Molecular Biology of Plant Pathogens

[illegible] pathogens [illegible]

The principle of hybridization analysis for indexing of viroids and viruses is [illegible]. The first requirement is to prepare highly radio active complementary DNA (cDNA) to the viroid or viral nucleic acid using ^{32}P or ^{3}H isotopes. The cDNA is prepared enzymatically on the viral or viroid nucleic acid so that its sequence is exactly complementary. When this ^{32}P or ^{3}H cDNA is incubated with a nucleic acid extract of a test plant under appropriate conditions, it hybridizes to any viral or viroid sequences present to produce a double strand ^{32}P cDNA-RNA hybrid in case of viroid and RNA viruses or ^{32}P cDNA-DNA hybrid in case of DNA viruses. The extent of hybrid formation is a measure of the concentration of viral or viroid sequences in the plant extract (Symons, 1984).

Nucleic acid specific hybridization techniques are being used for testing plant viruses and it involves the use of labeled complementary DNA (cDNA) or RNA (cRNA) prepared from purified viral nucleic acid as a probe or a recombinant clone of the viral nucleic acid as a probe. These labeled probes are used to detect the presence of plant viral nucleic acid by forming hybrid with them. Presently nucleic acid hybridization includes the formation of DNA-DNA, DNA-RNA, RNA-RNA complexes (Singh and Dhar, 1998). To detect the plant viruses and viroides three types of "Molecular hybridization" tests are being used: (a) Solution hybridization test (b) Filter hybridization test or Dot-blot hybridization or Nucleic acid spot hybridization (NASH) and (c) *in-situ* hybridization test.

Two procedures are available for the molecular hybridization analysis of [illegible] similar in principle differ in the way the ^{32}P cDNA:RNA hybrid [illegible]

[illegible]

[illegible] Hybridization Assay

In this method, nucleic acid hybridization takes place in solution [illegible] performed using both the target and probe DNAs in solution [illegible] liquid-liquid hybridization. There are also some data available for [illegible] DNA-DNA interaction in solution. Presently most of plant pathogen diagnosis was involved [illegible] hybridization [illegible] liquid hybridization [illegible] 2002). Liquid-liquid hybridization has been used [illegible] plant viruses [illegible]

[illegible] Palukaitis, 1979 [illegible]

Molecular Biology of Plant Pathogens

— *Editors* —

Dr. L.V. Gangawane
Professor Emeritus
Department of Botany
Dr. Babasaheb Ambedkar
Marathwada University
Aurangabad – 431 004, India

and

Dr. V.C. Khilare
Lecturer (Sr. Scale)
Department of Botany
Vasantrao Naik Mahavidyalaya
Aurangabad – 431 003, India

2010

DAYA PUBLISHING HOUSE

Delhi - 110 035

ISBN 81-7035-638-5
ISBN 978-81-7035-638-7

Published by : **Daya Publishing House**
1123/74, Deva Ram Park
Tri Nagar, Delhi - 110 035
Phone: (011) 27383999
Fax: (011) 23260116
e-mail : dayabooks@vsnl.com
website : www.dayabooks.com

Showroom : 4760-61/23, Ansari Road, Darya Ganj,
New Delhi - 110 002
Phone: 23245578, 23244987

Laser Typesetting : **Classic Computer Services**
Delhi - 110 035

Printed at : **Chawla Offset Printers**
Delhi - 110 052

PRINTED IN INDIA

Preface

Today we stand at the convergence of an incredible range of new science and technologies including recombinant DNA technology, information technology and nanotechnology, to facilitate our understanding of the structure and function of the genome and to harness this information for improvement of host plants. A product emanating from modern biotechnology offers the sustainable productivity in agriculture. Accurate identification and diagnosis of plant pathogens during early stages of infection can help a lot in better management of the diseases. All disease management programmes need a simple, cost-effective, safe and molecular detection of fungal pathogens is high than viral, bacterial and phytoplasmas. Analysis of molecular variability in plant pathogens by different markers used in RAPD, RFLP, AFLP, ISSR helps comparative analysis of pathogen race. Mapping, cloning and characterization of disease resistance genes opened a challenging field in the management of crop diseases. More than 70 diseases resistance genes have been cloned from different crops. Various defense response genes have been used for developing resistant plants to fungal pathogens using genetic approach. Genomics has emerged as one of the frontier technologies during this century using genome sequencing of many plant pathogens defense genes were validated. Sequencing of conserved genes has been used to develop PCR-based detection with varying levels of specificity for viruses, fungi, and bacteria. The application of these technologies in plant pathology has greatly improved our ability to detect plant pathogens and is increasing our understanding of, their ecology and epidemiology. We believe that molecular devices for plant pathogens became a routine tool for almost everyone involved in field research or in making disease management decisions. This book gives an idea of molecular approach towards plant pathogens, which helps us in their identification and management. This book would not have completed without the help of different scientists from various institutes. We are thankful to them. We are also grateful to Daya Publishing House, New Delhi for bringing this volume for student, researchers within stipulated time.

L.V. Gangawane
V.C. Khilare

Contents

Preface *v*

1. **Detection and Molecular Characterization of a Phytoplasma Associated with Frogskin Disease in Cassava** **1**
Elizabeth Alvarez, Juan F. Mejía and Germán A. Ilano

2. **Molecular Characterization of *Bipolaris sorokiniana* Causing Spot Blotch of Wheat** **18**
Rashmi Aggarwal, V.B. Singh, M.S. Gurjar and Sangeeta Gupta

3. ***Trichoderma*: Biology and its Potential for Biocontrol** **40**
Pranav Chettri, Mala C. Ganiger and P.U. Krishnaraj

4. **Molecular Biology of *Phytophthora infestans* Causing Late Blight of Potato** **58**
V.P. Chimote and S.J. Gawande

5. ***Macrophomina phaseolina*: An Overview on Molecular Biology and Host Specialization** **73**
I.K. Das and A.V. Gadewar

6. **Recent Advances in Diagnosis of Plant Pathogens** **81**
S.J. Gawande, V.P. Chimote and Nimisha Kaushal

7. **Molecular Characterization of *Fusarium* spp. Causing Wilt of Patchouli and Gladiolus** **104**
Yashoda R. Hegde, Sridevi Chavan, Sumitra Patil Kulkarni, Srikant Kulkarni and S.K. Prashanthi

8. **Reusing of Agarose in Molecular Plant Pathology Laboratories** 123
Kamel Ahmed Abd-Elsalam

9. **Rapid Detection of *Ganoderma* Disease of Coconut by Immunoassay and PCR** 131
A. Kandan, V. Rajendran, V. Raguchander and R. Samiyappan

10. **Novel Approaches for the Detection of Plant Viruses** 145
S.S. Kang and Abhishek K. Sharma

11. **Molecular Characterization of Formae Speciales of *Fusarium oxysporum*: A Review** 159
V.C. Khilare and L.V. Gangawane

12. **Molecular and Pathogenic Variability in *Plasmopora halstedii* Causing Downy Mildew of Sunflower** 178
Srikant Kulkarni, Yashoda R. Hegde and V. Ramesh Kota

13. **Molecular Variability of Rice Blast Pathogen, *Pyricularia grisea* and Resistant Gene Deployment in Karnataka** 191
S.K. Prashanthi, Srikant Kulkarni, B.S. Meena, Y.R. Hegde and N.G. Hanamaratti

14. **Diagnosis, Molecular Characterization and Management of Phytoplasma Diseases: A Review** 203
G.P. Rao and Smriti Mall

15. **Biochemical and Molecular Variability in *Pythium aphanidermatum* Causing Rhizome Rot of Ginger** 236
Shalini D. Sagar and Shrikant Kulkarni

16. **Molecular Based Techniques for Characterization, Detection and Diagnosis of Plant Viruses, Viroids and Phytoplasmas** 248
Satya Prakash and S.J. Singh

17. **Characterization of *Trichoderma* spp.** 302
Pratibha Sharma, A. Muthu Kumar and Monica Henry

18. **Vegetative Compatibility Groups (VCGs) in Fungi** 336
Pratibha Sharma, Susanta Banik, Mousa Najafinia and Monica Henry

19. **Molecular Methods for Detection of Plant Pathogen with Special Reference to DNA Based Methods** 360
Kahekashan Tanveer, Sairose S. Lalani and M.M.V. Baig

Index 377

Chapter 1

Detection and Molecular Characterization of a Phytoplasma Associated with Frogskin Disease in Cassava

Elizabeth Alvarez, Juan F. Mejía and Germán A. llano*

Plant Pathology Program,
International Center for Tropical Agriculture (CIAT),
P.O. Box 6713, Cali, Valle del Cauca, Colombia

ABSTRACT

Cassava Frogskin disease is a major disease of cassava roots that is spreading in some countries of Latin America, with yield losses of 90 per cent. In this work, it was detected a phytoplasma associated with the CFSD by using a nested PCR assay with the specific primers R16mF2/R16mR1 and R16F2n/R16R2. To classify the phytoplasma, the universal primers P1/P7 and R16F2n/R2 to amplify the 16S ribosomal DNA gene were used. Fragments measuring 1.2-kb were amplified only from samples collected from symptomatic plants. Sequence analysis of the cloned fragments showed that the phytoplasma was similar to *Cirsium* white leaf phytoplasma (GenBank acc. No. AF373106, 16SrIII X-disease group), with a 99 per cent sequence homology. The pathogen's transmission was achieved through *Cuscuta* from infected cassava plants to healthy periwinkle (*Catharantus roseus*) plants and from infected periwinkle plants to healthy cassava plants. This study is the first to report a phytoplasma in association with CFSD.

Keywords: *Manihot esculenta, Phytoplasma, Frogskin disease.*

* E-mail: g.llano@cgiar.org

Introduction

Cassava frogskin disease (CFSD) is an important disease affecting cassava (*Manihot esculenta* Crantz) roots, whose causal agent had remained unknown for many years despite its economic significance. Recently, CFSD has been reported with increasing frequency in Colombia, Brazil, Venezuela, Panama, and Costa Rica; incidence of up to 90 per cent has been recorded in Colombian commercial fields (Pineda *et al.*, 1983; Calvert *et al.*, 2004).

CFSD affects cassava roots, reducing their diameter. In many cassava varieties, symptoms are expressed in the roots. Some varieties may also show leaf symptoms such as mosaic, chlorosis, curling and / or curvature in leaf margins, however these symptoms are difficult to distinguish under field conditions, and could be confused with damage from mites, thrips, viruses, deficiencies of micro elements or they can be masked when temperatures are >30°C. Characteristic CFSD symptoms in the roots are also the woody aspect and the thick peel that is cork-like, fragile, and opaque. The peel also presents lip-like slits that, when they join, create a net-like or honeycomb pattern. When roots do not tuberize adequately, the stems tend to be thicker than normal. In contrast, the roots of healthy plants are well developed, with thin, brilliant, and flexible peel.

To develop appropriate management strategies for reducing the spreading of this disease, a proper identification of possible etiological agent(s) must be carried out. Phytoplasmas were recently detected in CFSD-affected cassava roots, leaf, midribs, petioles, and peduncles in susceptible commercial varieties, using nested-PCR on a fragment of the 16S rRNA gene. The phytoplasmas were not detected in healthy plants from the same varieties harvested from fields free of disease (Alvarez *et al.*, 2003). Since various diseases that induce symptoms in roots were reported as associated with phytoplasmas in diverse plant species (Crossley and Clark, 1996; Dyer and Sinclair, 1991; Orenstein *et al.*, 1999; Lee *et al.*, 2006), molecular tools were applied to verify phytoplasma identity and association with CFSD-infected roots, leaf midribs, petioles, and peduncles in different cassava varieties and genotypes from various regions of Colombia.

Materials and Methods

PCR Amplification

DNA was extracted according to described protocols (Gilbertson and Dellaporta, 1983; Prince *et al.*, 1993). Three pairs of universal primers P1/P7 (Deng and Hiruki, 1991; Smart *et al.*, 1996), R16mF2/R16mR1, and R16F2n/R16R2 (Gundersen and Lee, 1996) were used in direct PCR to amplify the region of the genes 16S rRNA and 23S rRNA plus spacer region between them and the 16S rRNA region respectively. The P1/P7 and R16mF2/R16mR1 amplicons were diluted at 1:29 with sterilized water to carry out nested PCR reaction with primers R16F2n/R16R2. Samples amplified in direct PCR with R16F2n/R16R2 primers were diluted at 1:50 with sterilized water and then subjected to nested PCR using group specific primer pair R16(III)F2/R16(III)R1 (Lee *et al.*, 1994), specifically designed for amplification of the 16SrIII group of phytoplasmas (X-disease) (Table 1.1).

Table 1.1: Direct and Nested PCR Assays Results for Phytoplasma Detection in Samples of 39 Cassava Varieties

Sample Code	*Symptom Intensity*[a]	*Cassava Genotype*	*Site*[b]	*Tissue*[c]	*PCR*[d]	*Primer Combinations*[e]
Y1	100	CM 6740-7	VC	LmP/R[f]	+/+	A-C
Y2	65	CIAT Parrita	VC	LmP/S/R	+/+/+	B-C
Y3	65	ICA Catumare	VC	LmP/R[f]	+/+	B-C
Y4	65	Manzana	VC	LmP/R[f]	+/+	B-C
Y5	90	M Bra 383	VC	LmP/R	+/+	B-C
Y6	100	CM 849-1	VC	LmP/R	+/+	B-C
Y7	100	CM 5460-10	VC	LmP	+/+	C
Y8	35	CM 2177-2	VC	LmP/R	+/+	B-C
Y9	35	CM 4919-1	VC	LmP/R	+/+	B-C
Y10	65	CM 3306-9	VC	LmP[f]	+	B-C
Y11	35	CM 3306-19	VC	LmP[f]	+	B-C
Y12	10	M Bra 856-54	VC	LmP[f]	+	B-C
Y13	35	M Per 335	VC	R	+	C
Y14	10	M Bra 856	C	LmP/R	-/+	C
Y15	90	SM 909-25	VC	LmP[f]/R[f]	+/+	A-B-C-D
Y16	90	CG 6119-5	VC	LmP/R	+/+	C
Y17	100	M Col 2063	VC	LmP[f]	+	A-B-C
Y18	90	ICA Nataima	VC	LmP/R	-/+	C
Y19	10	SM 1201-5	VC	LmP	-	C
Y20	10	GM 228-14	VC	LmP	-	C
Y21	90	CM 9582-64	VC	LmP/R	+/+	A-B-C
Y22	90	CM 9582-65	VC	LmP/R	+/+	A-B-C
Y23	100	CM 9582-24	VC	LmP/R	+/+	A-B-C
Y24	65	M CR 81	VC	LmP/R	+/+	A-B-C
Y25	90	Venezolana	S	R	+	A-B-C
Y26	65	M Per 16	C	LmP/R	+/+	C
Y27	65	M Col 634	C	LmP/R	+/+	C
Y28	10	M Bra 829	C	LmP/R	+/+	C
Y29	100	SM 1219-9	VC	LmP/R[f]	+/+	A-B-C-D
Y30	0	M Chn 2	C	LmP/R	-/-	C
Y31	35	HMC-1	C	LmP/R	+/+	C
Y32	0	M Arg 2	C	LmP/R	-/-	C
Y33	10	M Bra 325	C	LmP/R	+/+	C
Y34	10	M Bra 839	C	LmP/R	+/+	C
Y35	10	M Col 1178	C	LmP/R	+/+	C

Contd...

Table 1.1–Contd...

Sample Code	Symptom Intensity[a]	Cassava Genotype	Site[b]	Tissue[c]	PCR[d]	Primer Combinations[e]
Y36	10	M Col 1468	C	LmP/R	+/+	C
Y37	10	M Cub 74	C	LmP/R	-/+	C
Y38	10	M Bra 886	C	LmP/R	+/+	C
Y39	10	M Bra 882	C	LmP/R	+/+	C
Y40	90	M Bra 383	C	LmP/R	+/+	B
Y41	0	ICA Catumare	Q	LmP/R	-/-	B-C
Y42	0	Manzana	Q	LmP/R	-/-	B-C

a: Scale of severity (0-5) previously transformed according to (Little and Hills, 1978). (0), healthy plant; (10), very light; (35), light; (65), moderate; (90), severe; (100), very severe.

b: VC = Department of Valle del Cauca; C = Cauca; S = Sucre; Q = Quindío.

c: LmP = mix of leaf midrib and petioles; R = roots; S = stems.

d +: Amplification positive for phytoplasma; –: Amplification negative for phytoplasma.

e: Primers used for amplification were A = P1/P7–R16F2N/R16R2; B = R16mF2/R16mR1–R16F2N/R16R2; C = R16F2/R16R2–R16(III)F2/R16(III)R1. D = P1/P7.

f: Plants aslo showing foliar symptoms of chlorosis and deformed leaf blades.

A coffee crispiness phytoplasma strain AY525125 (Galvis *et al.*, 2007) was employed as positive control for PCR reactions, and sterile distilled water, together with health cassava samples were employed as negative controls. Each reaction was performed in a volume of 25 µL, using final concentrations of 100 ng of DNA, 1X buffer, 3 mM $MgCl_2$, 1 U *Taq* polymerase, 0.8 mM dNTPs, and 0.1 µM of each primer.

For primers P1/P7 and R16mF2/R1mR1, 35 cycles were carried out in a PTC-100 thermal cycler, following these conditions: 1 min (2 min for the first cycle) of denaturation at 94 °C, annealing for 2 min at 55°C, and extension of the primer for 3 min (10 min in the final cycle) at 72°C. The primer pairs R16F2n/R16R2 and R16(III)F2/R16(III)R1 were employed with same cycle at an annealing temperature of 50°C. The PCR products were visualized in a 1.5 per cent agarose gel, stained with 0.75 µg/mL ethidium bromide, and analyzed in a Stratagene Eagle Eye® II video system.

RFLP Analyses of Phytoplasma 16Sr Gene and Spacer Region

For identification of the detected phytoplasmas, the direct and nested-PCR products obtained with phytoplasma universal and specific primers from cassava samples were subjected to analyses of restriction fragment length polymorphism (RFLP). R16(III)F2/R16(III)R1 amplicons were digested with *Rsa*I, *Alu*I, *Mse*I, and *Taq*I restriction enzymes, while P1/P7 and R16F2n/R16R2 amplicons were digested with *Hpa*II, *Tru*I, and *Hha*I; the mixture was incubated for 16 h at 37°C, except for enzymes *Tru*I and *Taq*I, for which the incubation temperature was 65°C, following the instructions of the manufacturer. Visualization of RFLP products was performed in a 5 per cent polyacrylamide gel stained as described above for agarose gels.

Further RFLP analyses were performed using strains of 16SrIII subgroups from GenBank (Table 1.2); after their alignments using CLUSTALX (Thompson *et al.*, 1997) putative restriction site maps of the 16S rRNA gene sequences were generated by using the DNASTAR program MapDraw option (DNASTAR Inc.). The maps were manually aligned for comparison of CFSDY15 with endonucleases *Hpa*II, *Mse*I, *Hha*I, and *Sau*3AI to clover yellow edge (CYE-C), walnut witches' broom (WWB), Virginia grapevine yellows VGYIII (VGY), chayote witches' broom (ChWBIII-Ch10) (ChWB), strawberry leafy fruit (SLF), Dandelion virescence phytoplasma (Dan Vir), and black raspberry witches'-broom (BRWB7) phytoplasmas.

Table 1.2: Phytoplasma 16S rDNA Sequences Retrieved from GenBank, Including Cassava Sample CFSDY15 from Colombia and Employed for Phylogenetic and Putative Restriction Site Analyses

GenBank	*Phytoplasma Strain*	*Origin*	*Plant Host*	*16Sr RNA Group*
M30790	'*Ca.* Phytoplasma asteris'	Michigan (USA)	*Oenothera* spp.	16SrI-B
U15442	'*Ca.* P. aurantifolia'	Oman	Lime	16SrII-B
L04682	'*Ca.* P. pruni'*	California (USA)	Peach	16SrIII-A
U18747	'*Ca.* P. palmae'*	Florida (USA)	*Veitchia merrillii*	16SrIV
X80117	'*Ca.* P. cocostanzianae'*	Tanzania	*Cocos nucifera*	16SrIV
Y14175	'*Ca.* P. cocosnigerianae'*	Nigeria	*Cocos nucifera*	16SrIV
AF122910	'*Ca.* P. ulmi'	New York (USA)	Elm	16SrV-A
AF305240	'*Ca.* P. ziziphi'	South Korea	Jujube	16SrV-B
X76560	'*Ca.* P. vitis'*	France	Grapevine	16SrV-C
AY390261	'*Ca.* P. trifolii'	Canada	Clover	16SrVI-A
AF092209	'*Ca.* P. fraxini'	New York (USA)	Ash	16SrVII-A
AF086621	'*Ca.* P. luffae'*	Taiwan	Luffa	16SrVIII-A
AF515637	'*Ca.* P. phoenicium'	Iran	Almond	16SrIX-B
AJ542541	'*Ca.* P. mali'	Italy	Apple	16SrX-A
AJ542544	'*Ca.* P. prunorum'	Germany	Peach	16SrX-B
AJ542543	'*Ca.* P. pyri'	Germany	Pear	16SrX-C
X92869	'*Ca.* P. spartii'	Italy	*Spartium junceum*	16SrX-D
D12581	'*Ca.* P. oryzae'	Japan	Rice	16SrXI-A
AF248959	'*Ca.* P. solani'*	Serbia	Periwinkle from pepper	16SrXII-A
L76865	'*Ca.* P. australiense'	Australia	Grapevine	16SrXII-B
AF248960	Mexican periwinkle virescence (MPV)	Mexico	Periwinkle	16SrXIII
AJ550984	'*Ca.* P. cynodontis'	Italy	*Cynodon dactylon*	16SrXIV-A
AF147708	'*Ca.* P. brasiliense'	Brasil	Hibiscus	16SrXV-A
AY725228	'*Ca.* P. graminis'	Cuba	Sugarcane	16SrXVI
AY725234	'*Ca.* P. caricae'	Cuba	*Carica papaya*	16SrXVII

Contd...

Table 1.2–Contd...

GenBank	*Phytoplasma Strain*	*Origin*	*Plant Host*	*16Sr RNA Group*
DQ174122	'*Ca.* P. americanum'	Nebraska (USA)	Potato	16SrXVIII-A
DQ086423	'*Ca.* P. fragariae'	Lithuania	Strawberry	n.a.
AJ310849	'*Ca.* P. pini'	Germany	*Pinus sylvestris*	n.a.
AB010425	'*Ca.* P. japonicum'	Japan	Hydrangea	n.a.
X76431	'*Ca.* P. rhamni'	Germany	*Rhamnus frangula*	n.a.
AY135523	'*Ca.* P. allocasuarinae'	Australia	*Allocasuarina muelleriana*	n.a.
AB054986	'*Ca.* P. castaneae'	South Korea	Chestnut	n.a.
EU346761	Cassava frogskin disease strain (CFSDY15) [a]	Colombia	Cassava	16SrIII-L
AY737646	Cassava frogskin disease strain (CFSDY17) [a]	Colombia	Cassava	16SrIII-L
AY737647	Cassava frogskin disease strain (CFSDY29)[a]	Colombia	Cassava	16SrIII-L
AF147706	Chayote witches' broom (ChWB)	Brasil	Chayote	16SrIII-J
AF274876	Strawberry leafy fruit phytoplasma (SFL)	Maryland (USA)	Strawberry	16SrIII-K
AF373106	Cirsium white leaf phytoplasma (CWL)	Lithuania	Cirsium	16SrIII-U
AF189288	Clover yellow edge (CYE)	Oregon (USA)	Hazelnut	16SrIII-B
AF175304	Clover yellow edge (CYE-C)	Canada	Clover	16SrIII-B
AF302841	Black raspberry witches-broom (BRWB7)	Oregon (USA)	*Rubus occidentalis*	16SrIII-Q
AF060875	Virginia grapevine yellows VGYIII (VGY)	Virginia (USA)	Grapevine	16SrIII-I
AF510724	Milkweed yellows phytoplasma (MW1)	New York (USA)	milkweed	16SrIII-F
AF190227	Walnut witches' broom phytoplasma (WWB)	Georgia (USA)	Walnut	16SrIII-G
AF190223	Poinsettia branch-inducing phytoplasma (PoiBI)	USA	poinsettia	16SrIII-H
AF370119	Dandelion virescence phytoplasma (DanVir)	Lithuania	*Dandelion* sp.	16SrIII-P
M23932	*Acholeplasma laidlawii*			n.a.

a: Sequences determined in this study, an.a.: Group not available, *: '*Candidatus*' names proposed at the X International Congress of the International Organization of Mycoplasmology, 1994, held in Bordeaux, France, but not yet formally described, and are reported here as incidental citations which do not constitute prior citations, according to rule 28b of the bacteriological code (Lapage *et al.*, 1992).

Cloning of PCR Products and Sequencing

The R16F2n/R16R2 PCR products were purified from amplified samples CFSDY17, CFSDY29 (Table 1.1), using the QIAquick PCR Purification Kit, ligated in pGEM-T Easy vector, which was introduced into the *Escherichia coli* strain DH5-a by. Blue/white color screening for plasmid was performed followed by extraction with a Plasmid Miniprep System Kit (Gibco-BRL). Different-sized (1,200-1,300 bp) fragments were selected for sequencing by automated dideoxy sequencing (ABI Prism 377-96 DNA Sequencer), using a DNA-sequencing kit from Applied Biosystems, after plasmid restriction with *Eco*RI and electrophoresis in 1.5 per cent agarose gel. The PCR products from the same samples were also directly sequenced with a DNA-sequencing kit from Applied Biosystems.

Furthermore direct sequencing was performed from P1/P7 amplicon of sample SM909-25 (CFSDY15) after cleaning with QIAquick PCR Purification Kit; both directions sequencing with primers P1, F1 (Davis and Lee, 1993), and P7 was obtained using the BIG DYE sequencing terminator kit. The nucleotide sequences determined in this study were deposited in the GenBank data library (Table 1.1).

Phylogenetic Analyses

The public available 16S rDNA sequences of representative strains of the genus '*Candidatus* Phytoplasma' (IRPCM, 2004) and some additional strains (belonging to 16SrIII group) were retrieved from GenBank and aligned using CLUSTALX (Thompson *et al.*, 1997) and BioEdit (Hall, 1999). A phylogenetic tree was then constructed with MEGA version 4 (Tamura *et al.*, 2007) for aligning samples CFSDY15, CFSDY17, CFSDY29 16S ribosomal sequences with similar sequences from 43 phytoplasma strains (Table 1.2); bootstrap analysis was also performed and replicated 100 times. *Acholeplasma laidlawii* a cultivable *Mollicute* phylogenetically related to phytoplasmas was designated as the out-group to root the tree.

PCR/RFLP Analyses on Ribosomal Protein Gene

Further molecular characterization was performed by direct and nested-PCR using primers rpL2F3/rp(I)R1A (Martini *et al.*, 2007) followed by rpIIIF1/rpIIIR1 primer pair, that specifically amplify part of 16SrIII group rp operon (about 300 bp) (Davis *et al.*, 1998). The PCR reaction mix and negative controls were as described above. Thirty-eight PCR cycles were performed for both amplification under the following conditions: 1 min (2 min for the first cycle) for denaturation step at 94°C, 2 min for annealing at 50°C, and 3 min (7 min for the last cycle) for primer extension at 72°C. Six µl of PCR products were separated in 1 per cent agarose gel, stained with ethidium bromide and visualized with UV transilluminator. To further characterize the nested-PCR amplicons obtained with phytoplasma rpIIIF1/rpIIIR1 primers from cassava samples were subjected to RFLP analyses with restriction enzymes *Tru*I, *Alu*I and *Tsp509*I; visualization of RFLP products in a 5 per cent polyacrylamide gel was carried out as described above.

Transmission to Healthy Plants

Transmission of the detected phytoplasmas was carried out under insect-proof greenhouse at 20–25°C and 50 per cent–90 per cent RH from two cassava genotypes infected with CFSD. Dodder (*Cuscuta* sp.) transmission from naturally infected potted cassava plants to the healthy species, and grafts (clefts or chip-budding and splices whip-grafting) using the leaf midribs, petiols and shoots of infected cassava plants (Table 1.3) were carried out. For grafting the donor plants were selected with similar diameter to the stem of recipient plant where the graft was made at the top and in the middle. Two sources of inoculum were employed: cassava variety SM909-25 showing severe symptoms in roots and leaf chlorosis and curling under greenhouse conditions (23°C and 80 per cent RH), and CW94-21 cassava clone (CW family in CIAT's cassava genetic improvement program) showing characteristic disease symptoms under field conditions.

Table 1.3: Results Obtained with *Cuscuta* sp. and Grafts for Transmission of Phytoplasmas from Cassava Infected with CFSD Phytoplasma to Periwinkle and Healthy Cassava Plants

Treatment		*PCR (+)*[a]	
No.	*Description*	*SM 909-25*	*CW 94-21*
Cuscuta sp.–transmission period of about 2 months			
1	Infected cassava to healthy periwinkle	2/6	0/6
2	Infected periwinkle to healthy periwinkle	2/6	0/6
3	Infected cassava to healthy cassava	1/6	0/6
Grafts–union was carefully covered with parafilm® and plastic bag for 1-3 weeks			
1	Infected cassava to healthy periwinkle	5/6	3/6
2	Infected cassava to healthy cassava	5/6	4/6

a: Values refer to number of replications out of 6, where detection was successful.

Transmission was performed from these plants to young cassava plantlets (2 months after re-establishment form *in vitro* culture), and to periwinkle seedlings 5 to 6 weeks old as described in Table 1.3. Phytoplasma presence and identity after transmission experiments were determined 4 months after the grafting or the insertion of dodder bridges by nested-PCR using the phytoplasma 16Sr general and specific primers and the sequencing methods described above.

Results

PCR Amplification and RFLP Analyses of Phytoplasma 16S Gene and Spacer Region

Direct PCR with P1/P7 primers and nested-PCRs with R16F2n/R16R2 or R16(III)F2/R16(III)R1 primers resulted in the amplification of 1.7, 1.2, and 0.8 kb DNA fragments respectively from many of the tested samples (Table 1.1) indicating that the majority of cassava plants were infected by phytoplasmas.

The phytoplasmas were detected in 35 of the 39 cassava genotypes tested that exhibit mild to severe symptoms of CFSD (Table 1.1). Four genotypes resulted negative,

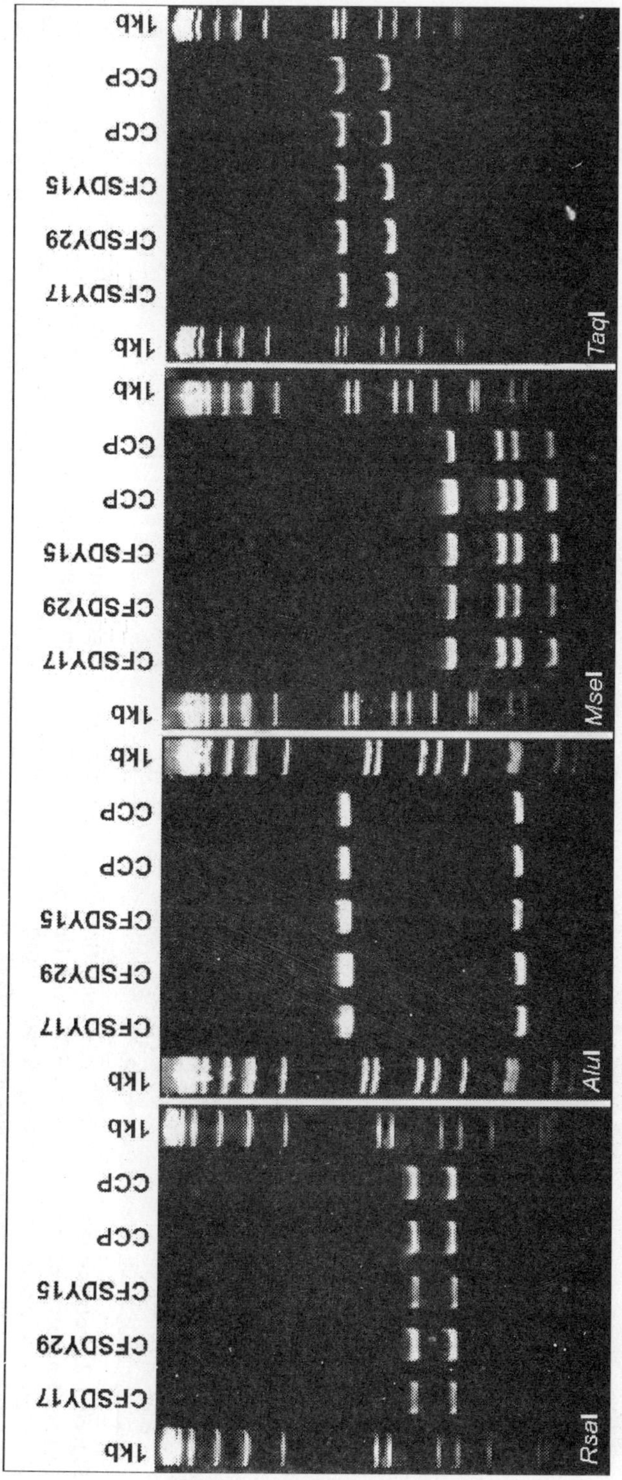

Figure 1.1: Restriction Fragment Length Polymorphism (RFLP) Profiles of Amplified 16S rDNA Products Obtained in Nested PCR with Primer Pairs R16(III)F2/R16(III)R1 on 5 per cent Polyacrylamide Gels. CFSDY17, Cassava frogskin diseases phytoplasma strain Y17; CFSDY29, strain Y29; CFSDY15, strain Y15; CCP, Coffee crispiness phytoplasma. 1 kb DNA ladder, fragment sizes in base pairs from top to bottom: 40,000; 10,000; 8,144; 6,108; 5,090; 4,072; 3,054; 2,036; 1,636; 1,018 and 517/506 (Invitrogen life technologies, CA). 117x49mm (518 x 518 DPI).

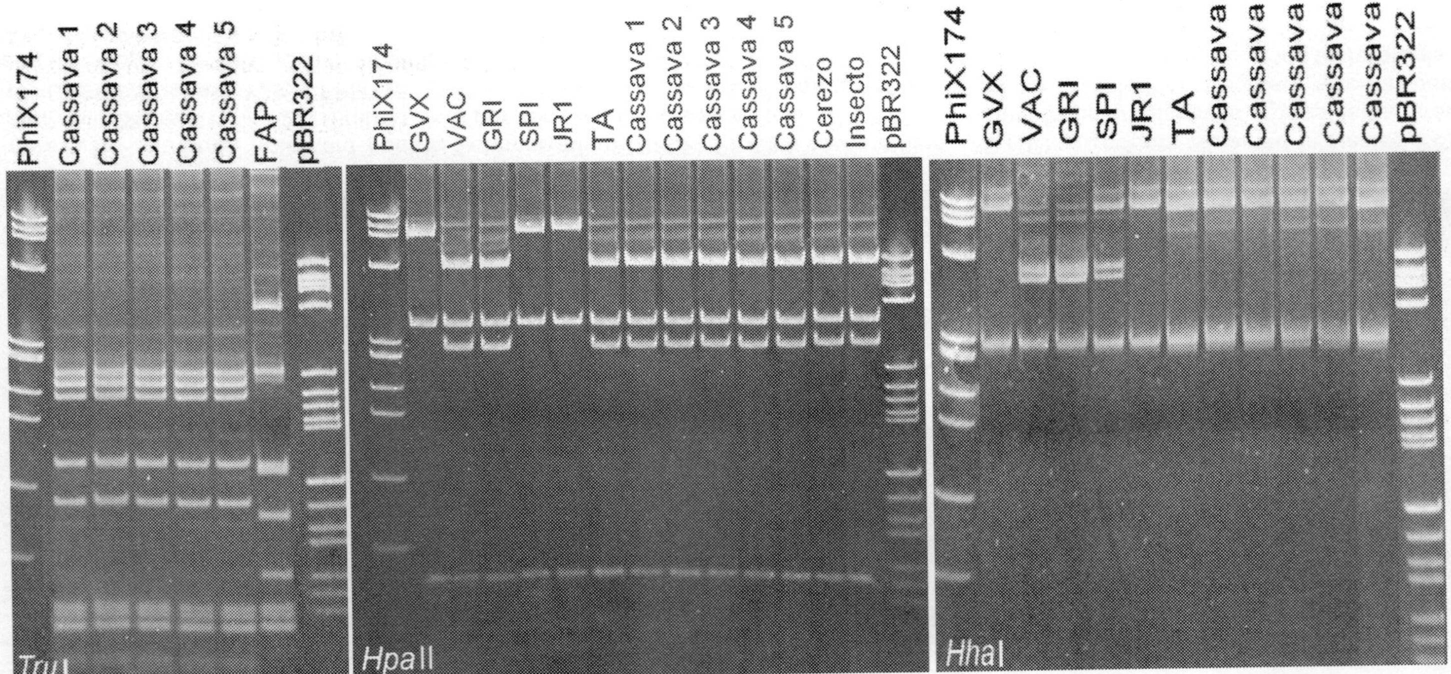

Figure 1.2: Polyacrylamide Gels (5 per cent) Showing the Restriction Fragment Length Polymorphism (RFLP) of 16S rDNA Plus Spacer Region Amplified in Nested-PCR with Primers R16F2n/R16R2 from Representative Phytoplasma Strain and from Cassava. Restriction endonucleases employed were *Hpa*II, *Tru*I, *Hha*I. PhiX174, marker FX174 *Hae*III digested and pBR322, marker pBR322 *Hae*I digest. PYLR, peach yellows leafroll; GVX, Green Valley X-disease; CX, X-disease; API, from Euscelidius variegates from Italy; RA, ranunculus virescence; GRI, golden rod yellows; SPI, spirea stunt; VAC, vaccinium witches' broom; MW1, Milkweed yellows; SBB, Solanum marginatum big bud from Equador; JRI, poinsettia branching factor. PhiX174, marker AX174 *Hae*III digested; fragment sizes in base pairs from top to bottom: 1,353; 1,078; 872, 603, 310, 281, 271, 234, 194, and 118; pBR322, marker pBR322 *Hae*I digest, fragment sizes in base pairs from top to bottom: 622, 527, 404, 307, 242, 238, 217, 201, 190, 180, 160, 147, 123, 110, 90, 76, and 67.

however two (CFSDY30 and CFSDY32) showed no symptoms and two (CFSDY19 and CFDY20) showed a rate of 10 per cent symptoms (Table 1.1). The control samples from *in vitro* cassava shoots were always negative in both direct and nested-PCR assays in all the systems employed.

Preliminary identification of phytoplasmas was obtained by the positive nested-PCR assays with primer pair R16(III)F2/R16(III)R1. RFLP analysis of these strains show that all of the cassava samples tested and control strain CCP yielded mutually indistinguishable collective RFLP patterns with the four restriction enzymes employed (Figure 1.1).

RFLP patterns obtained from P1/P7 and R16F2n/R16R2 amplicons were compared with reference phytoplasmas (Figure 1.2): collective profiles distinguished the CFSD phytoplasma strains from other employed phytoplasmas classified in group 16SrIII; *Hpa*II, *Tru*I, and *Hha*I distinguished the cassava strain CFSDY15 from reference strains employed.

Phylogenetic Analyses

Sequencing of two R16F2n/R16R2 amplicons from cassava samples CFSDY17, CFSDY29, and of one P1/P7 amplicon from CFSDY15 resulted in sequences of 1,260, 1,298 bp and of 1,679 bp that were deposited in GenBank with the accession numbers AY737646, AY737647 and EU346761 respectively (Table 1.2).

Phylogenetic comparison of the 16S rRNA gene of CFSD phytoplasmas with 43 representative strains of the genus '*Candidatus* Phytoplasma' confirmed that the cassava frogskin disease is associated with phytoplasmas very closely related to those belonging to 16SrIII group (Figure 1.3). Phylogenetic analyses of 16S rRNA gene delineated a separate subclade to which belong all the tree strains of CFSD.

PCR/RFLP Analyses on Ribosomal Protein

The use of ribosomal protein group III specific primers produced the expected length amplicons from several symptomatic cassava samples in nested-PCR, and from selected phytoplasma reference strains belonging to 16SrIII group (data not shown). RFLP analyses with *Tsp509*I and *Alu*I restriction enzymes showed no differences among cassava strains and most of the reference strains employed, however profiles obtained with *Tru*I clearly indicates that phytoplasmas infecting cassava differ from all reference strains employed (Figure 1.4).

Transmission to Healthy Plants

Molecular tests carried out on cassava and periwinkle plants after transmission trials allow detecting the presence of phytoplasmas related to group 16SrIII. Dodder transmission allows phytoplasma identification in periwinkle and cassava plants only when variety SM 909-25 was used as donor plant (Table 1.3). Graft transmission was able to transfer phytoplasmas from both genotypes SM 909-25 (83 per cent of phytoplasma transmission) and CW 94-21 (58 per cent of phytoplasma transmission) (Table 1.3).

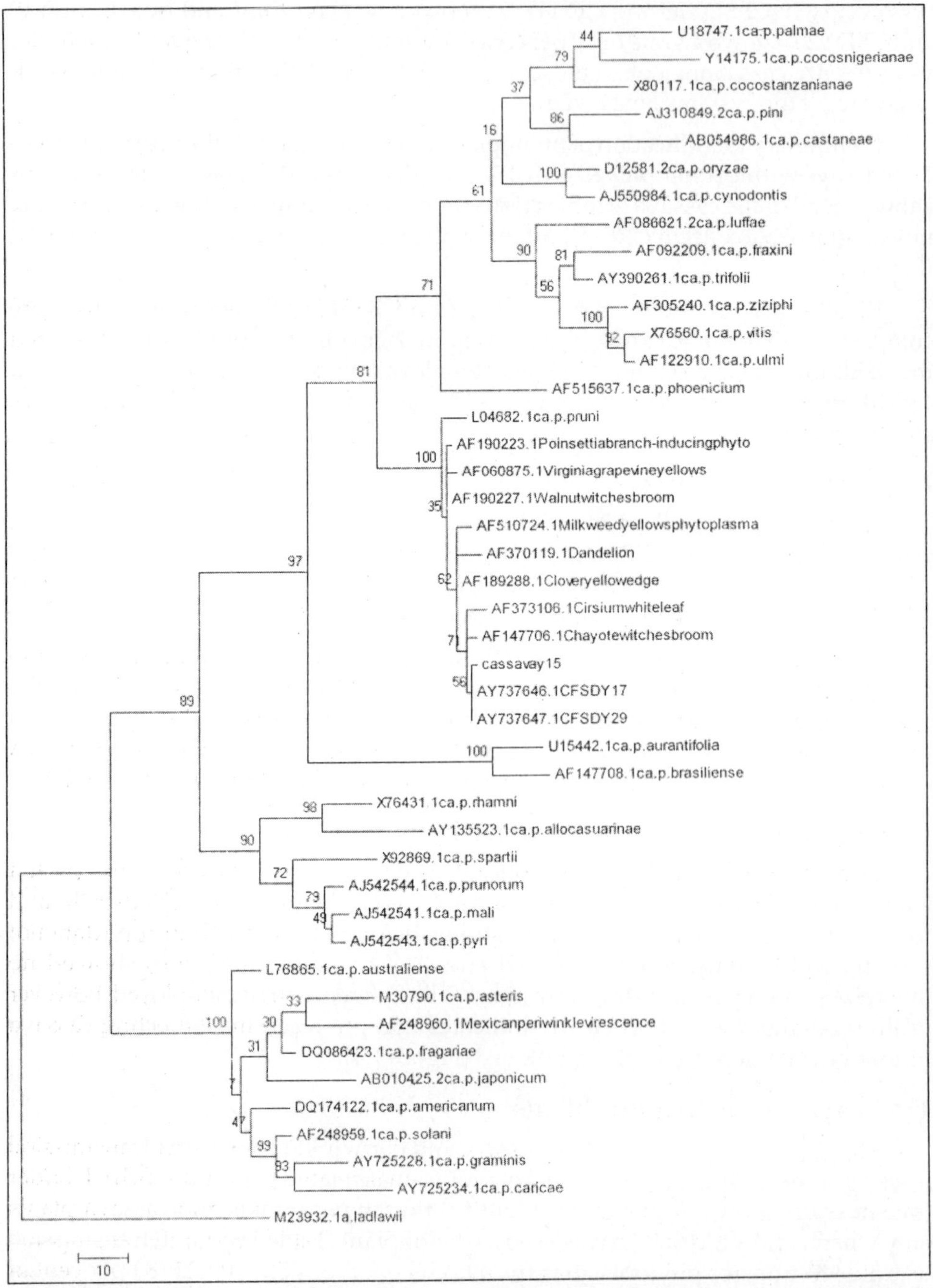

Figure 1.3: Phylogenetic Tree Constructed by Parsimony Analyses or Near Full Length 16SrDNA Sequences from Cassava Frogskin Diseases Phytoplasma Strain from Colombia, 40 Reference Strains of Phytoplasma and some Additional Strains from 16SrIII Group, Reporting the Placement of the CFSD Phytoplasma within the Genus '*Candidatus* Phytoplasma', Employing *Acholeplasma laidlawii* as the Outgroup

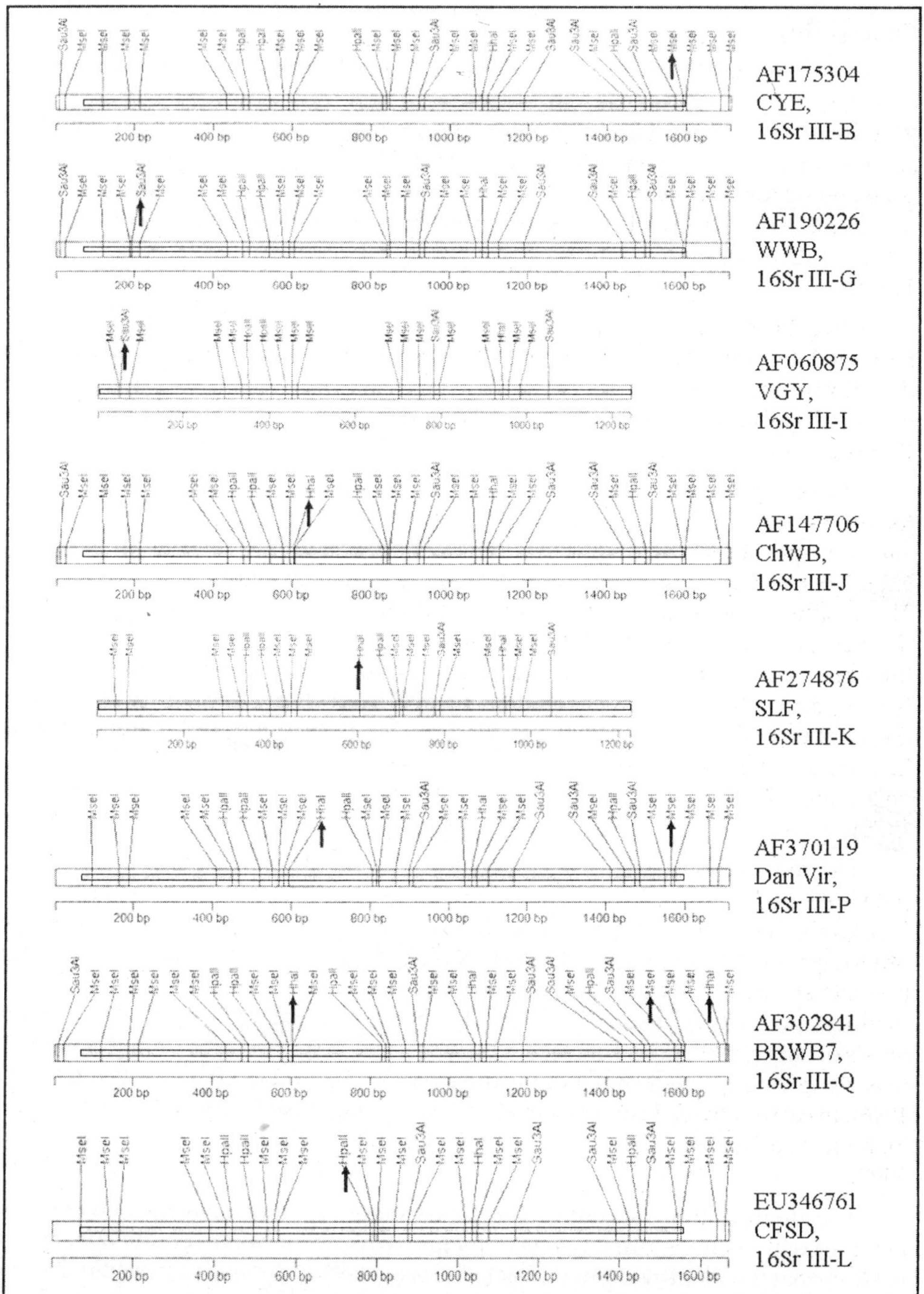

Figure 1.4: Putative Restriction Sites Analyses of 16S rRNA Gene Sequence of Cassava Strain CFSDY15 Compared with Other Phytoplasmas of 16SrIII Group. Arrows indicate differential restriction sites. Phytoplasma strains are described in Table 1.2.

Discussion

In this study the presence of a phytoplasma in tissues of plants exhibiting symptoms of cassava frogskin disease was confirmed by direct as well as nested-PCR assays on 16SrDNA and on phytoplasma ribosomal protein gene as well. RFLP analyses on both amplicons allowed classifying the phytoplasmas infecting cassava as belonging to group 16SrIII, and also allowed classifying the strain infecting cassava in Colombia into a new ribosomal subgroup in group 16SrIII, named 16SrIII-L; the results is also supported by the phylogenetical analyses of 16Sr RNA gene (Lee *et al.*, 1998).

Direct PCR detection was only possible in petioles and midribs of young leaves, for their detection in roots the nested PCR assay was necessary. The lack of pathogen detection in a few plants phenotypically evaluated as diseased could be related to the uneven distribution that phytoplasmas very often present in plants (EPPO/CABI, 1996) or to the lack of root sample testing (Table 1.1).

The adherence of phytoplasmas to the host's cellular membranes is necessary for successful colonization, which may then interfere with membrane function and alter transportation mechanisms as shown for human mycoplasmas (Rottem and Naot, 1998), restricting, in the case of cassava, sugar transfer from leaves to roots. This results in the accumulation of carbohydrates in leaves, and probably explains the increase of aerial part growth observed in infected cassava plants; the consequent low starch concentrations in roots lead to reduced tuberization (Maust *et al.*, 2003). Increased carbohydrate accumulation in leaves also occurs in periwinkle (*Catharanthus roseus*) (Lepka *et al.*, 1999) and pears (Catlin *et al.*, 1975) infected by phytoplasmas. Low starch concentrations in roots are also found in palms suffering from lethal yellowing (Montano *et al.*, 2000) and in pears infected by pear decline (Catlin *et al.*, 1975).

The reduced transfer of sugars could be caused by the physical blockade of phloem tubes by phytoplasmas and/or by the deposition of calluses or other material in response to phytoplasma infection (Maust *et al.*, 2003). In sweet potato infected by phytoplasmas (sweet potato little leaf), the whole plant, including the root system, becomes stunted with a pronounced proliferation of axillary shoots; latex production in vines and roots is also noticeably reduced. Depending on the time of infection, yields of harvestable tubers can be severely reduced, to the extent that plants infected in early growing stages may not produce harvestable tubers (Crossley and Clark, 1996). In carrot phytoplasma presence induce among other symptoms the reduction in the size and quality of taproots (Duduk *et al.*, 2007; Lee *et al.*, 2006; Orenstein *et al.*, 1999).

The identification of a new phytoplasma subgroup in cassava affected by CFSD together with its transmission to healthy periwinkle and cassava (data not shown) indicate that the identified phytoplasma is associated with cassava frogskin disease.

References

Alvarez, E., Mejía, J. F., Loke, J. B., Hernández, L. and Llano, G. A. (2003). Detecting the phytoplasma-frogskin disease association in cassava (*Manihot esculenta* Crantz) in Colombia. *Phytopathology* 93: S4.

Crossley, S. J. and Clark, M. F. (1996). A plate capture PCR method for epidemiological studies with sweet potato little leaf and other phytoplasma diseases. Brighton Crop Protection Conference: *Pests and Diseases* 2: 18-21.

Calvert, L. A., Cuervo, M., Lozano, I., Villareal, N. and Arroyave, J. (2004). Identification of a reolike virus infecting *Manihot esculenta* and associated with cassava frog-skin disease. Sixth international scientific meeting of the cassava biotechnology network. PS4: 68.

Catlin, P. B., Olsson, E. A. and Beutel, J. A. (1975). Reduced translocation of carbon and nitrogen from leaves with symptoms of pear curl. *J. Am. Soc. Hortic. Sci.* 100: 184-187.

Davis, R.E. and Lee, I.-M. (1993). Cluster-specific polymerase chain reaction amplification of 16S rDNA sequences for detection and identification of mycoplasma like organisms. *Phytopathology* 83: 1008-1001.

Davis, R. E., Jomantiene, R., Dally, E. L. and Wolf, T. K. (1998). Phytoplasmas associated with grapevine yellows in Virginia belong to group 16SrI, subgroup A (tomato big bud phytoplasma subgroup), and group 16SrIII, new subgroup I. *Vitis*, 37(3): 131-137.

Deng, S. and Hiruki, C. (1991). Amplification of 16S rRNA genes from culturable and non-culturable mollicutes. *J. Microbiol. Meth.* 14: 53-61.

Dyer, A. T. and Sinclair, W. A. (1991). Root necrosis and histological changes in surviving roots of white ash infected with mycoplasma-like organisms. *Plant Dis.* 75: 814-819.

Duduk, B., Bulajiæ, A., Duduk, N., Calari, A., Paltrinieri, S., Krstiæ, B. and Bertaccini, A. (2007). Identification of phytoplasmas belonging to aster yellows ribosomal group (16SrI) in vegetables in Serbia. Bullettin of Insectology 60: 341-342.

EPPO/CABI (European and Mediterranean Plant Protection Organization and CAB International). (1996). Apple proliferation phytoplasma. In: Quarantine pests for Europe, 2nd ed. Edited by IM Smith; DG McNamara; PR Scott; M Holderness). CAB International, Wallingford, UK.

Galvis, C. A., Leguizamón, J. E., Gaitán, A. L., Mejía, J. F., Alvarez, E. and Arroyave, J. (2007). Detection and Identification of a Group 16SrIII-Related Phytoplasma Associated with Coffee Crispiness Disease in Colombia. *Plant Dis.* 91: 248-252.

Gilbertson, L. and Dellaporta, S. L. (1983). Molecular extraction DNA protocols. In: Molecular biology of plants. Cold Spring Harbor Laboratory, Cold Spring Harbor, NY. pp 395–397.

Gundersen, D. E. and Lee, I-M. (1996). Ultrasensitive detection of phytoplasma by nested-PCR assays using two universal primer pairs. *Phytopath Medit.*, 35: 114–151.

Hall, T. A. (1999). Bio Edit: a user-friendly biological sequence alignment editor and analysis program for Windows 95/98/NT. Nucleic Acids Symposium Series, 41: 95-98.

Lapage S. P., Sneath P. H. A., Lessel E. F., Skerman V. B. D., Seeliger H. P. R. and Clarck W. A. (1992). International code of nomenclature of bacteria: bacteriological code. 1990 Revision. Am. Soc. Microbiol., Washington, D.C.

Lee, I.-M., Gundersen, D. E., Hammond, R. W. and Davis, R. E. (1994). Use of mycoplasmalike organism (MLO) group-specific oligonucleotide primers for nested-PCR assays to detect mixed-MLO infections in a single host plant. *Phytopathology* 84: 559-566.

Lee, I.-M., Gundersen-Rindal, D. E., Davis, R. E. and Bartoszyk, I.M. (1998). Revised classification scheme of phytoplasma based on RFLP analyses of 16SrRNA and ribosomal protein gene sequences. *Int. J. Syst. Bacteriol.*, 48: 1153-1169.

Lee, I.-M., Bottner, K. D., Munyaneza, J. E., Davis, R. E., Crosslin, J. M. du Toit, L. J. and Crosby, Z. (2006). Carrot purple leaf: A new spiroplasmal disease associated with carrots in Washington State. *Plant Disease* 90: 989-993.

Lepka, P., Stitt, M., Moll, E. and Seemuller, E. (1999). Effect of phytoplasmal infection on concentration and translocation of carbohydrates and amino acids in periwinkle and tobacco. *Physiol. Mol. Plant Pathol.* 55: 59-68.

Little, T. M. and F. J. Hills. (1978). Agricultural Experimentation: Design and Analysis. John Wiley and Sons, New York, NY.

Martini, M., Lee, I.-M., Bottner, K. D., Zhao, Y., Botti, S., Bertaccini, A., Harrison, N. A., Carraro, L., Marcone, C., Khan, A. J. and Osler, R. (2007). Ribosomal protein gene-based phylogeny for finer differentiation and classification of phytoplasmas. *Int. J. Syst. Bacteriol.*, 57: 2037-2051.

Maust, B. E., Espadas, F., Talavaera, C., Aguilar, M., Santamaría, J.M. and Oropeza, C. (2003). Changes in carbohydrate metabolism in coconut palms infected with the Lethal Yellowing Phytoplasma. *Phytopathology* 93: 976-981.

Montano, H. G., Davis, R. E., Dally, E. L., Pimentel, J. P. and Brioso, P. S. T. (2000). Identification and phylogenetic analysis of a new phytoplasma from diseased chayote in Brazil. *Plant Dis.*, 84: 429-436.

Orenstein, S., Franck, A., Kuznetzova, L., Sela, I. and Tanne, E. (1999). Association of phytoplasmas with a yellows disease of carrot in Israel. *Journal of Plant Pathology* 81(3): 193-199.

Pineda, B., Jayasinghe, U. and Lozano, J. C. (1983). La enfermedad "Cuero de Sapo" en yuca (Manihot esculenta Crantz). ASIAVA 4: 10-12.

Prince, J. P., Davis, R. E., Wolf, T. K., Lee, I-M., Mogen, B., Dally, E., Bertaccini, A., Credi, R. and Barba, M. (1993). Molecular detection of diverse mycoplasmalike organisms (MLOs) associated with grapevine yellows and their classification with aster yellows, X–disease, and elm yellows MLOs. *Phytopathology*, 83(10): 1130–1137.

Rottem, S. and Naot, Y. (1998). Subversion and exploitation of host cells by mycoplasmas. *Trends Microbiol.* 6: 436-440.

Smart, C. D., Schneider, B., Blomquist, C. L., Guerra, L. J., Harrison, N. A., Ahrens, U., Lorenz, K. H., Seemuller, E. and Kirkpatrick, B. C. (1996). Phytoplasma-specific PCR primers based on sequences of the 16S-23S rRNA spacer region. *Appl. Environ. Microbiol.* 62: 2988-2993.

Tamura, K., Dudley, J., Nei. M. and Kumar, S. (2007). MEGA4: Molecular Evolutionary Genetics Analysis (MEGA) software version 4.0. *Molecular Biology and Evolution*, 24: 1596-1599.

Thompson, J. D., Gibson, T. J., Plewniak, F., Jeanmougin, F. and Higgins, D.G. (1997). The Clustal X windows interface: flexible strategies for multiple sequence alignment aided by quality analysis tools. *Nucleic Acids Research*, 24: 4876-4882.

Chapter 2

Molecular Characterization of *Bipolaris sorokiniana* Causing Spot Blotch of Wheat

Rashmi Aggarwal*, V.B. Singh, M.S. Gurjar and Sangeeta Gupta

Division of Plant Pathology,
Indian Agricultural Research Institute, New Delhi, India

ABSTRACT

One hundred and three isolates of *Bipolaris sorokiniana* infecting wheat collected from different agro-climatic zones of India were studied for their morphological characters, pathogenicity and DNA fingerprinting. All the isolates were categorized into five groups based on their colony characteristics. The frequency of the dull white/greenish black colony type was maximum (38.83 per cent), while both black, suppressed type and white fluffy type colonies showed minimum frequency (11.65 per cent) in the population studied. A total of 40 isolates were selected out of the five identified groups for further studies on pathogenic and molecular variability. Although all the isolates were pathogenic but their pathogenicity on susceptible genotype, Agra local varied. Isolate BS-49 showed 4.5 average infection index (ADI) and BS-75 exhibited 63.4 ADI. Based on the disease response on differential hosts, five pathotypes were identified. Pathotype 1 constituted maximum isolates and was least virulent as it produced R/MR response on most of the differential hosts. Pathotype 5 was categorized highly virulent as isolates belonging to this category produced S/MS response on most of the differential hosts. RAPD analysis of isolates of these identified pathotype groups showed 100 per cent polymorphism. The amplified DNA band ranged

* E-mail: rashmi.aggarwal2@gmail.com

between 200 bp to 3.5 Kb. The dendrogram generated based on polymorphic data revealed a considerable amount of diversity among the isolates. OPB 18 specifically amplified a DNA band of 260 bp in isolate BS-53 collected from Malan (HP) which could be used as a specific marker to identify strainal differences. (URPs' are universal rice primers, which are derived from DNA repeat sequences in the rice genome) URP-PCR high polymorphism analysis of forty pathogenic isolates showed high polymorphism. Out of the 12 URPs used in the study, 10 markers were effective in producing polymorphic fingerprint patterns from DNA of *B. sorokiniana* isolates. The statistic analysis of entire fingerprint profile differentiated *B. sorokiniana* isolates obtained from different geographic regions. One isolate BS-53 from northern hill zone was different showing less than 50 per cent similarity with rest of the isolates. Broadly, three major clusters are formed, one cluster consisted of isolates from North western plain zone; second cluster having isolates from North eastern plain zone and third cluster consisted of isolates from Peninsular zone showing more than 75 per cent similarity among them. One of the primers, URP-2F (5'GTGTGCGATCAGTTGCTGGG 3') gave three monomorphic bands of size 0.60kb, 0.80kb and 0.90kb which could be used as specific markers for identification of *B. sorokiniana*. Further, based on URP-PCR analysis, the grouping of the isolates according to the geographic origin was possible. This analysis also provided important information on the degree of genetic variability and relationship between the isolates investigated, revealing polymorphism and establishing electrophoretic profiles useful to characterize this phytopathogen.

Keywords: Bipolaris sorokiniana, Wheat, Spot blotch, Molecular variability, RAPD, URP, PCR.

Introduction

Wheat (*Triticum aestivum* L.) is the world's most extensively grown crop and important staple food for about one billion people in as many as 43 countries and provides about 20 per cent of the total food calories (Anon, 1997). In India, bread wheat (*T. aestivum*) is the major species occupying over 90 per cent of wheat area followed by durum wheat (*T. turgidum* var. *durum*) and *T. dicoccum* or khapali wheat. Considerable economic losses occur chiefly due to diseases caused by fungi. Spot blotch, caused by *Bipolaris sorokiniana* (teleomorph: *Chochliobolus sativus*) is one of the most important foliar diseases limiting wheat production in warmer, non-traditional growing areas (Dubin and Ginkel, 1991; Duveiller and Gilchrist, 1994). Yield losses caused by *B. sorokiniana* have been reported to vary from 2.7 to 100 per cent in various countries depending on varieties (Anon, 1997; Villareal *et al.*, 1995; Mehta, 1985; Duveiller and Gilchrist, 1994). In some locations, the disease is preventing wheat from becoming a commercial crop (Duveiller *et al.*, 1998). *B. sorokiniana* besides causing spot blotch also causes common root rot in wheat in Australia and southern Brazil (Diehl *et al.*, 1982, Tinline *et al.*, 1988). In Indian sub-continent, such environments are encountered in the warmer parts of eastern India, most of Bangladesh and tarai region of Nepal. The disease was first recorded in Bihar as early as 1914 by Mohy (HCIO No. 12508) in India. In India the disease causes up to 36 per cent loss in the favorable conditions (Singh and Srivastava, 1997). The pathogen shows enormous morphological and physiological variability due to multinucleate condition of mycelium and conidia, with subsequent heterokaryosis (Day, 1974; Mitra, 1931), This occurs chiefly through the anastomosis between adjacent hyphae, enables the parasexual cycle, which is the main source of genetic diversity in fungi with asexual reproduction (Day, 1974). Fungal species that are pathogenic on plants or animals exhibit variation in many important traits. Variability in morphology can make pathogen identification difficult while physiological variability can affect the evaluation of the damage potential (virulence) of a particular pathogen. The discrimination of species using these traits can be very difficult and gives erroneous results (Meyer *et al.*, 1992). Molecular markers, however can be applied at these levels with great reliability, and they allow simultaneous measurement of variability at multi loci in each individual tested. The development and refinement of molecular techniques have provided tools to advance the study of genetic diversity, of the intra- and inter-specific phylogenetic relationships, and also the identification of races and pathotypes (Mullins and Faloona, 1987; Bruns *et al.*, 1991; Gallego and Martinez, 1997).

PCR-based markers have been employed extensively for confirming genotypes of organisms at level of species and population. The PCR method requires little biological material and provides a rapid method for screening large sample sizes. PCR markers have been developed using arbitrary primers or specifically designed primers from known DNA information such as repetitive sequences. Recently, repeat sequences from Korean weedy rice, originally referred to as universal rice primer (URP) have been used for the fingerprinting of diverse genomes of plants, animals and microbes (Kang *et al.*, 2002). This URP-PCR has been used earlier in molecular analyses of only a very few fungi (Kang *et al.*, 2002; Jana *et al.*, 2005; Kang *et al.*, 2001; Aggarwal *et al.*, 2008). The aim of present investigations is to generate DNA fingerprint profile of isolates of *B. sorokiniana* collected from different geographic locations of India using RAPD and URP primers and to establish genetic variability among the isolates.

Materials and Methods

Fungal Isolation from Infected Wheat Leaves and Maintenance of Monoconidial Cultures

A total of 103 samples of wheat leaves infected with *Bipolaris sorokiniana* were collected from different geographical regions of India divided into 6 zones (Table 2.1). The isolation of fungus, *B. sorokiniana* was made and all cultures were maintained on PDA slants at 4°C. Monoconidial cultures were established and the pathogenicity of isolates was tested.

Morphological Variability and Pathogenicity of Isolates

The colony morphology with respect to colour and growth behaviour were studied among all the 112 isolates of *B. sorokiniana* on potato dextrose agar medium. These isolates were also tested for their pathogenicity on the susceptible genotype, Agra local at seedling stage under glass house conditions. The disease severity was measured in terms of lesion number per leaf and infection index was calculated (Adlakha *et al.*, 1984).

Table 2.1: Isolates of *Bipolaris sorokiniana* Collected from Different Zones of India

Sl.No.	*Zone*	*No. of Isolates*	*Isolate Designation*
1.	NWPZ (North-western plain zone)	46	BS 1-26, BS 86-103
2.	NEPZ (North-eastern plain zone)	30	BS 27-52
3.	NHZ (Northern hill zone)	11	BS 53-60
4.	CZ (Central zone)	4	BS 61-64
5.	PZ (Peninsular zone)	9	BS 65-73
6.	SHZ (Southern hill zone)	12	BS 74-85

Pathogenic Variability

Pathogenic variability among 40 selected isolates was studied by inoculating them on a set of wheat differential hosts. Fourteen wheat genotypes obtained from Division of Genetics, IARI, New Delhi were used as differential hosts (Table 2.4). The experiment was conducted in the green house. A mixture of sandy loam and FYM in the ratio of 1:1 was prepared for sowing of fourteen differentials. Spore suspension of isolates was sprayed on the seedlings @ 10^4 conidia/ml using hand atomizer. The inoculated seedlings were kept under humid conditions for 48 hours and then incubated in alkathene house for 1 week. The leaves showing typical spot blotch lesions were colleted and infection index was calculated after disease severity scoring as per the scale given below:

0= free of spots; 1= necrotic spots without chlorosis, up to 5 per cent leaf area involved; 2=necrotic spots with light chlorosis, 6-20 per cent of the leaf area involved; 3=necrotic spots with pronounced chlorosis, 21-40 per cent of the leaf area involved; 4=41-60 per cent of leaf area involved; 5= spots merging more than 60 per cent of the leaf area involved (Adlakha *et al.*, 1984). Average infection index (AIDX) was transformed into disease responses *viz.*, 0 = No infection; 0-10.0- resistant response (R); 10.1-20.0 =Moderately resistance (MR); 20.1-30.0 =Moderately susceptible(MS); 30.1- 50.0=Susceptible(S); and more than 50.0= Highly susceptible(HS).

Molecular Variability Using RAPD and URP Markers

Extraction of Genomic DNA

DNA was extracted using cetyltrimethyl ammonium bromide (CTAB) method (Murray and Thompson, 1980). Isolates of *B. sorokiniana* were grown in 100 ml potato dextrose broth [20 per cent potato, 2 per cent dextrose (w/v)] taken in 250 ml flask for 5 days at 24±2°C with 12 h photoperiod and 100 rpm shaking in shaker incubator [Kuhner, Biogenetek (I) Pvt. Ltd.]. The mycelia (20g) were harvested and separated by filtration through Whatman No.1, washed thrice with sterile water, and freeze dried at –80°C. The mycelium was finally ground in liquid nitrogen and to this powder 20 ml of DNA extraction buffer (0.1 M Tris, 1.5 M NaCl, 0.01 M EDTA) was added. The suspension was incubated at 65°C for one hour with occasional stirring. Equal volume of Chloroform:Isoamyl alcohol (24:1) was added to each tube followed by centrifugation. The upper aqueous phase so obtained by precipitation with 0.6 volume of ice-cold isopropanol was again centrifuged. The pellet was washed with 70 per cent ethanol and dried at room temperature. Finally, the nucleic acid was dissolved in TE buffer (10 mM Tris-Cl, pH 8.0; 1 mM EDTA); quantified spectrophotometrically and stored at –20°C.

Polymerase Chain Reaction Using RAPD Primers and URP Primers

Thirty Operon primers each of 10 nucleotide in length were used for this study. Amplification was carried out in a thermalcycler (BioRAD, USA). The RAPD-PCR reaction were performed in 25 µl volumes containing 75 ng of genomic DNA, 0.6 mM each dNTP (dATP, dGTP, dCTP and dTTP), 9 µM primer, 3.5 mM $MgCl_2$, 1.5 U *Taq* DNA polymerase and 1 X Taq buffer. Thermalcycler programmed for one cycle of denaturation at 94°C for 5 min followed by 35 cycles of denaturation at 94°C for 1 min, annealing at 35°C for 1 min and extension at 72°C for 2 min. A final extension step at 72°C for 5 min was also performed.

Twelve Universal Rice Primers each of 20 nucleotide in length, derived from repeat sequences of weedy rice (Kang *et al.*, 2002) were synthesized by Genuine Chemical Corporation (GCC), India. Amplification was carried out in a Gradient thermalcycler (BioRAD, USA).

The URP-PCR reactions were performed in 25 µl volume containing 75 ng of genomic DNA of each isolate, 200 µm each dNTP (dATP, dGTP, dCTP and dTTP), 0.2 µM primer, 1.5 mM $MgCl_2$, 2.5 U Taq DNA polymerase and 1 X Taq buffer. Thermal cycler was programmed for one cycle of denaturation at 94°C for 4 min followed by 35 cycles of denaturation at 94°C for 1 min, annealing at 55°C for 1 min and extension at 72°C for 2 min. A final extension step at 72°C for 7 min was also performed.

The URP-PCR products were separated by electrophoresis on a 1.2 per cent (w/v) agarose gel in TBE buffer, stained with ethidium bromide (@0.5ug/ml) and photographed under Gene Genius Gel Documentation System (Syngene Inc, Cambridge, UK). All the amplifications were repeated three times for each primer.

Statistical Analysis

RAPD and URP-PCR data were assessed using the statistical package of the NTsys.pc (v.2.01; Rohlf, 1998). Relation among the 40 isolates of *B. sorokiniana* was estimated by means of scorable DNA bands amplified from different RAPD and URP primers. Each band was considered as character and was scored as either present (coded as 1) or absent (coded as 0). Similarity between isolates was assessed by calculating the simple association coefficient and cluster analysis using the unweighted pair group method with arithmetical averages (UPGMA).

Results

Colony Characteristics

The colony morphology with respect to colour and growth pattern of 103 isolates of *B. sorokiniana* were studied and these were characterized in five different groups. These groups were: (I) black suppressed growth; (II) brown/dull black suppressed growth; (III) gray with white spots cottony growth; (IV) dull white and/greenish black fluffy growth and (V) white fluffy growth (Figure 2.1, Table 2.2). The frequency

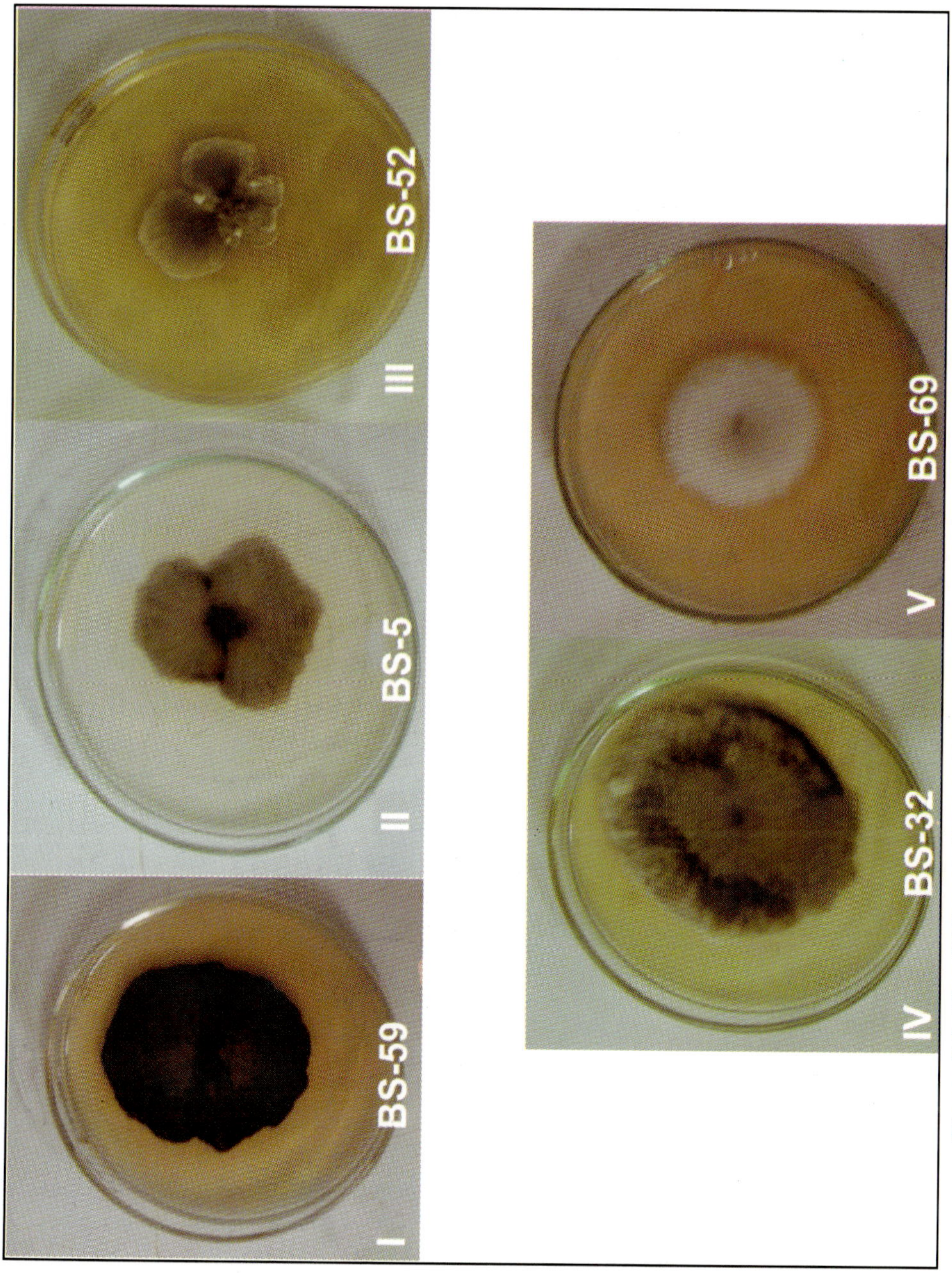

Figure 2.1: Five Different Groups Based on Colony Morphology of *Bipolaris sorokiniana*

of the dull white/greenish black colony type was maximum (38.83 per cent), while both black, suppressed type and white fluffy type colonies showed minimum frequency (11.65 per cent) in the population studied (Table 2.2).

Table 2.2: Morphological Characteristics, Pathogenicity, Infection Index on the Susceptible (Agra Local) Genotype and Frequency Distribution of 103 Isolates to Different Code Groups of *B. sorokiniana*

Group	*Characteristics*	*Isolate Popⁿ (Per cent)*	*Isolates No.*	*Isolates Employed for RAPD Study*	*AIDX (Agra Local)*
I	Black suppressed growth	12 (11.65%)	9, 21, 45, 46, 58, 59, 60, 63, 67, 71, 88, 98	9, 45, 59, 60, 63, 88	15.91
II	Brown/dull black suppressed growth	20 (19.41%)	5, 6, 11, 14, 20, 25, 26, 28, 31, 42, 53, 66, 68, 75, 76, 93, 94, 95, 96, 97	5, 6, 11, 14, 28, 31, 42, 53, 66, 68, 75, 94, 95	28.65
III	Gray with white spots cottony growth	19 (18.44%)	1, 2, 3, 16, 29, 36, 38, 40, 41, 44, 48, 51, 52, 62, 72, 78, 79, 87, 101	1, 3, 29, 41, 48, 52, 72	24.62
IV	Dull white/greenish black fluffy growth	40 (38.83%)	4, 7, 8, 10, 12, 15, 17, 19, 18, 22, 23, 24, 27, 30, 32, 35, 47, 49, 50, 54, 55, 56, 61, 64, 70, 73, 74, 77, 80, 81, 82, 83, 84, 85, 86, 89, 91, 92, 99, 100	27, 30, 31, 32, 49, 54, 55, 74, 77, 91, 92	36.77
V	White fluffy growth	12 (11.65%)	13, 33, 34, 37, 39, 43, 57, 65, 69, 90, 102, 103	34, 65, 69,	15.43

All the isolates of *B. sorokiniana* were pathogenic on susceptible genotype, Agra local although the isolates showed differences in no. of lesions/leaf and infection index. The average infection index of colony type IV was maximum (36.77) and colony type V exhibited minimum infection index (15.43).

Forty representative isolates from each group mentioned above were selected for studies on pathogenic and molecular variability.

Pathogenic Variability

Forty isolates of *B. sorokiniana* were tested for their differential behavior on a set of 14 wheat genotypes at seedling stage. Based on average infection index and disease response on differential hosts; five pathotypes were identified. Pathotype 1 was least virulent having maximum number of isolates, and pathotype 5 was most virulent constituting isolates BS-55,60,63,75 which showed 'S' to 'HS' response on most of the differential hosts (Tables 2.3 a,b).

Table 2.3(a): Grouping of 40 Isolates of *B. sorokiniana* Based on Disease Response on Wheat Genotypes/Differentials

Wheat Genotypes/ Differentials	*Disease Response of Isolates*																								
	Pathotype 1																								
	1	*5*	*6*	*9*	*11*	*14*	*29*	*31*	*32*	*34*	*41*	*42*	*48*	*52*	*53*	*54*	*59*	*65*	*68*	*69*	*72*	*88*	*92*	*94*	*95*
1.	R	R	MR	R	R	R	R	R	R	R	R	R	R	MR	R	R	R	R	R	R	R	R	R	R	R
2.	R	MR	MR	R	R	MR	R	R	MR	R	R	R	R	R	R	R	MR	MR	R	R	R	R	R	R	R
3.	R	MR	MR	R	R	MR	R	R	R	MR	R	R	R	MR	R	R	MR	R	R	R	R	R	R	R	R
4.	R	R	MR	R	R	MR	R	MR	MR	MR	R	R	R	MR	R	R	R	R	R	R	R	R	R	R	R
5.	R	MR	MR	R	R	R	R	MR	R	R	R	R	R	MR	R	R	MR	MR	R	R	R	R	R	R	R
6.	R	MR	MR	R	R	R	R	MR	R	R	R	R	R	R	MR	R	MR	R	R	R	R	R	R	R	R
7.	R	R	R	R	R	R	R	R	R	R	R	R	R	R	R	R	MR	R	R	R	R	R	R	R	R
8.	R	R	MR	R	R	R	R	MR	R	R	R	R	R	MR	R	R	R	R	R	MR	MR	R	R	R	R
9.	MR	R	R	MR	R	R	R	R	R	MR	R	R	R	MR	MR	R	MR	R	R	MR	MR	R	R	R	R
10.	R	R	R	R	R	R	R	MR	R	MR	R	R	R	MR	R	R	MR	R	R	R	MR	R	R	R	R
11.	R	R	R	R	R	R	R	R	R	R	R	R	MR	R	R	R	MR	MS	R	R	MR	R	R	R	R
12.	R	R	R	R	R	MR	R	MR	R	R	R	R	MR	MR	MR	MR	MR	MR	R	MR	MR	R	R	R	R
13.	R	R	R	R	R	R	R	MR	R	R	R	R	MR	MR	MR	R	MR	MR	R	R	MR	R	R	R	R
14.	S	MS	MS	S	MS	MS	MS	S	S	MS	MS	MS	S	MS	HS	S	S	MS	MS	MS	S	S	MS	S	S

Table 2.3(b): Grouping of 40 Isolates of *B. sorokiniana* Based on Disease Response on Wheat Genotypes/Differentials

	Pathotype 2							*Pathotype 3*		*Pathotype 4*		*Pathotype 5*			
	27	*30*	*49*	*66*	*74*	*77*	*79*	*28*	*91*	*45*	*3*	*55*	*60*	*63*	*75*
1.	R	R	MS	R	R	R	MR	R	R	R	R	MR	MR	MR	R
2.	R	R	MS	MR	R	R	MS	R	R	MR	MR	MS	S	S	S
3.	R	R	R	R	R	R	MR	MR	R	R	MR	MS	MS	MS	S
4.	R	R	MR	MR	R	R	MS	MS	MR	S	MS	MS	MS	MS	HS
5.	MR	R	MR	MR	R	MR	MR	MS	S	MR	MR	MS	S	MS	HS
6.	MR	R	R	R	R	R	MR	MR	MR	R	MR	MS	MR	MS	S
7.	MS	MS	R	MR	R	R	R	MR	MR	R	MR	MR	S	MS	HS
8.	MR	MR	R	R	R	R	R	S	MS	R	MR	S	S	HS	HS
9.	MR	R	MR	R	R	R	MR	MR	R	S	MS	S	MS	HS	S
10.	MR	R	R	R	R	R	R	S	MS	S	MS	S	MS	MS	MS
11.	MS	MR	R	R	R	R	R	R	R	MS	MS	MS	MS	S	S
12.	MR	R	MR	R	R	R	MR	MS	MS	MS	S	MR	MR	MR	MS
13.	MR	MR	R	R	R	R	R	MS	MS	MR	MS	MR	MR	MS	MS
14.	HS	MS	R	MS	S	S	MS	S	S	MS	MS	MS	S	S	HS

Genetic Variations

RAPD- PCR

DNA amplification with only 10 random primers out of 30 primers of OPA and OPB series produced reproducible and scorable bands. The fingerprint profile of 40 isolates of *B. sorokiniana* showing polymorphic bands is presented in Figure 2.2. The statistical analysis showed a high degree of polymorphism among the isolates and a total 160 bands were scored among the selected 40 isolates and all of which were polymorphic showing 100 per cent polymorphism (Table 2.4). Some isolates like, BS 34, BS 42, BS 48, BS 49, BS 66 and BS 91 produced a unique band of 2400 bp with primer OPA 10 while a unique band of 1600 bp was produced in isolate BS 32 and BS-34 by primer OPA 20 and a band of 260 bp was obtained by amplification with primer OPB 18 in case of isolate BS 53. Primer OPB 11 produced a DNA band of 2500

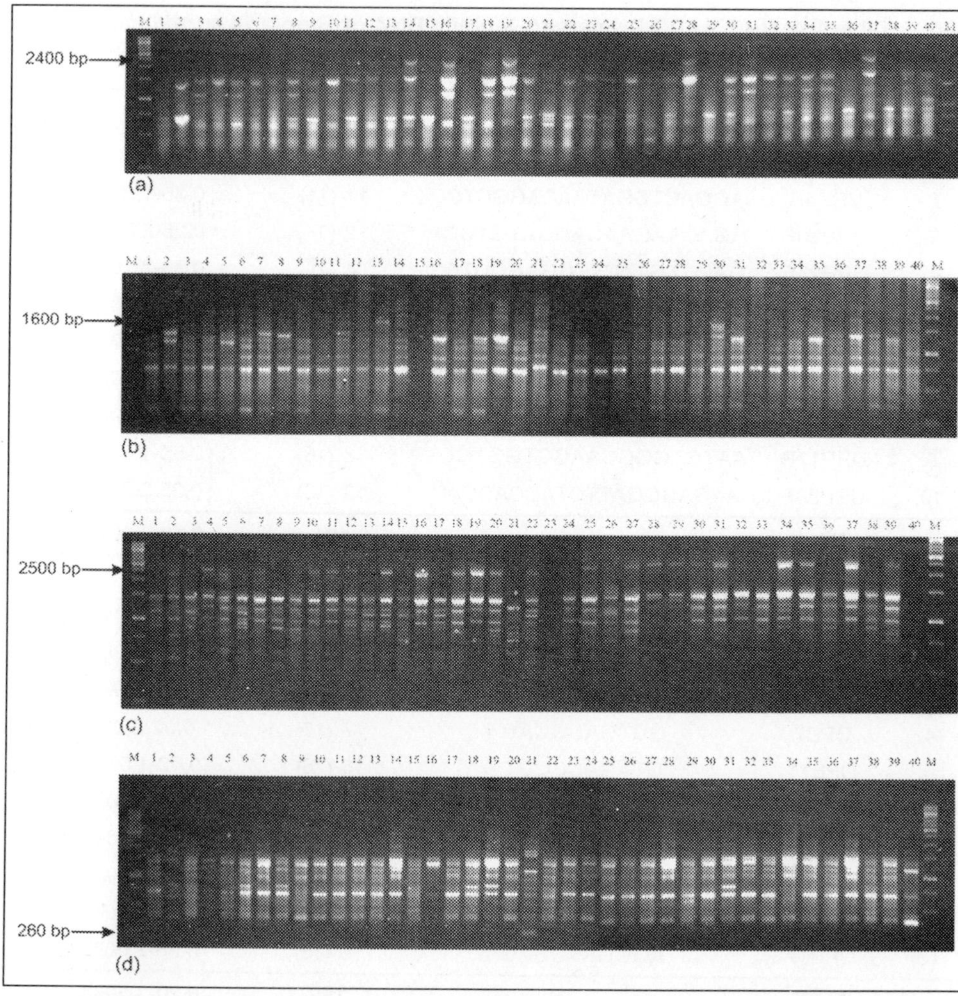

Figure 2.2: Amplification Profiles of 40 Isolates of *Bipolaris sorokiniana* Using Random Primers (a) OPA 10; (b) OPA 20; (c) OPB 11 and (D) OPB 18

bp in isolates BS 6, BS 28, BS 29, BS 30, BS 31, BS 34, BS 42, BS 48, BS 49, BS 52, BS 59, BS 60, BS 65, BS 66, BS 68, BS 69, BS 72, BS 77, BS 79, BS 91, and BS 94. There was no band uniformly amplified in all the isolates. A dandrogram was constructed by the NTSYS-pc programme for all the selected isolates and it was found that the overall similarity indices ranged from 0.91 to 0.46. Maximum similarity was observed between isolates BS 29 and BS 11 both from different agroclimatic zones; and BS 92 and BS 88 both from SHZ, while minimum similarity index was between BS 49 and BS 41 both from NEPZ. Forty isolates belonging to 5 patho-groups of *B. sorokiniana* were divided into a number of clusters by UPGMA cluster analysis. There were total 9 clusters formed. Isolate BS 53 separated out from rest of the clusters showing 59 per cent similarity with all other isolates. BS 48, BS 91, BS 95, BS 69 and BS 53 formed individual branches with less similarity with the isolates of the same group. Among all the isolates BS 53 was the most diverse (Table 2.3).

Table 2.4: URP/RAPD Primers to Amplify DNA of *Bipolaris sorokiniana*

Sl.No.		Primer Sequence	No. of Bands	Fragment Size (Range)
	URP Markers			
1.	URP6R	GGCAAGCTGGTGGGAGGTAC	16 (16)	0.30-2.40kb
2.	URP4R	AGGACTCGATAACAGGCTCC	11 (11)	0.30-1.50kb
3.	URP30F	GGACAAGAAGAGGATGTGGA	9 (9)	0.25-2.00kb
4.	URP25F	GATGTGTTCTTGGAGCCTGT	4 (4)	0.75-2.10kb
5.	URP1F	ATCCAAGGTCCGAGACAACC	11 (11)	0.10-2.00kb
6.	URP2F	GTGTGCGATCAGTTGCTGGG	12 (9)*	0.10-2.00kb
7.	URP9F	ATGTGTGCGATCAGTTGCTG	14 (14)	0.20-1.50kb
8.	URP2R	CCCAGCAACTGATCGCACAC	13 (13)	0.10-1.70kb
9.	URP17R	AATGTGGGCAAGCTGGTGGT	18 (18)	0.50-1.20kb
10.	URP38F	AAGAGGCATTCTACCACCAC	13 (13)	0.25-2.50kb
	Total		**121**	**0.10-2.50kb**
		RAPD Markers		
1.	OPA10	GTGATCGCGC	12 (12)	0.50-1.0kb
2.	OPA18	AGGTGACCGT	21 (21)	0.20-1.0kb
3.	OPB5	TGCGCCCTTC	20 (20)	0.32-3.0kb
4.	OPB7	GGTGACGCAG	17 (17)	0.32-2.0kb
5.	OPB11	GTAGACCCGT	16 (16)	0.40-1.00kb
6.	OPA20	GTTGCGATCC	13 (13)	0.30-3.50kb
7.	OPB2	TGATCCCTGG	13 (13)	0.30-3.00kb
8.	OPB18	CCACAGCAGT	17 (17)	0.28-1.00kb
9.	OPB19	ACCCCCGAAG	15 (15)	0.25-3.50kb
10.	OPB17	AGGGAACGAG	16 (16)	0.35-2.00kb
	Total		**160**	**0.20-3.50kb**

*: Monomorphic bands of 0.60 kb, 80 kb and 0.90 kb.

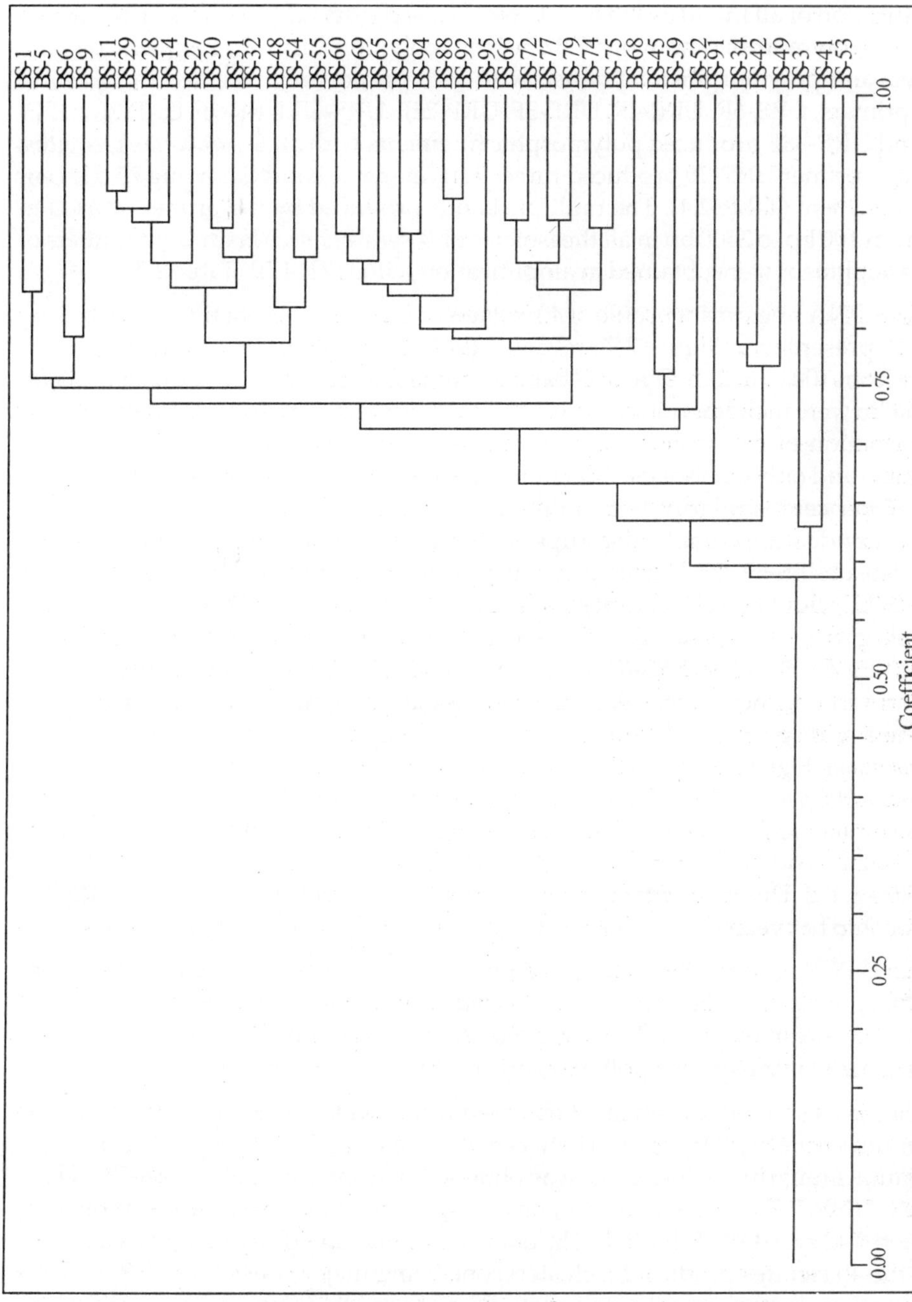

Figure 2.3: Dendrogram of 40 Isolates of *Bipolaris sorokiniana* of Wheat Revealed by UPGMA Cluster Analysis Based on RAPD-PCR Amplification

URP-PCR

Out of 12 URP primers tested, 10 produced reproducible well resolved DNA bands and remaining 2 primers did not show amplification. Genomic DNA amplification of all the 40 isolates of *B. sorokiniana* produced 121 bands out of which 118 were polymorphic showing 97.52 per cent polymorphism. The sequences of the primers and polymorphism obtained with each primer are presented in Table 2.3. Nine primers, URP-6R, URP-4R, URP-3F, URP-25F, URP-1F, URP-9F, URP-2R, URP-17R and URP-38F produced polymorphic fragments with all *B. sorokiniana* isolates, whereas, primer URP-2F produced three monomorphic bands showing 75 per cent polymorphism (Table 2.4). The multiple bands produced by the primers varied in size from 100 bp to 2500 bp in all the isolates of *B. sorokiniana*. Maximum numbers of bands (eighteen) were obtained in amplification with URP-17R (Table 2.4).

The DNA fingerprint profile of 40 isolates of *B. sorokiniana* obtained with primer URP-2F presented in Figure 2.4a revealed that, the fragment size of the amplicons varied from 100bp to 2kb. Out of 12 bands obtained, three bands of size 600bp, 800bp and 900bp were monomorphic (Table 2.4). The dandrogram based on the data showed the formation of one major cluster comprising of 39 isolates having 75 per cent similarity and only one isolate, BS-3 from Asandh (Haryana) separated out showing only 57 per cent similarity with all the other isolates (Figure 2.4b). Major cluster further divided into small subgroups showing only 25 per cent variations among them. Isolates BS-69, BS-72 from peninsular zone (PZ), and BS-88 from southern hill zone (SHZ) showing 100 per cent similarity were grouped with BS-6 from NWPZ in this sub-group (a). Isolates, BS-29, BS-30, BS-31 and BS-42 from Northeastern plain zone (NEPZ); BS-66 and BS-68 (PZ), and BS-92 (SHZ) also showed 100 per cent similarity among them in the other major sub-group (b). Similarly, DNA amplification obtained with primer URP-6R also exhibited variability among the isolates of *B. sorokiniana* (Figure 2.5a). All the amplicons were polymorphic and band size ranged between 300bp to 2400bp. The dendrogram obtained with UPGMA based similarity coefficient showed formation of two clusters (Figure 2.5b). One major cluster consisting of 38 isolates and other minor cluster with only two isolates, BS-53 from NHZ and BS-63 from PZ. This data obtained with individual primers on analysis indicated no relationship between DNA fingerprinting and geographical origin of isolates.

URP-PCR results obtained in the present study enabled a fast and efficient variability analysis for *B. sorokiniana*, collected from different geographical regions of India. These results also led to the generation of electrophoretic profiles, which discriminated intra-specific polymorphism in the isolates studied.

Jaccard's similarity coefficient obtained from combined data analysis showed maximum similarity (99 per cent) between BS-77 and BS-79 both from SHZ whereas minimum similarity (37 per cent) was obtained between the isolates BS-53 (NHZ) and BS-3 (NWPZ). All the other isolates presented a similarity coefficient ranging between 0.42 and 0.97 (Table 2.5). The dandrogram obtained with UPGMA analysis of all the 40 isolates produced 2 clusters; one being major consisting of 39 isolates and the other cluster had only one isolate, BS-53 from NHZ (Figure 2.6). Further four different sub-clusters were identified in the major cluster, out of which 3 sub-groups

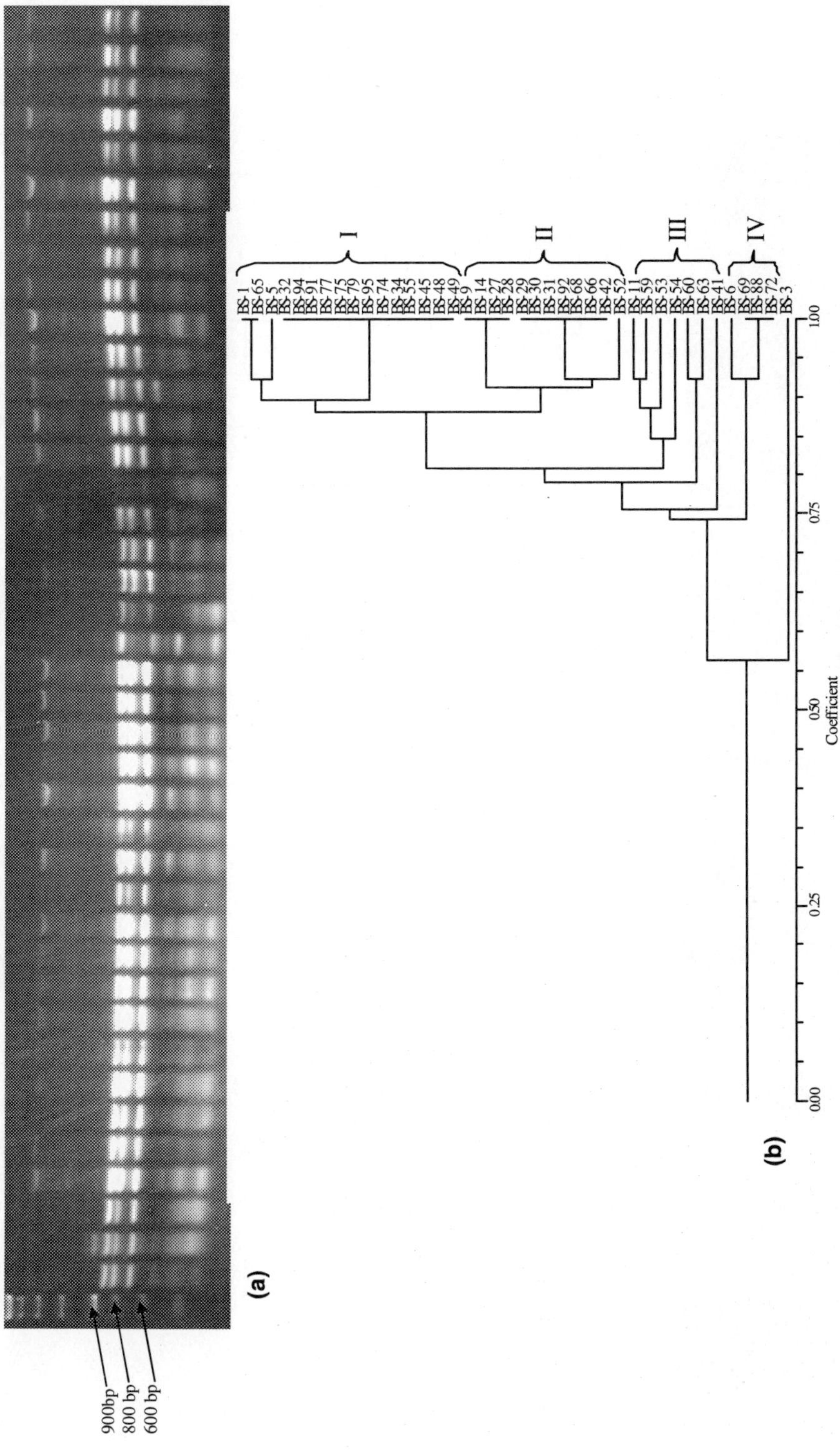

Figure 2.4: DNA Fingerprint Profiles of Different Isolates of *Bipolaris sorokiniana* Obtained with Primer URP-2F
(a) M is 1 Kb ladder marker (MBI, Fermentas), Lanes 1-40, isolates of *B. sorokiniana.*
(b) Dendrogram obtained from 40 isolates of *Bipolaris sorokiniana* with UPGMA based similarity coefficient with URP-2F.

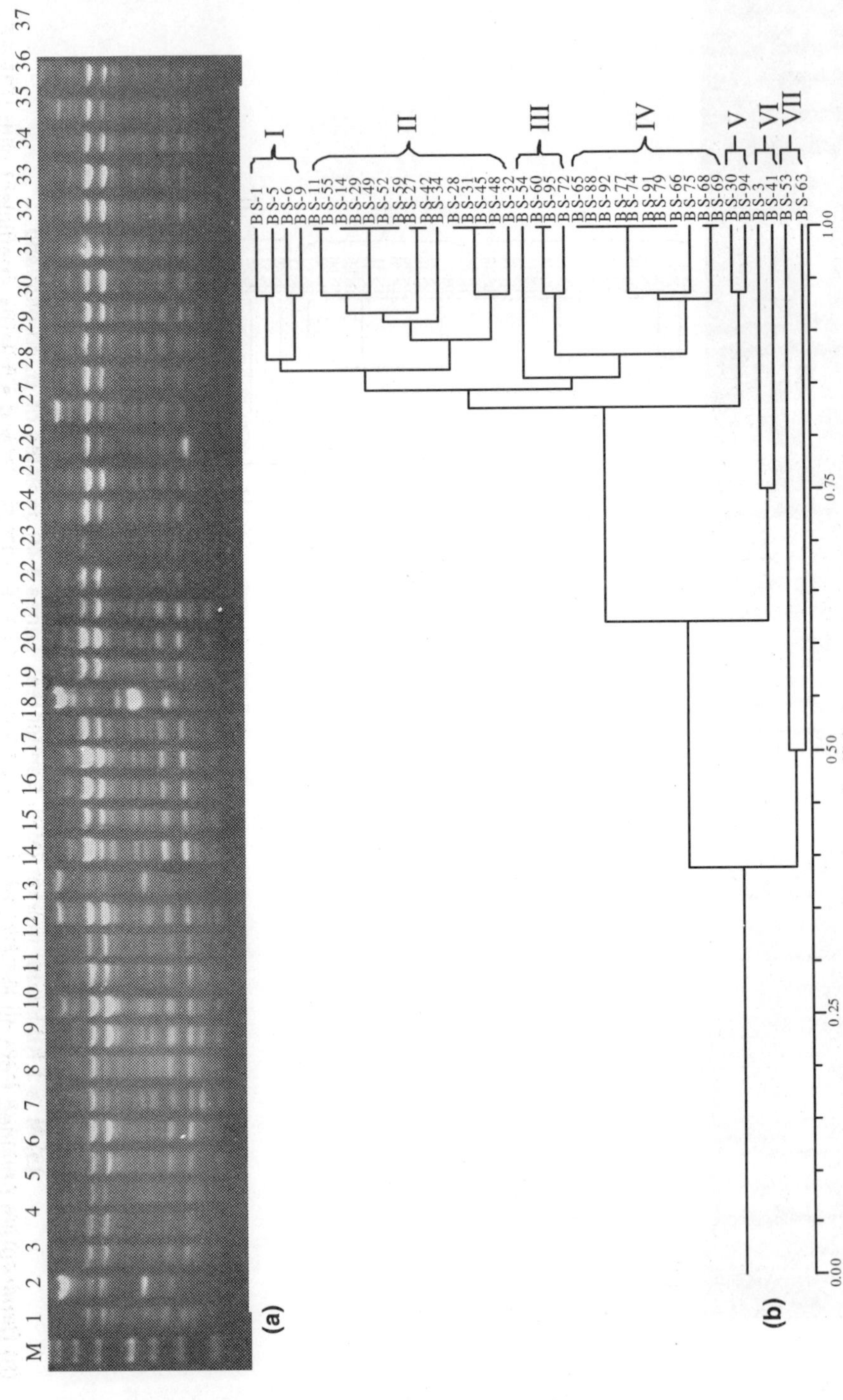

Figure 2.5: DNA Fingerprint Profiles of Different Isolates of *Bipolaris sorokiniana* Obtained with Primer URP-6R
(a) M is 1 Kb ladder marker (MBI, Fermentas), Lanes 1-40, isolates of *B. sorokiniana*.
(b) Dendrogram obtained from 40 isolates of *Bipolaris sorokiniana* with UPGMA based similarity coefficient with URP-6R.

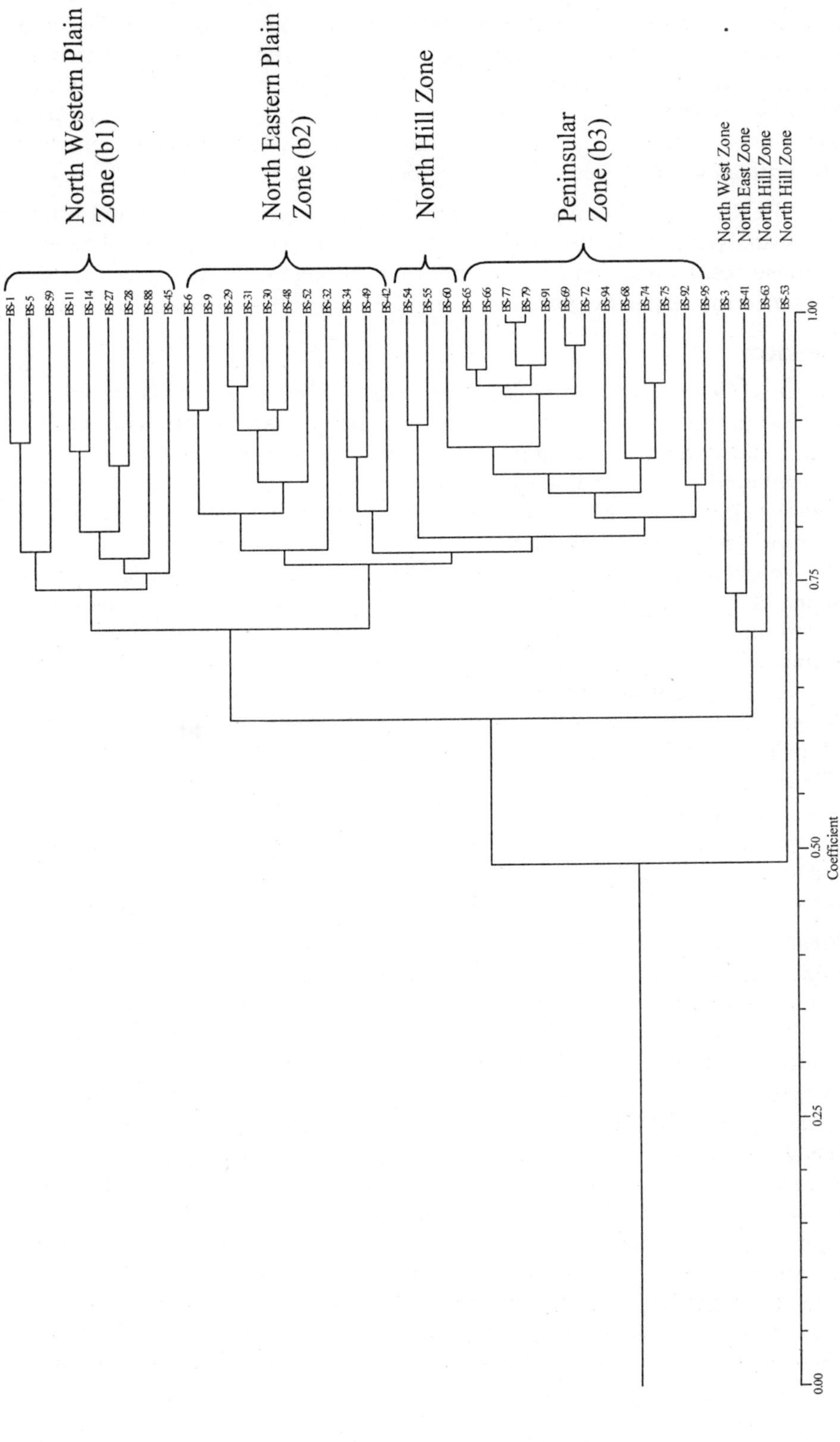

Figure 2.6: Dandrogram Obtained from Combined Data of URP-PCR of Forty Isolates of *Bipolaris sorokiniana*

could be broadly correlated with their zone of origin showing up to 75 per cent similarity among them. Considering the dandrogram prepared based on combined data analysis, a correlation between genetic variability and geographic origin of isolates was observed. The sub-cluster b consisted of isolates mainly from NWPZ (b1); sub-cluster b2 had isolates from NEPZ and in sub-cluster b3 isolates were mainly from PZ with a few exceptions, like BS-54, BS-55 and BS-60 which were from NHZ.

URP-PCR results obtained in the present study enabled a fast and efficient variability analysis for *B. sorokiniana*, collected from different geographical regions of India. These results also led to the generation of electrophoretic profiles, which discriminated intra-specific polymorphism in the isolates studied.

Discussion

Bipolaris sorokiniana is a hemibiotrophic pathogen which infects both wheat and barley. Host pathogen interaction studies have shown that the pathogen is initially biotrophic subsequently turning into necrotrophic phase (Aggarwal *et al.*, 2008 a). Our present work on morphological, pathological and molecular variations has been done taking a large number of isolates from different agro-ecological zones of India. Five groups based on colony characteristics have been defined out of 103 isolates studied. The frequency of the dull white/greenish black colony type was maximum (38.83 per cent), while both black, suppressed type and white fluffy type colonies showed minimum frequency (11.65 per cent) in the population studied. Our data on pathogenicity of these groups on susceptible genotype, Agra local indicated variations in terms of infection index. Average infection index of colony type IV isolates was maximum (36.77). Earlier studies have also indicated a high level of morpho-pathological variability in the pathogen (Chand *et al.*, 2003; Mishra, 1981; Nelson, 1960). Based on the reaction on differential hosts five pathogenic groups were identified which were different from groups formed based on morphological variations. Patho-group 5 having 4 isolates, BS55, 60, 63, and BS 75 was highly virulent, therefore, can be used for creating artificial epiphytotics for evaluating resistance in the breeding programme.

Bipolaris sorokiniana reproduces asexually; therefore, the cause of genetic variability is parasexual recombination (Tinline, 1962). Another possible cause of variability in spot blotch pathogen was suggested to be the variable rearrangements of one to six nuclei per cell (Chand *et al.*, 2003). According to Burdon and Silk (1997), plant pathogenic fungi most commonly rely on mutation and recombination as the main source of genetically based variations. Within a species, gene flow between populations supplement these processes, as propagules spread from one epidemiological area to another. Gene flow along with other evolutionary forces can result in the spread of single gene or DNA sequences and even in the establishment of the whole populations in different regions (McDermott and McDonald, 1993). Although morphological and pathological traits can be used for preliminary screening and grouping of isolates (Gordon and Martyn, 1998), molecular markers are more precise in characterizing the isolates for genetic variability. Genetic variability through RAPD markers in 40 isolates showed high level of polymorphism. Nine clusters were identified indicating more variations at molecular level. The genetic similarity coefficient varied between 0.46 and 0.91 indicating much variation among the isolates. A unique band of 1600 bp was amplified in isolate BS 32 and BS-34 by primer OPA 20 and a band of 250 bp was obtained by amplification with primer OPB 18 in case of isolate BS 53. These primers could be used as markers for location specific strains, as BS 32 and BS 34 both are from Northwestern zones and BS 53 is from Hill zone. The result of the present study revealed that the isolates of *B. sorokiniana* could be distinguished based on culture characters, physiological characters and RAPD analysis, however, RAPD analysis illustrated that identification of *B. sorokiniana* isolates based on physiological and morphological data is not sufficient due to similarities among the isolates as was evident from the five groups formed, but genetic variations based on RAPD analysis grouped the isolates into nine clusters. Therefore, utility of PCR-RAPD as a specific and sensitive method for genetic variations is emphasized (Henson and French, 1993). Earlier, molecular polymorphisms have been successfully used in characterizing the population linkages within a species (Aggarwal *et al.*, 2008 b; Crawford *et al.*, 1993; Soltis *et al.*, 1992). Our results would be useful in characterizing the Indian isolates of spot blotch pathogen and the use of specific isolates in the resistance breeding programmes of wheat and barley that are highly prone to this disease in warm and humid environments.

Our URP-PCR based results showed that 39 isolates out of 40 collected from different regions of India originating from different wheat cultivars showed >50 per cent variability, suggesting high level of genetic variability among the isolates. Although RAPD method has been successfully used to characterize fungi at the intra and inter-specific level (Gulthrie *et al.*, 1992; Assigbetse *et al.*, 1994; Nicholson and Rezanoor, 1994; Lehman *et al.*, 1992) but this is the first report of use of URP markers to characterize this fungus. There are very few reports on URP-PCR to characterize variability in phytopathogenic fungi such as, *Macrophomina phaseolina* (Jana *et al.*, 2005); *Pleurotus* spp. (Kang *et al.*, 2001) and *Chaetomium* spp. (Aggarwal *et al.*, 2008 b). Our studies have indicated that these rice repeat sequences, from where URP primers have been designed, are conserved in the fungi also, making us enable to detect variations at intraspecific levels. Earlier, genetic characterization of this fungus has been done using RAPD markers on Brazilian isolates by Muller *et al.*, 2005. They observed that 19 isolates out of 20 tested formed a single group having a similarity coefficient of >78 per cent. In our study we observed that all the isolates showed > 50 per cent variability and isolate BS-53 separated out of the major cluster, which is from NHZ. It may be noted here that NHZ is not a hot spot area for this disease, as spot blotch infection requires warm and humid conditions which are more prevalent in western, eastern and peninsular zones of India. Zhong and Steffenson (1999) assessed genetic diversity of *Cochliobolus sativus* based on virulence and AFLP markers, and no close correlation between genetic similarity measured by AFLP markers and geographic origin was observed. But, in our study, we could correlate the URP-PCR data with geographic origin of isolates. Our results also agree with the earlier findings of Oliveira *et al.* (1998) who observed extensive polymorphism among different isolates of *B. sorokiniana* and *B. oryzae*. Nucleotide alterations, insertions and deletions at initiation sites may result in polymorphic DNA, which is detectable by the PCR based technique such as RAPD (Williams *et al.*, 1990) or URP-PCR (Kang *et al.*, 2001).

In the present study, the multinucleated conditions of *B. sorokiniana*, mycelial cells and conidia, with subsequent heterokaryosis which in turn could lead to mitotic recombination and new haploidization arrangements may account for a likely contribution to the DNA polymorphism detected for this pathogen. Only one of the URP primers, URP-2F gave monomorphic bands (0.60kb; 0.80 kb and 0.90kb). These fragments can be used as specific molecular markers to identify *B. sorokiniana* isolates.

Acknowledgements

The authors are thankful to Department of Biotechnology, Government of India for financial support in the form of project funding (BT/PR4436/AGR/02/216/2003) and Head, Division of Plant Pathology, IARI, New Delhi-110012, India for providing facilities.

References

Adlakha, K.L., Wilcoxson, H.D. and Raychaudhury, S.P. (1984). Resistance of wheat to leaf spot caused by *Bipolaris sorokiniana*. *Plant Dis.*, 68: 320-321.

Aggarwal, R., Sharma, Vandana, Kharbikar, Lalit, L. and Renu. (2008). Genetic differentiation of *Chaetomium species* using URP-PCR. *Genetics and Molecular Biology* (Brazil) (communicated).

Aggarwal, R., Das, Soma, Jahani, Mehdi, and Singh, D.V. (2008). Histopathology of spot blotch [*Bipolaris sorokiniana* (teleomorph: *Cochliobolus sativus*)] infection in wheat. *Acta Phytopathologica* 43: 23-30.

Ahmed, A.V., Rahman, M.Z., Bhuiyan, K.A. and Mian, I.H. (1997). Variation in isolates of *Bipolaris sorokiniana* from wheat. *Bangladesh Journal of Plant Pathology* 13 (1-2): 29-35.

Anonymous. (1997). Vision 2020: Perspective Plan. Directorate of Wheat Research, Karnal, India. pp.104.

Assigbetse, K.B., Fernandez, D., Dubois, M.P. and Geiger, J.P. (1994). Differentiation of *Fusarium oxysporum* f. sp. *vasinfectum* races on cotton by random amplified polymorphic DNA (RAPD) analysis. *Phytopathology* 84: 622-626.

Batista, M.F. (1993). Métodos moleculares para identificação de patógenos de plantas. *Rev. Anu. Patol. Plantas.* 1: 165-196.

Bruns, T.D., White, T.J. and Taylor, J.W. (1991). Fungal molecular systematics. *Ann. Ver. Ecol. Syst.* 22: 525-564.

Burdon, J.J., and Silk, J. (1997). Sources and patterns of diversity in plant pathogenic fungi. *Phytopathology*, 87: 664-669.

Chand, R., Pandey, S.P., Sandeep, Kumar. and Joshi, A.K. (2003). Variability and its probable cause in the natural populations of spotblotch pathogen *Bipolaris sorokiniana* of wheat (*T. aestivum*). *Ind. J.Plant Dis. Protect.* 110: 27-35

Crawford, D., Brauner, S., Cosner, M.B. and Striessy, T.F. (1993). Use of RAPD markers to document the origin of the intergeneric hybrids *Margyracaena scottsbergii* (Rosaceae) on the Jaun Frenadez Island. *American J. Botany*, 80: 89-92.

Duveiller, E.L., and Gilchrist, L. (1994). Production constraints due to *Bipolaris sorokiniana* in wheat: current situation and future prospects. In: *Wheat in Heat Stressed Environments: Irrigated Dry Areas and Rice-Wheat Farming System* (Eds. Saunders, D.A. and Hettel, G.P.). CIMMYT, Mexico. pp. 343-352.

Duveiller, E., Garcý´a, I., Toledo, J., Franco, J., Crossa, J. and Lopez, F. (1998). Evaluating spot blotch resistance of wheat: improving disease assessment under controlled conditions and in the field. In: *Proceedings of the International Workshop on Helminthosporium Diseases of Wheat: Spot Blotch and Tan Spot* (Duveiller, E., Dubin, H.J., Reeves, J., Mcnab, A., eds.) CIMMYT, El Bata´ n, Mexico, February 9–14, 1997 171–81.

Day, P.R. (1974). Genetics and host-pathogen interaction W.H. Freeman and Company, San Francisco, CA, USA.

Diehl, J.A., Tinline, R.D., Kochhann, R.A., Shipton, P.J. and Rovira, A.D. (1982). The effect of fallow periods on common root rot of wheat in Rio Grande do Sul, Brazil. *Phytopathology*.72: 1297–301.

Dubin, J. and Ginkel, M. (1991). The status of wheat diseases and disease research in warmer areas. In: Wheat for the Non-traditional Warm Areas (Eds. Saunders, D.A.). pp. 125-151, CIMMYT, Mexico, pp. 549.

Gallego, F.J. and Martinez, I. (1997). Method to improve reliability of random-amplified polymorphic DNA markers. *Biotechniques*. 23: 663-664.

Gordon, T.R. and Martya, R.D. (1998). The evolutionary biology of Fusarium oxysporum. *Annu. Rev. Phytopathology*, 35: 111-128.

Gulthrie, P.A.L., Magill, C.W., Frederkisen, R.A. and Odvordy, G.N. (1992). Random amplified DNA markers: a system for identifying and differentiating isolates of *Colletotrichum graminicola*. *Phytopathology*, 82: 832-835.

Henson, J.M. and French, R. (1993). The polymerase chain reaction and plant disease diagnosis. *Ann. Rev. Phytopathology*, 31: 81-109.

.Jana, T.K., Singh, N.K., Koundal, K.R. and Sharma, T.R. (2005). Genetic differentiation of charcoal rot pathogen, *Macrophomina phaseolina* into specific groups using URP-PCR. *Cand. J. of Microbiol.*, 51(2): 159-164.

Kang, H.W., Park, D.S., Go, S.J. and Eun, M.Y. (2001). Genomic differentiation among oyster mushrooms (*Pleurotus* spp.) cultivars released in Korea by URP-PCR. *Mycobiology*. 29: 85-89.

Kang, H.W., Park, D.S., Park, Y.J., You, C.H., Lee, B.M., Eun, M.Y., and Go, S.J. (2002). Fingerprinting of diverse genomes using PCR with universal rice primers generated from repetitive sequence of Korean weedy rice. *Molecules and Cells*. 13: 281–287.

Lehmann, P.F., Lin, D. and Lasker, B.A. (1992). Genotypic identification and characterization of species and strains within the genus *Candida* by using random amplified polymorphic DNA. *Journal of Clinical Microbiology*. 30: 3249-3254.

Mc Dermott, J.M. and Mc Donald, B.A. (1993). Gene flow in plant pathosystems. *Annual Review of Phytopathology*, 31: 335-373.

Mehta, Y.R. (1985). Breeding wheats for resistance to spot blotch. In: *Wheat for More Tropical Environments* (Eds. Villareal, R.L. and Klatt, A.R.). pp. 135-144. Proceeding of the international symposium. CIMMYT, Mexico. pp. 354.

Meyer, W., Morawetz, R., Börner, T. and Kubicek, C.P. (1992). The use of DNA fingerprinting analysis in classification of some species of the *Trichoderma* aggregate. *Current Genetics*. 21: 27-30.

Mitra, M. (1931). Saltation in the genus *Helminthosporium*. *Trans. Brit. Mycol. Soc.* 16: 115-117.

Mishra, A.P. (1981). Variability, physiological specialization and genetics of pathogenesity in graminicolus *Helminthosporium* affecting cereal crops. *Indian Phytopath.* 34: 1-22.

Muller, M.V.G., Germani, J.C. and Van Der Sand, S.T. (2005). The use of RAPD to characterize *Bipolaris sorokiniana* isolates. *Genet. Mol. Res.* 4 (4): 642-652.

Mullins, K.B. and Faloona, F.A. (1987). Specific synthesis of DNA *in vitro* via a polymerase-catalised chain reaction. *Methods Enzymol.* 155: 335-350.

Murray, M.G. and Thompson, W.F. (1980). Rapid isolation of high molecular weight plant DNA. *Nucleic Acids Research*. 8 (19): 4321-4326.

Nelson, R.R. (1960). Evolution of sexuality and pathogenicity in interspecific crosses in genus *Helminthosporium*. *Phytopathology*, 50: 375-377.

Nicholson, P. and Rezanoor, H.N. (1994). The use of random amplified polymorphic DNA to identify pathotype and detect variation in *Pseudocercosporella herpotrichoides*. *Mycol. Res.* 98: 13-21.

Oliveira, A.M.R., Matsumura, A.T.S., Prestesi, A.M., Matos, C.S. and Sand, S.T. (1998). Morphological variability and pathogenicity in isolates of *Bipolaris sorokiniana*. *Phytopathologia-Brasileira* 33 (3): 319-353.

Rohlf, F.J. (1998). *NTSYS–PC. Numerical Taxonomy and Multivariate Analysis System, Version 2.02.* Exeter Software: Setauket (New York).

Singh, D.V. and Srivastava, K.D. (1997). Foliar blights and Fusarium scab of wheat. Present status and strategies for management. In: *Management of Threatening Plant Diseases of National Importance*. Malhotra Publishing House, New Delhi. pp. 1-16.

Soltis, P.S., Soltis, D.E. and Doyle, J.J. (1992). Molecular Systematics of Plants. Chapman and Hall, Inc. New York.

Tinline, R.D. (1962). *Cochliobolus sativus* V. *Heterokaryosis* and parasexuality. *Can. J. Bot.* 40: 425-437.

Tinline, R.D., Wildermuth, G.B. and Spurr, D.T. (1988). Inoculum density of *Cochliobolus sativus* in soil and common root rot of wheat cultivars in Queensland. *Australian Journal of Agricultural Research*. 39: 569–77.

Villareal, R.L., Mujeeb-Kazi, A., Gilchrist, L.I. and Del Toro, E. (1995). Yield loss to spot blotch in spring bread wheat in warm non-traditional wheat production areas. *Plant Disease*, 79: 893-897.

Williams, J.G.K., Kuberlik, A.R. and Kenneth, J. L. (1990). DNA polymorphisms amplified by arbitrary primers are useful as genetic markers. *Nucleic Acids Research.* 18: 6231-6235.

Zhong, S. and Steffenson, B.J. (1999). Molecular mapping of loci conferring virulence on barley and wheat. *Fungal Genetics newsletter*: 46 (Suppl.), 122.

Chapter 3

Trichoderma: Biology and its Potential for Biocontrol

Pranav Chettri*, Mala C. Ganiger and P.U. Krishnaraj

Institute of Agri-Biotechnology, University of Agricultural Sciences, Krishinagar, Dharwad – 580 005, Karnataka, India

ABSTRACT

The ultimate goal of agriculture is to feed the world's growing population. Achievement of higher productivity and production to meet the everlasting demand of food commodities is the need of the hour. Loss due to the pests is one of the major limiting factors for sustained increase in crop productivity and production but growth in this sector is influenced by various external factors like monsoon, pest and diseases, economic and social factors. More than 1,00,000 kinds of fungi exist, of which about 8,000 fungi can cause diseases in plants, and a relatively small number of them cause disease in humans and livestock. Almost all the agricultural and horticultural crop species suffer severe yield losses due to fungal diseases. In the Indian context, fungal diseases are rated as the most important factor causing yield losses in major cereal, pulse, fruit and oilseed crops. Various strategies are being employed to control fungal plant pathogens. So this article is mainly focused on the biology and the use of *Trichoderma* as a potential biocontrol agent.

Keywords: Trichoderma, Biological control, Chitinases, Glucanases, Transgenic.

Strategies to Control Plant Diseases

The efforts on effective disease control are being made since the mid 1600, when it was reported that some species and varieties were more resistant to a disease than

* E-mail: pronovoo3@gmail.com

the others. Knowingly or unknowingly, farmers had been selecting these resistant plants (Fokunang *et al.*, 2004). New technology in all areas has improved agricultural production, but at the cost of environment. So the recent challenge faced by modern agriculture is to achieve satisfactory control of plant diseases in an environment friendly manner (Lorito and Scala, 1999). Strategies include biological control, breeding of resistant varieties, improved cultural practices, storage conditions that are less favorable for pathogen attack and survival and integrated pest management (IPM). Pesticides and organic compounds are widely used to control plant pathogens in many countries. However, the non-degradable components of these compounds have accumulated over the years and entered the food chain, causing toxicity in animals (Cigdem and Merih, 2003; Chet, 1987; Lynch, 1990). Hence, it is compelling to look for alternative disease management practices, which include the use of biocontrol agents, pathogen-resistant crop cultivars and other strategies.

Biological Control Agents (BCAs) for Plant Diseases

Biological control of plant pathogens is an attractive proposition to decrease heavy dependence of modern agriculture on costly chemical fungicides, which not only cause environmental pollution but also lead to the development of resistant strains (Harjono and Widyastuti, 2001). Biological control of plant disease is defined as the involvement and the use of beneficial microorganisms, such as specialized fungi and bacteria, to attack and control plant pathogens and the diseases they cause (Lewis and Papavizas, 1991). Different biological control agents (BCAs) can be used for the control of diseases. These include bacteria, fungi and actinomycetes. The most important BCAs belong to the genus *Trichoderma.*

Trichoderma as a Biocontrol Agent (BCA)

The success of *Trichoderma* as BCAs is due to their high reproductive capacity, ability to survive under very unfavorable conditions, efficiency in the utilization of nutrients, capacity to modify the rhizosphere, aggressiveness against phytopathogenic fungi, and efficiency in promoting plant growth and defense mechanisms. These properties have made *Trichoderma* a ubiquitous genus present in any habitat and at high population densities (Chet *et al.*, 1997). *Trichoderma* BCAs control ascomycete, deuteromycete and basidiomycete fungi, which are mainly soil-borne but control airborne pathogens (Monte, E., 2001). *Trichoderma* is more efficient in acidic than alkaline soils. Excellent results of integrated control have been attained with strains of *T. virens* and metalaxyl against *Pythium ultimum,* infecting cotton (Chet *et al.*, 1997), *T. harzianum* and captan against *Verticillium dahliae,* infecting potato (Chet *et al.*, 1994), *T. virens* and thiram against *Rhizoctonia solani* infecting tobacco and others (Chet *et al.*, 1997).

Trichoderma spp. has proved effective and selective enough to eliminate many plant diseases. The mode of action of *Trichoderma* is suggested to be either by mycoparasitism, competition, antibiosis or a combination of all these (Elad, 2000).

Biocontrol by Competition for Nutrients

Starvation is the most common cause of death for microorganisms; here competition for limiting nutrients results in biological control of fungal phytopathogens (Chet *et al.*, 1997). For instance, in most filamentous fungi, iron uptake is essential for viability, and under iron starvation, most fungi excrete low molecular weight ferric-iron specific chelators, termed siderophores, to mobilize environmental iron. Subsequently, iron from the ferri-siderophore complexes is recovered via specific uptake mechanisms. In *Aspergillus fumigatus* and *Aspergillus nidulans*, siderophore biosynthesis is negatively regulated by carbon source (Esindle *et al.*, 2004). In *Ustilago maydis*, gene products related to iron uptake affect the development of plant disease (Mcintyre *et al.*, 2004). Some *Trichoderma* BCAs produce highly efficient siderophores that chelate iron and stop the growth of other fungi (Chet *et al.*, 1994). In addition, *T. harzianum* T35 controls *Fusarium oxysporum* by competing for both rhizosphere colonization and nutrients, with biocontrol becoming more effective as the nutrient concentration decreases (Tjamos *et al.*, 1992). Competition has proved to be particularly important for the biocontrol of phytopathogens such as *Botrytis cinerea*, the main pathogenic agent during the pre- and post-harvest in many countries (Latorre *et al.*, 2001). *Trichoderma* has a superior capacity to mobilize and take up soil nutrients compared to other organisms. The efficient use of available nutrients is based on the ability of *Trichoderma* to obtain ATP from the metabolism of different sugars, such as those derived from polymers wide-spread in fungal environments: cellulose, glucan and chitin among others, all of them rendering glucose (Chet *et al.*, 1997).

Biofertilization and Stimulation of Plant Defense Mechanisms

Plant Root Colonization and Defense Mechanism

Trichoderma strains must colonize plant roots before stimulation of plant growth and protection against infections. Colonization implies the ability to adhere and recognize plant roots, penetrate the plant, and withstand toxic metabolites produced by the plants in response to invasion by a foreign organism, whether pathogen or not. Mycorrhizal interaction is modulated by plant flavonoids and fungal auxins, followed by morphogenetic events that include appressorium development. In addition, genes that encode hydrophobins and other cell-wall structural proteins are specifically expressed, or their expression is up regulated (Franken *et al.*, 2002). Hydrophobins and repellents are small, functionally similar to hydrophobic proteins, that play fundamental roles in fungal morphogenesis, including infection structures, hyphal aggregation, cell to cell communication, and attachment of hyphae to hydrophobic surfaces and adhesion (Kershaw and Talbot, 1998). Some *Trichoderma* strains establish long-lasting colonization of plant roots and penetrate into the epidermis, where they release compounds that induce localized or systemic plant resistance responses. Plants react against fungal invasion by synthesizing and accumulating phytoalexins, flavonoids and terpenoids, phenolic derivatives, aglycones and other antimicrobial compounds. *Trichoderma* strains are generally more resistant to these compounds than most fungi, their ability to colonize plant roots strongly would then depend on the capacity of each strain to tolerate them. This resistance is considered as an essential requirement for plant colonization (Harman *et al.*, 2004). In another mechanism proposed to explain biocontrol activity of *Trichoderma* species (Yedida *et al.*, 2000) showed that inoculation of cucumber roots with *T. harzianum* induced an array of pathogenesis-related proteins, including a number of hydrolytic enzymes. Plants treated with a chemical inducer (2, 6-dichloro isonicotinic acid) of disease resistance displayed defense responses that were similar to those of plants inoculated with the biocontrol agent. Howell *et al.* (2000) demonstrated that seed treatment of cotton with biocontrol preparations of *T. virens* or application of *T. virens* culture filtrate to cotton seedling radical induced synthesis of much higher concentrations of the terpenoids desoxyhemigossypol, hemi gossypol and gossypol in developing roots than those found in untreated controls.

Biofertilization

Root colonization by *Trichoderma* strains frequently enhances root growth and development, crop productivity, resistance to abiotic stresses and the uptake and use of nutrients (Arora *et al.*, 1992). Crop productivity in fields can increase up to 300 per cent after the addition of *Trichoderma hamatum* or *Trichoderma koningii.* In experiments carried out in greenhouses, there was also a considerable yield increase when plant seeds were previously treated with spores from *Trichoderma* (Chet *et al.*, 1997). The same increase was observed when seeds were separated from *Trichoderma* by a cellophane membrane, which indicates that *Trichoderma* produces growth factors that increased the rate of seed germination (Benitez *et al.*, 1998). However, there are very few reports on growth factors which have been detected and identified in the laboratory, despite the identification of many filamentous fungi that produce phytohormones such as indole acetic acid (IAA) and ethylene (Arora *et al.*, 1992). *Trichoderma* strains that produce zeatin, gibberellin or related cytokinin-like molecules have been recently detected. The controlled production of these compounds could improve biofertilization (Osiewacz, 2002). Together with the synthesis or stimulation of phytohormone production, most *Trichoderma* strains acidify their surrounding environment by secreting organic acids, such as gluconic, citric or fumaric acid (Gomez-Alarcon *et al.*, 1994).

Rhizosphere Modification

One of the mechanisms of *Trichoderma* strains for achieving colonization and pathogen control in a dynamic pH environment is that they appropriately respond to each given pH condition. Some strains of *T. harzianum* control external pH strictly, ensuring optimal conditions for their own, secreted enzymes (McIntyre *et al.*, 2004). Different extracellular proteins are synthesized at different pH conditions. In addition, at the transcriptional level, several proteases, glucanases, cell-wall proteins and a glucose transporter are pH-controlled, which suggests a pH dependent transcriptionally controlled response of different enzymes. External pH is also important to pathogens because their pathogenicity factors are produced or effective only within a very narrow pH range (Prusky and Yakoby, 2003) so that pH modification determines the pathogen's ability to successfully colonize and invade the targeted host. *Trichoderma* strains, which are able to modify external pH and adapt their own metabolism to the surrounding growth conditions, would consequently reduce the virulence of phytopathogens because most pathogenicity factors could not be synthesized.

Antibiosis in Biocontrol

Antibiosis is the antagonism mediated by specific or non-specific metabolites of microbial origin, *viz.*, lytic enzymes, volatile compounds or other toxic substances (Jackson, 1965). Elad *et al.* (1983) observed that, degradation and lysis of fungal cell and *sclerotial* wall were mainly due to key enzymes β-1, 3-glucanase and chitinase. The relationship between mycolytic enzymes produced by *Trichoderma* and their significance in host cell degradation has been studied by Cook and Baker (1983).

Dennis and Webster (1971a) studied the production of non-volatile (diffusible) antibiotics by *Trichoderma spp.* by an "agar layer technique". They noticed that, many isolates produced non-volatile antibiotics, active against a range of fungi. Upadhyay and Mukhopadhyay (1983) studied the effect of volatile compounds of *T. harzianum* on the growth of *S. rolfsii*. When *Trichoderma* isolates were grown for three days and *S. rolfsii* was exposed for 48 hours to its vapour, the growth of *S. rolfsii* was reduced by some isolates and stimulated by others. The vapour action of *T. harzianum* on 6th day was most inhibitory and thereafter it declined and reached zero on 15th day.

Mycoparasitism and Hyphal Interactions

Following the discovery of *Trichoderma* as a biocontrol agent (Weindling. 1934), many researches dealing with *Trichoderma* have noticed that, hyphae of the antagonists parasitized the hyphae of other fungi *in vitro* and brought about several morphological changes (coiling, haustoria formation, disorganization of host cell contents and penetration). Dennis and Webster (1971) studied in detail the ability of isolates of different species of *Trichoderma* and found that they coiled around or penetrated the hyphae of other fungi, in dual cultures. Hyphae of majority of the *Trichoderma* isolates coiled around the hyphae of majority of its host fungi. Penetration of hyphae seldom occurred. The hyphae of test fungi *R. solani* and *F. annosus* nearest to the *Trichoderma* colony showed vacuolation and coagulation of cytoplasm. The host fungus generally ceased growth soon after contact. The *Trichoderma* isolates continued to grow and eventually covered the whole plate and attempts to reisolate the test fungi generally failed. Under scanning electron microscope, Elad *et al.* (1983) observed that *T. harzianum* attached to the hyphae of *S. rolfsii* by coiling, hooks or appressoria. Benhamou and Chet. (1996) studied the interaction between *T. harzianum* and *S. rolfsii* by scanning electron microscopy (SEM) and transmission electron microscopy (TSM). SEM investigation revealed that the hyphae of *T. harzianum* grew abundantly on the sclerotial surface and formed a densely branching mycelium. Cross sections of parasitized sclerotia showed that the hyphae of the antagonist multiplied on the sclerotial surface and penetrated into the rind.

Cell-Wall-Degrading Enzymes

Trichoderma produces several enzymes such as chitinases, glucanases and proteases, which degrade the cell wall of several phytopathogenic fungi.

Chitinases

Trichoderma, which are efficient producers of chitinase, can be exploited to get plant resistance to various pathogens. *Trichoderma* chitinase (EC 3.2.1.1.14) belongs to hydrolase group can hydrolyze chitin, a polymer of β-1, 4 linked N-acetyl glucoseamine (Kas, 1997). The Chitinases are classified into two families *viz.*, family 18 and 19, based on amino acid sequence similarities (Henrissat and Bairoch, 1993). Family 18 includes chitinases found in bacteria, fungi, viruses, and animals, and class III and V of plant chitinases. Family 19 includes class I, II and IV chitinases of plant origin only, with the exception of chitinase C from *Streptomyces griseus* HUT 6037 (Ohno *et al.*, 1996) and chitinases F and G from *Streptomyces coelicolor* (Saito *et al.*, 1999). The *Trichoderma* chitinase belongs to family 18 and are mainly classified into three categories (Sahai and Manocha, 1993). Endochitinases (EC 3.2.1.14) cleave internal bonds within chitin, releasing chitotetraose, chitotriose, and chitobiose. Exochitinase (chitobiosidase) catalyzes the release of chitobiose without the formation of oligo- or monosaccharides. β-N-Acetylhexosaminidases (N-acetyl-β–D-glucosaminidases, EC 3.2.1.52) cleave chitobiose, chitotriose, and chitotetraose to N -acetylglucosamine monomers in a manner similar to exochitinase (Henrissat and Bairoch, 1993).

Glucanases

Chitin and β-1, 3 glucan form the main structural components of the fungal cell wall, chitinases and β-1, 3 glucanases have been proposed as the key enzymes in the degradation of cell wall during mycoparasitism against phytopathogenic fungi (Elad *et al.*, 1983). Glucanases are PR-2 proteins, which are active *in vitro* at micromolar levels (~50 mg/ml) against a wide number of fungi, including human and plant pathogens (Coutos *et al.*, 1993).

Though the glucan-degrading enzymes are distributed among a number of families of glycosyl hydrolases, glucanase genes from *Trichoderma* species characterized to date occur only in glycosyl hydrolase family 5 and 55 (Dong-Jinkim *et al.*, 2002).

Proteases

Proteases have emerged as a class of antifungal proteins that have potent activity against plant and animal pathogens (Blanco-Labra *et al.*, 1980). Biocontrol of *B. cinerea* by *T. harzianum* has been attributed in part to the action of proteases produced by the BCA that inactivate hydrolytic enzymes produced by this pathogen on bean leaves (Howell, 2003). Proteases involved in the degradation of heterologously produced proteins have been characterized (Delgado-Jarana *et al.*, 2000). For example, alkaline protease *Prb1* from *T. harzianum* IMI 206040 has been demonstrated to play an important role in biological control (Benitez *et al.*, 1998), and *prb1* transformants showed an increase of up to five-fold in the biocontrol efficiency of *Trichoderma* strains against *R. solani*. Protease *Pra1* from *T. harzianum* has affinity for fungal cell walls (Elad *et al.*, 2000). The gene for an extracellular serine protease (*tvsp1*) has been cloned from *T. virens* (Pozo *et al.*, 2004) and it's over expression significantly increased protection of cotton seedlings against *R. solani*. This gene shows great potential in improving biocontrol ability, as serine proteases are effective against oomycetes and nematodes (Dunne *et al.*, 2000). A serine protease of 28 kDa with trypsin activity isolated from strain 2413 also reduced the number of hatched eggs of root-knot nematodes and showed synergistic effects with other proteins produced during antagonistic activity of the strain (Surez *et al.*, 2004). Antal *et al.* (1995) screened cold-tolerant strains and found that most of them antagonized phytopathogens and produced chitinases, β-glucosidases and trypsin-like, and chymotrypsin-like proteases, active at low temperatures. The role of proteases in mycoparasitism has been reinforced with the isolation of new protease-overproducing strains of *T. harzianum* (Szekeres *et al.*, 2004).

Metabolism of Germination Stimulants

In a study on the mechanisms involved in the biocontrol of pre-emergence damping-off of cotton seedlings incited by *Rhizopus oryzae*, Howell (2002) found that control by *T. virens* was due to metabolism of germination stimulants released by the cotton seed. The importance of metabolism of stimulatory compounds by the biocontrol agent is further supported by the fact that cotton cultivars that did not produce pathogen propagule stimulants during germination were virtually immune to the disease. Again, artificial induction of pathogen propagule germination rendered these cotton cultivars susceptible to disease (Howell. 2003).

Problems in Using *Trichoderma* spp. as Biological Control Agents (BCAs) Directly in the Field

Although *Trichoderma* spp. is an effective biocontrol agent against several fungal soilborne plant pathogens, possible adverse effects of this fungus on arbuscular mycorrhizal (AM) fungi might be a drawback in its use in plant protection. AM fungi are obligate biotrophic endosymbionts in roots of most herbaceous plants. These fungi grow from the roots out into the surrounding soil, forming an external hyphal network, which increases uptake of mineral nutrients (Smith and Read, 1997) and consequently promotes plant growth. The results from pot experiments suggest that *Trichoderma* species suppress AM root colonization (Siddiqui and Mohmood, 1996). The presence of *T. harzianum* in soil reduced root colonization by *G. intraradices*. The external hyphal length and density of *G. intraradices* was reduced by the presence of *T. harzianum* in combination with wheat bran. On the other hand, adverse effects of AM fungi on the population density of *Trichoderma koningii* also have been observed (McAllister *et al.*, 1994). Another problem of *Trichoderma* spp while directly applying as biocontrol to the field was low field performance (Graham and Sticklen, 1994). Therefore, genes encoding cell wall degrading enzymes such as chitinase and glucanase are being cloned and transferred to plants to impart resistance against several plant pathogens. Plants have been transformed with plant chitinase and glucanase encoding genes as a means to alter plant's resistance to fungal pathogens. In addition, better control was observed when genes encoding chitinase was used in combination with genes encoding glucanases (Cornelissen, 1996).

Molecular Tools for the Characterization and Improvement of *Trichoderma* Species

The taxonomy of *Trichoderma* was based largely on morphological character (Bissett, 1991) such as conidial form, size, color, branching pattern with short side branches, short inflated phialides and the formation of sterile or fertile hyphal elongations from conidiophores. However, these descriptions are based on study of a limited number of strains where the morphological differences are clear but these differences become less clear as bulk strains are studied. This result suggests that there are not enough morphological and cultural characters to reliable define species level. Molecular characterization and identification has gained popularity over the past decade and authenticated differentiation between and among *Trichoderma* isolates has been reported. It should be noted, however, that more than one molecular method should be used in combination to attain reproducible and accurate results.

With the advancement in the molecular biology field and novel techniques in analyzing DNA sequence polymorphisms and their increasing use in fungal systematics, the evolutionary relationships among species in Trichoderma are now emerging. The molecular techniques include microsatellite-primed PCR based on the phage M13 core sequence, restriction fragment length polymorphism (RFLP) and random amplified polymorphic DNA (RAPD) sequence analysis using a single gene (Kindermann *et al.*, 1998) to multiple genes (Kullnig *et al.*, 2002; Samuels *et al.*, 2006; Chaverri, *et al.*, 2003). Genes that have been used to study the phylogenetic relationships among species of Trichoderma include the nuclear ribosomal internal transcribed spacers (ITS1 and 2) and partial or complete sequences of several genes [the 28S rRNA gene, the translation elongation factor (tef1) gene, the gene coding for endochitinase 42 (ech42), the calmodulin gene (cal), the actin gene (act), and the RNA polymerase subunit II (RPB2) gene]. For the identification of new species in the genus Trichoderma, most authors have used the combination of ITS and tef1 (Kubicek *et al.*, 2003; Bissett *et al.*, 2003; Kraus *et al.*, 2004; Zhang *et al.*, 2005; Lu *et al.*, 2004). Till date the number of phylogenetically distinct species of the Hypocrea/Trichoderma has increased to 100 from 27 morphological species (Bissett, 1991; Druzhinina *et al.*, 2006).

Furthermore the oligonucleotide barcode for *Trichoderma* and *Hypocrea* *Trich*OKEY (http://www.isth.info/tools/molkey/index.php) has been established and are used for molecular identification.

Genetically improved antagonistic hyperparasites microbes tend to increase their effectiveness as biological control agents however; effective performance requires either improvement of the environment to favour their activity or genetic improvement of the agent (Hornok, 2000). Genetic improvement here implies to the development of one or more of its desired characters through genetic technique(s). Genetic improvement can be achieved by chemical and physical mutation, sexual hybrids, homokaryons, and genetic manipulation *e.g.*, directed mutagenesis, protoplast fusion, recombination, transformation or isolation of useful genes from biocontrol fungi without functional sexual stages (Mohamed *et al.*, 2004 and Palumbo *et al.*, 2005)

In one of the study protoplast fusion techniques were used to combine genetic traits desirable for improving biocontrol activity by *T. harzianum* and increased amounts of specific proteins. Fusion of protoplast derived from two efficient biocontrol strains of *T. harzianum* resulted in the recovery of a progeny strain with greatly improved biocontrol ability. A mitogen-activated protein kinase encoding gene, *tvk1*, from

Trichoderma virens was cloned, and its role during the mycoparasitism, conidiation, and biocontrol was examined in *tvk1* null mutants. These mutants showed a clear increase in the level of the expression of mycoparasitism-related genes when confronted with the plant pathogen *Rhizoctonia solani* (Mukherjee *et al.*, 2003). The null mutants displayed an increased protein secretion phenotype as measured by the production of lytic enzymes in culture supernatant compared to the wild-type (Zaldívar *et al.*, 2001). A marker-assisted strategy is being developed to improve selection of antagonistic *Trichoderma* spp. for biological control of soil-borne diseases. In order to find key genes of *T. harzianum* predominantly expressed during the antagonistic interaction with the plant pathogenic fungus *Rhizoctonia solani* a cDNA subtraction library technique (Rapid Subtraction Hybridisation, is also being used. cDNA from *T. harzianum* grown on complete medium was subtracted from cDNA derived from *T. harzianum* grown as antagonist in confrontation with *R. solani*. Differential gene expressions of 200 clones were verified by Reverse Northern blot analysis. About 50 clones were sequenced and estimated for similarities by BLAST search of the JGI *T.reesei* and NCBI genome data bases (JGI *T. reesei* genbank: (http://gsphere.lanl.gov/cgibin/runBlast?db=trirel). A total of 13 fragments showed no hits on the databases; for 37 gene fragments, representing 25 different genes, corresponding counterpart sequences were found (with similarities to *T. reesei*. The 14 most interesting genes were characterised for their differential expression patterns during the time course of interaction between *T. harzianum* and *R. solani* by Northern blot analysis. Four genes seem to be significantly up-regulated during antagonism between *T. harzianum* and *R. solani*.

Biotechnological Approaches for Exploiting Chitinases and Glucanase Gene from *Trichoderma* Species

For cloning chitinase genes from *Trichoderma* mainly two approaches *viz.*, genomic DNA approach and cDNA approach were used. In genomic DNA approach, mostly PCR amplification with specific primers was followed. Genes encoding endochitinase were cloned from different *Trichoderma* spp such as *T. harzianum ech 42* (Baek *et al.*, 1999, Woo *et al.*, 1999). *T. asperellum* (*chit36*) (Viterbo *et al.*, 2002) and *T. atroviride* (*chit36*) (Viterbo *et al.*, 2002). *T. hamatum* (Steyaert *et al.*, 2004). *T. reesei* (Ike *et al.*, 2006) and *T. virens* (Kim *et al.*, 2002; Baek *et al.*, 1999). In other cases cDNA approach was used for cloning the gene coding for endochitinase from *T. harzianum* (Garcia *et al.*, 1994). *T. reesei* (Ike *et al.*, 2006), *T. virens* (Kim *et al.*, 2002).

The amino acid analysis of endochitinase shows that it belongs to glycosyl hydrolases family 18 which contains two conserved motifs, chitinase family active site ([LIVMFY]-[DN]-G-[LIVMF]-[DN]-[LIVMF]-[DN]-X-E) and chitin binding domain (XXXSXGG) (Terwisscha *et al.*, 1996). This gene also contains signal peptide which is not highly conserved and varies in length, and the possible site for the cleavage is Ala-Ser site. The presence of N-terminal signal peptide indicated that they were targeted for the secondary pathway (Emanuelsson *et al.*, 2000).

Glucanase genes have been cloned from many fungal and bacterial strains. Jesus *et al.* (1995) purified and biochemically characterized endo-β-1, 3-glucanase BGN 13.1 from *Trichoderma harzianum* CECT 2413. Attempt was made to clone β-1, 3-

glucanase (*bgn13.1*) gene by PCR techniques on the basis of the amino acid sequence of an internal peptide of the protein. For PCR, DNA from a unidirectional cDNA library in *λgt11-Sfi-Not* vector was used as the template and a degenerate oligonucleotide corresponding to the internal sequence and the *λgt11* reverse primers were used. A single 250 bp fragment of the C-terminal region, corresponding mostly to the 3′ untranslated regions was repetitively amplified and was further sub cloned and sequenced. On screening cDNA library with amplified DNA as probe, the longest insert was chosen and cloned in the Bluescript SKII (+) vector and sequenced. The cloned cDNA has a size of 2484 bp, of which 32bp corresponds to the 3′ untranslated region and 2288bp corresponds to ORF which translated to give product of 762 amino acids with molecular mass 81246 Da. Comparison of nucleotide and amino acid sequences of BHN13.1 with sequences present in the database did not show any significant homology to any α- or β-glucanases or other related protein families. On alignment of BGN13.1 with known β-1, 3-glucanase sequence from bacteria, yeast strains and plants showed that BGN13.1 has none of the conserved domains present in the other proteins.

Soledad *et al.* (1998) analysed the β-1, 3-glucanolytic system of the biocontrol agent *T. harzianum*. It was found that *T. harzianum* (IM 1206040) secretes β-1, 3-glucanases in the presence of different glucose polymers and fungal cell walls. The level of secreted β-1, 3-glucanase activity was found to be proportional to the amount of glucan present in the inducer. The fungus produces at least seven extracellular β-1, 3-glucanases upon induction with laminarin, a soluble β-1, 3-glucan. The molecular weights of five of these enzymes fall in the range 60,000-80,000 kDa. Glucose appears to inhibit the formation of all of the inducible β-1, 3-glucanases detected. A 77kDa glucanase was partially purified from the laminarin culture filterate. This enzyme is glycosylated and belongs to the exo-β-1, 3-glucanase group. The properties of this complex group of enzymes suggest that the enzyme might play different roles in host cell wall lysis during mycoparasitism. Gabor *et al.* (2001) identified glucan degrading fungi *Coniothyrium minitans* strain CM2. PCR based strategy was used to clone *cmg1*-encoding β-1, 3-glucanase of 100 kDa. The nucleotide and deduced amino acid sequence of this gene showed high levels of similarity to the sequences of other fungal exo-β-1, 3-glucanase genes. Hirofumi *et al.* (1998) reported the sequence and analysis of cDNA and genomic clones encoding low molecular mass EGIII (endo–β-1, 4-glucanase) from *Trichoderma reesei*. For the cloning of *egl3* cDNA by PCR, N-terminal and C-terminal sequences of the protein were predicted from the similarity of the deduced sequence from the genomic *egl3* gene to that of F1-CMcase from *A. aculeatus* (Ooi *et al.*, 1990). The nucleotide sequence of the cDNA fragment was verified to contain a 702bp open reading frame that encoded a 234-aminoacid propeptide. The deduced protein sequence had significant homology with family H endo-β-1, 4-glucanases. The 16-aminoacid N-terminal sequence was shown to function as a leader peptide for possible reaction. Rika *et al.* (2004) cloned the gene *lamAI*, which encodes a novel laminarinase (β-1, 3-glucanase) AI of *Trichoderma viride* U-1 using RT-PCR in conjunction with the rapid amplification of cDNA ends (RACE) technique. The open reading frame consisted of 2,277 bp encoding a protein of 759 amino acid residues, including a 32-residue signal prepropeptide. The protein showed

91 per cent sequence similarity to the putative *Trichoderma virens* β-1, 3-glucanase BGN1, but no significant similarity to fungal β-1, 6-glucanases or β-1, 3-glucanases from other organisms. Shaikh, (2005) has cloned and characterized the two genes from *T. reesie* β-1, 4-endoglucanase and *T. virens* β-1, 6-endoglucanase.

Transgenic Plants with Genes Encoding Chitinases and Glucanases

Several efforts have been made to test the efficiency of *Trichoderma* endochitinase in plant. Lorito *et al.* (1999) expressed an endochitinase-encoding gene from *T. harzianum* in tobacco and potato. High expression levels of the fungal gene were obtained in different plant tissues, which had no visible effect on plant growth and development. Substantial differences in endochitinase activity were detected among transformants. Selected transgenic lines were highly tolerant or completely resistant to the foliar pathogens; *A. alternata*, *A. solani*, *B. cinerea*, and the soilborne pathogen *R. solani*. The high level and the broad spectrum of resistance obtained with a single chitinase gene from *Trichoderma* overcame the limited efficacy of transgenic expression of chitinase genes isolated from plants and bacteria in plants. Liu *et al.* (2004) transformed rice plant with a gene encoding endochitinase (*ech42*) from the biocontrol fungus *T. atroviride*. The transformed plants showed increased resistance to sheath blight caused by *R. solani* and rice blast caused by *Magnaporthe grisea*. In case of apple, endochitinase gene from *T. harzianum* was transferred through *Agrobacterium* mediated transformation. The presence of the gene in apple was confirmed by southern analysis. Eight plants were used for the bioassay with *Venturia inaequalis* which causes the apple scab. The disease severity was compared with the control and there was a reduction in the number of lesions (0-99.7 per cent) and per cent leaf area infected (0-90 per cent). However, the endochitinase had negative effect on the growth of the plant (Jyoti *et al.*, 2000). Chandrakanth *et al.* (2003) transferred the 42 kDa endochitinase from *T. virens* to cotton. Transgenic plants showed resistance to the pathogens *i.e.*, *A. alternata* and *R. solani*. Recently, Shah *et al.* (2005) transformed tobacco plants with *ech42* gene cloned from an Indian isolate of *T. virens* and confirmed the integration of endochitinase into tobacco genome by PCR amplification using specific primers. Movsesyan *et al.* (2001) developed tobacco plants expressing bacterial genes for thermostable glucanase. It was shown that bacterial genes for thermostable β-glucanase were expressed in tobacco retained their activity and substrate specificity. The transgenic plants obtained in the study are proposed as model objects for investigating the role of glucanases in plants. Vijaya *et al.* (1999) transformed mulberry with gene encoding β-1, 3-glucanase through *Agrobacterium* mediated transformation by using leaf explants and shoot meristems for developing resistance to leaf spot pathogens. Potrykus *et al.* (1995) transformed rice with β-1, 3-glucanase from tobacco and barley to develop resistance against *M. grisea* and *R. solani* which cause blast and sheath blight. Transgenic plants expressing more than one PR proteins (Chitinase and glucanase) genes in a constitutive manner were developed and transgenic showed better resistance level than transgenic having a single gene (Cornelissen, 1996). Antifungal activity of chitinase *in vitro* is enhanced when applied in combination with β-1, 3-glucanase. Genes encoding different chitinases and β-1, 3-glucanases have been cloned and used to transform tobacco. The transgenic tobacco plants showed increased survival relative to control plants in the presence of pathogenic fungi, such as *Rhizoctonia solani* (Collinge *et al.*, 1993). Genes encoding chitinase and glucanase from *Trichoderma atroviride have* been cloned and expressed in rice, which showed increased resistance to blast disease (Liu *et al.*, 2003). In recent years, considerable progress has been made in producing disease resistant and high-yielding transgenic plants. It may be necessary to integrate different resistance genes together in order to extend host defense.

References

Antal, Z., Manczinger, L., Szakacs, G., Tengerdy, R. P. and Ferency, L. (2000). Colony growth, *in vitro* antagonism and secretion of extracellular enzymes in cold-tolerant strains of *Trichoderma* species. *Mycol. Res.*, 104: 545-549.

Arora, D. K., Elander, R. P. and Mukerji, K. G. (1992). Handbook of applied mycology. *Fungal Biotech.*, 4: 40-45.

Baek, J. M., Howell, C. R. and Kenerley, C. M. (1999). The role of an extracellular chitinase from *Trichoderma virens* Gv29-8 in the biocontrol of *Rhizoctonia solani. Current Genetics*, 35 (1): 41-50.

Benhamou, N. and Chet, I. (1996). Parasitism of sclerotia of *Sclerotium rolfsii* ultra structural and cytochemical aspects of the interaction. *Phytopathol.*, 86: 405-416.

Benitez, T., Delgado-Jarana, J., Rincon, M., Rey, M. and Limon, M, C. (1998). Biofungicides: *Trichoderma* as a biocontrol agent against phytopathogenic fungi. *Recent Res. Develop. in Microbiol.*, 2: 129-150.

Bissett, J. (1991). A revision of the genus *Trichoderma*. III. Sect. *Pachybasium. Canadian Journal of Botany*, 69: 2373-2417.

Bissett, J., Szakacs, G. and Nolan, C. A. (2003). New species of Trichoderma from Asia. *Can J Bot*, 81: 570–586

Blanco-labra, A., Iturbe-chinas, F. A. and Summerbell, R. (1980). Purification and characterization of an amylase inhibitor from maize (*Zea maize*). *J. Food Biochem.*, 5: 1-17.

Chandrakanth, E., Juan, M. G., Emily L. F., Mariajose, P., Pedro, U., Dong, J. K., Ganesan, S., Cook, D.R., Charles, M. K. and Keerti, S. R. (2003). Enhanced fungal resistance in transgenic cotton expressing an endochitinase gene from *Trichoderma virens. Plant Biotechnology Journal*, 1 (5): 321-328.

Chet, I. and Inbar, J. (1994). Biological control of fungal pathogens. *Appl. Biochem. Biotechnol.*, 48: 37-43.

Chet, I. (1987). *Trichoderma* application, mode of action and potential as biocontrol agent of soil borne plant pathogenic fungi. In: *Innovative Approaches to plant disease control*, John Wiley and Sons Publishers, New York, pp 137-160.

Chet, I., Inbar, J. and Hadar, I. (1997). Fungal antagonists and mycoparasites. In: *The Mycota IV: Environmental and microbial relationships*, pp 165-184.

Cigdem, K.K. and Merih, K.K. (2003). Isolation of *Trichoderma* spp. and determination of their antifungal, biochemical and physiological features. *Turkish Journal of Biology*, 27: 247-253.

Collinge, D. B., Kragh, K. M., Mikkdern, J. D., Nielsen, K. K., Rasmussen, U. and Vad, K. (1993). Plant Chitinases. *Pl. J.*, 3: 31-40.

Cook, R. J. and Baker, K. F. (1983). The Nature and Practice of Biological control of Plant Pathogens. *American Phyto pathol. Soc.*, 539-545.

Cornelissen, B. J. C., Does, M. P. and Melchers, L. S. (1996). *Rhizoctonia* species: Taxonomy, Molecular Biology, Ecology, Pathology and Control (eds. Sneh, B., Jabaji-Hare, S., Neate, S. and Deist, G.) Kluwer, Dordrecht, pp. 529-536.

Coutos, B., Coutos, J., Keller, C., Beven, L. and Delattre, C. (1993). Purification and characterization of a novel glucuronan lyase from *Trichoderma* sp. *J. Biol. Chem.*, 260: 1-11.

Delgado, J. J., Pintor, T, J. A. and Benitez, T, (2000). Overproduction of β-1, 6-glucanase in *Trichoderma harzianum* is controlled by extracellular acidic proteases and pH. *Biochem. Biophys Acta.*, 1481: 289-296.

Dennis, C. and Webster, J. (1971a). Antagonistic properties of species of groups of *Trichoderma* I. Production of non-volatile antibiotics. *Transactions of British Mycol. Soc.*, 57: 25-29.

Dong, J., Jong, M., Baek, P., Charles, M., Kenerley D. R. and Cook. (2002). Cloning and characterization of multiple glycosyl hydrolase genes from *Trichoderma virens*. *Curr. Genetics.*, 40: 374–384.

Druzhinina, I. S., Kopchinskiy, A. and Kubicek, C. P. (2006). The first 100 *Trichoderma* species characterized by molecular data. Mycoscience, 47: 55–64

Dunne, C., Moenne-Loccoz, Y., DE Bruijn, F. J. and O'gara, F. (2000). Overproduction of an inducible extracellular serine protease improves biological control of *Pythium ultimum* by *Stenotrophomonas maltophilia* strain W81. *Microbiol.*, 146: 2069-2078. ms involved in the biological control of *Botrytis cinerea* incited diseases. *Eur. J. of Plant Path.*, 102: 719-732.

Elad, Y., Chet, I., Boyle, P. and Henis, Y. (1983). Parasitism of *Trichoderma* spp. on *Rhizoctonia solani* and *Sclerotium rolfsii* scanning electron microscopy. *Phytopathol.*, 73: 85-88.

Elad, Y., Freeman, S. and Monte, E. (2000). Biocontrol agents: Mode of action and interaction with other means of control. *IOBC wprs Bulletin.*, 24: 58-68.

Emanuelsson, H., Nielsen, H., Brunak, S. and Heijne, G. (2000). Predicting sub cellular localization of proteins based on their N-terminal amino acid sequence. *Journal of Molecular Biology*, 300: 1005-1016.

Esindle, M., Oberegger, H., Buttinger, R., Illmer, P. and Haas, H. (2004). Biosynthesis and uptake of siderophores is controlled by the Pac C mediated ambient-pH regulatory system in *Aspergillus nidulans*. *Eukaryot. Cell.*, 3: 561-563.

Fokunang, C. N., Beynon, J. L., Watson, K. A., Battey, N. H., Dunwell, J. M. and Tembe-fokunang, E. A. (2004). Advancement in Genetic Modification Technologies Towards Disease Resistance and Food Crop Production. *Biotech.*, 3(1): 1-20.

Franken, P., Khun, G. and Gianinazzi-pearson, V. (2002). Development and molecular biology of arbuscular mycorrhizal fungi. In: Osiewacz HD (ed) *Mol. Boil. of fungal developt.*, 325-348.

Gabor, Z, K. and Larzlo, F. (2001). Expression of *emg-1* an Exo-b-1, 3-glucanase gene from *Comiothurium minitants* increase during sclerotial parasitism. *App. and Environ. Microbiol.*, 67: 865-871.

Garcia, J., Lora, J.M., Cruz J. De la., Benitez, T., Liobell, A. and Pintor, T. J. (1994). Cloning and characterization of a chitinase (CHIT42) cDNA from the mycoparasitic fungus *Trichoderma harzianum*. *Current Genetics*, 27: 83-89.

Gomez, A. G. and De la Torre ma (1994). Mecanismos de corrosion microbiana sobre los materiales pétreos. *Microbiología*, 10: 111-120.

Graham, L. S. and Sticklen, M. B. (1994). Plant chitinases. *Can. J. Bot.*, 72: 1057-1083.

Harjono and Widyastuti, S. M. (2001). Antifungal activity of purified endochitinase produced by biocontrol agent *Trichoderma reesei* against *Ganoderma philippii*. *Pakistan Journal of Biological Sciences*, 4(10): 1232-1234.

Harman, G. E., Howell, C. R., Viterbo, A., Chet, I. and Lorito, M. (2004). *Trichoderma* species-opportunistic, avirulent plant symbionts. *Nature Rev.*, 2: 43-56.

Henrissat, B. and Bacrioch, A. (1993). New families in the classification of glycosyl hydrolases based on amino acid sequence similarities. *Biochem. J.*, 293(3): 781-788.

Hirofumi, Okada., Kohjii, Tada., Tadashi, Sekuiya., Kengoyokoyama., Akinori-takahashi., Hideki, Tohda., Hiromichi, Kumagai, Yasushi and Morikawa. (1998). Molecular characterization and heterologous expression of the gene encoding a low molecular mass endoglucanase from *Trichoderma reesei* QM 9414. *App. and Environ. Microbio.*, 64(2): 555-563.

Hornok, L. (2000). Genetically modified microorganisms in biological control. Novenyvedelem, 36: 229-237.

Howell, C. R. (2002). Cotton seedling pre-emergence damping-off incited by *Rhizopus oryzae* and *Pythium* spp. and its biological control with *Trichoderma* spp. *Phytopathol.*, 92: 177-180.

Howell, C. R. (2003). Mechanisms employed by *Trichoderma* species in the biological control of plant diseases. The History and evolution of current concepts. *Pl. Disease*, 87(1): 4-10.

Howell, C. R., Hanson, L. E., Stipanovic, R. D. and Puckhaber, L. S. (2000). Induction of terpenoid synthesis in cotton roots and control of *Rhizoctonia solani* by seed treatment with *Trichoderma virens*. *Phytopathol.*, 90: 248-252.

Ike, M., Kazuhisa, N., Akiko, S., Masahiro, N., Wataru, O., Hirofumi, O. and Yasuchi M. (2006). Purification and characterization and gene cloning of 46 kDa chitinase (*chi46*) from *Trichoderma reesei* PC-3-7 and its expression in *Escherichia coli*. *Applied Microbiology and Biotechnology*, 71: 294-303.

Jackson, R. M. (1965). Antibiosis and Fungistasis of soil microorganisms. In: *Ecology of Soil Borne Plant Pathogens* (eds. Baker, K. F. and Synder, W. C), *University of California Press, Berkley.*, 363-369.

Jesus De La Cruz., Jose, A., Pintor, T., Jahia, Benitez and Anomia, L. (1995). Purification and characterization of an endo-b-1, 3-glucanase from *Trichoderma harzianum* that is related to its Mycoparasitism. *J. of Bacteriol.*, 177: 1864-1871.

Jyoti, B., John, L. N., Kwai Weng, W., Cristopher, K. H., Harman, G.E. and Herb, S. A. (2000). Expression of endochitinase from *Trichoderma harzianum* in transgenic apple increase resistance to apple scab and reduces vigor. *Phytopathology*, 90(1): 72-77.

Kas, H.S. (1997). Chitosan: Properties preparations and application to micro-encapsulation system. 14: 687-711.

Kershaw, M. J. and Talbot, N. J. (1998). Hydrophobins and repellents: proteins with fundamental roles in fungal morphogenesis. *Fungal Genet Biol.*, 23: 18-33.

Kim, D. J., Jong, M. B., Pedro, U., Charles, M. Kener L.Y., Douglas, R. and Crook (2002). Cloning and characterization of multiple glycosyl hydrolase gene from *Trichoderma virens. Current Genetics*, 40: 374-384.

Kindermann, J., El-Ayouti Y. and Samuels, G. J. (1998). Phylogeny of the genus *Trichoderma* based on sequence analysis of the internal transcribed spacer region 1 of the rDNA clade. *Fungal Genet Biol*, 24: 298–309.

Kraus, G. F., Druzhinina, I. and Gams, W. (2004). *Trichoderma brevicompactum* sp. nov. *Mycologia*, 96: 1059–1073.

Kubicek, C. P., Bissett, J. and Druzhinina, I. (2003). Genetic and metabolic diversity of *Trichoderma*: a case study on South–East Asian isolates. *Fungal Genet Biol*, 38: 310–319.

Kullnig, C. M., Szakacs, G. and Kubicek, C. P. (2002). Phylogeny and evolution of the genus *Trichoderma*: a multigene approach. *Mycol Res*, 106: 757–767.

Latorre, B. A., Lillo, C. and Rioja, M. E. (2001). Eficacia de los tratamientos fungicidas para el control de Botrytis cinerea de la vid en funcion de la epoca de aplicacion. *Cien Inv Agr.*, 28: 61-66.

Lewis, J.A. and Papavizas, G.C. (1991). Biocontrol of plant diseases: the approach for tomorrow. *Crop Protection*, 10: 95-105.

Liu, M., Sun Zong-Xiu, Zhu, Xu Tong, Harman, G. E. and Lorito, M. (2004). Enhancing rice resistance to fungal pathogens by transformation with cell wall degrading enzyme genes from *Trichoderma atroviride. Journal of Zhejiang University Science*, 5(2): 133-136.

Lorito, M. and Scala, F. (1999). Microbial genes expressed in transgenic plants to improve disease resistance. *Journal of Plant Pathology*, 81 (2): 73-88.

Lu, B. S., Druzhinina, I. S. and Fallah, P. (2004). *Hypocrea/Trichoderma* species with pachybasium-like conidiophores: teleomorphs for *T. minutisporum* and *T. polysporum* and their newly discovered relatives. *Mycologia*, 96: 310–342.

Lynch, J. M. (1990). Fungi as antagonists. In New directions in biological control: alternatives for suppressing agricultural pests and diseases, Liss, New York, pp. 243-253.

Mcallister, C. B., Garcia, R. I., Godeas, A. and Ocampo, J. A. (1994). Interactions between *Trichoderma koningii, Fusarium solani* and *Glomus mosseae*: effects on plant growth, arbuscular mycorrhiza and the saprophyte inoculants. *Soil Bio and Biochem.*, 26: 1363-1367.

McIntyre, M., Nielsen, J., Arnau, J., Brink, H., Hansen, K. and Madrid, S. (2004). Proceedings of the 7th European Conference on Fungal Genetics. Copenhagen, Denmark, pp. 125-130.

Mohamed, H. A. A., Haggag, M. and Abo-Aba, S. M. (2004). Influence of salt stress on *Pseudomonas fluorescens* plasmids, some phenotypic traits and antibiosis against *Diplodia theobromae. J. Gentic Eng. and Biotechnol*, 2(2): 265-281.

Monte, E. (2001). Understanding *Trichoderma*: between biotechnology and microbial ecology. *Int. Microbiol.*, 4: 1-4.

Movsesyan, N. R., Alizade, K., Muriychuk, K. A., Popov, Y. G. and Piruzian, E. S. (2001). Transgenic tobacco plants expressing bacterial genes for thermostable glucanases. *Russian J. Genetics.*, 37 (6): 610-616.

Mukherjee, M., Hadar, R., Mukherjee, P. K., and Horwitz, B. A. (2003). Homologous expression of a mutated betatubulin gene does not confer benomyl resistance on *Trichoderma virens. J. Applied Microbiology*, 95: 861-867.

Ohno, T., Armand, S., Hata, T., Nikaidou, N., Henrissat, B., Mitsutomi, M. and Watanable, T. (1996). A modular family 19 chitinase found in the prokaryotic organism *Streptomyces griseus* HUT 6037. *Journal of Bacteriology*, 178: 5065-5070.

Ooi, T., Shinmyo, A., Okada, H., Murao, S., Kawaguchi, T. and Arai, M. (1990). Complete nucleotide sequence of a gene coding for *Aspergillus aculeatus* cellulase (F1-CMCase). *Nuclei Acid Res.*, 18: 5884.

Osiewacz, H. D. (2002). Molecular biology of fungal development. Marcel Dekker, New York.

Palumbo, J. D., Yuen, G. Y., Jochum, C. C., Tatum, K. and Kobayashi, D. Y. (2005). Mutagenesis of beta-1,3-glucanase genes in *Lysobacter* enzymogenes strain C3 results in reduced biological control activity toward *Bipolaris* leaf spot of tall fescue and *Pythium* damping-off of sugar beet. *Phytopathology*, 95: 701-707.

Potrykus, I., Armstrong, P., Bieri, P. K., Bnekhardt, H. and Ding, C. (1995). Transgenic indica rice for the benefit of less developed countries: toward fungal, insect, and viral resistance and accumulation of β-carotene in the endosperm. *Pl. Diseases.* 5: 521-526.

Pozo, M. J., Baek, J. M., Garcia, J. M. and Kenerley, C. M. (2004). Functional analysis of *tvsp1*, a serine protease-encoding gene in the biocontrol agent *Trichoderma virens. Fungal Genet. Biol.*, 41: 336-348.

Prusky, D. and Yakoby, N. (2003). Pathogenic fungi: leading or led by ambient pH? *Mol. Pl. Pathol.*, 4: 509-516.

Rika, N., Yoichi, S., Kihachiro, O. and Masahito, S. (2004). Cloning and Expression of a Novel *Trichoderma viride* Laminarinase AI Gene (*lam AI*). *Biosci. Biotechnol. Biochem.*, 68 (10): 2111-2119.

Sahai, A. S. and Manocha, M. S. (1993). Chitinases of fungi and plants: their involvement in morphogenesis and host-parasite interaction. *FEMS Microbiological Reviews*, 11: 317-338.

Saito, A., Fujii, T., Yoneyama, T., Redenbach, M., Ohno, T., Watanable, T. and Miyashita, K. (1999). High multiplicity of chitinase genes in *Streptomyces coelicolor*. *Bioscience, Biotechnology and Biochemistry*, 63: 710-718.

Samuels, G. J., Suarez, C. and Solis, K. (2006). *Trichoderma theobromicola* and *T. paucisporum*: two new species isolated from cacao in South America. *Mycol Res*, 110: 381–392.

Shah, M. R., Mukherjee, P. K. and Eapen, S. (2005). Transformation of Tobacco with *Endochitinase ech42* Gene from *Trichoderma virens*. Second Global Conference. "Plant Health-Global Wealth". Abstracts, 96-97.

Shaikh, Z. (2005). Cloning and Characterization of Glucanase gene from *Trichoderma* spp. MSc Thesis, UAS Dharwad, Karnataka.

Siddiqui, Z. A. and Mohmood, I. (1996). Biological control of *Heterodera cajani* and *Fusarium udum* on pigeonpea by *Glomus mosseae, Trichoderma harzianum*, and *Verticillium chlamydosporium*. *Israel J. Pl. Sci.*, 44: 49-56.

Smith, S. E. and Read, D. J. (1997). Mycorrhizal symbiosis. Academic Press, San Diego.

Soledad, V., Carlos, A. L. and Alfredo, H. (1998). Analysis of the b-1, 3-Glucanolytic system of the Biocontrol Agent *Trichoderma harzianum*. *App. and Environ. Microbiol.*, pp 1442-1446.

Steyaert, J. M., Stewart, A. and Ridgway,H. J. (2004). Co-expression of two genes, a chitinase (*chit42*) and proteinase (*prb1*), implicated in mycoparasitism by *Trichoderma hamatum*. *Mycologia*, 96 (6): 1245-1252.

Suarez, B., Rey, M., Castillo, P., Monte, E. and Llobell, A. (2004). Isolation and characterization of *PRA1*, a trypsin-like protease from the biocontrol agent *Trichoderma harzianum* CECT 2413 displaying nematicidal activity. *Appl. Microbiol. Biotechnol.* (in press).

Szekeres, A., Kredics, L., Antal, Z., Kevei, F. and Manczinger, L. (2004,). Isolation and characterization of protease overproducing mutants of *Trichoderma harzianum*. FEMS *Microbiol. Let.*, 233: 215-222.

Terwisscha, V.S., Henning, D. and Dijkstra, B. W. (1996). The 1.8 OA resolution structure of hevamine, a plant chitinase lysozyme and analysis of the conserved sequence and structure motif of glycoxyl hydrolase family 18. *Journal of Molecular Biology*, 262: 243-257.

Tjamos, E. C., Papavizas, G. C. and Cook, R. J. (1992). Biological control of plant diseases. Progress and challenges for the future. Plenum Press, New York, pp. 58-65.

Upadhyay, J. P. and Mukhopadhyya, A. N. (1983). Effects of non-volatile and volatile antibiotics of *Trichoderma harzianum* on the growth of *Sclerotium rolfsii*. *Indian J. of Mycol and Pl. Pathol.*, 13: 232-233.

Vijaya, D. S., Chitra. and Padmaja, G. (1999). Clonal propagation of mulberry (*Morus indica* L. Cultivar M-5) through in vitro culture of nodal explants. *Scientia Hortic*, 80: 289-298.

Viterbo, A., Manuel, M., Ramot, O., Dana, F., Enrique, M., Antonio, L. and Chet, I. (2002). Expression regulation of the endochitinase *chit36* from *Trichoderma asperellum* (*T. harzianum* T-203). *Current Genetics*, 42: 114–122.

Weindling, R. (1934). Studies on a lethal principle effective in the parasitic action of *Trichoderma lignorum* on *Rhizoctonia solani* and other soil fungi. *Phytopathol.*, 24: 1153-1179.

Woo, S. L., Donzelli, B., Scala, F., Harman, G. E., Kubicek, C. P. and Lorito, M. (1999). Disruption of *ech42* (endochitinase encoding) gene affects biocontrol activity in *T. harzianum* strain P1. *Plant-Microbe interaction*, 12: 419-429.

Yedida, I., Benhamou, N., Kapulink, Y. and Chet, I. (2000). Induction and accumulation of PR proteins activity during early stages of root colonization by the mycoparasite *Trichoderma harzianum* strain T-203. *Pl. Physiol. and Biochem.*, 38: 863-873.

Zaldívar, M., Velásquez, J. C., Contreras, I., and Pérez, L. M. (2001). *Trichoderma aureoviride* 7-121, a mutant with enhanced production of lytic enzymes: its potential use in waste cellulose degradation and/or biocontrol. *Electronic Journal of Biotechnology*, 4: 1-9.

Zhang, C. L., Druzhinina, I. S. and Kubicek, C.P. (2005). *Trichoderma* biodiversity in China: Evidence for a North to South distribution of species in East Asia. *FEMS Microbiol Lett* 251: 251–257.

Chapter 4

Molecular Biology of *Phytophthora infestans* Causing Late Blight of Potato

V.P. Chimote[1]* *and S.J. Gawande*[2]**

[1]*Biotechnology Research Centre, Mahatma Phule Krishi Vidyapeeth, Rahuri – 413 722, India*

[2]*Division of Crop Protection, Central Institute of Cotton Research, Nagpur, India*

Introduction

Late blight caused by *Phytophthora infestans*, an oomycete, is the most devastating diseases of potato and to some extent in tomato. The genus *Phytophthora* (meaning plant destroyer) is an oomycete and *P. infestans*, is perhaps its most notorious species, costing annually on a global basis in billions of dollors in terms of losses of the potato crop and control measures. In Indian hills, potato is grown as a temperate summer crop and accounts for about 20 per cent total potato acreage, with late blight recurring every year in severe epiphytotic form causing 40-75 per cent yield losses. It is particularly more damaging in North-Eastern hills, where the moist weather is more congenial for buildup and prevalence over longer periods. In Indo-Gangetic plains where 80 per cent of total potato is grown as sub-tropical winter crop, late blight infection is mild to moderate, but occasionally becomes epiphytotic once in every 2-3 years (15-75 per cent losses) (Bhattacharyya *et al.*, 1990).

Potato is the main host of the late blight pathogen. It appears as spots on leaf tips and margins that enlarge rapidly into necrotic brownish black spots on its leaves,

E-mail: *vivekchimote@rediffmail.com; **sureshgawande76@gmail.com stems and tubers. Under favorable conditions entire field populations may be killed within few days and blackened with lesions showing cottony mildew around dead areas. Tubers are readily infected while in soil by rain borne spores that drops from infected aerial foliage cover.

The oomycete fungi, *Phytophthora* species are responsible for economically important diseases of a wide range of agronomic and ornamental crops. *Phytophthora infestans* (Mont.) de Bary, the causal agent of potato late blight is best known for its role in the great Irish famine of 1845-1849 which resulted in the deaths of over 1 million people. The western world became aware of *P. infestans* with the devastating late blight epidemics in the Europe and North-eastern United States in the 1840's. Tomato late blight was detected sometime later and has also been a persistent problem. Since then, the disease has become established in all potato growing countries and is the most important pathogen of potatoes worldwide.

Molecular Markers in Population Studies

Identification of various strains of *P. infestans* has traditionally been based upon microscopic examination of morphological characters and growth characteristics of the pathogen on specific media. Variations in the morphological characters of both the sexual and asexual stages of this group of pathogens exist, leading to difficulties in accurate identification by traditional methods. In addition, identification based on pathogenicity assays or growth characteristics are time-consuming. Accurate and rapid characterization of *P. infestans* is important for precise diagnosis of related plant pathogens and understanding of the population structure and genetics of the pathogen for several reasons.

The markers available for *P. infestans* can be classified in three categories: (1) biological markers, (2) cytoplasmic markers, (3) neutral or molecular markers. Biological markers include mating types, fungicide resistance and virulence. Some of them have been discussed elsewhere in this chapter in greater details. Cooke and Lees (2004) have extensively reviewed both conventional and molecular markers being used in *P. infestans* characterization.

Molecular markers on the other hand are highly reliable for cataloguing variability in the plant pathogens since they are not related to biologically importance characteristic. Molecular techniques have also been used to study genetic diversity and evolutionary origins in populations of many different fungal genera. They can be subdivided into two categories: (1) Allozyme markers, (2) Polymorphic DNA markers. Molecular tools including isozyme analysis, restriction fragment length polymorphisms in nuclear and mitochondrial DNA, randomly amplified polymorphic DNA PCRs, serological assays, DNA probes, and PCR of internal transcribed spacer (ITS) regions and nuclear small- and large-subunit ribosomal DNA (rDNA) have been used to evaluate intraspecific and interspecific variation in *Phytophthora* species.

Isozymes

Isozymes are based on affordable technology and are codominant, yielding data amenable to population genetic analysis However, of the many isozymes tested, only glucosephosphate isomerase and peptidase which have been used extensively world over for cataloguing variability in *P. infestans* (Tooley *et al.*, 1985; Spielman *et al.*, 1990). In isozyme analysis there are problems with throughput, band nomination and level of polymorphism. In Isozyme analyses, migration distance is expressed in relative terms and can be difficult to interpret, a different stain is required for each enzyme, the precise nature of the genetic change that alters migration distances is unknown and the assays are time-consuming.

RFLP Using Repetitive Probes

Use of respective probes or short oligonucleotide repeat in one of them. RG-57 which is moderately repetitively DNA probe has been used extensively to characterize *P. infestans* across the globe (Goodwin *et al.*, 1992a; 1992b; Fry *et al.*, 1993; Drenth *et al.*, 1994; Perez *et al.*, 2001). The moderately repetitive RFLP probe RG57 (Goodwin *et al.*, 1992b) yields a genetic fingerprint of 25–29 bands (Forbes *et al.*, 1998) and has proved a valuable tool in monitoring *P. infestans* genetic diversity. However, the method does have disadvantages, however; large amounts of pure DNA are required, often radiolabelling of probes, it is time consuming, the banding patterns can be difficult to interpret and the resultant data are dominant.

AFLP

AFLPs (Amplified fragment length polymorphisms) (Vos *et al.*, 1995) are very powerful markers, since they yield many loci per primer combination. They have been used in genetic mapping (van der Lee *et al.*, 2004) and to various population divergence studies (Knapova and Gisi, 2002; Cooke *et al.*, 2003; Flier *et al.*, 2003) in *P. infestans*. Purvis *et al.* (2002) used AFLPs to increase marker resolution as a result of which certain isolates which had the same RG-57 fingerprint could be resolved into distinct isolates based on AFLP fingerprints.

Microsatellite or Simple Sequence Repeat (SSR) Analysis

SSR markers are highly polymorphic markers based on loci specific primers designed from sequences flanking tandemly repeated 1-6 base repeats. Knapova and Gisi (2002) developed three *P. infestans* SSR markers for the analysis of *P. infestans* populations on potato and tomato in France and Switzerland. Lees *et al.* (2006) developed 12 SSR markers that amplified 2–9 highly reproducible and polymorphic alleles per locus. Microsatellite markers being single locus, co-dominant markers have advantages over dominant multilocus (RFLP and AFLP) markers because each allele of the markers can be detected and data on frequency of various homozygotes and heterozygotes can be used to characterize the breeding system.

Mitochondrial DNA Haplotype Studies

Polymorphisms in mitochondrial DNA (mtDNA) of *P. infestans* are particularly useful for monitoring pathogen populations; they are easily detected and, because they are uniparentally (and probably maternally) inherited, ideal for tracing lines of descent. These polymorphisms have previously been studied by hybridization of digested total DNA with labelled, cloned mtDNA (Goodwin, 1991) and digestion of isolated mtDNA (Klimczak and Prell, 1984). Carter *et al.* (1990) defined two mitochondrial types, type I and type II, by digestion of total DNA with the frequently cutting restriction enzymes *Msp*I or *Cfo*I (which produce bands of mtDNA on a background smear of nuclear DNA upon separation). Type II differed from type I by an insert of 1.6 kb and rearrangement of flanking sequences. Type I was further differentiated into haplotypes Ia and Ib, the latter possessing an additional *Msp*I site; similarly, type II was subdivided into haplotypes IIa and IIb, the latter possessing an additional *Cfo*I site. The recent sequencing of the mitochondrial genome of *P. infestans* (Chesnik *et al.*, 1996, Paquin *et al.*, 1997) allows the design of primers to amplify by PCR the known polymorphic sequences of the genome.

Some correlation has been made between the Carter haplotypes and the six mitochondrial haplotypes (A to F) designated by Goodwin, 1991), and it is known that the haplotype Ib isolates also have a nuclear DNA fingerprint (with probe RG57) identical to the US-1 clonal lineage. In some countries only a single haplotype was detected (*e.g.*, Costa Rica, all Ia; Bolivia, all haplotype IIa), while in others (*e.g.*, Russia, United Kingdom) two or three haplotypes were detected. Lack of diversity may be due to the limited number of samples collected or sites sampled and/or to restricted sampling dates. In Western Europe, recent collections are predominantly of haplotype Ia, with much lower frequencies of IIa (Day and Shattock, 1997; Lebreton *et al.*, 1998). In this study, haplotype IIb was detected only in California and British Columbia (Vancouver), but it has since been detected in one isolate from The Netherlands (Flier, W., unpublished).

In our mtDNA haplotype studies on *Phytophthora infestans* strains from India there were 42 Ib, 25 Ia and 3 IIb mtDNA haplotype isolates. Prior to 2002 all isolates had Ib mtDNA haplotype when first Ia mtDNA haplotype were detected. mtDNA haplotype analysis suggested large scale drift during 2004 to 2006 from Ib mtDNA haplotype to Ia mtDNA haplotype all over India except in north eastern hills.

Single Nucleotide Polymorphisms (SNP)

SNPs are single base-pair differences in DNA due to point mutations (substitutions or insertions/deletions). With recent developments in *P. infestans* genome sequencing projects and available sequences in public domain it will most likely be the ideal marker for *P. infestans* population analysis for the immediate future. They represent most common variation and provide genome-wide view of population. However, SNP development is time consuming and requires technically skilled manpower.

Other Markers

A range of new co-dominant markers are becoming increasing available which include or cleaved amplified polymorphic sequences (CAPS) which is a combination of PCR and RFLP or by the single stranded conformational polymorphism (SSCP) method.

Molecular Studies on *Phytophthora* Taxonomy

Phytophthora belong to kingdom Stramenopila (=Chromista), class Oomycetes. This group of organisms is characterized by the absence of chitin in the cell walls (true fungi contain chitin), zoospores with heterokon flagella, diploid nuclei in vegetative cells, and sexual reproduction via antheridia and oogonia. Chesnick *et al.* (1996) studied the phylogenetic position of *Phytophthora* using the derived protein sequences of the mitochondrial *nad*4L genes. The bootstrap support (in percentage) they found that that *Phytophthora* is not a member of the true fungi. Nucleic acid data have now shown that they are more closely related to the diatoms, brown and golden algae in the Stramenopila lineage than to true fungi.

The most important plant pathogenic oomycetes belongs to two orders namely Pernosporales and Saprolegniales. The order Pernosporales includes several of the most important genera of plant pathogenic known like *Phytophthora, Pythium, Bremia, Peronospora, Plasmopara, Pseudoperonospora, Albugo, Peronosclerospora, Sclerospora and Sclerophthora.*

Molecular Characterization of *P. infestans* Mating Types

Most of late blight infections start from asexual zoospores, but oospores play a significant role. With oospores for surviving host-free periods, sporangia for long distance dispersal, and zoospores that actively seek out host over short distance, it is not surprising that *Phytophthora* species are such formidable pathogen.

Phytophthora species have a sexual cycle that return in the production of oospores. *P. infestans* is heterothallic and requires two mating (compatibility) types, A1 and A2 mating types to produce oospores. Following nearly a century in which only the A1 mating type was distributed widely in the old world, A1 and A2 types are now worldwide. These mating types represent compatibility types differing in the production and response to mating hormones, rather than dimorphic sexual forms.

Oospores are usually thick walled and survive environmental extremes to initiate epidemics in subsequent seasons. Oospores provide a long-lived source of inoculums that can allow the otherwise near-obligate pathogens to survive outside their host. The long-term survival of oomycetes is also enhanced by genetic exchanges during mating with recombinants may be more fit or adapted to certain environments. Host specificity, aggressiveness, temperature optima, fungicide resistance, and other traits vary between isolates of *P. infestans.*

Sexual Reproduction Outside Mexico

Before 1984, A2 mating type had only been detected in Mexico (Smoot *et al.*, 1958). The A1 mating type prevailed in the rest of the world, including the United States, Canada, West Europe, South Africa, and West India (Smoot *et al.*, 1958). Detection of the A2 mating type outside Mexico was first reported in 1984 from Switzerland (Hohl and Iselin, 1984), and subsequently from other places world including India (Singh *et al.*, 1994). Mating type analysis at Central Potato Research Institute, Shimla revealed that A1 type is predominant (90 per cent) in subtropical Indo-Gangetic plains; on the contrary A2 mating type is dominant (93 per cent) in high altitude hills.

Two hypotheses have been proposed to explain the new distribution of the A2 mating type.

1. Fry *et al.* (1993) suggested that migration was the cause of the new occurrence of the A2 mating type, while
2. Ko concluded that sexual offspring or mutation from the descendants of A1 mating type pioneers could be the alternative origin (Ko, 1994).

Goodwin and Drenth (1997) re-analysed the published genotypic data to test the hypotheses on the origin of the A2 mating type outside Mexico. They concluded that the migration hypothesis was strongly supported and rejected Ko's mating type change hypothesis.

Mating Type Specific Markers

Here have been limited studies on the genetics of mating types in *P. infestans*, by using a DNA marker linked to the A2 mating type in *P. infestans*. Markers linked to A2 mating types are genetically and physically linked to one region (Judelson *et al.*, 1995). They developed W16 CAPS (Cleaved Amplified Polymorphic Sequences) marker which differentiated A1 and A2 mating types. Although all of the strains of *P. infestans* amplified a 600 bp PCR product were detected, digestion with *Hae*III revealed a 600 bp band linked to the A1 mating type and a 550 bp band linked to the A2 mating type. Another DNA marker linked to the A2 mating type in *P. infestans* was detected by a powerful AFLP fingerprinting technique (Vos *et al.*, 1995). AFLP polymorphism detected a 347 bp fragment that was specific in the A2 mating type of *P. infestans* segment, but not in the A1 mating type of *P. infestans*. Kim and Lee (2002) also observe that similar mating type pattern and converted this mating type specific fragment into a mating type specific SCAR (Sequence Characterized Amplified Region) marker named PHYB.

Molecular Studies in Revealing Centre of Origin of *P. infestans*

Speculation about the origin of *P. infestans* and the source of inoculum for the epidemics began soon after the Irish famine catastrophe and remains the subject of debate. There are three theories regarding origin of this pathogen.

1. Till recently, most scientists had an opinion that the center of origin of *P. infestans* is in the highlands of central Mexico's Toluca Valley because high nuclear genetic diversity and the presence of sexual reproduction of the pathogen occurs there (Goodwin *et al.*, 1994; Andrivon, 1996) and it had been considered the ultimate source for all known migrations. The cool-short day climate in central Mexican highlands favours sexual cycle and co-existence of wild tuber bearing *Solanum* species in these highlands supports it. It was only in Central Mexico that both mating types of the pathogen were common prior to the 1980's. However, high levels of nuclear genetic variability found in central Mexico could be the result of sexual reproduction and not of ancestry.
2. Nineteenth-century scientists thought that *P. infestans* originated in the South American Andes (currently Bolivia, Ecuador, and Peru) (Berkeley, 1846; de Bary, 1876), the center of origin of the cultivated potatoes and other Solanaceous species, and assumed cospeciation of the pathogen with

its host. Historical records of the potato disease in the Andes suggest that it was endemic in the region (Abad, 1997). However, only a few clonal lineages of *P. infestans* have been described from the Andes, so the Andean hypothesis has not been generally accepted (Goodwin *et al.*, 1994). Sexual reproduction is less common in Andean populations and evidence for host adaptation and reproductive isolation has been reported (Oliva *et al.*, 2002).

Greater diversity in a place may be due to a particular history of founder effects, extinctions, and expansions of local populations. In contrast, there is less mitochondrial diversity in Toluca Mexico and the predominance of one maternal lineage suggests either a single maternal origin for this population or selection (Gavino and Fry, 2002). The mitochondrial genome is inherited maternally as a unit in *P. infestans*, without genetic recombination (Goodwin *et al.*, 1994; Gavino and Fry, 2002).

3. A third theory, known as the Three-Step or Hybrid theory, suggests Mexico as the center of origin of the pathogen, but that the source of inoculum for the 19th century epidemics originated from the South American Andes (Andrivon, 1996). It was speculated that *P. infestans* migrated first from Mexico to the South American Andes centuries before the 1840's and was subsequently dispersed from the Andean region to the U.S. and Europe (Andrivon, 1996).

Gomez-Alpizar *et al.* (2007) assessed the genealogical history of *P. infestans* using sequences from portions of two nuclear genes (*tubulin* and *ras*) and several mitochondrial loci P3, (*rpl*14, *rpl*5, tRNA) and P4 (*cox1*) from 94 isolates from South, Central, and North America, as well as Ireland. Summary statistics, migration analyses and the genealogy of current populations of *P. infestans* for both nuclear and mitochondrial loci are consistent with an "out of South America" origin for *P. infestans*.

Mexican populations of *P. infestans* from the putative center of origin in Toluca Mexico harbored less nucleotide and haplotype diversity than Andean populations. Coalescent-based genealogies of all loci were congruent and demonstrate the existence of two lineages leading to present day haplotypes of *P. infestans* on potatoes. The oldest lineage associated with isolates from the section *Anarrhichomenun* including *Solanum tetrapetalum* from Ecuador was identified as *Phytophthora andina* and evolved from a common ancestor of *P. infestans*. Nuclear and mitochondrial haplotypes found in Toluca Mexico were derived from only one of the two lineages, whereas haplotypes from Andean populations in Peru and Ecuador were derived from both lineages. Haplotypes found in populations from the U.S. and Ireland was derived from both ancestral lineages that occur in South America suggesting a common ancestry among these populations.

The geographic distribution of mutations on the rooted gene genealogies demonstrate that the oldest mutations in analysis of the mitochondrial and nuclear loci of *P. infestans* strongly supports a South American center of origin of this pathogen (Gomez-Alpizar *et al.*, 2007). *P. infestans* originated in South America and are consistent with a South American origin. They proposed evolutionary history of *P. infestans* which is as follows:

- An ancestral population of *Phytophthora* diverged into different lineages in the South American Andes in association with wild *Solanum* species.
- Two of the divergent lineages gave rise to the extant haplotypes of *P. infestans* capable of infecting potato, tomato, and some wild *Solanum* species.
- Other lineages evolved into distinct species, closely related to *P. infestans* and morphologically identical to it (section *Anarrhichomenum* isolates). Host specificity became the driving force for maintaining the divergent lineages in Ecuador and Peru. An Andean source of inoculum initiated epidemics first in the U.S. and then Ireland that led to the famine.
- Their data provided strong evidence for an "out of South America" origin of this destructive plant pathogen and clearly demonstrate that the oldest mutations in the ancestral strains occurred in South America.

Phytophthora infestans: Population Dynamics

P. infestans exists as distinct populations divided based on physiological race, fungicide resistance, mating type and genetic markers. Physiological races are identified by the presence of one or more of the 11 known virulence genes. The appearance of resistance to the phenylamide fungicides in the late 1970's indicated that populations of *P. infestans* were changing. An anti-resistance strategy was developed for growers in an effort to reduce the spread of resistant strains. Strains can also be divided into different genotypes by genetic fingerprinting which is based on the presence or absence of different genetic markers. Prior to the 1980s no reliable methods were available for adequate identification of genotypes. Development of molecular markers specific to *P. infestans* has made this possible

The introduction of the A2 mating type into Europe resulted in a shift from low to high nuclear genetic diversity in the Netherlands, particularly in places where both mating types were found together, mirroring the diversity found in central Mexico (Zwankhuizen *et al.*, 2000). It was suggested that *P. infestans* originally migrated from Mexico to the United States in infected wild potato tubers in the 19th century to cause famine-era epidemics (Goodwin *et al.*, 1994).

In the U.S., the pathogen infected potatoes and then spread to Europe and the rest of the world (Goodwin *et al.*, 1994). Spread of a single clonal lineage, the US-1 (Ib mtDNA haplotype) was proposed (Goodwin *et al.*, 1994). The US-1 lineage is not found widely in extant Mexican populations of *P. infestans* (Gavino and Fry, 2002; Flier *et al.*, 2003), whereas this lineage is still found in other populations around the world including the Andes. Sequencing of mtDNA from historic specimens of *P. infestans* from the Irish famine revealed that the Ia haplotype was common (Ristaino *et al.*, 2001; May and Ristaino, 2004). The US-1 lineage (Ib mtDNA haplotype) did not cause the famine, but was identified in more recent samples from the Andean region in Ecuador and Bolivia (May and Ristaino, 2004). This finding suggests either extinction of the US-1 lineage from the Mexican population or a non-Mexican origin of this lineage. Avila-Adame *et al.* (2006) published the mitochondrial genome sequences of extant mtDNA haplotypes of *P. infestans*. Two independent ancestral lineages, the type I (Ia, Ib) and type II (IIa, IIb), are derived from a common ancestor and the type I lineage is more closely related to the common ancestor.

Long-distance migration of *P. infestans* frequently appears to have resulted from the inadvertent movement of infected plant material (potato tubers, tomatoes) during trade. The appearance of new populations of *P. infestans* has often been accompanied by devastating results: loss of resistant varieties of host and the appearance of fungicide resistant strains.

Sources of Variation

In *P. infestans*, wherever sexual recombination is new variants may develop by segregation and recombination of genes. Hybridization between races gives rise to new biotype and races. Within potatoes different varieties/genotypes (with different set of resistance genes) are infected by pathogen variants thereby leading to race based classification based on 'Gene for Gene hypothesis". However there are many other factors that also contribute significantly to pathogen's diversity.

Migration

Migration is a strong evolutionary force that has a visible effect on the genetic structures of population of *Phytophthora* species. Migration is the force that units all members of species in a common gene pool, relatively low level of migration can prevent population from diverging. There are two components to migration, movement of dispersal of individuals and the contribution of migrants to the gene pool, of the recipient population in subsequent generations. Gene flow only occurs if the migrant individuals become established and reproduce in the new population. Large migrations can result in little gene flow, and limited migration can have a large effect on gene flow, depending on the fate of the immigrant individuals after dispersal. The effect of migration also depends on the degree of differentiation among populations. Migrations that start new population in previously unoccupied territories are called founder events. Founder population typically contains only a small usually small, genetic drift.

Mutation

Mutation is the primary source of new genetic variation. Unfortunately, little is known about mutation or mutation rate in any oomycete. However, clonal reproduction is predominant in population of many *Phytophthora* species, and some information about mutation can be gleaned from analyses of genetic variation within clonal lineages.

The concept of a clonal lineage is important to understanding the population biology of *P. infestans*. A clonal lineage includes the asexual descendants of a single genotype and the conclusion emerging from population genetic studies a *P. infestans* is that prior to the 1980's, old world populations were dominated and consisted solely of a single clonal linage. All members of the clonal lineages have descended from a single individual; any variation within a lineage must arise by mutation (or possibly mitotic recombination). So far, clonal lineages have been identified in *P. infestans* and *P. sojae*. The most commonly detected clonal lineage of *P. infestans* has been designated US-1. All members of US-1 clonal lineage of *P. infestans* worldwide most likely were derived from a single Mexican ancestor during the past 157 years.

Analyses of US-1 isolated worldwide have detected a number of probable mutation both in nuclear and mitochondrial (mt) DNA. Mutations also were detected in the mtDNA of *P. infestans* among various mtDNA haplotypes. The changes involve insertion and deletion, additional variants undoubtedly would be found in a systematic survey with more enzymes. Although these results are very limited, they do indicate that mutation in both nuclear and mtDNA occur at a high enough frequency to be detected.

Mitotic Recombination

The changes described above involved the appearance of new alleles that must have been due to mutation. However, most of the variation within clonal lineage of *P. infestans* did not involve new alleles but, rather apparent changes to homozygous at loci that were heterozygous in most common genotype. These changes could have been caused by mitotic recombination. A cross over during mitosis causes a change to homozygosity at all heterozygous loci on the same chromosome that are distal to the recombination break point. Although mitotic recombination does not generate new variation, it can reveal recessive variation that previously was hidden heterozygotes.

Putative mitotic recombination of *P. infestans* was detected more often than mutant's. All of the observed allozyme variation within clonal lineage involved changes from heterozygosity to homozygosity. However, very specific types of mutation would be required to effect these changes, which would probably occur much more rarely than mitotic recombination. Therefore, mitotic recombination is much more likely explanation.

Parasexual Recombination

Another potential explanation for new variation within clonal lineages is recombination. There is some evidence of parasexual recombination in *Phytophthora* species in laboratory although it has not been confirmed with molecular markers. Somatic fusion followed by a parasexual cycle has proposed to explain the recovery of new races of *P. infestans* in co-inoculation studies. Although parasexual recombination is a possibility, it is not most likely explanation for the majority of the observed changes.

Interspecific Hybridization

One final potential source of genetic variation is hybridization with other species. Interspecific hybrids between *P. infestans* and *P. mirabilis* have been made in the laboratory, and it has been hypothesized that *P. meadii* arose by interspecific hybridization, possible between *P. botryosa* and another species. This is an interesting area of research that should be pursued more intensively in the future.

Phytophthora infestans Genome Sequencing Projects

While great advances have been made in the area of human genomics, we know little about the genomes of plant pathogens. The development of automated sequencing techniques have opened the way for rapidly identifying genes from organisms of interest, and *P. infestans* is no exception. *P. infestans* was ranked number two by the American Phytopathological Society on the list of plant pathogen species targeted for DNA sequencing and whole genome sequencing is now underway.

Currently two major *P. infestans* genome sequencing projects are going in the world. Sequencing of *P. infestans* isolate T30-4 is in assembly phase at Broad Institute, MIT, USA; while sequencing work is also in progress at Sanger Institute, USA. There is a large database of genetic information available on *P. infestans* including a genetic linkage map and expressed sequence tag (EST) libraries and genetic systems are tractable.

Broad Institute, MIT, USA produced whole genome shotgun sequence from two plasmid libraries (4kb and 10kb inserts) and a Fosmid library. A preliminary 5X assembly was made public on 7/19/2006. The full-coverage (8X) assembly is now available. T30-4 used in this genome sequencing project is an F1 of two aggressive strains of *P. infestans* originally isolated from potato in the Netherlands, and is considered the reference isolate for most genetic studies. Significant resources have been generated in this strain: (1) A well characterized large insert BAC library has been constructed (Whisson *et al.*, 2001) and is available for use in that project; (2) it originates from the cross that was used to construct a genetic map based on AFLP markers (van der Lee *et al.*, 2004) and that will be used for high density mapping in this project; and (3) It was the source for the 1X draft genome sequence produced previously by Syngenta (Randall *et al.*, 2005). The Sanger Institute sequenced and annotated two BAC clones derived from *P. infestans*. They also fingerprinted additional *P. infestans* BAC clones.

The genome size of *P. infestans* is estimated to be 237 Mb (Sogin and Silberman, 1998), although (preliminary assembly data suggest it is slightly larger, 240-245 Mb. Two independent analyses indicate that the genome has about 52 per cent GC content. Microscopic analyses indicate that *P. infestans* has 8-10 chromosomes (Sansome and Brasier, 1973). Several studies suggest that genes are often tightly clustered in *Phytophthora* genomes (Kamoun, 2003; Randall *et al.*, 2003; Cvitanich and Judelson, 2003).

Randall *et al.* (2005) aligned EST sequences from the Syngenta *Phytophthora* Consortium to the assembly. Of ~80,000 ESTs available in GenBank, ~65,000 align to the assembly. The remainder contains a high fraction of host transcripts as well as sequences (such as ribosomal genes) which are difficult to assemble by whole genome shotgun.

mtDNA Genome

Complete mitochondrial genome sequencing of various haplotypes of the mitochondrial genome *P. infestans* has been completed by collaborating North Carolina State University; The Institute of Genome Research (TIGR) and Universite de Montreal, Canada. *Phytophthora* mitochondrial DNA is smaller (37.914kb) in size as compared to larger size ranges in plants (195kb in *Oenothera* to 2400 kb in cucurbits). Completely sequenced the circular 37.914 kbp mitochondrial (mt) DNA is very A+T-rich (76 per cent) and very tightly packed with genes.

Most of the genes in the mt DNA of *P. infestans* do usually not occur in mt genomes of animals and plants, including three subunits of the NADH dehydrogenase complex, eleven small-subunit and six large-subunit ribosomal protein genes (*rps* and *rpl*, respectively), two open reading frames (*orf*) with a remote similarity to *rps*2 and *rps*3 sequences, the gene encoding the ATPase-α subunit (*atp*A) and at least four unique *orfs*. No introns have been identified in this mt genome. The set of genes resembles more plant than fungal mitochondria.

Mitochondrial genome sequencing was proposed to help in detection of base-level differences amongst various mtDNA haplotypes (Paquin *et al.*, 1997). It is a tool that would be of use to scientists and growers would be the ability to rapidly genotype isolates of the pathogen. Haplotypes could be correlated with virulence and resistance to pesticides and used in epidemiological studies. In addition, whole mitochondrial genome sequence analysis will be useful in studies to examine molecular evolution in the pathogen.

References

Abad, Z.G. and Abad, J.A. (1997). Another look at the origin of late blight of potatoes, tomatoes and pear melon in the Andes of South America. *Plant Disease.* 81: 682–688.

Andrivon, D. (1996). The origin of *Phytophthora infestans* present in Europe in the 1840s: A critical review of the historical and scientific evidence. *Plant Pathology,* 45: 1027–1035.

Avila-Adame, C., Gomez-Alpizar, L., Zismann, V., Jones, K.M., Buell, C.R. and Ristaino, J.B. (2006). Mitochondrial genome sequences and molecular evolution of the Irish potato famine pathogen, *Phytophthora infestans. Current Genetics.* 49: 39-46.

Berkeley, M.J. (1846). Observations, botanical and physiological, on the potato murrain. *Journal of the Horticultural Society of London* 1: 9-34. (Reprinted in *Phytopathological Classics,* no. 8, American Phytopathological Society, East Lansing, Michigan, USA, 1975).

Bhattacharyya, S.K., Shekhawat, G.S. and Singh, B.P. (1990). Potato late blight, Central Potato Research Institute, Shimla, India. *Central Potato Research Institute Technical Bulletin* No. 27: pp. 46.

Carter, D.A., Archer, S.A., Buck, K.W., Shaw, D.S. and Shattock, R.C. (1990) Restriction fragment length polymorphisms of mitochondrial DNA of *Phytophthora infestans. Mycological Research.* 94: 1123-1128.

Chesnick, J.M., Tuxbury, K., Coleman, A., Burger, G. and Lang, B.F. (1996). Utility of the mitochondrial *nad*4L gene for algal and protistan phylogenetic analysis. *Journal of Phycology.* 32: 452-456.

Cooke, D.E.L. and Lees, A.K. (2004) Markers, old and new, for examining *Phytophthora infestans* diversity *Plant Pathology.* 53: 692–704.

Cooke, D.E.L., Young, V., Birch, P.R.J., Toth, R., Gourlay, F., Day, J.P., Carnegie, S. and Duncan, J.M. (2003). Phenotypic and genotypic diversity of *Phytophthora infestans* populations in Scotland (1995–97). *Plant Pathology.* 52: 181–92.

Cvitanich, C. and Judelson, H.S. (2003). Stable transformation of the oomycete, *Phytophthora infestans*, using microprojectile bombardment. *Current Genetics*. 42: 228-235.

Day, J.P. and Shattock, R.C. (1997). Aggressiveness and other factors relating to displacement of populations of Phytophthora infestans in England. *Euro. J. Plant Pathology*. 103: 379-391.

deBary, H.A. (1876). Researches into the nature of the potato-fungus, *Phytophthora infestans. Journal of Royal Agricultural Society*. 12: 239–268.

Drenth, A., Tas, I.C.Q., Govers, F. (1994). DNA fingerprinting uncovers a new sexually reproducing population of *Phytophthora infestans* in the Netherlands. *European Journal of Plant Pathology*. 100: 97–107.

Flier, W.G., Grundwald, N.K., Kroon, L.P.N.M., Sturbaum, A.K., van den Bosch, T.B.M., Garay-Serrano, E., Lozoya-Saldana, H., Fry, W.E., Turkensteen, L.J. (2003) *Phytopathology*. 93: 382–390.

Forbes, G.A., Goodwin, S.B., Drenth, A., Oyarzun, P., Ordoñez, M.E., Fry, W.E. (1998). A global marker database for *Phytophthora infestans*. *Plant Disease*. 82: 811–818.

Fry,W.E., Goodwin, S.B., Dyer, A.T., Matuszak, J.M., Drenth, A., Tooley, P.W., Sujkowski, L.S., Koh, Y.J., Cohen, B.A., Spielman, L.J., Deahl, K.L., Inglis, D.A., Sandlan, K.P. (1993) Historical and recent migration of *Phytophthora infestans*: Chronology, Pathways and implications. *Plant Disease*, 77: 653-661.

Gavino, P.D. and Fry, W.E. (2002) Diversity in and evidence for selection on the mitochondrial genome of *Phytophthora infestans*. *Mycologia*. 94: 781-793.

Gomez-Alpizar, L., Carbone, I. and Ristaino, J.B. (2007) An Andean origin of *Phytophthora infestans* inferred from mitochondrial and nuclear gene genealogies *Proceedings of National Academy of Sciences, USA* 104: 93306-93311.

Goodwin, S.B. and Drenth, A. (1997). Origin of A2 mating type of *Phytophthora infestans* outside Mexico. *Phytopathology*. 87: 992-999.

Goodwin, S.B., Cohen, B.A., Fry, W.E. (1994) Panglobal distribution of a single clonal lineage of the Irish potato famine fungus. *Proceedings of National Academy of Science, USA*. 91: 11591-11595.

Goodwin, S.B., Drenth, A., Fry, W.E. (1992a). Cloning and genetic analysis of two highly polymorphic, moderately repetitive nuclear DNAs from *Phytophthora infestans*. *Current Genetics*. 22: 107–15.

Goodwin, S.B., Spielman, L.J., Matuszak, J.M., Bergeron, S.N., Fry, W.E. (1992b). Clonal diversity and genetic differentiation of *Phytophthora infestans* populations in northern and central Mexico. *Phytopathology*. 82: 955–61.

Goodwin, S.B. (1991). DNA polymorphisms in *Phytophthora infestans*: the Cornell Experience. In: Lucas JA, Shattock RC, Shaw DS, Cooke LR, eds. Phytophthora. Cambridge, UK: Cambridge University Press. 256-271.

Hohl. H., Iselin. K. (1984) Strains of *Phytophthora infestans* from Switzerland with A2 mating type behaviour. *T Brit Mycol Soc*. 83: 529-53.

Judelson, H.S., Spielman, L.J. and Shattock, R.C. (1995). Genetic mapping and non-Mendelian segregation of mating type loci in the oomycete, *Phytophthora infestans. Genetics*. 141: 503-512.

Kamoun, S. (2003). Molecular genetics of pathogenic oomycetes. *Eukaryotic Cell* 2: 191-199.

Kim, K.J. and Lee, Y.S. (2002): Genetic DNA Marker for A2 mating type in *Phytophthora infestans. Journal of Microbiology*. 40: 254–259.

Klimczak, L.J. and Prell, H.H. (1988) Isolation and characterisation of mitochondrial DNA of the oomycetous fungus *Phytophthora infestans. Current Genetics*. 8: 323–326.

Knapova, G., Tenzer, I., Gessler, C. and Gisi, U. (2001). Characterization of *Phytophthora infestans* from potato and tomato with molecular markers. *Proceedings of the 5th Congress of the European Foundation for Plant Pathology (Biodiversity in Plant Pathology)*. Taormina, Italy: SIVP, 6–9.

Knapova, G. and Gisi, U. (2002). Phenotypic and genotypic structure of *Phytophthora infestans* populations on potato and tomato in France and Switzerland. *Plant Pathology*. 51: 641–53.

Ko, W.H. (1994). An alternative possible origin of the A2 mating type of *Phytophthora infestans* outside Mexico. *Phytopathology*. 84: 1224-1227.

Lebreton, L., Laurent, C. and Andrivon, D. (1998). Evolution of *Phytophthora infestans* populations in the two most important potato production areas of France during 1992-1996. *Plant Pathology*. 47: 427-439.

Lees,A.K., Wattier, R., Shaw, D.S., Sullivan, L., Williams, N.A. and Cooke, D.E.L. (2006). Novel microsatellite markers for the analysis of *Phytophthora infestans* populations. *Plant Pathology*. 55: 311–319.

May, K.J. and Ristaino, J.B. (2004) Identity of the mtDNA haplotype(s) of *Phytophthora infestans* in historical specimens from the Irish potato famine. *Mycological Research*. 108: 471-479.

Oliva, R.F., Erselius, L. J., Adler, N.E., Forbes, G.A.(2002) Potential of sexual reproduction among host adapted populations of Phytophthora infestans sensu lato in Ecuador. *Plant Pathology*. 51: 710–719.

Paquin, B., Laforest, M.J., Forget, L., Roewer, I., Wang, Z., Longcore, J., Lang, B.F. (1997). The fungal mitochondrial genome project: evolution of fungal mitochondrial genomes and their gene expression. *Current Genetics*. 31: 380–95.

Perez,W.G., Gamboa, J.S., Falcon, Y.V., Coca, M., Raymundo, R.M., Nelson, R.J. (2001). Genetic structure of Peruvian populations of *Phytophthora infestans*. *Phytopathology*. 91: 956-965.

Purvis, A.L., Pipe, N.D., Day, J.P., Shattock, R.C., Shaw, D.S., Assinder, S.J. (2001). AFLP and RFLP (RG57) fingerprints can give conflicting evidence about the relatedness of isolates of *Phytophthora infestans. Mycological Research*. 105: 1321–1330.

Randall, T.A., Fong, A., Judelson, H.S. (2003). Chromosomal heteromorphism and an apparent translocation detected using a BAC contig spanning the mating type locus of *Phytophthora infestans*. *Fungal Genet Biol.* 38: 75-84.

Randall, T.A., Dwyer, R.A., Huitema, E., Beyer, K. and Judelson, H.S. (2005). Large-scale gene discovery in the oomycete *Phytophthora infestans* reveals likely components of phytopathogenicity shared with true fungi. *Mol Plant Microbe Interact.* 18: 229-243.

Ristaino, J.B., Groves, C.T. and Parra, G.R. (2001). PCR amplifications the Irish potato famine pathogen from historic specimens. *Nature.* 411: 695–697.

Sansome, E. and Brasier, C.M. (1973). Diploidy and chromosomal structural hybridity in *Phytophthora infestans*. *Nature.* 241: 344-345.

Singh, B.P., Roy, S., Bhattacharyya, S.K. (1994) Occurrence of the A2 mating type of *Phytophthora infestans* in India. *Potato Res.* 37: 227-231.

Smoot, J.J., Gough, F.J., Lamey, H.A., Eichenmuller, J.J., Gallegly, M.E. (1958). Production and germination of oospores of *Phytophthora infestans*. *Phytopathology.* 48: 165-171.

Sogin, M.L. and Silberman, J.D. (1998). Evolution of the protists and protistan parasites from the perspective of molecular systematics. *International Journal of Parasitology.* 28: 11-20.

Spielman, L.J., Sweigard, J.A., Shattock, R.C., Fry, W.E. (1990). The genetics of *Phytophthora infestans*: segregation of allozyme markers in F2 and backcross progeny and the inheritance of virulence against potato resistance genes *R2* and *R4* in F1 progeny. *Experimental Mycology.* 14: 57–69.

Tooley PW, Fry WE, Villerreal Gonzalez MJ. (1985). Isozyme characterization of sexual and asexual *Phytophthora infestans* populations. *Journal of Heredity.* 76: 431–435.

van der, Lee. T., Testa, A., Robold, A., van't Klooster, J.W. and Govers, F. (2004). High density genetic linkage maps of *Phytophthora infestans* reveal trisomic progeny and chromosomal rearrangements. *Genetics.* 167: 111-129.

Vos. R., Hoger, R., Bleeker, M., Reijans, M., van der Lee, T., Hornes, M., Frijters, A., Pot, J., Peleman, J., Kuiper, M. and Zabeau, M. (1995). AFLP: a new fingerprinting technique for DNA fingerprinting. *Nucleic acids Research.* 23: 4407-4414.

Whisson, S., Lee, T., Bryan, G., Waugh, R., Govers, F. P. (2001). Physical mapping across an avirulence locus of *Phytophthora infestans* using a highly representative, large-insert bacterial artificial chromosome library. *Molecular Genetics and Genomics.* 266: 289-295.

Zwankhuizen, M.J., Govers, F., Zadoks, J,C. (2000). Inoculum sources and genotypic diversity of *Phytophthora infestans* in Southern Flevoland, the Netherlands. *European Journal of Plant Pathology.* 106: 667–680.

Chapter 5

Macrophomina phaseolina: An Overview on Molecular Biology and Host Specialization

I.K. Das[1] and A.V. Gadewar[2]***

[1]*National Research Centre for Sorghum, Rajendranagar, Hyderabad – 500 030, A.P., India*
[2]*Centre on Rabi Sorghum, Solapur – 413 006, Maharashtra, India*

ABSTRACT

Macrophomina phaseolina inflicts losses on many agriculturally important crops worldwide. Application of molecular tools like RFLP, AFLP, RAPD and SSR markers have detected wide range of genetic variations among isolates from different hosts and narrow range among isolates from a given host. As host resistance is highly influenced by environment and host factors, precise assessment of virulence/aggressiveness of isolates, becomes a real challenge. Molecular biology tools like marker assisted selection (MAS) have been attempted for enhancement of resistance against charcoal rot. Available information on host specialization in this pathogen is scanty. Cross-inoculation studies suggested that there was no host specialization in isolates from soybean, sorghum and cotton but it appeared to occur with corn. Chlorate phenotype might not be useful for studying host specialization in this pathogen as this phenotype is not specific to host, and possible shift in chlorate phenotype may occur in relation to host

E-mail: *das@nrcsorghum.res.in; ** gadewar@nrcsorghum.res.in.

interaction. This chapter presents an overview on recent research developments with special emphasis on molecular diversity, pathogenic variations and host specialization.

Keywords: *Macrophomina phaseolina, RAPD, Genetic diversity.*

Introduction

Macrophomina phaseolina (Tassi) Goid, is a soil-borne plant pathogen of global importance. It is present in most cultivated soils and can infect many important crops, particularly in the warm and tropical environments. About 500 plant species of agriculturally important crops including pulses, cereals, oilseeds, vegetables, fruits, fiber crops, medicinal plants and many others are affected worldwide. In India sorghum, maize, soybean, jute, and pulses are major crops which are severely affected by this pathogen. In sorghum the pathogen causes charcoal rot and causes significant yield loss especially on high yielding cultivars (Rosenow and Clark, 1995; Das *et al.*, 2008b). Ecology, biology, and management aspects of this pathogen have been reviewed in the past (Tarr, 1962; Dhingra and Sinclair, 1978; Sinclair, 1984). Recent developments in the field of molecular biology facilitated studies on genetic diversities, pathogenic variations and host specialization in this plurivorous pathogen. Herein an overview on recent research developments on molecular biology and host specialization in *M. phaseolina*.

Genetic Diversity Studies

M. phaseolina has aerial, superficial or immersed, hyaline to brown, septate, profusely branched or dendroid mycelium. The fungus is highly variable in size of sclerotia and presence or absence of pycnidia. Sclerotia are loose, brown to black, scale like, irregular in shape, size highly variable within an isolate. Pycnidia production in PDA is rare except under some specific incubation conditions (Gaetan *et al.*, 2006) and specific to some host crops (Mihail and Taylor, 1995). Pycnidia (or pycnidial) stage is widespread in jute and garden beans but uncommon in soybean, maize and sorghum. *Macrophomina* is a monotypic genus with a single species. Efforts to divide it in to sub-species were unsuccessful largely due to the extreme intra-specific variations in morphology and pathogenicity (Dhingra and Sinclair, 1973; Echavez-Badel and Perdomo, 1991). Recently, molecular techniques like amplified fragment length polymorphism (AFLP), restriction fragment length polymorphism (RFLP), random amplified polymorphic DNA (RAPD) and simple sequence repeats (SSR) have been used to unveil genetic variability in this soil-borne plant pathogen (Vandemark *et al.*, 2000; Su *et al.*, 2001; Jana *et al.*, 2003; Jana *et al.*, 2005).

Genetic diversity in populations of *M. phaseolina* originating in different countries such as Australia, Brazil, India, and USA have been analyzed using RAPD (Fuhlbohm, 1997; Alvaro *et al.*, 2003; Jana *et al.*, 2003; Purkayastha *et al.*, 2006; Das *et al.*, 2008a). Wide range of variability was detected among isolates, isolated from different hosts. Isolates from a given host were, however, genetically closer (Su *et al.*, 2001; Jana *et al.*, 2003). Forty-three isolates from four countries (India, USA, Pakistan and Iraq) formed five groups in RAPD cluster analysis (Jana *et al.*, 2003). Cropping system seemed to have some influence on genetic makeup of *M. phaseolina*. Working on Brazilian isolates of this pathogen, Alvaro *et al.* (2003) observed that genetically more diverse isolates originated from areas with a single crop than areas practicing crop rotation. RFLP analysis of the ITS region of *M. phaseolina* was used for detecting variability among isolates. No variations were observed among isolates in restriction patterns of DNA fragments amplified by polymerase chain reaction covering the internal transcribed spacer region, 5.8S rRNA and part of 25S rRNA, suggesting that *M. phaseolina* constitutes a single species (Su *et al.*, 2001). Single primers of simple sequence repeats (SSR) or microsatellite markers have been used for the characterization of genetic variability of different populations of *M. phaseolina* from soybean and cotton. The variability found within closely related isolates indicated that such microsatellites are useful in population studies in *M. phaseolina* (Jana *et al.*, 2005). Jana *et al.* (2003) developed a single RAPD primer OPA-13 that distinguishes isolates of *M. phaseolina* from soybean, sesame, groundnut, chickpea, cotton, common bean, okra, and 13 other hosts and this was claimed to be useful as taxonomic marker for population studies.

Pathogenic Variations

M. phaseolina is highly variable in morphology and pathogenicity. Wide variation in pathogenic properties has been demonstrated in soybean, sunflower, groundnut, common bean and sorghum (Dhingra and Sinclair, 1973; Manici *et al.*, 1992; Sobti and Sharma, 1992; Mayek-perez *et al.*, 2001; Das *et al.*, 2008a). Host resistance against *M. phaseolina* is polygenic and governed by many quantitative trait loci (QTL). Major factors that influence expression of disease and its severity include physiological conditions of the host (vigor, water content, sugar content, mineral content, balance between current photosynthates and its translocation) (Liebhardt and Munson, 1976; Dodd, 1977; Pedgaonkar and Mayee, 1990), root and stalk morphology and anatomy (stalk diameter and strength in cereals, vascular bundle number, diameter of xylem vessel) (Maranville and Clegg, 1984; Das and Prabhakar, 2003), soil moisture (Seetharama *et al.*, 1987) and senescence (Rosenow, 1980). Therefore, resistance level is highly influenced by environment and host factors and precise assessment of virulence/aggressiveness of isolates, become a real challenge. Isolates from a single sorghum cultivar vary in pathogenicity when inoculated on other cultivars (Diourte *et al.*, 1995). As many as 43 pathotypes were obtained from 84 common bean isolates in Mexico (Mayek-perez *et al.*, 2001). Eight sorghum isolates from India showed six types of disease reactions (pathotypes) on a set of 5 sorghum lines, and further supported the existence of a high degree of variation in the pathogen (Das, unpublished data).

Molecular Approaches for Management

Use of host resistance has been the major focus for management of charcoal rot especially in sorghum. But genetic improvement of such polygenic traits through classical approaches has been rather slow. Molecular biology tools have been tried for enhancement of resistance against quantitative traits like charcoal rot resistance. For traits greatly influenced by environment and no reliable genetic clues have been available for recombination breeding, the marker-assisted selection (MAS) approach is gaining importance. Combining MAS methods with conventional breeding schemes can increase the overall selection gain and, therefore, the efficiency of a breeding programme (Bhat *et al.*, 2004). Mapping of quantitative trait loci (QTL) in the mapping population, E36-1 x IS22380 identified five QTLs for the component traits of charcoal rot resistance in sorghum (Srinivasa *et al.*, 2008). Drought often predisposes *Macrophomina* infection in many crop plants. Incorporation of non-senescent properties into cultivar by using molecular techniques will possibly help combating drought and charcoal rot. Molecular tools have been tried for inactivation of pathogen weapons. A cell-wall degrading enzyme, endoglucanase (egl II) has been cloned from *M. phaseolina*. This enzyme enables *M. phaseolina* to penetrate the cell wall of living plant. Production of antibody for this enzyme can disable *M. phaseolina* from infection (Wang and Jones, 1995).

Host Specialization

Fungal pathogens provide a good model for scientific studies of evolution because changes can occur over relatively short time scales. Though evolution of sibling fungal plant pathogens is not well understood, morphological and genetic similarities between individuals suggest that they have evolved from a common ancestor. There are many examples of sibling pathogens that exhibit distinct host ranges. As for example, 3 forma specials have been distinguished in *Phytophthora megasperma*, based on host specificity (Kuan and Erwin, 1980). There are more than 20 closely related sibling species of *Botrytis* confined to one or a few closely related host species (Jarvis, 1977). While evidence is growing for host specialization in pathogen populations attacking genetically distant plant species (Gudelj *et al.*, 2004), the situation for sorghum is less clear. *M. phaseolina*, in general, is non-specific and can infect wide range of hosts (Holliday and Punithalingam, 1970; Singh and Nene, 1990). Isolates are more aggressive towards the host species from which they were recovered than other hosts (Diourte, 1987). Except in few earlier reports on the occurrence of host specific isolates in jute, available information on host specialization in this pathogen is scanty (Ahmed and Ahmed, 1969). Cloud and Rupe (1991) observed significantly greater infection of soybean roots in plants grown in soil with the soybean isolates compared with soil either infested with both the isolates or with the sorghum isolate, and concluded that host specialization does occur in soybean. However, cross-inoculation study with a large number of isolates revealed no host specialization in isolates from soybean, sorghum and cotton but it appeared to occur with corn (Su *et al.*, 2001).

Pearson *et al.* (1986) suggested the use of chlorate phenotypes (colony morphologies on media supplemented with 120 mM potassium chlorate) as the marker for identifying host-specific isolates in *M. phaseolina*. It was found that isolates from corn stalks was chlorate resistant (120 mM) and had dense phenotype (growth like that on medium lacking chlorate), whereas isolates from soybean root tissue and field soil were chlorate sensitive with feathery or restricted phenotypes (inhibited growth). Later studies reported that chlorate phenotype was not specific to host, and possible shift in chlorate phenotype occurs in relation to host interaction (Cloud and

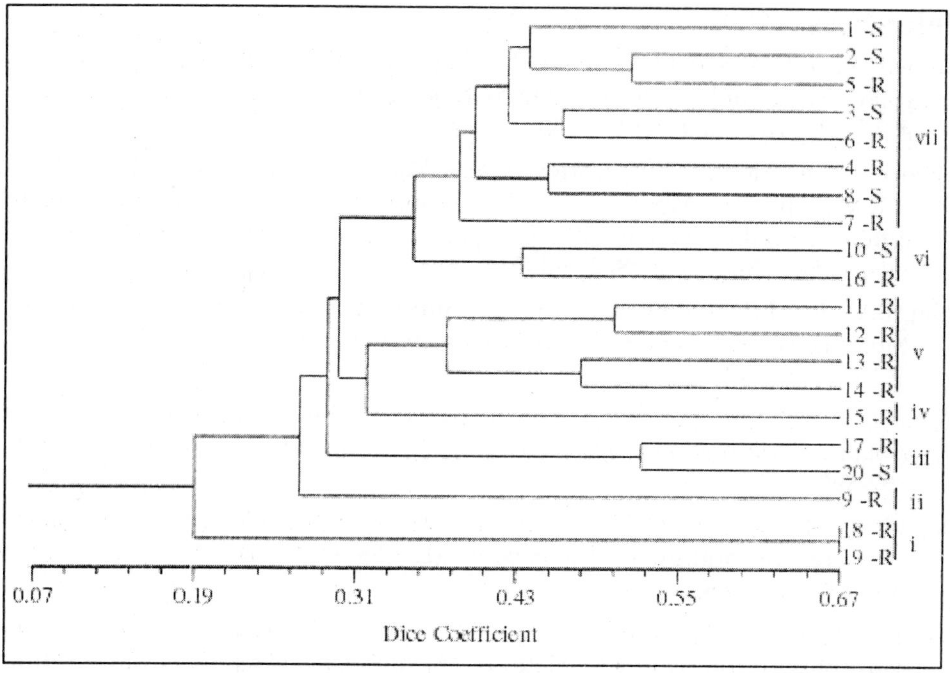

Figure 5.1: Dendrogram Constructed with the UPGMA Clustering Method for 20 Isolates of *M. phaseolina* Infecting Sorghum. Distance is based on genetic similarity coefficient. S or R followed by isolate name indicates chlorate sensitivity of the isolate.

Rupe, 1991). Moreover, in some hosts such as sorghum, all three phenotypes were noted and no specific chlorate phenotype could be ascribed to sorghum isolates. Chlorate phenotype may not be useful for studying host specialization in such cases. It was also observed that chlorate sensitive isolates were distinct from chlorate resistant isolates within a given host (Figure 5.1). Chlorate sensitivity in *Macrophomina* had some relation with charcoal rot severity in sorghum (Das *et al.*, 2008a). Further studies on the mechanism of chlorate assimilation and genetic basis of chlorate sensitivity in *M. phaseolina* can generate more information.

It is concluded that *M. phaseolina* is a pathogen of global concern. Genetic diversity is wide among isolates from different hosts and narrow among isolates from a given host. Available information on host specialization is scanty, and it seems to occur in corn isolates and not in those from soybean, sorghum and cotton. The mechanism of chlorate assimilation and genetic basis of chlorate sensitivity in *M. phaseolina* can generate more information. For management of *Macrophomina* diseases, resistant cultivars have been the major focus. Molecular biology tools like marker assisted selection (MAS) have also been attempted in a few crops and success is likely to come in future.

References

Ahmed, N. and Ahmed, Q.A. (1969). Physiological specialization in *Macrophomina phaseolina* (Maubl.) Ashby causing root rot in jute, *Corchorus* species. *Mycopathol et Mycolog Applicata*. 39: 129-138.

Alvaro, M.R.A., Ricardo, V.A., Carlos, A.A.A., Valdemar, P.C., David, S.J.F., Silvana, R.R.M., Luis, C.B., Mauro, C.P. and Claudio, G.P.C. (2003). Genotypic diversity among Brazilian isolates of *Macrophomina phaseolina* revealed by RAPD. *Fitopatologia Brasileira*. 28: 279-285.

Bhat, B.V., Balakrishna, D., Sateesh, K., Srinivas, G. and Seetharama, N. (2004). Molecular marker-aided approaches for sorghum improvement. *AgBiotechNet*. 6: 1–13, ABN 127.

Cloud, G.L. and Rupe, J.C. (1991). Morphological instability in a chlorate medium of isolates of *Macrophomina phaseolina* from soybean and sorghum. *Phytopath.* 81: 892-895.

Das, I.K. and Prabhakar, (2003). Identification of stable morphological and anatomical characters of sorghum (*Sorghum bicolor* (L) Moench) stalk. *Indian J Genet Plant Breed.* 63: 347-348.

Das, I.K., Fakrudin, B. and Arora, D.K. (2008a). RAPD cluster analysis and chlorate sensitivity of some Indian isolates of *Macrophomina phaseolina* from sorghum and their relationships with pathogenicity. *Microbiol Res*, 163: 215-224.

Das, I.K., Prabhakar. and Indira, S. (2008b). Role of stalk-anatomy and yield parameters in development of charcoal rot caused by *Macrophomina phaseolina* in winter sorghum. *Phytoparasitica*. 36(2): 199-208.

Dhingra, O.D. and Sinclair, J.B. (1973). Location of *Macrophomina phaseolina* on soybean plant related to culture characteristics and virulence. *Phytopath*. 63: 934-936.

Dhingra, O.D. and Sinclair, J.B. (1978). Biology and pathology of *Macrophomina phaseolina* Vicosa, Brazil: Imprensia Universitaria, Universidade Federal de Vicosa. 166 pp.

Diourte, M., Starr, J.L., Jeger, M.J., Stack, J.P. and Rosenow, D.T. (1995). Charcoal rot (*Macrophomina phaseolina*) resistance and the effects of water stress on disease development in sorghum. *Plant Pathol.* 44: 196-202.

Diourte, M. (1987). Pathogenic variations and morphological studies of *Macrophomina phaseolina* (Tassi) Goid. M Sc Thesis, Texas A and m Univ., College Station, Texas.

Dodd, J.L. (1977). A photosynthetic stress-translocation balance concept of corn stalk rot. In: HD, Loden and D, Wilkinson ed. Proceedings, 32nd Annual Corn and Sorghum Research Conference, Washington DC, USA, American seed trade association. pp 122-130.

Echavez-Badel, R. and Perdomo, A. (1991). Characterization and comparative pathogenicity of two *Macrophomina phaseolina* isolates from Puerto Rico. *J Agric Univ P R* 75: 419-421.

Fuhlbohm, M. (1997). Genotypic diversity among Australian isolates of *Macrophomina phaseolina*. XX Biennial Australian Plant Pathology Society Conference, Lincoln University, New Zeeland. p52.

Gaetan, S.A., Fernandez, L. and Madia, M. (2006). Occurrence of charcoal rot caused by *Macrophomina phaseolina* on canola in Argentina. *Plant Dis.* 90: 524.

Fill, G.B.D.L. and Bosch, F van den. (2004). Evolution of sibling fungal plant pathogens in relation to host specialization. *Analytical Theoritical Plant Pathology.* 94: 789-795.

Holliday, P. and Punithalingam, E. (1970). *Macrophomina phaseolina.* No. 275 in CMI (Commonwealth Mycological Institute), Description of pathogenic fungi and bacteria. Kew, Surrey, UK: CMI

Jana, T.K., Sharma, T.R., Prasad, R.D. and Arora, D.K. (2003). Molecular characterization of *Macrophomina phaseolina* and *Fusarium* species by a single primer RAPD technique. *Microbiol Res.*158: 249-257.

Jana, T.K., Sharma, T.R. and Singh, N.K. (2005). SSR-based detection of genetic variability in the charcoal root rot pathogen *Macrophomina phaseolina. Mycol Res.*109: 81-86.

Jarvis, W.R. (1977). Botryotinia and Botrytis specis: Taxonomy, Physiology and Pathogenecity; A guide to the literature. Tech. Rep Res Branch, Can Dep. Agric. Monog. N. 15

Kuan, T.L. and Erwin, D.C. (1980). Differentiation of soybean and alfalfa isolates of *Phytophthora megasperma. Pytopathology* 70: 333-338.

Liebhardt, W.C. and Munson, R.D. (1976). Effect of chloride and potassium on corn lodging. *Agron J.* 68: 425-328.

Manici, L.M., Cerato, C. and Caputo, F. (1992). Pathogenic and biologic variability of *Macrophomina phaseolina* (Tassi) Goid. isolates in different areas of sunflower cultivation in Italy. *Proc. Sunflower Conf.* 13th Vol. 1. Pisa, Italy. p779-784.

Maranville, J.W. and Clegg, M.D. (1984). Morphological and physiological factors associated with stalk strength. In: Sorghum root and stalk rots-a critical review, *Proceedings of the Consultative Group Discussion on Research Needs and Strategies for Control of Sorghum Root and Stalk Rot Diseases*, 27 Nov–2 Dec 1983, Bellagio, Italy, pp 111-118.

Mayek-perez, N., Lopez-caataneda, C., Gonzalez-Chavira, M., Garch-Espinosa, R., Acosta-Gallegos, J., de la Vega, O.M. and Simpson, J. (2001). Variability of Mexican isolates of *Macrophomina phaseolina* based on pathogenesis and AFLP genotype. *Physiol Mol Plant Pathol.* 59: 257-264.

Mihail, J.D. and Taylor, S.J. (1995). Interpreting variability among isolates of *Macrophomina phaseolina* in pathogenicity, pycnidium production, and chlorate utilization. *Can J Bot.* 73: 1596–1603.

Pearson, C.A.S., Leslie, J.F. and Schwenk, F.W. (1986). Variable resistance in *Macrophomina phaseolina* from corn, soybean and soil. *Phytopath.* 76: 646-649.

Pedgaonkar, S.M. and Mayee, C.D. (1990). Stalk water potential in relation to charcoal rot of sorghum. *Indian Phytopath.* 43 (2): 192-196.

Purkayastha, S., Kaur, B., Dilbaghi, N. and Chaudhury, A. (2006). Characterization of *Macrophomina phaseolina*, the charcoal rot pathogen of cluster bean, using conventional techniques and PCR-based molecular markers. *Plant Pathology.* 55(1): 106.

Rosenow, D.T. (1980). Stalk rot resistance breeding in Texas, In: Sorghum Diseases-a world review, *Proceedings of the International Workshop on Sorghum Diseases,* 11-15 Dec 1978, ICRISAT, Patancheru, Hyderabad, India, pp 306-314:

Rosenow D.T and Clark L.E. (1995). Drought and lodging research for quality sorghum crop pp82-97. In: Proceedings of the 5th ANN Corn and Sorghum Industry Research Conference, Dec6-7, 1995, Illinois, American Seed Trade Association.

Seetharama, N., Bidinger, F.R., Rao, K.N., Gill, K.S. and Mulgund, M. (1987). Effect of pattern and severity of moisture deficit stress on stalk rot incidence in sorghum I. Use of line source irrigation technique, and the effect of time of inoculation. *Field Crop Res* 15: 289-308.

Sinclair, J.B. (1984). Root and stalk rots caused by *Macrophomina phaseolina* in legume and other crops, In: Sorghum root and stalk rots-a critical review, pp 173-182: Proceedings of the Consultative Group Discussion on Research Needs and Strategies for Control of Sorghum Root and Stalk Rot Diseases, 27 Nov–2 Dec 1983, Bellagio, Italy, Patancheru, A. P. 502324, ICRISAT, India.

Singh, S.K. and Nene, Y.L. (1990). Cross inoculation studies on *Rhizoctonia bataticola* isolates from different crops. *Indian Phytopath.* 43: 446-448.

Sobti, A.K. and Sharma, L.C. (1992). Cultural and pathogenic variations in isolates of *Rhizoctonia bataticola* from groundnut in Rajasthan. *Indian Phytopath.* 45: 117-119.

Srinivasa, P.R., Fakrudin, B., Rajkumar, Punnuri, S.M., Arun, S.S., Kuruvinashetti, M.S., Das, I.K. and Seetharama, N. (2008). Molecular mapping of genomic regions harboring QTLs for stalk rot resistance in sorghum. *Euphytica.* 159: 191-198.

Su, G., Suh, S.O., Schneider, R.W. and Russin, J.S. (2001). Host specialization in the charcoal rot fungus, *Macrophomina phaseolina. Phytopath.*91: 120-126.

Tarr, S.A.J. (1962). Diseases of sorghum, sudan grass and broom corn. Kew, Surrey, UK: *Commonwealth Mycological Institute,* 380 pp.

Vandemark, G., Martnez, O., Pecina, V. and Alvardo, M. de J. (2000). Assessment of genetic relationships among isolates of *Macrophomina phaseolina* using simplified AFLP technique and two different methods of analysis. *Mycologia.* 92: 656–664.

Wang, H. and Jones, R.W. (1995). A Unique endoglucanase-encoding gene cloned from the phytopathogenic Fungus *Macrophomina phaseolina. Appl Environ Microbiol.* 61: 2004–2006.

Chapter 6

Recent Advances in Diagnosis of Plant Pathogens

S.J. Gawande, V.P. Chimote* and Nimisha Kaushal*

Division of Plant Protection, Central Institute for Cotton Research, Shankarnagar, P.B. No. 2, Nagpur – 440 010, India

ABSTRACT

Effective plant pathogen diagnosis is critical for efficient plant disease management. Traditional pathogen detection methods have been either modified or replaced by techniques having higher speed, reliability, robustness, ease of use and cost-effectiveness. As the result of biotechnological advances, new products and techniques are becoming available that will complement or replace time-consuming laboratory procedures. Most of these developments have been made in serological and molecular technologies, mainly due to the developments in monoclonal antibodies and PCR technology. Serological and molecular techniques are frequently used in viral and bacterial diagnosis; however their use in fungal pathogen detection becomes necessary in few cases. Most recent improvements have been time resolved fluoroimmunoassay, flow cytometry, real time PCR, multiplex PCR and microarray, which are more automated thereby have the potential to increase diagnostic throughput and sensitivity.

Keywords: *Pathogen, Detection techniques, Serology, PCR, ISEM*

Introduction

Different kinds of pathogens namely fungi, bacteria, viruses, phytoplasmas etc. are capable of infecting various types of crop plants and induce diseases that cause

---

* E-mail: sureshgawande76@gmail.com; **vivekchimote@rediffmail.com

considerable economic loss. The first and most important step in effective plant disease management lies in rapid correct diagnosis of the disease and detection of its causative agent. The prevalence, the extent of spread and damage is assessed through surveys on the basis of symptoms followed by detection of pathogen in laboratories. This also helps us to assess, alter and modify the effectiveness of the plant protection measures that are being followed.

Pathogen diagnostic methodology can be divided into four categories: (1) Disease symptoms based diagnosis, (2) Microbial visualization methods, including microbial cultivations, and/or light-microscopic examinations, (3) Protein detection methods, mostly based on immunodiagnostics, and (4) Nucleic acid based diagnostics. Traditionally, plant pathologists detect and diagnose some plant diseases by visually examination of the characteristic symptoms wherever possible; others require laboratory testing for diagnosis. They involve microscopic observation and/or by isolating the disease causing organism from infected plant tissues. These laboratory procedures may take days or even weeks to complete and are, in some cases, relatively insensitive.

Classical techniques employed to analyze maladies caused by microbes other than viruses and mycoplasma have relied on sampling plant tissue, soil or other relevant material, and incubating and growing them on synthetic growth media under the controlled condition in the laboratory. Pathogens that grow from these samples can be observed for telltale visual characteristics such as colony/mycelium appearance and colour. Alternatively they can be viewed under the microscope for distinctive cell shapes or reproductive structures, or subjected to other growth tests that differentiate particular strains. These techniques have been useful in successfully describing the causal agents of numerous diseases of various crops; however, they are too variable, slow, and labor-intensive to be of much use in most of the investigations. Conventional broth- and agar-based anti-microbial susceptibility testing methods provide a phenotypic profile of the response of a given microbe to an array of agents. However, recently it has been estimated that less than 1 per cent of microbial species will be detected under such circumstances, due to the selectivity of growth media and conditions. Hence, the vast majority of pathogens from the environment that are not stimulated to grow under laboratory conditions will be overlooked.

A part from the role of detection and diagnosis in plant disease management, detection of pathogen(s) can act as important tool to localize and prevent the spread of the disease(s). In the era of globalization and WTO regime, detection of plant pathogens has immense role to play because of free movement of materials across the countries creates chances for introduction of new pathogens in to the importing countries. Therefore we need to have techniques that are rapid, accurate, sensitive, robust and easy to adopt. The emerging technologies, both serological and molecular, which satisfy many of these requirements are discussed and reviewed in this chapter.

Serological Detection Based on Protein Components of Pathogens

Serological detection involves application of serological tests to diagnose disease infection by studying antibody-antigen interaction. In practice, the term serology usually refers to the diagnostic identification of antibodies in the serum. Antibodies are specific serum proteins, which are elicited in the animal in response to alien proteins or polysaccharides called antigen. Antigens include parts (coats, capsules, cell walls, flagella, fimbrae, and toxins) of bacteria, viruses, and other microorganisms. Lipids and nucleic acids are antigenic only when combined with proteins and polysaccharides. Antibodies react specially with the antigen that stimulated their production. The most common type of antibodies in blood serum is α-globlulin, also called as IgG. Binding sites on surface of antibodies are called paratopes, which are complementary in shape, size and hydrophobicity to the antigenic sites called epitopes on the surface of homologous antigens.

Serological techniques involve specific antibody-antigen recognition and their application is extended to detection of wide range of plant pathogens including virus, bacteria and fungi. Earlier serological techniques relied on the clumping of pathogen particles by specific antisera (serum containing antibodies) that are visible to the naked eye. Various improvements in immunological techniques have resulted from increased specificity. The different forms of serological assay include agglutination, gel diffusion, complement fixation, enzyme linked immunosorbent assay (ELISA), immune-fluorescence (IF) and immuno-fluorescent colony staining.

1. *Agglutination*: Insoluble aggregate formed on antigen-antibody/cell interaction; detected by naked eyes as clumps; earliest method; fibrinogen is the clumping factor.
2. *Precipitation*: Precipitate formed when antibody (precipitin) and soluble antigen (precipitinogen) react in suitable proportion. Gel diffusion assay is needed.
3. *Complement fixation*: Antigen-antibody interacts in presence of complement; complement used up is estimated by titration of serial dilutions along with known concentration standard.
4. *Fluorescent antibodies based detection*: ELISA; Immunosorbent electron-microscopy; Colony staining; Flow cytometry.

Polyclonal antiserum is a combination of many different antibodies recognizing specific antigen. Each β-lymphocyte cell produces specific antibody; mixture of them together produce polyclonal antibodies. Since antigens usually possess several different epitopes, multiple β-lymphocyte cell clones are involved in immune response. Usually the ability of animal immune systems to produce antibodies capable of recognizing specific antigens is used.

Polyclonal antibody production involves rearing of fresh laboratory or farm animals under aseptic conditions and injecting them (sometimes weekly) with specific antigen samples to evoke high antibody expression in serum. Blood serum is collected from bleeding and polyclonal antibodies are recovered. Antisera purification method depends upon its intended application. Crude method involves precipitation of total proteins involving antibodies present. During affinity purification, crude protein extract is passed through resin coupled with immobilized antigen, unbound nonspecific proteins washed out as flow through and specific antibodies are eluted using low pH elution buffer.

The antibody producing B cells do not multiply *in-vitro* but they can be multiplied by fusing them with carcinoma cells. Somatic hybrids of β-lymphocyte and myeloma (HGPRT$^-$) cells are screened on HAT medium (Hypoxanthine + Aminopterin + Thymidine). Resulting hybridoma cell grow indefinitely on HAT medium; each one producing only single specific antibody that can be can be singled out and mass cultured for producing produced it in large amount. Such hybridoma clones secrete antibodies for only a single epitope and are called as monoclonal antibodies (Monoclonal antibodies) which is collected as culture supernatant. Purified antibodies are further modified by antibody fragmentation (cleaved into smaller antigen binding domains); conjugation with enzyme or other detectable markers for quick detection (Alkaline phosphatase/horse radish peroxidase/penicillinase) or immobilized to solid support for antigen binding.

Here we discuss more popular and robust techniques that are useful in detection and diagnosis of plant pathogens.

ELISA

ELISA tests are handy to perform and can be carried in simple glass/polystyrene microtitre plates where the enzyme conjugated antibody is made to bind its complementary antigen. The antigen-antibody complex is probed by adding enzyme substrate, which upon enzymatic breakdown produces a colored product. Most of antibodies and antigen are adsorbed onto a polystyrene surface by hydrophobic interaction when incubated in a suitable buffer solution. Sometimes the polysterene surface is modified chemically by replacing benzene ring or modified with carboxyl or amine group to allow covalent binding to a molecule. Intensity of the colored product is proportional to the amount of enzyme trapped, which in turn is the function of antigen titer and time. Polyclonal and monoclonal antisera for many viruses and bacteria have been developed for commercial use. Commercial kits (agencies: Agdia, Adgen and Ayr) are available for detection of many viruses, bacteria and fungi. These antibodies have been put to use to detect viruses, bacteria using different protocols *viz.*, immune-diffusion assay, western blots, dot blot immuno binding assay and serologically specific electron microscopy (SSEM). However, due to its specificity and robustness.

ELISA still remains standard protocol of choice for viruses and bacteria in diagnostic laboratories since its inception in 1970s (Clarks and Adams, 1977). The sensitivity of an ELISA assay varies depending on the organism and precaution needs to be taken as these tests sometime show false negative or false positive detection. Many formats of ELISA are in use and are referred to as Enzyme Immuno Assay (EIA). EIA can be grouped under two categories of direct and indirect ELISA.

1. *Direct ELISA*: In direct ELISA, antigen is trapped on surface and detected with antigen specific antibodies (primary antibodies) labeled with the enzyme whereupon the enzyme substrate is added. Double antibody sandwich ELISA (DAS-ELISA) is the most common method among direct ELISA and it is highly specific detecting closely related strains of virus/bacteria.

2. *Indirect ELISA*: The trapped antigen is first combined with nonconjugated homologous antibodies (primary antibodies). These in turn are detected in another step using enzyme conjugated anti-IgG antibodies (secondary antibodies). Various formats of indirect ELISA are, TAS ELISA and Fab/2-ELISA (Adams and Barbara, 1982). TAS ELISA *i.e.* triple antibody sandwich ELISA though less specific but it has broad spectrum and is more sensitive. This helps even in detecting strains of same virus/or of different viruses (Koenig and Paul, 1982).

   Alkaline phosphatase is common enzyme used for conjugation with antibodies. Another variant, penicillinase based ELISA, uses materials that are cheaper and more easily available in developing countries. Sensitivity of ELISA can be increased by bridging between pathogen bound antibody and enzyme. This is possible by coupling biotin to the antiserum globulin and the protein (Streptavidin) to the enzyme, exploiting the high bridging affinity of avidin for biotin (biotinavidin ELISA). Major improvements like fluorogenic substrates, enzyme amplification or time resolved fluorometry came during 1980s. Besides these the following protocols are also being applied for detection of various plant pathogens:

3. *Immunoblotting*: It is a variant of ELISA that employ labeled antibodies including enzyme conjugates but use nitrocellulose/nylon membranes for binding the antigen instead of polystyrene ELISA plates.
4. *Voltametric Enzyme Immunoassay*: detects the change in electrical conductivity of the substrate, rather than a colour change, when acted upon by an enzyme attached to a secondary antibody.

   This method is claimed to be an order of magnitude more sensitive than ELISA. It was used to detect *Cucumber mosaic virus*3.

5. *Tissue Blot Immunoassay (TBIA)*: In this technique fresh surface of infected tissue (leaf/stem) is directly pressed on to dry membrane.
6. *Squash Blot Immunoassay (SMB)*: Seeds/insects are squashed onto membrane.
7. *Electroblot Immunoassay (EBIA)*: Antigen is transferred to a membrane through electrophoresis.
8. *Microarray multiplex ELISA*: It involves performing many ELISA assays simultaneously in the same assay plate, by using different colored fluorescent dyes. Now a day, microassay based multiplex ELISA is being adopted for disease diagnosis. It is advantageous over other forms of ELISA as it can detect numerous antigens simultaneously.

Reliability of ELISA depends upon antibodies of high affinity and specificity. This objective can be obtained through use of monoclonal antibodies. Monoclonal antibodies are expensive to produce and some hybridoma cell lines either die or use antibody production during storage. The latest approach is use of recombinant antibody produced in bacteria. Recombinant antibodies comprise of variable heavy and light chain regions of antibodies (ScFv). For viruses, for which specific poly Abs could not be produced, ScFv R-ELISA has proved effective. Production of recombinant antibodies in *Escherichia coli* takes 2-4 weeks.

Immunosorbent Electron Microscopy (IEM/ISEM)

Techniques involve the detection and identification of viruses/bacteria by union of electron microscopy with serology and the technique is highly sensitive. The first electron microscopic observation of antigen antibody complex was made by Lafferty and Oertelis (1961). Derrick (1973) introduced the present form of ISEM technique that involves coating of an electron microscope grid with specific antiserum. Grids are floated on purified antigen or crude sap from an infected plant, and antibodies on the grid surface trap antigens from solution. Improved detection is achieved by combining ISEM with a decoration step wherein the antigens are coated with a second layer of antibodies. Decoration is useful for examining relationships between pathogens and looking at mixed infection especially when pathogens have similar morphology. However, the technique has major limitation as it requires initial high cost electron microscope, expertise and not suitable to scan large number of sample.

Flow Cytometry

Efficiency of immunodiagnostic detection has been greatly enhanced by flow cytometry. It has applications in detection of plant pathogenic bacteria and viruses; and in viability assessment of spores and bacteria (Bergervoet *et al.*, 2007). Flow cytometry is based on simultaneous multiparametric evaluation of physio/chemical properties of single cells flowing through a stream of fluid. The technique enables rapid identification and quantification of thousands of cells/particles per second. Fluorescent microspheres are conjugated to different antigens/antibodies, thereby constituting the solid phase for detecting antibodies or antigens in biological samples. These particles pass individual through a sensor in liquid stream, get sorted individually and therefore they are detected. Multiple parameters (relative size, granularity, and up to three/nine colors of emitted fluorescence) can be measured simultaneously and correlated particle by particle. The signals can be instantly collected as histograms for immediate results and/or stored as raw data for experimental analysis later.

Various uses of flow cytometry include measurements of auto-fluorescent proteins, DNA and RNA content, antigen or ligand density, enzyme activity, membrane potential, viability obtained from cells, isolated nuclei, organelles or microorganisms. These assays seem to be more sensitive than traditional immunoassays, have a high throughput capacity and provide a wide analytical dynamic range. Additional, they have multiplexing ability, *i.e.* capable of measuring multiple antibodies or antigens simultaneously. Compared to a common green organic dye, semiconductor quantum dots (QDs) composed of CdSe/ZnS core/shell bioconjugates display brighter fluorescence intensities, lower detection thresholds, and better accuracy in analyzing bacterial cell mixtures using flow cytometry (Hahn *et al.*, 2008). In near future the small, affordable flow cytometers are likely to replace ELISA and RIA testing.

Molecular Detection by Nucleic Acids Based Techniques

As we proceed, in the 21st century, scientists are becoming increasingly able to diagnose and manage diseases at the molecular level. Molecular methods offer an entirely new approach to the plant disease diagnosis, however many a times molecular methods may be an improvement over conventional microbiology testing in many ways. The rapid development in the fields of molecular plant pathology has provided new insights into the genetic and structural features of a large number of pathogens. These results obtained through intensive basic research are further leading to improvement in diagnostic procedures. Automation and high-density oligonucleotide probe arrays (DNA chips) also hold great promise for characterizing microbial pathogens.

Molecular strain typing is now being recognized to be an important component of a comprehensive disease control program. The most widely used molecular typing methods include plasmid profiling, restriction endonuclease analysis of plasmid and genomic DNA, Southern hybridization analysis using specific DNA probes, and chromosomal DNA profiling using either pulsed-field gel electrophoresis (PFGE) or those exploiting polymerase chain reaction (PCR) based methods. All these methods use electric fields to separate DNA fragments, whole chromosomes, or plasmids into unique patterns or fingerprint that are visualized by staining with ethidium bromide or by nucleic acid probe hybridization.

Nucleic Acid Hybridization (NAH)

The use of nucleic acid probes for identifying cultured organisms and for direct detection of pathogen(s) in disease infected material was the first exposure that most laboratories had to molecular tests. Nucleic acid-based tests used in diagnosing diseases use standard methods for isolating nucleic acids from organisms and host plant material, and restriction endonuclease enzymes, gel electrophoresis, and nucleic acid hybridization techniques are used to analyze DNA or RNA. Nucleic acid hybridization involves specific interaction in in vitro conditions between two polynucleotide sequences (RNA: RNA, RNA: DNA or DNA: DNA) by complementary base pairing. Depending on the type of nucleic acid separation and their transfer to solid supports, the hybridization method can be termed as Southern transfer hybridization (Southern, 1975) for DNA and Northern transfer hybridization (Alwine *et al.*, 1977) for RNA. This binding can take place in solution (liquid phase hybridization) or on membrane (Solid support hybridization) or in the tissue itself (*in situ* hybridization).

Although direct detection of organisms in field specimens by nucleic acid probes is rapid and simple, it suffers from lack of sensitivity. Most direct probe detection assays require at least 10^4 copies of nucleic acid per microliter for reliable detection, a requirement rarely met in samples without some form of amplification. There is need to develop user friendly probe systems that use solution-phase hybridization and chemiluminescence for direct detection of infectious agents in field material. The main steps of nucleic acid hybridizations are preparation of probes, sample preparations, sample denaturation, application and its immobilization on the membranes, prehybridization and hybridization washing of the membranes and detection of hybridized probes.

Hybridization Signal Detection

Radioactive membranes are visualized through autoradiography at –70°C using an X-ray film with an intensifying screen. For nonradioactive labels colorimetric visualization of immuno-detected nucleic acid is carried out with nitroblue tetrazolium (NBT) and 5-bromo-4-chloro-3-inaolyl-phosphate (BCIP) reagents. For chemiluminiscent detection, CSDD or CDP star is added slowly to the membranes and incubated for half an hour at room temperature followed by autoradiography.

Nucleic Acid Spot Hybridization (NASH)

A solid support hybridization technique called as nucleic acid spot hybridization (NASH) is very useful for the detection of viroids (virus without protein coat), which cannot be detected through immunological approaches. The main steps involved in NASH include preparation of probes, sample preparations, sample denaturation, application and its immobilization on the membranes, prehybridization and hybridization washing of the membranes and detection of hybridized probes.

First the target nucleic acid (DNA/RNA) is spotted onto a solid support, commonly nitrocellulose/nylon membrane and immobilized by baking (at 80°C for 2 h), UV cross-linking or by electrostatic interaction on a charged membrane. Depending upon the virus/bacterium involved, different buffered sap extracts or nucleic acid extracts can be denatured prior or after application to the membranes by heating or by alkali treatment. Several chemical denaturants *viz.*, dimethyl sulphoxide, methyl mercuric hydroxide, formaldehyde and glyoxal have been described. Free binding sites on the membrane are then blocked with non-homologous DNA (usually Salmon sperm or calf thymus DNA) and protein source (bovine serum albumin or nonfat dried milk). This step is designed to block all the sites that could nonspecifically bind the probe. Thereafter hybridization with a labeled probe is carried out. It is necessary to ensure that the added probe is single stranded.

This is achieved by boiling or by denaturing the dsDNA probes in alkali. In NASH technique, probes can be dsDNA, ssDNA, RNA or synthetic oligonucleotides specific for the pathogen to be detected. Majority of the plant viruses and viroids has ssRNA genome. Therefore, for the detection of viruses/viroids, reverse transcription of RNA genome is needed, which is performed by RNA dependent DNA polymerase (reverse transcriptase). The cDNA thus formed is inserted into a bacterial plasmid DNA for making probes.

Polymerase Chain Reaction (PCR)

DNA amplification technique known as the polymerase chain reaction (PCR) was originally introduced by Saiki *et al.* (1985), subsequently automated by Mullis and Fallona, (1987) and with time the PCR applications have grown in an exponential manner. It can be used to multiply specific regions of the DNA of any microorganism with the help of primers. This method uses the essential part of cellular DNA replication machinery (DNA polymerase) in combination with two synthetic target segments of the DNA in vitro. The reaction is cycled to produce an exponential increase in the target sequences *i.e.* around 10^6 copies within few hours. In the diagnostic field, the importance of PCR stems from its enormous amplification power, thus allowing the synthesis of a detectable product from a single target molecule.

Another very important quality of PCR unit is high level of specificity, which when using carefully designed primers, allows for the discrimination of sequences differing by single nucleotide innovations. It also provides the ability to selectively amplify specific targets present in low concentrations to detectable levels; thus, amplification-based methods offer superior performance, in terms of sensitivity, over the direct (non-amplified) probe-based tests. Since the target DNA or RNA may be present in very small amounts in various specimens, various signal amplification and target amplification techniques are used in plant pathogen diagnostic laboratories. Because of the flexibility and ease of performance, PCR amplification remains the most widely used molecular diagnostic technique in research laboratories. Amplification-based methods are also valuable for identifying cultured and non-cultivable organisms. Amplification reactions may be designed to amplify either a genus-specific or "universal" target, which then is characterized by using restriction endonuclease digestion, hybridization with multiple probes, or sequence determination to provide species or even subspecies delineation.

The amplification method involves three basic steps:

1. Melting of target DNA;
2. Annealing of the oligonucleotide primers to the denatured DNA strands; and
3. Primer extension by DNA polymerase.

Although a number of thermo-resistant DNA polymerases are now available, Taq DNA polymerase isolated from *Thermus aquaticus* is by far the most frequently used enzyme for PCR assays. The DNA polymerase must have 5'-3' exonuclease (proof-reading) activity, which results in an increased fidelity of the amplification process.

Most of the plant viruses and viroids have RNA genome, which require the introduction of a preliminary reverse transcription (RT) step before PCR amplification process (RT-PCR). Enzymes isolated from Avian Myoblastosis Virus (AMV) or from Moloney Marine Leukemia virus (M-MuLV) are generally used for reverse transcription. Some thermo resistant enzymes (*Thermus thermiphilus*) have both reverse transcriptase (RT) and DNA polymerase activity, allowing the use of a single enzyme is RT-PCR assays. All components of RT and of PCR reaction can thus be added together. This form of RT-PCR is also referred as one-step RT-PCR (Sellner *et al.*, 1992). This format has advantage of requiring fewer hands on time to set up an assay and also to reduce possibility of false positive results through contamination, since it involves only a single series of pipeting. Besides this, many variants of original PCR protocols have been described which are described below.

Nested PCR

It involves two rounds of PCR amplification, the second of which is performed using primers that hybridize within the fragment amplified during the first round of PCR. This approach offers the advantage of increasing both the sensitivity and specificity of PCR. A large number of pathogen specific primers are available through web. Specificity of PCR depends upon the uniqueness of the sequences selected for primers and probes. Improvements in sequencing technique are making selection of reliable primers a routine (Schaad and Frederick, 2002).

Immunocapture (IC-RT-PCR) and Print Capture Polymerase Chain Reaction (PC-RT-PCR)

They are TaqMan variants of PCR are being used for the detection of plant viruses. These methods are simple, sensitive, rapid and cost effective also (Gawande *et al.*, 2005). Advantage of IC-PCR and PC PCR is that there is no need to go for RNA extraction and crude sap can be used directly.

Duplex and Multiplex PCR

They involve parallel amplification of different targeted fragments in one-tube reactions for the multiple pathogen detection in a single reaction as described by various workers (Singh *et al.,* 2000). Though duplex and multiplex PCR are advantageous in terms of time and cost, but difficulties arise as more than one primer pairs are added in one tube. One has to modify the reaction conditions specially annealing temperature, amount of $MgCl_2$ and polymerase extension time.

Real-Time PCR

Real-time PCR can detect plant pathogens accurately and rapidly has been reviewed (Schaad and Fredreick, 2002). Real-time PCR is a kinetics-based quantitative PCR technique, where the amount of newly synthesized DNA is measured after each cycle throughout the PCR amplification process. It may be based on fluorescent light emission that is proportional to amplicon amounts generated. These fluorescent signals maybe generated by the use of either non-specific double-strand DNA (dsDNA) intercalating SYBR Green dye or fluorescently labeled target sequence-specific probes. In terms of signal intensity curves, ten times the standard deviation from the average baseline signal is defined as the threshold cycle (C.T. value). Real-time PCR can use Taqman probes, fluorescent resonance energy transfer (FRET) probes or molecular beacons to detect the production rate of the amplicons. These methods are based upon the hybridization of fluorescently labeled oligonucleotide probes sequences to a specific region within the target amplicon that is amplified using traditional forward and reverse primer.

Taqman Probes

This method utilizes 5'-3' exonuclease activity of *Taq* DNA polymerase to clean a dual labeled fluorogenic hybridization probe during extension that is quantitatively measured by using a combined thermocycler fluorescence detector. Three primers are used together in a single real-time PCR reaction, a primer pair anneal to sites flanking the sequence of interest and the third fluorescently labelled primer anneals in between them. As PCR elongation step takes place the flanking primers extend, the labelled primer is released and fluorescence detection occurs. This system eliminates the time consuming process after PCR, and is also relatively resistant to carry over contamination of PCR amplification.

FRET

Principle relies on the non-radioactive transfer of energy from an excited donar (fluorophore) to an acceptor (fluorophore) by means of intermolecular long range dipole-dipole coupling.

Ligase-Mediated (LM) Amplification

It involves the use of a T4 DNA ligase to ligate two same-strand targeting proximal adjacent labeled primers, a detection scheme known as Oligonucleotide Ligation Assay (OLA). This procedure involves use of reverse orientated internal primers for amplification of external regions for inverse PCR combined with restriction digestion of template DNA, and fragment circularization via DNA ligation. The Ligase Chain Reaction.(LCR) utilized a thermostable DNA ligase to exponentially amplify short DNA stretches utilizing OLA-primers, in a thermal cycling process similar to PCR. Modifications were made to amplify longer DNA stretches (gapLCR) and for a ligase mediated probe circularization (The Padlock Probe).

Isothermal Amplification Strategies

Isothermal amplification strategies are carried out at a uniform temperature, without any need for repeated thermocycling. They involve use of strand displacing polymerases that are capable of displacing downstream duplex DNA segments during primer extension, which constitute a key strategy in numerous isothermal amplification schematics. There are three techniques for isothernal amplification *i.e.* Strand Displacement Amplification (SDA); Rolling Circle Amplification (RCA) and Circle-to-Circle Amplification (C2CA).

DNA Sequence Data Based Analysis

Molecular methods that utilize DNA sequence information can thus be employed to broadly survey the range of organisms infecting crop plants or living in their phyllosphere. Once a region of DNA is multiplied using the PCR, the sequential order of the four constituent molecules or 'bases' (designated by the letters A, C, G or T) in the DNA can be determined. Usually Sanger's dideoxy chain termination method is used for sequencing to generate a range of fragments differing by single base. Fluorescently labeled ddNTP (4 different colour dyes for 4 different ddNTP) are used for sequencing reaction, followed by capillary electrophoresis and laser based detection.

Known as the 'DNA sequence', the order of bases in a given piece of DNA is usually unique to an organism, and can therefore act like a signature to confirm its identity. The size or sequence of a portion of its DNA can thus identify an unknown microbe, and this knowledge can in turn suggest whether that organism might cause disease. Furthermore, such methods can be fine-tuned for assays, which can diagnose the presence, and even estimate the abundance, of specific plant pathogens. The

DNA sequence data has been used to design very specific molecular diagnostic assays that can be used to distinguish single or dual infections by these important pathogens.

With recent developments in high throughput sequencing there has been tremendous increase in sequenced data with over 700 microbial genomes been completely sequenced and made available at on-line databases. With these developments in DNA sequencing technology the multidisciplinary field of bioinformatics emerged as an important discipline.

Biochips and Microarray

Microarrays with key features of speed, reliability, robustness, ease of use and cost-effectiveness, have already made a marked impact on many fields of plant biology and are likely to become a standard tool of the microbiology/pathology laboratory. They may be based on nucleic acid detection (either cDNA or oligo based) or protein detection array with potential of recognizing thousands of spots on a single array chip/slide. They have potential to increase diagnostics throughput, automate the process while simultaneously reducing unit cost. At present, DNA microarray technology is the most suitable technique for high-throughput detection and identification, as well as quantification, of multiple pathogens in a single assay. Lievens *et al.* (2006) demonstrated the utility of DNA array technology to detect single nucleotide polymorphisms (SNPs) that may be targeted for pathogen identification and should be pursued in any diagnostic assay. Pathogen diagnosis by microarray technique has been reviewed by Yoo and Lee (2008).

Continued commercial interest in microarray technology promises increasing array element density, better detection sensitivity, and cheaper, faster methods. The exponential growth of pathogen nucleic acid sequences available in public domain databases has invited their direct use in pathogen detection, identification, and surveillance strategies. DNA microarray technology has offered the potential for the direct analysis of a broad spectrum of pathogens of interest. However, to achieve the practical attainment of this potential, numerous technical problems need to be addressed.

The key unifying principle of all microarray technique is that labeled nucleic acid/antibodies molecules in solution hybridize, with high sensitivity and specificity, to complementary sequences immobilized on a solid substrate, thus facilitating parallel quantitative measurement of many different sequences in a complex mixture. They are composed of densely packed probe arrays based either on nucleic acid or antibodies coated on optically flat glass plates containing 96 wells. Main steps involve generation of ssDNA fragment to be spotted on glass slides or membranes and subsequently hybridized with labelled probes. Resulting hybridization image is scanned and data analyzed. Although several methods for building microarrays have been developed, two have prevailed.

In one method, DNA microarrays comprise of physically attached DNA fragments such as library clones or polymerase chain reaction (PCR) products to a solid substrate. The main advantage of this method is relatively low cost and substantial flexibility (which explain its wide implementation in the academic setting) in addition, to no prior requirement of primary sequence information to print a DNA element. By using a robotic array and capillary printing tips, we can print above 40,000 elements on a microscope slide. In this system scanning is done by dual lasers capable of recognizing Cy3 and Cy5 dyes.

In the other method, arrays are constructed by synthesizing single stranded 25-mer oligonucleotides *in situ* by use of photolithographic techniques. Advantages of the latter method include higher density (>280,000 features on a 1.28 × 1.28-cm array) and elimination of the need to collect and store cloned DNA or PCR products. This microarray scanning system is based on single laser baser detection of pyroerthyrin dye.

Application of Diagnostic Techniques

Diagnosis of Plant Viruses

Since the introduction of ELISA, it has been in routine use for the detection of nepoviruses (Clark and Adams, 1977) as these viruses are relatively easy to purify and are highly immunogenic. There are about 800 different antisera available for plant viruses through American Type Culture Collection (ATCC). Polyclonal and monoclonal antibodies have been produced against Grapevine leafroll associated closteroviruses and successfully used in ELISA (Zimmermann *et al.*, 1990; Monis and Bestwick, 1997). Potyviruses can be easily detected in leaves and tubers through various forms of ELISA of which DAS-ELISA (Gugerli, 1979; Gugerli and Fries, 1983; Singh and Somerville, 1992), dot-ELISA (Weidemann, 1980) and direct tissue blotting (Samson *et al.*, 1993) are commonly used. ELISA is being used routinely during the past two decades for bulk testing of infected samples (Barker *et al.*, 1993; Singh *et al.*, 1993). DIBA has great potential for mass screening (Singh *et al.*, 1993). With the introduction of Monoclonal antibodies, serological differentiation of strains improved greatly (Fernandez Northcote and Gugarli, 1988; Ellis *et al.*, 1996; Carovska, 1998). PVX can be easily detected in leaves and tubers using commercial ELISA kits and immunoelectron microscopy (Koenig, 1988). TAS-ELISA has been used for the detection of potato moptop virus (Torrance *et al.*, 1993).

For the detection of carlaviruses, earlier precipitation and microprecipitation tests were used (Shepard, 1970, 1972) but these tests were superseded by mere sensitive method such as latex agglutination (Fribourg and Nakashima, 1984). ISEM (Milne and Luisoni, 1977) and ELISA (de Bokx *et al.*, 1980; Dedic, 1995) are now a days the methods of choice for detection of carlaviruses (PVM and PVS). Voltametric Enzyme Immunoassay provided an order of magnitude more sensitivity than ELISA in detection of *Cucumber mosaic virus*3 (Sun *et al.*, 2001). Direct tissue blotting which is more convenient and cheaper than ELISA is also being used (Samson *et al.*, 1993). Time resolved fluoroimmunoassay using monoclonal antibodies which is 10-20 times more sensitive than ELISA is being preferred (Jarvekuly *et al.*, 1989). TAS-ELISA is also used to detect some carlaviruses (Goth *et al.*, 1999). Various temperate fruit crop viruses *viz.*, plumpox virus (PPV), arabis mosaic virus, apple mosaic virus, apple chlorotic leafspot virus are detected using polyclonal antisera through das-ELISA. Monoclonal antibodiess have been produced against prunus dwarf virus, prunus necrotic ring spot and raspberry busty dwarf. Recombinant ELISA has been found very effective for the detection of citrus tristeza virus. Among the various serodiagnostic tests, ELISA remains the most preferred tool for large scale indexing due to its specificity and relative cheapness.

Nucleic Acid Based Detection of Viruses

Polymerase Chain Reaction

Within two decades of inception polymerase chain reaction technique, it has been used frequently to detect viruses. One important feature of viral genome is the absence of a DNA repair system. Viral genomes have mutation rates 1000 times that of the bacterial genomes. Genomic sequences of most of the viruses are available and with the rapidly advancing techniques of PCR, it has become a routine for virus detection. With the availability of sequence data it has become possible to detect viruses at different levels. Alignment of groups of viral sequence are valuable for identifying conserved regions in primers that will amplify numerous viruses as well as variable regions that are used to distinguish them at strains and species levels (Ding *et al.*, 1990; Gibbs and Mackenzie, 1997).

Since most of the viruses have +SS RNA genome, so reverse transcription has to be performed before PCR amplification. For large scale testing, RT-PCR is not very suitable due to cost associated with it. To overcome these problems Jansen *et al.* (1990) developed IC-PCR for detection of Hepatitis A virus. In plant virology, Whetzel *et al.* (1991) applied this technique for the detection of Plum pox potyvirus. Immunocapture (Varveri, 2000) and modified print capture PCR (Gawande *et al.*, 2005) have been used to detect potato virus Y in leaves and aphids respectively. Immuno-capture and fluorogenic and nuclease RT-PCR has been used to detect PLRV in tubers (Schoen *et al.*, 1996; Russo *et al.*, 1999). For the differentiation of strains in PVY, RT and IC-PCR have been used (Barker *et al.*, 1993; Weidemann and Maiss, 1996). Combination of immunocapture with multiplex PCR, where several viruses can be detected in a single reaction and RNA extraction step also eliminated has been used to detect many viruses can also reduce the cost of detection considerably. Further, these days real time PCR is in vogue due to its speed, sensitivity and quantitative nature (Boonham *et al.*, 2002; Eun *et al.*, 2000, Korimbocus *et al.*, 2002; Louws and Couples, 2001; Schoen *et al.*, 1996). This technique is so sensitive that it can detect Tomato spotted wilt virus in thrips also (Boonham *et al.*, 2002).

Nucleic Acid Hybridization

Nucleic acid hybridization have been used successfully to detect a number of potyviruses (Hopp *et al.*, 1991), potex virus (Dhar and Singh, 1994; Singh and Singh, 1995; Eweida *et al.*, 1990), tomato spotted wilt virus (De Haan *et al.*, 1991) and PLRV (Lobenstein *et al.*, 1997). Using biotinylated acid/or digoxigenin labeled probes using a nonradioactive fluorescein labeled cDNA probe, have detected PVY in tubers and leaves of potato. Simultaneous detection of two viruses (PVY, PLRV) and viroid (PSTVd) was achieved using digoxigenin labeled probe (Welnicki *et al.*, 1994). Radioactive probes have been used to detect and differentiate the strains of PVY.

In India potato viruses *viz.*, PVY, PVX, PLRV, PVA, PVS, PVM damage potato crop widely and more recently reported one is potato apical leaf curl virus. These viruses are transmitted through planting material *i.e.* tubers. Their elimination is an important part of potato seed production programme. ELISA is still the method of choice for virus detection. However it has certain limitations like inability to detect low concentration of virus, particularly in the dormant tubers/aphids and as it gives false results few times. PCR has advantage of sensitivity over ELISA but huge cost factor is involved when it comes large scale screening of viruses. To overcome these limitations, RT-PCR protocol for IC-RT-PCR and PC-RT-PCR was developed at CPRI (Gawande *et al.*, 2005). Both of these methods are simple, cost effective, time saving and have ability to detect virus even in single aphid. CPRI is well equipped with facilities to produce antisera and ELISA kits against all these six major potato viruses. Screening of viruses and viroid through DAS-ELISA, radioactive/non-radioactive NASH and RT-PCR is routinely done at the institute.

Diagnosis of Plant Pathogenic Bacteria/Prokaryotes Serodiagnosis of Bacteria

Some bacterial taxa can be easily identified using polyclonal antibodies *e.g. Clavibacter michiganensis* subsp. *michiganensis* (Alverez *et al.*, 1993) and *Xanthomonas axonopodis* pv *pelargonii* (Benedict *et al.*, 1990). Polyclonal antiserum raised to a bacterial species is a mixture of antibodies with multiple specificities since bacterial cell surface represents a range of epitopes. Therefore cross-reaction with unrelated species is common. The development in production of monoclonal antibodies in hybridoma cells has brought immediate improvement and revolution in bacterial serology. Monoclonal antibodies have been used to distinguish a variety of bacterial pathogens upto genera, species, subspecies and pathogens level (Alvarez *et al.*, 1994; 1996; 1997; Franken, 1990). Alvarez, (2004) listed various monoclonal antibodies used for taxon specific detection and diagnosis of plant pathogenic bacteria.

Polyclonal antisera and Monoclonal antibodies both are available in various format *viz.*, agglutination assays, ELISA, Western blot immunofluorescence (IF), IF colony staining (IFC) and lateral flow devices. ELISA is the most established technique for testing large number of bacterial isolates. ELISA kits are available for large number of phytopathogenic bacteria commercially. Sensitivity of ELISA can be increased over ten-fold using extraction buffer containing EDTA and lysozyme. Flow cytometry is a technique suitable for the rapid identification of bacterial cells.

Molecular bacterial detection techniques have been reviewed extensively by Louws *et al.* (1999), and all PCR primers are now available for phytobacterial diagnostic applications. These primers may be neutral markers unrelated to host pathogen interaction for example rRNA gene based primers or may be related to pathogenicity and avirulence like *hrc* or *hrp* based primers. The resolving power of various techniques varies. AFLP (amplified fragment length polymorphism), Rep-PCR (repetitive DNA-PCR), RAPD (Random Amplification of Polymorphic-DNA) can resolve any sample upto strain level, while ITS-PCR and tRNA-PCR resolve upto subspecies level. DNA sequencing of conserved region gives the clearest picture among all molecular techniques. A key marker in the diagnostics of bacteria is the

16S rRNA gene, as it is present in all bacteria and encodes for the same product. The Ribosomal Database Project (RDP) since March 2008 catalogued over 480,000 16 S rRNA sequences. Phylogeny can be defined based on its nucleic acid differences, which can be sufficient for species identification.

One of the most highly touted benefits of molecular testing for diseases is the promise of earliest/incipient detection of certain pathogens, thereby playing crucial role in plant quarantine. For example, in Australia using RFLP (Restriction fragment length polymorphism) technique the early identification of Moko disease of banana caused by *Burklolderia (Ralstonia) solanacearum* race 2 on *Heliconia* could lead to its quick eradication. Even in uncultivable phytoplasmas, PCR amplification of 16S rRNA genes has been successfully used in their detection and establishment of phylogenetic relationship. For the detection of bacterial plant pathogens in heterogeneous mixtures, only PCR based techniques are being applied (Louws *et al.*, 1999). A new cooperational PCR has also been developed to detect *Ralstonia solanacearum* in water (Caruso *et al.*, 2003). For the detection of bacteria, bio-PCR is preferred over simple PCR (Schaad *et al.*, 2002).

Sequencing of 1540 bases of 16S rRNA gene has confirmed and extended our understanding of the relationships among the pseudomonads. All cultivable Gram negative plant pathogenic prokaryotes occur within the α, β and γ subdivisions of phylum Proteobacteria. Homology groups II and III pseudomonads are contained within β proteobacteria, whereas homology groups I and V (the Xanthomonads) are members of the γ proteobacteria subdivision. DNA sequencing shows that some of the presently recognized genera are not true taxonomic entities but phylogenetically heterogeneous. For example complex Pseudomonads were split into five homology groups on the basis of rRNA: DNA hybridization (Palleroni *et al.*, 1973). As per DNA sequencing results those in homology group III are now assigned to the genus *Acidovorax*, most of those with homology group II to the genus *Burkholderia*, true pseudomonads including all those producing fluorescent pigment, are contained homology group I, including *Pseudomonas aeruginosa*, the type species of the genus *Pseudomonas* as originally defined.

Diagnosis of Plant Pathogenic Fungi

Different fungal pathogens are capable of infecting various crops, and the diseases they cause can be of considerable economic importance. Fungi are easily diagnosed by conventional techniques of microbial examination of morphological characters and growth characteristics on specific media, therefore serological and molecular techniques are not as frequently used as in case of viral and bacterial diagnoses. However, many a times this process is time consuming, needing pathogenicity assays and variations in morphological characters of both sexual and non-sexual stages of fungi, leads to difficulties in accurate identification by traditional methods. Current diagnostic kits give indirect evidence for the presence of fungal pathogen by detecting toxic byproducts. Since many of these byproducts are a food safety concern, it is advantageous to directly detect the fungi at low levels before the toxins accumulate. New fungal detection kits are coming which can detect the pathogen before the visual disease symptoms are apparent.

Isozyme assays have been extensively used for strain differentiation, as they are quick and easy to apply. Isozymes like glucose phosphate isomerase, peptidase, and mannose phosphate isomerase are the favorites. PCR diagnosis helped in effective rapid diagnosis of Karnal bunt (c.o. *Tilletia indica*) of wheat in USA in 1996 followed by its control (using various methods) and also enforcing quarantine regulations (Bonde *et al.*, 1997). Real time PCR assay techniques has been described for many plant pathogenic fungi and reviewed (Schaad *et al.*, 2003).

With the advent of molecular biology and the ability to compare conserved regions, the development of fungal diagnostics has improved at amazing rate. The internal transcribed spacer (ITS) region of the ribosomal DNA (rRNA genes) of fungi is one region that can be PCR amplified for the identification of potential plant pathogens. The comparison of sequences of ITS region of ribosomal genes of various fungi has been used for taxonomic studies and to develop PCR based detection techniques.

The nuclear small sub-unit rRNA genes evolves relatively slowly and are useful for studying distinctly related organisms, whereas ITS regions and intergenic repeat evolves fastest and may vary among species and populations (White *et al.*, 1990). New data is being added to international DNA databases at an exponential rate, but already around thousands of sequences for the well-studied ITS DNA region exist for fungi. Many times as in case of Genus *Phytophthora*, ITS-rRNA amplification is combined with restriction endonuclease digestion for species identification and to further develop species-specific primers (Ristaino *et al.*, 1998). Zeng *et al.* (2005) developed species specific primers based on the 18S rDNA sequence variation pattern for detection of conifer pathogens of genus Gremmeniella.

Mitochondrial haplotypes based on mitochondrial DNA RFLP polymorphism are also used to study biotypes of *Phytophthora infestans*. This is done after PCR amplification of specific mitochondrial sequences (Griffith and Shaw, 1998). AFLP kits and multiloci probes like RG 57 have also been used for population studies analysis in *P. infestans* (Goodwin *et al.*, 1992; Shaw, 1999).

Financial Considerations for Molecular Diagnostics

Most of the plant pathologists and microbiologists enthusiastically welcome molecular techniques for identifying and detecting microorganisms and the techniques have become common in use. However, the cost of these techniques and their contribution to agriculture needs to be addressed (Kant, 1995). Although technical issues such as ease of performance, reproducibility, sensitivity, and specificity of molecular tests are important, cost and potential contribution to agriculture are also of concern. The true financial impact of molecular testing will only be realized when testing procedures are integrated into total disease assessment. More expensive testing procedures may be justified if they reduce the use of less-sensitive and less-specific tests and eliminate unnecessary diagnostic procedures and ineffective therapies. Some of the molecular diagnostic tests are not highly expensive and the cost varies depending on the test's complexity and sophistication. Inexpensive molecular tests are generally kit based and use methods that require little instrumentation or technologist experience. DNA probe methods that detect pathogens are examples of low-cost molecular tests. Although the more sophisticated tests may require expensive equipment (*e.g.*, DNA sequencer) and reagents, advances in automation and the production of less-expensive reagents promise to decrease these costs as well as technician time. Much of the justification for expenditures on molecular testing is speculative; however, the cost of equipment, reagents, and trained personnel is real and substantial, and reimbursement issues are problematic. Given these concerns, facilities needed for molecular diagnostic testing for diseases should be examined critically by the concerned authorities.

Conclusions and Future Needs

An accurate, rapid and cost-effective diagnosis is the cornerstone of efficient field disease management. Rapid detection of the pathogen is important to take preventive steps. Leading laboratories should provide a mechanism for standardization of diagnostic techniques and certification. Basic research is the backbone to open new directions for applied research. Future scope in basic research on microbial diagnosis includes (*a*) sequencing of a wider representation of microbial pathogens, and (*b*) functional genomics, proteomics, bioinformatics, and microarray technology. There are also promising developments towards interdisciplinary methodology usage, such as the Proximity Ligation Assay (PLA), where antibody-antigen interaction is detected via coupled nucleic acid amplification. Alternatively, direct antibody detection of nucleic acid heteroduplexes was developed with the Hybrid Capture assay (HCA). All these are needed for the development of new tools for accurate and rapid diagnosis of plant diseases and for determining global relationships of select plant pathogens.

References

Adams, A.N.X. and Barabara, D.J. (1982). The use of F(ab2)-based ELISA to detect serological relationships among carlaviruses. *Ann. Appl. Biol.* 101: 495-500.

Alvarez, A.M.(2004). Integrated approaches for detection of plant pathogenic bacteria and diagnosis of bacterial diseases. *Annu. Rev. Phytopathol.* 42: 339-366

Alvarez, A.M., Benedict, A.A., Mizumoto, C.Y., Hunter, J.E. and Gabriel, D.D. (1994). Serological, pathological and genetic diversity among strains of *Xanthomonas campestris* infecting crucifers. *Phytopathology*. 84: 1449-1457.

Alvarez, A.M., Rehman, F.V. and Leach, J. (1997). Comparison of serological and molecular methods for detection of *Xanthomonas oryzae* pv. *oryzae* in rice. Presented at Seed Health Test: Progress towards the 21'st century, Cambridge, UK.

Alvarez, A.M., Schenck, S. and Benedict, A.A. (1996). Differentiation of *Xanthomonas albilineans* strains with monoclonal antibody reaction patterns and DNA fingerprints. *Plant Pathol.* 45: 358-366.

Alverez, A.M., Benedict, A.A. and Granamanickam, S.S. (1993). Identification of seed borne bacterial pathogens of rice with monoclonal antibodies. Presented at Proc. Symp Seed Health Test. Ist, Ottawa, Ont. Canada.

Alwine, J.C., Kemp, D.J. and Stark, G.R. (1977). Method for detection of specific RNAs in agarose gels by transfer to diazobenzyloxymethol-paper and hybridization with DNA probes. *Proc. Natl. Acad. Sci. USA.* 74: 5350-5354.

Barker, H, Webster, KD and Reary, B. (1993). Detection of potato virus Y in potato tubers. A comparison of polymerase chain reaction and enzyme linked immunosorbent assay. *Potato Res.* 36: 13-20.

Benedict, A., Alverez, A.M. and Polland, L.W. (1990). Pathovar-specific antigens of *X. carnpestris* pv *begoniae* and *X. campestris* pv *pelargonii* detected with monoclonal antibodies. *Appl. Environ. Microbiol.* 56: 572-574.

Bergervoet, J. H.W., van der Wolf, L. M., and Peters J. (2007). Detection and viability assessment of plant pathogenic microorganisms using flow cytometry. In 'Flow Cytometry with Plant Cells'. Eds. Dole Zel J., Greiihuber J. and Suda J. pp 217-229.

Bonde, M., Peterson, G.L., Schaad, N.W. and Smilanick, J.L. (1997). Karnal bunt of wheat. *Plant Dis.* 81: 1370-1377.

Boonham, N., Smith, P., Walsh, K., Tame, J., Marris, J. (2002). The detection of tomato spotted wilt virus (TWV) in individual thrips vectors using real-time fluorescent RT-PCR (TaqMan). *J. Virol. Methods.* 101: 37-48.

Caruso, P., Bertolini, E., Cambra, M. and Lopez, M.M. (2003). A new and sensitive cooperational polymerase chain reaction (Co-PCR) for rapid detection of *Ralstonia solanacearum* in water. *J. Microbiol. Methods.* 55: 257-272.

Cerovska, N. (1998). Production of monoclonal antibodies to potato virus YNT strains and their use for strain differentiation. *Plant Pathol.* 4: 505-509.

Clark, M.F. and Adams, A.N. (1977). Characteristics of the microplate method of enzyme linked immunosorbent assay for the detection of plant viruses. *J. Gen. Viro.*134: 475-483.

de Bokx, JA, Piron, P.G.M. and Cother, E.(1980). Enzyme linked immunosorbent assay (ELISA) for the detection of potato viruses S and M in potato tubers. *Neth. J. Plant Path.* 86: 285-290.

De Haan, P.T., Gielen, J., Van Grinsven, M., Peters, D., Goldback, R. (1991). Detection of tomato spotted wilt virus in infected plants by molecular hybridization and PCR. In Exploring and Exploiting the RNA Genome of Tomato spotted wilt virus, Ph.D. Thesis Agricultural University, Wageningen, The Netherlands, pp.93-106.

Dedic, P. (1995). The reliability of potato virus M detection in progeny of primarily infected potato plants. *Ochi Rost.* 131: 277-285.

Derrick, K.S. (1973). Quantitative assay for plant viruses using serologically specific electron microscopy. *Virology*, 56: 652-653.

Dhar, A.K. and Singh, R.P. (1994). Improvement in the sensitivity of PVY detection by increasing the cDNA probe size. *J. Virol. Methods.* 50: 197-210.

Ding, S. W., Howe, J., Keese, P., Mackenzie, A., Meek D., *et al.* (1990.) The tymobox, a sequence shared by most tymoviruses: its use in molecular studies of tyomoviruses. *Nucleic Acids Res.* 18: 1181-1187.

Ellis, P., Stace-Smith, R., Bowler, G. and Mackenzie, D.J. (1996). Production of monoclonal antibodies for detection and identification of strains of potato virus Y. *Can J. Plant Path.* 18: 64-70.

Eun, A.J.C., Seoh, M.L. and Wong, S.W. (2000). Simultaneous quantitation of two carlaviruses by the TaqMan, real-time fluorescent RT-PCR. *J. Virol Methods.*87: 151-160.

Eweida, M., Xu, H., Singh, R.P. and Abon Haider, M.G. (1990). Comparison between ELISA and biotin labeled probes from cloned cDNA of potato virus X for the detection of virus A crude tuber extracts. *Plant Pathology* 39: 623-628.

Feinberg, A.P. and Vogelstein, B. (1983). A technique of radiolabelling DNA restriction fragments to high specific activity. *Ann. Biochem.* 132: 6-15.

Fernandez-Northcote, E.N. and Gugerli, P. (1988). Reaction of a broad spectrum of potato virus Y isolates to monoclonal antibodies in ELISA. *Fitopatologia* 22: 33-36.

Fribourg, C.E. and Nakashima, J. (1984). An improved latex agglutination test for routine detection of potato viruses. *Potato Res.* 27: 237-249.

Gawande, S.J., Shukla, A., Garg, I.D. and Khurana, S.M.P. (2005). Development of PCR based technique for the detection of immobilized virion. National Symposium on Crop Disease Management in Dryland Agriculture. Jan 12-14,2005, MAU, Parbhani, pp. 88.

Gibbs, A. and Mackenzie, A. (1997). A primer pair for amplifying part of the genome of all potyviride by RT-PCR. *J. Virol. Methods.* 63: 9-16.

Goodwin, S.B., Smart, C.D., Sandrock, R.W., Deahl; K.L., Punja, Z.K., Fry W.E. (1992). Cloning and genetic analyses of two highly polymorphic, moderately repetitive nuclear DNAs from *Phytophthora infestans*. *Current Genet.* 22: 107-115.

Goth, R.W., Ellis, P.J., Villievs, G., Goine, E.W. and Wright, N.S. (1999). Characteristics and distribution of potato latent carlavirus (Red Lasoda Virus) in North America. *Plant Dis.* 83: 751-753.

Griffith, G. W. and Shaw, D.S. (1998). Polymorphism of *Phytophthora infestans*: four mitochondrial haplotypes are detected after PCR amplification of DNA from pure cultures or from host lesions. *Appl. Environ. Microbiol.* 64: 4007-4014.

Gugerli, P. (1979). Potato virus A and potato leafroll virus: Purification, antiserum production and serological detection in potato and test plants by enzyme linked immuno-sorbent assay (ELISA). *Phytopath. Z.* 96: 97-107.

Gugerli, P. and Fries, P. (1983). Characterization of antibodies to potato virus Y and their use for virus detection. *J. Gen. Virol.* 64: 2471-2477.

Hahn, M.A., Keng, P.C. and Krauss, T. D. (2008). Flow cytometric analysis to detect pathogens in bacterial cell mixtures using semiconductor quantum dots. *Anal. Chem.* 80 (3): 864-872.

Hopp, H.E., Hain, L., Bravo-Almonacid, F., Tazzini, A.C., Orman, B. and Mentaberry, A.N. (1991). Development and application of non-radioactive nucleic acid hybridization system for simultaneous detection of potato pathogens. *J. Virol. Methods.* 31: 11-30.

Jansen, R.W., Siegl, G., Lemon, S.M. (1990). Molecular epidemiology of human Hepatitis A virus defined by an antigen capture polymerase chain reaction method. *Proc. Natl. Acad. Sci. USA.* 87: 2867-2871.

Jarvekuly, L., Sober, J., Sinijaru, R., Toots, I. and Saarme, M. (1989). Time-resolved fluorimmunoassay of potato virus 14 with monoclonal antibodies. *Ann. Appl. Biol.* 114: 279-291.

Kant, J.A. (1995). Molecular diagnostics: Reimbursement and other selected financial issues. *Diagnostic Mol Pathol.* 4: 79-81.

Koenig, R. and Paul H.L. (1982). Variants of ELISA in plant virus diagnosis. *J. Virol. Methods.* 5: 113-125.

Koenig, R. (1988). Serology and immunochemistry. In: The Plant Viruses. RG Milne (ed.) Vol.4, pp. 111-158. Plenum Press, New York.

Korimbocus, J., Coates, D., Barker, I. and Boonham, N. (2002). Improved detection of sugarcane yellow leaf virus using a real-time fluorescent (Taqman) RTPCR assay. *J. Virol. Methods* 103: 109-120.

Lafferty, K.J. and Oertalis, S.J. (1961). Attachment of antibody to influenza virus. *Nature* 192: 764-765.

Lievens, B., Claes, L., Vanachter, A.C.R.C., Cammue, B.P.A. and Thomma, B.P.H.J. (2006). Detecting single nucleotide polymorphisms using DNA arrays for plant pathogen diagnosis. *FEMS Microbiology Letters.* 255 (1): 129–139.

Mackay, I.M., Arden, K.E. and Nitsche A. (2002). Real-time PCR in virology. Nucl. *Acids Res.* 30(6): 1292-1305.

Schaad, N. W. and Frederick, RD. (2002). Real time PCR and its application for rapid plant disease diagnostics. *Can. J. Plant Pathol.* 24: 250-258.

Schaad, N.W., Frederick, R.D., Shaw, J., Schneider, W.L., Hickson, R., Petrillo, M.D. and Luster, D.G. (2003). Advances in molecular-based diagnostics in meeting crop biosecurity and phytosanitary issues. *Annu. Rev. Phytopathol.* 41: 306-324.

Schaad, N.W., Jones, J.B. and Chun, W. (2002). Laboratory Guide for Identification of Plant Pathogenic Bacteria. St.Paul, MN: APS Press.

Schoen, C.D., Knor, D. and Leone, G. (1996). Detection of potato leafroll virus in dormant potato tubers by immunocapture and fluorogenic 5' nuclease RTPCR assay. *Phytopathology.* 86: 993-999.

Sellner, L.N., Coelen, R.J. and Mackenzie, J.S. (1992). A one-tube, one manipulation RT-PCR reaction for detection of Ross river viruses. *J. Virol. Methods.* 40: 255-264.

Shaw, D. (1999). Molecular approaches-Fingerprinting made easy in Late Blight-A threat to global food security. Volume-1, in "Proceedings of Global Initiative on Late Blight" conference. March 16-17, 1999. Quito, Ecuador. pp. 75-76.

Shepard, J.F. (1972). Gel-diffusion methods for the serological detection of potato viruses X, S and M. Bull. Montana Agric. Res. 564, 562, 72PP.

Shepard, J.F. (1970). A radial immunodiffusion test for the simultaneous diagnosis of potato viruses S and X. *Phytopathology*.60: 1669-1671.

Singh, M and Singh, R.P. (1995). Digoxigenin labeled cDNA probes for the detection of potato virus Y in dormant potato tubers. *J. Virol. Methods*. 52: 133-143.

Singh, R.P., Boucher, A., Somerville, T.H. and Dhar, A.K. (1993). Selection of a monoclonal antibody to detect PVYJN and its use in ELISA and DIBA. *Can. J. Plant Path*. 15: 273-300.

Singh, RP. and Somerville T.H. (1992). Evaluation of the enzyme amplified ELISA for the detection of potato virus A, M, S, X, Y and leafroll. Amer. *Potato J*. 69: 21-30.

Singh, RP., Nie, X. and Singh, M. (2000). Duplex RT-PCR: urgent concentrations at reverse transcription stage affect the PCR performance. *J. Virol. Methods*. 86: 121-129.

Southern, E.M. (1975). Detection of specific sequences among DNA fragments separated by gel electrophoresis. *J. Mol. Biol*. 98: 503-517.

Sun, W., Jiao, K., Zhang, S., Zhang, C. and Zhang, Z. (2001). Electrochemical detection for horseradish peroxidase-based enzyme immunoassay using *p*-aminophenol as substrate and its application to detection of plant virus. *Anal. Chim. Acta*. 434: 43–50.

Torrance, L., Cowan, G.H, and Pereira, L.G. (1993). Monoclonal antibodies specific for moptop virus and some properties of the coat protein. *Ann. Appl. Biol*. 122: 311-322.

Varveri, C. (2000). Potato Y Potyvirus detection by immunological and molecular techniques in plants and aphids. *Phytoparasitica* 28: 141-148.

Weidemann, H.L. (1988). Rapid detection of potato viruses by dot-ELISA. *Potato Res*. 31: 485-492.

Weidemann, H.L. and Maiss, E. (1996). Detection of the potato necrotic ringspot strain of potato virus Y (PVY"T") by reverse transcription and immunocapture polymerase chain reaction. *J. Plant Dis. Prot*. 103: 337–345.

Welnicki, M., Zekanowski, C. and Zagorski, W. (1994). Digoxigenin labeled molecular probe for the simultaneous detection of three potato pathogens, potato spindle tuber viroid (PSTVd), potato virus Y (PVY) and potato leaf roll virus (PLRV). *Acta Biochem. Polonica* 41: 473-475.

Wetzel, T., Candresse, T., Macquaire, G., Ravelonandro, M. and Dunez, J, (1992). A highly sensitive immunocapture polymerase chain reaction method for plumpox potyvirus detection. *J. Virol. Methods* 39: 27-37.

White, T.J., Bruns, T., Lee, S. and Taylor, J. (1990). Amplification and direct sequencing of fungal ribosomal RNA genes for phylogenetics, pp. 315322. In: (ed. MA Innis, DH Gelfand, JJ Sninsky and TJ White, PCR protocols: a guide to methods and applications, Academic Press, Inc., New York, N.Y).

Yo, S.M. and Lee, S.Y. (2008). Diagnosis of pathogens using DNA microarray. *Recent Patents on Biotech*. 2: 124-129.

Zeng Q.-Y., Hansson, P., and Xiao-Ru Wang X.R. (2005). Specific and sensitive detection of the conifer pathogen Gremmeniella abietina by nested PCR. *BMC Microbiol*. 5

Zimmerman, D., Sommermeyer, G., Walter, B. and Van Regenmortel, M.H.V. (1990). Production and characterization of monoclonal antibodies specific to closterovirus like particles associated with grapevine leafroll disease. *J. Phytopathol*. 130: 277-288.

Chapter 7

Molecular Characterization of *Fusarium* spp. Causing Wilt of Patchouli and Gladiolus

Yashoda R. Hegde, Sridevi Chavan, Sumitra Patil Kulkarni, Srikant Kulkarni and S.K. Prashanthi*

Department of Plant Pathology, University of Agricultural Sciences, Dharwad – 580 005, Karnataka, India

ABSTRACT

The cultural and morphological features of eight isolates of *F. solani*, showed four groups when cultured on potato dextrose agar. Of the 20 primers used for amplification, OPA3, OPA4, OPA7, OPA10, OPA11, OPA15, OPB4 showed 100 per cent polymorphism. The coefficient ranged from 0.63 to 0.92. The maximum genetic similarity of 92 per cent was observed between *Fs*2 and *Fs1*. Least genetic similarity was observed between *Fs6* and *Fs2* isolates. Amongst *F. solani* isolates of same geographical locations were closely related. Morphological and cultural variation could not help to group among the isolates of *F. oxysporum* f.sp. *gladioli*. The suitability of random amplified polymorphic DNA (RAPD) was used to detect similarity coefficient. It was ranged from 7 to 58 per cent. There was 53 per cent similarity between *Fog 3* and *Fog 4*. About 36 per cent similarity between *Fog 1* and *Fog* 2. RAPD data distinguished the six isolates into two major clusters A and B. Cluster A was classified up to sub cluster A6. Cluster B comprising only one isolate *Fog 6* (Bangalore isolate). Cluster A was sub grouped into A1 and A2. A1

*E-mail: uasyashoda@rediffmail.com

was comprised of only one isolate *i.e. Fog 1* (Arabhavi isolate). A2 was sub grouped into A3 and A4. A4 comprising the only one isolate *i.e., Fog 5* (Belgaum isolate). Sub cluster A3 comprised of A5 and A6. A5 comprising of two isolates *viz., Fog 2* and *Fog 4* (Hubli and Saidapur farm isolate, respectively). A6 comprising of *Fog 3* isolate (Agriculture College, Dharwad isolate). The isolates *Fog 2* and *Fog* 4 are from the same geographical origin which appeared close to each other with respect to similarity degree. Bangalore is geographically far away from all these locations and it is grouped in separate cluster.

Keywords: *Fusarium solani, F. oxysporum f.sp. gladioli, RAPD, Patchouli, Gladiolus, Molecular variability, wilt.*

Introduction

Patchouli (*Pogostimon patchouli* Pellet.) belonging to family Lamiaceae is one of the important aromatic and medicinal plants. It is native of Philippines. It is highly valued for its essential oils-patchouli oil–which is obtained by steam distillation and is perfume by itself. Patchoulene is the important component of essential oil of patchouli. Due to lack of any synthetic substitute for patchouli oil, it is used in a wide range of toilet soaps, scents and body lotions. Wilt caused by *Fusarium solani* is one of the important diseases of patchouli resulting in yellowing of leaves, brown to black discoloration of stem and roots, resulting in complete death of plants (Plate 7.1). Gladiolus (*Gladiolus hybridus* Hort.) is one of the most popular cut flowers, at both national and international level and can rightly be called the "Queen of bulbous flower crops". Gladiolus is commonly propagated by corms and cormels. *Fusarium oxysporum* Schlecht Fr f. sp. *gladioli* (Massey) Synd. and Hans. is one of the most important fungal pathogens which cause rotting of corms, yellowing of leaves, malformation of flowers, stunting and wilting of plants (Plate 7.2). The suitability of random amplified polymorphic DNA (RAPD) was used to detect the variations among the isolates of *Fusarium solani* and *F. oxysporum* f.sp. *gladioli.*

Eight isolates of *Fusarium solani* were collected from different places of Karnataka for variability studies.

Sl.No.	Locality of Collection of Isolates	Isolate No.
1	Arabhavi	*Fs1*
2	Belgaum	*Fs2*
3	Mangenkoppa	*Fs3*
4	Adavisomapur	*Fs4*
5	Saidapur	*Fs5*
6	Shimoga	*Fs 6*
7	Uttar Kannada	*Fs7*
8	Adur	*Fs 8*

Cultural and Morphological Variability

Isolates of *F. solani* were grown on potato dextrose agar for variability studies. Growth of each isolate, *viz.*, colony colour, colony diameter, sporulation and spore size were recorded. Sporulation and spore size were recorded by using Motic Images Software.

Molecular Variability

With the advent of molecular biology, isozyme analysis, restriction analysis of mitochondrial DNA and RAPD analysis have been successfully used to study the genetic variation of different pathogens. Keeping these intricacies in view, we have made an attempt to study genetic variation of *Fusarium* isolates of Karnataka, using RAPD markers.

Total Genomic DNA Extraction (Good Win and Lee, 1992)

Five days old fungal mycelial mat grown on potato dextrose broth was placed in pre cooled pestle and mortar. The pestle and mortar was filled with liquid nitrogen and was allowed to evaporate for 2 minutes. Immediately 500 µl of 65°C lysis buffer was added and transferred to 1.5 ml microcentrifuge tube. The tube was vortexed and placed in 65°C water bath for one hour by vortexing after 30 and 60 minutes. 500 µl phenol was added to extract and tube was spinned for 5 minutes at 8000 rpm to separate phases. 450 µl of aqueous phase was taken in a fresh tube and 450 µl of buffered phenol was added and spinned for 5 min at 8000 rpm. The aqueous phase of 400 µl was placed in a fresh tube. Further 400 µl of chloroform: isoamyl alcohol (24:1) was added and spinned for 5 min at 8000 rpm. 350µl of aqueous phase was removed and placed in a fresh tube. 50 µl Ammonium acetate was added and gently mixed. Immediately 880 µl of 95 per cent ethanol was added and placed in 20°C freezer for overnight. The DNA pellet obtained was spinned down for 20 min at 13,000 rpm and pellet was rinsed in 70 per cent ethanol. The pellet obtained was allowed to dry and resuspended in 20 µl of T_{10} E_1. Further, the DNA template was quantified by agarose gel electrophoresis.

Random Primers

Commercial kit OPA and OPB of decamer DNA primers were obtained from M/s Integrated DNA technologies supplied by Sigma Industrial and laboratory Equipments Inc., Bangalore, India.

The Thermoprofile for PCR

The PCR amplification for RAPD analysis was performed according to Williams *et al.* (1990) with certain modifications. The optimum conditions for DNA amplifications used were as follows.

Sl.No.	*Step*	*Temperature (ºC)*	*Duration (min)*	*Number of Cycles*
1.	Denaturation	94	4	1
2.	Denaturation	94	1	40
3.	Annealing	36	1	

Contd...

Contd...

Sl.No.	Step	Temperature (ºC)	Duration (min)	Number of Cycles
4	Extension	72	2	
5	Final extension	72	5	1
6	Hold temperature	-4	-	-

Plate 7.1: Symptoms of Fusarium Wilt of Patchouli

General view of infected field

Yellowing of leaves

Wilted plants

Contd...

Plate 7.1–Contd...

Black discolouration of roots

Brown discolouration

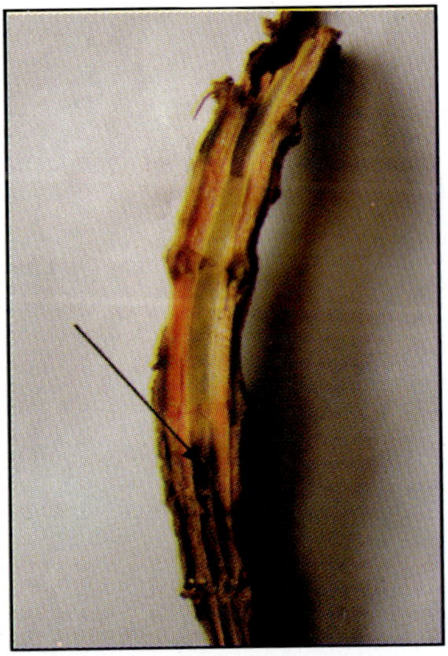

Vascular Discolouration

Master Mix for PCR

Amplification reaction mixture was prepared in 0.2 ml thin walled PCR tubes containing following components. The total volume of each reaction mixture was 20µl. The following reaction mixture was found to be optimum for PCR amplification.

1. 10 x assay buffer with 15 mM $MgCl_2$: 2.50 µl
2. dNTPs mix (2.5 mM each): 1.0 µl
3. Primer (5pM/µl): 1.0 µl
4. Template DNA (25ng/µl): 1.0 µl
5. Sterile distilled water: 14.30 µl
6. Taq DNA polymerase (3.0U μl^{-1}): 0.20 µl

Except template, the master mix was distributed to PCR tubes (19 µl/tube) and later 1 µl of template DNA from the respective isolates was added making the final volume of 20 µl. After the completion of the PCR, the products were stored at –4ºC until the gel electrophoresis was done. DNA bands were viewed under UV transilluminator and photographed using Alpha DigiDoc documentaton system (Alpha Innotech Corporation, USA).

Plate 7.2: Fusarium Wilt of Gladiolous

General view of infected field

Stunting

Yellowing of leaves

Unopened flower

Contd...

Plate 7.2–Contd...

Infected corms

Bending of spike

Healthy plant

RAPD Analysis

The amplified fragments were scored as '1' for the presence and '0' for the absence of a band generating the '0' and '1' matrix. The data obtained was analysed by obtaining the pair wise genetic similarities using DICE similarity coefficient. Clustering was done using the symmetric matrix of similarity coefficient and cluster obtained based on unweighted pair group arithmetic mean (UPGMA) using Sequential agglomerative hierarical nested (SAHN) cluster analysis of NTSYS-PC version 2.0 (Rohlf, 1998).

$$\text{Per cent polymorphism} = \frac{\text{Number of polymorphic bands}}{\text{Total number of bands}} \times 100$$

Results and Discussion

Cultural and Morphological Variability

The results presented in Table 7.1 indicate studies on cultural and morphological features of eight isolates of *F. solani,* carried on potato dextrose agar. On potato dextrose agar, the colony diameter ranged from 84 mm (*Fs7*) to 90 mm (*Fs* 1). Isolates *Fs1, Fs2,* and *Fs4* showed good mycelial growth and abundant sporulation, whereas *Fs5* and *Fs8* showed good growth and sporulation. However, sporulation was sparse in *Fs3* isolate. Isolate *Fs6* registered good growth but sporulation was sparse (Plate 7.3) These findings are in accordance with reports of Patel (1991) who categorized 19 *Fusarium* chickpea wilt isolates into six groups on the basis of morphology and cultural characters.

Genetic Variability

Of the 20 primers used for amplification, OPA3, OPA4, OPA7, OPA10, OPA11, OPA15, OPB4 showed 100 per cent polymorphism (Table 7.2 and Figure 7.1). A total of 175 amplicon levels resulted from 20 primers were available for analysis. Information on banding pattern for all primers was used to determine genetic distance between isolates and to construct dendrogram. Similarity coefficient ranged from 0.63 to 0.92. The maximum genetic similarity of 92 per cent was observed between *Fs2* and *Fs1*. Least genetic similarity was observed between *Fs6* and *Fs2* isolates (Table 7.3).

The dendrogram by RAPD study revealed that eight isolates were differentiated into two major clusters A and B (Figure 7.2). Cluster A was classified upto sub sub cluster A6, whereas cluster B was classified upto sub sub cluster B4. Major cluster A composed of isolates belonging to Shimoga, Uttar Kannada, Adur and Adavisomapur. Major cluster B composed of isolates belonging to Belgaum, Saidapur, Mangenkoppa and Arabhavi. In the present investigation, the results revealed that isolates of same geographical locations were closely related. So results obtained from the cluster analysis revealed that sub cluster group was composed of isolates belonging to same geographical locations with very less variability. Similar work was carried out by Sau Paulo (2004) who used PCR based methods to investigate the molecular variability among isolates of *F. solani* from four Brazilian states.

Table 7.1: Cultural and Morphological Variability in Isolates of *Fusarium solani* on Potato Dextrose Agar

Isolates	*Colony Diameter (mm)*	*Growth Characters*	*Sporulation*	*Size (μm)*		
				Macroconidia	*Microconidia*	*Chlamydospores*
Fs1	90.00	White colour cottony growth with smooth margin	++++	27.3-33.5x6.9-8.8	13.0-18.1x7.8x8.4	14
Fs2	88.20	Pink coloured pluffy colony	++++	21.5-26.3x7.0-9.1	14.2-15.7x7.1-7.9	22
Fs3	89.00	Light pink cottony colony	++	21.2-31.8x6.7-8.5	11.4-13.8x5.0-6.5	18
Fs4	87.75	Light pink coloured colony with yellowish pigmentation	++++	21.9-25.76x8.5x9.7	10.2-11.8x6.3-7.1	20
Fs5	85.60	Yellowish pluffy colony	+++	24.4-31.7x7.6-8.9	11.4-13.4x4.5x6.4	16
Fs6	89.75	Light pink cottony growth with irregular margin	+	23.5-26.9x6.0-9.0	11.4-14.6x5.4-6.9	18
Fs7	84.00	Pink coloured pluffy growth with irregular margin	++	26.9-32.8x6.7-9.0	13.0-14.7x5.0x7.0	20
Fs8	89.25	Light pink coloured colony with orange pigmentation	+++	30.4-32.9x5.4-9.4	16.6-17.2x5.8x7.5	24

++++: Abundant sporulation >80 number of conidia/microscopic field; ++: Moderate sporulation 25-50 number of conidia/microscopic field; +++: Good sporulation 50-80 number of conidia/microscopic field; +: Sparse sporulation <25 number of conidia/microscopic field.

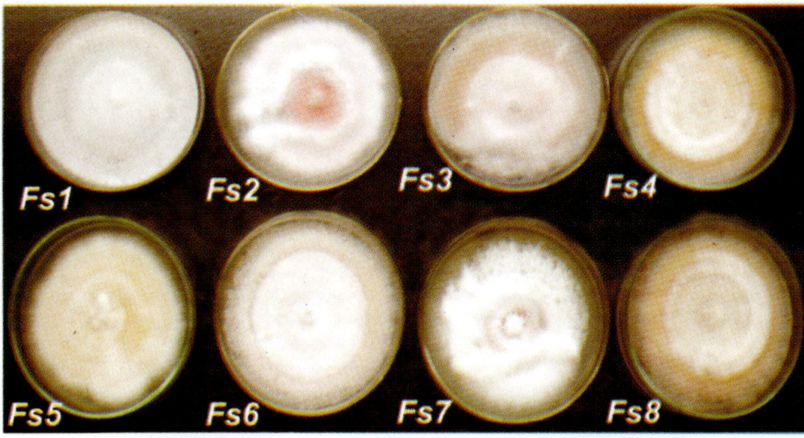

***Fs1*. Arabhavi**
***Fs2*. Belgaum**
***Fs3*. Mangenkoppa**
***Fs4*. Adavisomapur**
***Fs5*. Saidapur**
***Fs6*. Shimoga**
***Fs7*. Uttar Kannada**
***Fs8*. Adur**

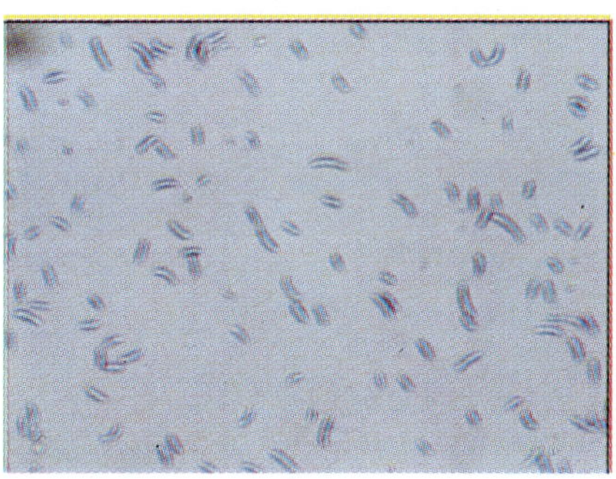

Fs8

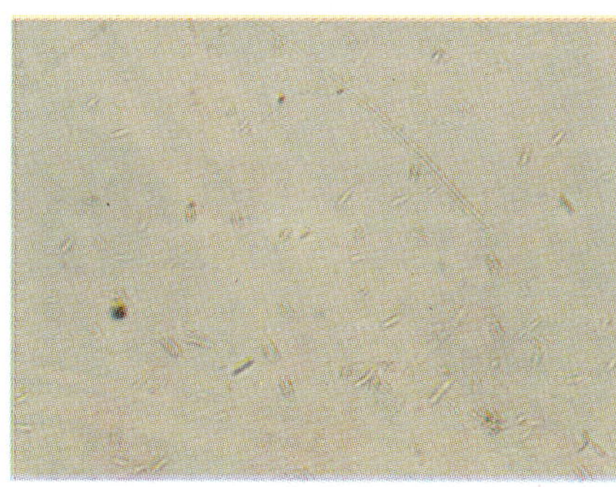

Fs3

Size of conidia

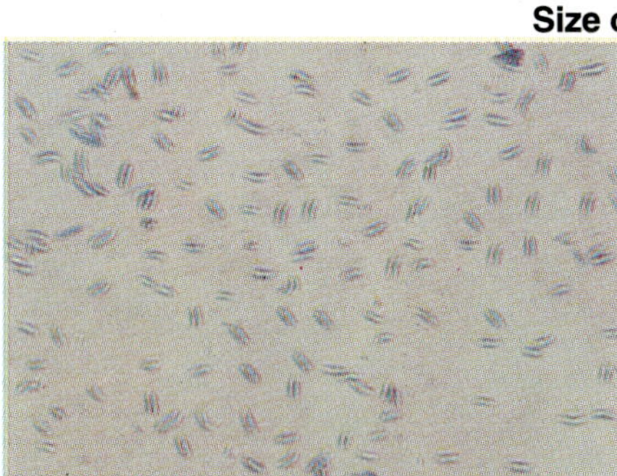

Fs1

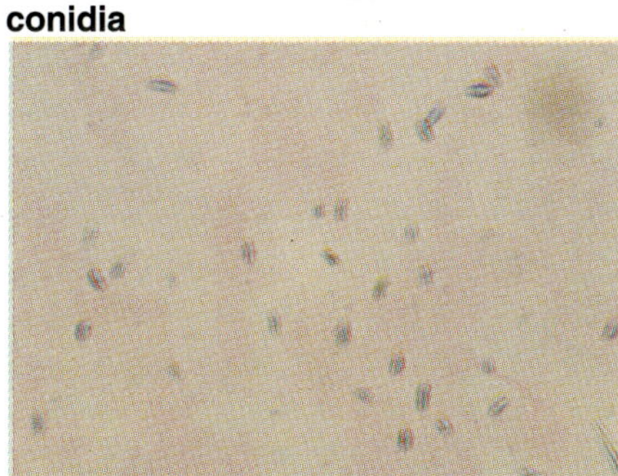

Fs6

Number of conidia

Plate 7.3: Cultural Variability of *Fusarium solani* on Potato Dextrose Agar

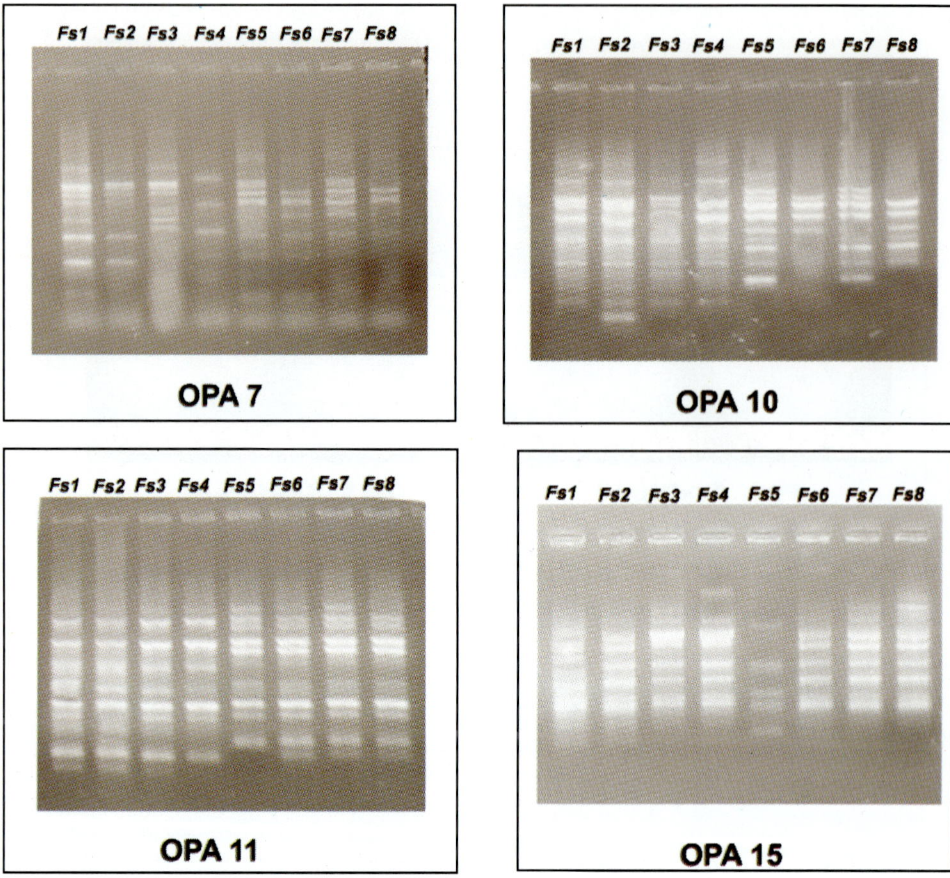

Figure 7.1: Gel Electrophoresis of RAPD Amplified from Eight Isolates of *F. solani* with Random Primers

Wilt of Gladiolus

For variability studies, the isolates collected are as follows:

Sl.No.	*Location*	*Designation*
1	Arabhavi	Fog 1
2	Hubli	Fog 2
3	Agriculture College, Dharwad	Fog 3
4	Saidapur, Dharwad	Fog 4
5	Belgaum	Fog 5
6	Bangalore	Fog 6

Table 7.2: Banding Profile of Different Primers for Different Isolates of *Fusarium solani*

Primers	*Total Bands*	*Polymorphic Bands*	*Per cent Polymorphism*
OPA-01	7	5	71.4
OPA-02	7	6	85.7
OPA-03	9	9	100
OPA-04	8	8	100
OPA-05	8	7	87.5
OPA-06	7	6	85.7
OPA-07	8	6	100
OPA-08	9	8	88.8
OPA-09	9	7	77.7
OPA-10	12	12	100
OPA-11	13	13	100
OPA-12	12	12	75.0
OPA-13	11	9	81.8
OPA-14	10	9	90.0
OPA-15	11	11	100
OPA-16	7	6	85.7
OPB-02	7	4	57.1
OPB-03	2	1	50
OPB-04	10	10	100
OPB-05	8	7	87.5

Table 7.3: Similarity Co-efficient of Eight Isolates of *Fusarium solani*

Isolates	*Fs1*	*Fs2*	*Fs3*	*Fs4*	*Fs5*	*Fs6*	*Fs7*	*Fs8*
Fs1	1.00							
Fs2	0.92	1.00						
Fs3	0.91	0.86	1.00					
Fs4	0.72	0.73	0.70	1.00				
Fs5	0.86	0.82	0.87	0.70	1.00			
Fs6	0.65	0.63	0.64	0.72	0.66	1.00		
Fs7	0.72	0.72	0.72	0.83	0.70	0.74	1.00	
Fs8	0.78	0.78	0.76	0.86	0.74	0.76	0.85	1.00

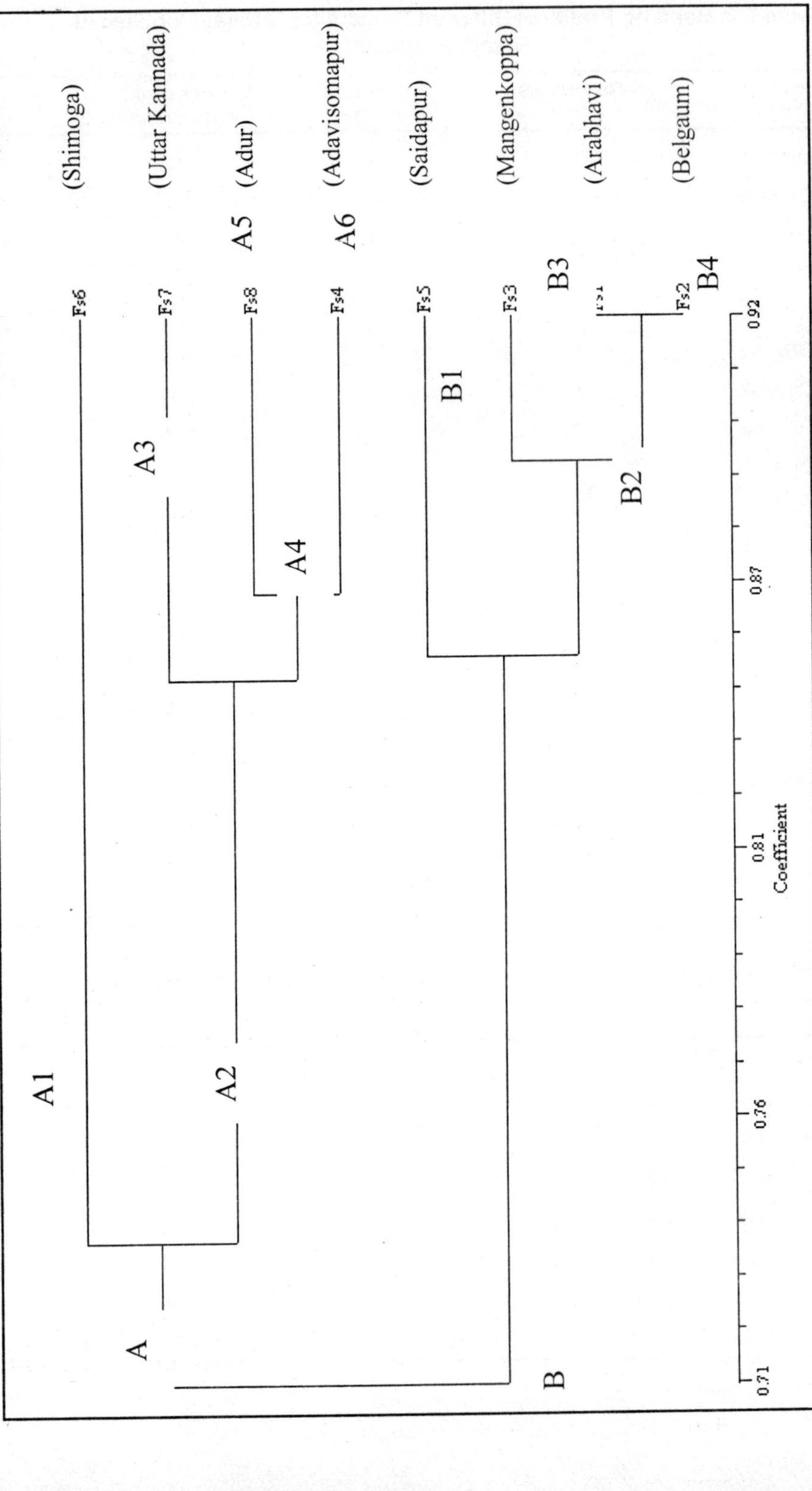

Figure 7.2: Dendrogram Based on RAPD Analysis of Eight Isolates of *Fusarium solani* OPB 10

Isolation of DNA of *F. oxysporum* f.sp. *gladioli* (Good Win and Lee, 1992)

25 mg of five days old fungal mycelial mat covered with agar was placed in pre cooled pestle and mortar. The pestle and mortar was filled with liquid nitrogen and was allowed to evaporate for 2 minutes. Immediately 500 µl of 65°C lysis buffer was added and transferred to 1.5 ml microcentrifuge tube. The tube was vortexed and placed in 65°C water bath for one hour by vortexing after 30 minutes and 60 minutes. 500 µl phenol was added to extract and tube was spinned for 5 minutes at 8000 rpm to separate phases. 450 µl of aqueous phase was taken in a fresh tube. Again, 450 µl of buffered phenol was added and spinned for 5 min at 8000 rpm. The aqueous phase of 400 µl was placed in a fresh tube. Further 400 µl of chloroform: isoamyl alcohol (24:1) was added and spinned for 5 min at 8000 rpm. 350µl of aqueous phase was removed and placed in a fresh tube. 50 µl of 7.5 M Ammonium acetate was added and gently mixed. Immediately 880 µl of 95 per cent ethanol was added and placed in 20°C freezer for overnight. The DNA pellet obtained was spinned down for 20 min at 13,000 rpm and pellet was rinsed in 70 per cent ethanol. The pellet obtained was allowed to dry and resuspended in 20 µl of TE buffer. Further, the DNA template was quantified by agarose gel electrophoresis.

Results and Discussion

Cultural variation regarding colony diameter ranged from 83.81 mm (*Fog 6)* to 89.5 mm *(Fog 1).* Isolates *viz., Fog 1, Fog 2, Fog 3* and *Fog 4* showed good mycelial growth whereas *Fog 5* and *Fog 6* produced moderate growth. Isolates like *Fog1, Fog 3* and *Fog 5* showed the pink colour cottony mycelial growth whereas isolates *Fog 2, Fog 4* and *Fog 6* showed white cottony growth. Sporulation was good in *Fog 1, Fog 2* and *Fog 4,* whereas it was moderate in *Fog 3, Fog 5* and *Fog 6* (Table 7.4). However in the present investigation, morphological and cultural variation could not help to group among the isolates of *F. oxysporum* f.sp. *gladioli.* It is difficult to distinguish these isolates using traditional morphological differences. The suitability of random amplified polymorphic DNA (RAPD) was used to detect the variations among the isolates of *F. oxysporum* f. sp. *gladioli.* OPB and OPF series primers were used to determine genetic distance between isolates and to construct a dendrogram. Banding profile of different primers for six isolates of *F. oxysporum* f.sp. *gladioli* is given in Table 7.5. Of the 16 primers used for amplification OPB1, OPB6, OPB9, OPB12, OPB13, OPF3, OPF12 and OPF17 showed 100 per cent polymorphism (Figure 7.3). Information on banding pattern for all the primers was used to determine genetic distance between isolates and to construct dendrogram. Similarity coefficient of six isolates of *F. oxysporum* f.sp. *gladioli* based on RAPD analysis is given in the Table 7.6. Similarity coefficient ranged from 7 to 58 per cent. There was 53 per cent similarity between *Fog* 3 and *Fog 4.* About 36 per cent similarity between *Fog 1* and *Fog 2*. RAPD data distinguished the six isolates into two major clusters A and B (Figure 7.4). Cluster A was classified up to sub sub cluster A6. Cluster B comprising only one isolate *Fog 6* (Bangalore isolate). Cluster A was sub grouped into A1 and A2. A1 was comprised of only one isolate *i.e. Fog 1* (Arabhavi isolate). A2 was sub grouped into A3 and A4. A4 comprising the only one isolate *i.e., Fog 5* (Belgaum isolate). Sub cluster A3 comprised

Table 7.4: Cultural and Morphological Variability in Isolates of *F. oxysporum* f.sp. *gladioli* on Potato Dextrose Agar

Sl.No.	*Isolates*	*Colony Diameter (mm)*	*Growth characters*	*Sporulation*	*Microconidia*	*Macroconidia*	*Dry Mycelial Weight (mg)*
1.	*Fog 1*	89.50	Light pink color cottony growth	+++	3.4-4.5x1.5-2.7	17.5-20x5.0-6.1	384.61
2.	*Fog 2*	88.60	White cottony and pluffy growth	+++	3.0-4.0x1.2-2.0	16.6-18.8x4.0-5.2	373.28
3.	*Fog 3*	85.75	Pink cottony growth with smooth mycelium	++	4.2-6.1x2.5x3.2	19.0-20.8x4.2-5.3	366.78
4.	*Fog 4*	88.25	White cottony growth	+++	3.1-4.2x2.0-3.2	16.0-18.5x4.8-5.9	376.81
5.	*Fog 5*	84.16	Pink cottony growth	++	3.1-4.5x2.2-2.9	16.219x3.8-4.2	354.08
6.	*Fog 6*	83.81	White cottony growth with irregular margin	++	2.5-3.0x1.3-2.2	14-17.2x3.0-4.5	338.46

+++: Good sporulation >50 number of conidia/microscopic field; ++: Moderate sporulation 30-50 number of conidia/microscopic field.

of A5 and A6. A5 comprising of two isolates *viz.*, *Fog 2* and *Fog 4* (Hubli and Saidapur farm isolate, respectively). A6 comprising of *Fog 3* isolate (Agriculture College, Dharwad isolate). The isolates *Fog 2* and *Fog 4* are from the same geographical origin which appeared close to each other with respect to similarity degree. Bangalore is geographically far away from all these locations and it is grouped in separate cluster.

Table 7.5: Banding Profile of Primers for Different Isolates of *F. oxysporum* f.sp. *gladioli*

Sl.No.	*Primer*	*Total Bands*	*Polymorphic Bands*	*Per cent Polymorphism*
1	OPB-01	7	7	100.0
2	OPB–05	6	5	83.30
3	OPB-06	7	7	100.0
4	OPB-07	9	8	88.88
5	OPB-08	7	6	85.68
6	OPB-09	8	8	100.00
7	OPB-10	10	8	80.00
8	OPB-11	7	6	85.68
9	OPB-12	6	6	100.00
10	OPB-13	8	8	100.00
11	OPB-14	10	8	80.00
12	OPB-15	6	5	83.30
13	OPF-03	4	4	100.0
14	OPF-09	5	4	80.00
15	OPF-12	4	4	100.00
16	OPF 17	5	5	100.00

Similar work was carried out by De Haan (2000) who tested 160 arbitrary 10-mer oligonucleotide primers on *F. oxysporum* f.sp. *gladioli* by PCR to find RAPD specific marker. He found that RAPD primer G-12 amplified two discriminating DNA fragments. This study indicates that RAPD is a simple and convenient tool to measure the extent of genetic variability of *Fusarium* isolates at the molecular level and identify the lineages.

Table 7.6: Similarity Co-efficient of Six Isolates of *F. oxysporum* f.sp. *gladioli*

Isolates	*Fog 1*	*Fog 2*	*Fog 3*	*Fog 4*	*Fog 5*	*Fog 6*
Fog 1	1.00					
Fog 2	0.36	1.00				
Fog 3	0.29	0.48	1.00			
Fog 4	0.30	0.58	0.53	1.00		
Fog 5	0.07	0.30	0.41	0.33	1.00	
Fog 6	0.08	0.26	0.11	0.27	0.28	1.00

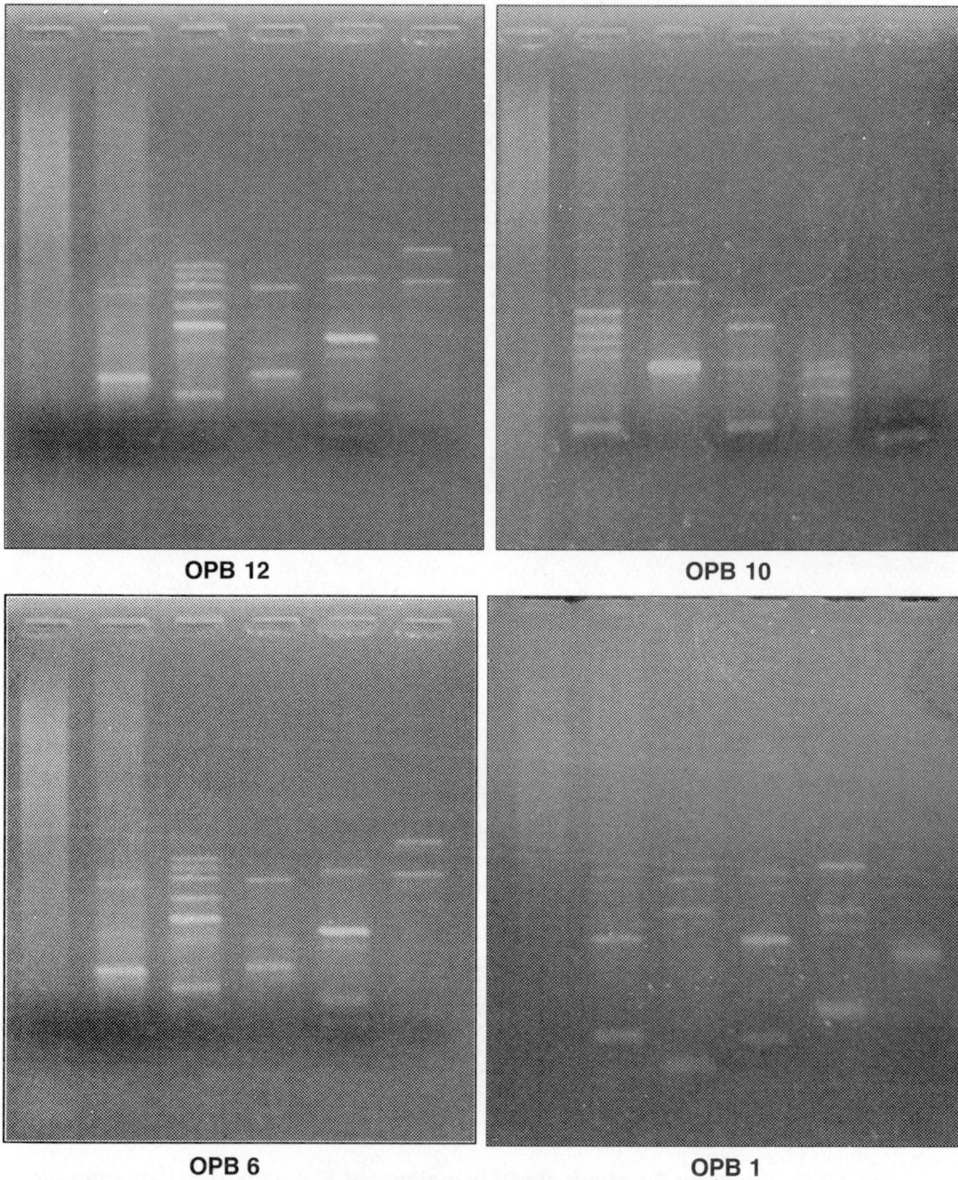

Figure 7.3: Gel Electrophoresis of RAPD Amplified from Six Isolates with Random Primers

Molecular biology has brought many powerful new tools to fungal taxonomists including the potential for rapid identification of isolates, methods for rapid determination of virulence or toxicity of strains, and the means to elucidate the relationships among fungal species. Molecular methods have also been used to distinguish between closely related species with few morphological differences and to distinguish strains within a species (Bhim Pratap Singh *et al.*, 2006).

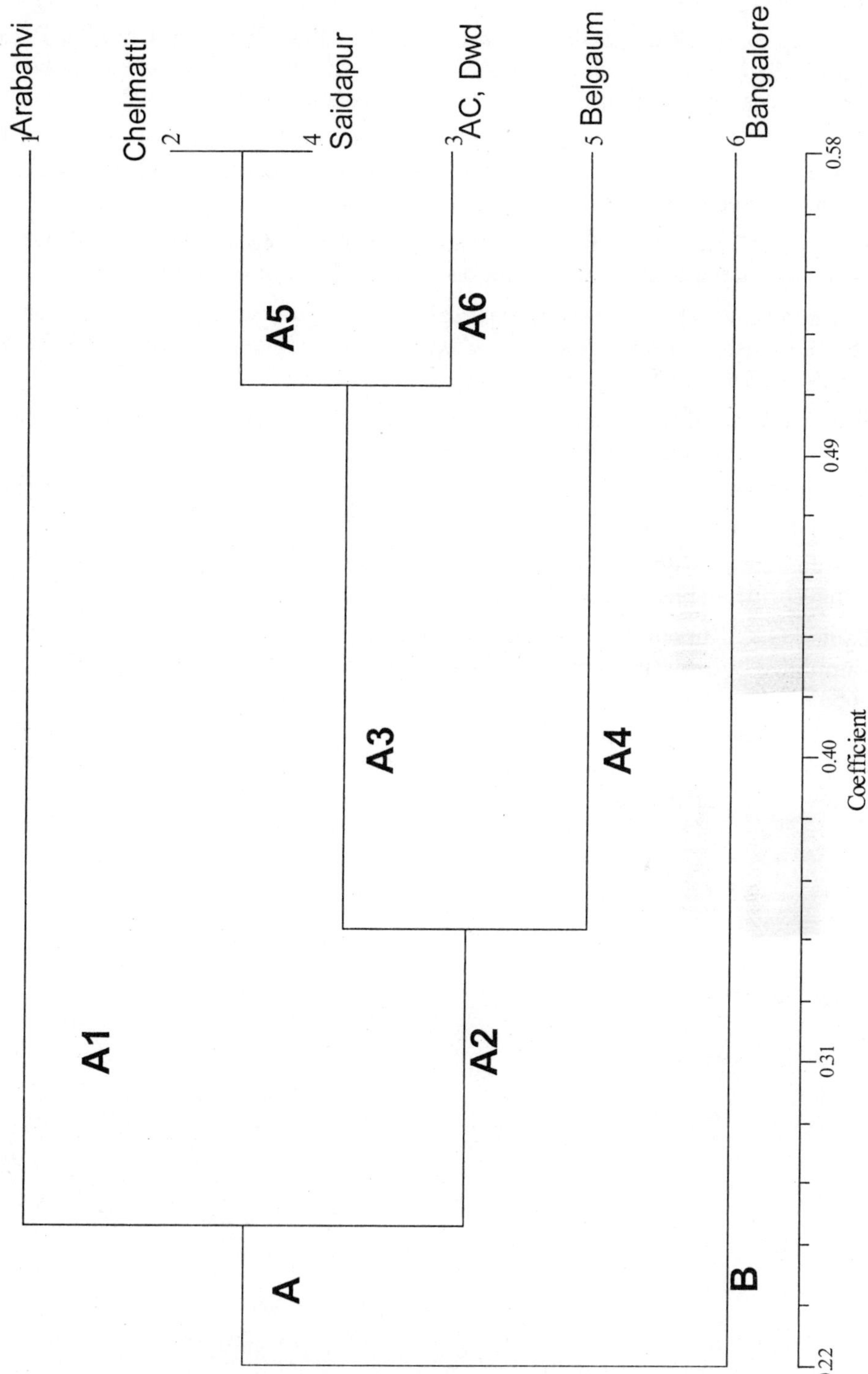

Figure 7.4: Dendrogram Based on RAPD Analysis of Six Isolates of *F. oxysporum* f. sp. *Gladioli*

References

Bhim Pratap, Singh., Ralut, Saikai., Mukesh, Yadhav., Rakesh, Singh., Chouhan, V. S. and Dilip, K. Arora. (2006). Molecular characterization of *Fusarium oxysporum* f. sp. *ciceri* causing wilt of chickpea. *African J. Biotech.* 105(6): 497–502.

De Haan, L.A.M., Numansen, A., Reobroek, E.J.A. and Van Doorn, J. (2000). PCR detection of *F. oxysporum.* f.sp. *gladioli* causal agent of gladiolus yellows disease, from infected corm. *Plant Pathology*. 49: 438-444.

Good win, D. C. and Lee, S. B. (1992). Microwave mini prep of total genomic DNA from fungi, plants protests and aniols for PCR. *Biotechniques.* 15: 438-444.

Nei, M. and Li, W. H. (1979), Mathematical model of studying genetic variation in terms of restriction endonucleases. *Proceedings of the National Academy of Sceinces,* USA, 76: 5269-5273.

Patel, S. T. (1991). Studies on some aspects of wilt of chickpea. *Ph.D. Thesis,* Uni. of Agric.Sci. Dharwad (India).

Rohlf, F. J. (1998) *NTSYS-PC Numerical Taxonomy and Multivariate Analysis Version* 2.0. Applied Biostatics Inc., New York.

Sau Paulo. (2004). Genetic variability within *Fusarium solani,* as revealed by PCR fingerprinting based on PCR markers. *Braz. J. Microbiol.* 35: 3

Williams, J.G.K., Kubelik, A.R., Livak, K.J., Rafalsti, J.A. and Tingey, S. V. (1990). DNA amplified by arbitrary primers are useful genetic markers. *Nuc. acid Res.* 18: 6531–6535.

Chapter 8

Reusing of Agarose in Molecular Plant Pathology Laboratories

Kamel Ahmed Abd-Elsalam*

King Saud University, College of Science, Botany and Microbiology Department, P.O. Box No. 2455, Riyadh 1145, Saudi

ABSTRACT

Because the agarose undergoes much commercial processing it is very expensive. To save expensive high-resolution agarose gel, agarose can be reused for some time by remelting it. To reuse agarose gel, simply re-melt it with occasional swirling until the solution is uniformly liquid. Just run the bands out the bottom and reload the gel for further uses. This agarose can be reused numerous times (4-6) and prior to repouring a little more ethidium (10 ul of a 10 mg/ml solution) was added after each remelt. The efficiency of reused agarose gel was tested by using microsatellite-primed PCR amplifications of 6 different *Fusarium* species and DNA marker on common DNA electrophoretic conductive media. The recycled gels showed different resolution performance with two Tris-based conductive media. Tris boric acid EDTA (TBE) gives better resolution and sharper bands with recycling agarose, and is particularly recommended for analyzing fragments <1 kb. To check the first results, recycled agarose recommended for PCR-based marker analysis, or other applications.

Keywords: *DNA support media, Re-melt, Tris-acetate, Tris-borate.*

---

* E-mail: abdelsalamka@gmail.com.

Introduction

Agarose gel electrophoresis, first introduced by Daniel Nathans in 1970, is a simple and efficient method for resolving DNA fragments of different sizes. DNA molecules are negatively charged due to the presence of phosphate groups. Consequently, the DNA molecules migrate towards the anode (positive electrode) in an electric field. The agarose gel matrix acts as a molecular sieve to sort fragments based on size. Electrophoretic mobility of the DNA fragment is inversely proportional to the $\log_{10}$ of the number of base pairs (Helling *et al.*, 1974). Walter Schaffner invented the currently used configuration of horizontal agarose gel electrophoresis. McDonell *et al.* (1977) and Southern (1977) described details of this technique. Tris acetate (TA) was used in DNA electrophoresis in 1971 (Danna and Nathans, 1971), where EDTA remained restricted to the loaded sample. Tris acetate EDTA (TAE) was used in 1972 as the gel conductive medium by two groups (Aaij and Borst, 1972). Tris boric acid EDTA (TBE) was used for RNA electrophoresis in 1968 (Peacock, 1968) and for RNA sequencing in 1973 (Loewen and Khorana, 1973). Tris Borate EDTA (TBE) Buffer is the traditional buffer used for agarose electrophoresis. Tris Acetate EDTA (TAE) is also a popular selection. These buffers provide good results, producing acceptable resolution of bands and reasonable size separation (Miura *et al.*, 1999).

It is possible that the 30-year arrested maturation of agarose gel electrophoresis reagents was responsible for a significant loss of time and money. Brody and Kern, (2004a) estimated that for agarose electrophoresis alone in the United States, the scientific community would have saved over 30 million dollars last year in reduced reagents expenses. Molecular biologist have compared and analyzed TAE and TBE buffers in DNA electrophoresis; however, to our knowledge no one has substantially investigated the efficiency of two conductive mediums on band resolution of recycling agarose. The purpose of the current research was to modify a procedure for recycling agarose gel, after carrying out cleansing of DNA or RNA loading dye and ethidium bromide; to study the effect of conductive media on resolution performance of recycling agarose.

Materials and Methods

Fusarium Species Isolates and DNA Extraction

Six isolates belonging to 6 *Fusarium* species was used in this experiment: *F. solani* (Fs), *F. chylamdiosporum* (Fc), *F. oxysporum* f. sp. *vasinfectum* (Fov), *F. oxysporum* (Fo) *F. poae* (Fp), and *F. gramanirium* (Fg). Fungal isolates used in the following study were obtained from the fungal collection of Plant Pathology Research Institute, Agricultural Research Center, Giza, Egypt. Isolates of *Fusarium* spp. were grown in 75 ml Czapek-Dox broth in 500 ml Erlenmeyer flasks for 4-5 days at 26°C without shaking. A hyphal mat was filtered through a double layer of sterile cheesecloth and was washed with sterile distilled water. Total DNA was extracted basically as described by Abd-Elsalam *et al.* (2007).

Microsatellite-Primed PCR

PCR products were obtained in a total volume of 25 ml with 20 ng of template DNA, 0.2 mM of (CAG) 5 primer (MWG, Germany), 200 mM of each dNTP, 1 U Taq Polymerase (JenaBioscience, Germany) and 1X reaction buffer for used polymerase. DNA and PCR mixture were amplified in Techne TC-312 (Techne, Stone, UK) under the following conditions: initial denaturation at 94ºC for 3 min; 40 cycles: denaturation at 94ºC for 20 s, annealing at 50ºC for 1 min, extension at 72ºC for 20 s and a final extension at 72ºC for 7 min.

Storage Conditions of Used Gel

Agarose gels already analyzed, was classified and separated based on gel like concentrations (1, 1.5, 2, and 2.5, per cent ...etc). And then stored in four plastic trays containing 1X TBE or TAE buffer at 8 ºC for more than four months. Be sure that the buffer covers all of the gels during storage. The container kept closed when not in use (Figure 8.1). Stored agarose that is ready for melting and reuse without further buffer adjustment.

Figure 8.1: Storage Container of Reused Agarose

How to Multiply Reuse Agarose Gels?

The method, which was initially suggested by Palacios, *et al.* (2000) and the International Potato Center protocol, Perú (unpublished protocol) was modified by balancing the agarose directly in TBE or TAE buffer, then repouring it. To remelt the gels, the bands was run out the bottom to removes most of the loading dye (for instance, bromophenol blue or orange G for us) and probably some of the ethidium bromide (EtBr). The stored agarose gel was divided into small pieces and the transferred into beaker that is 2-4 times of the gel volume. The flask was placed in a microwave and heated on high power for 30 sec to 1 min. The flask was taken out wearing appropriate hand protection (a thermal glove) and swirled it gently. The flask putted back into the microwave and heated again until the agarose comes to a rolling boil. The microwave power was turned off and the flask sited in the microwave for at least 30 sec. The flask was removed, again wearing proper hand protection, and transferred into a fume hood. The agarose solution was cooled to 60°C prior to casting and 10 ul of 10 mg/ml ethidium bromide solution was added and stirred until ready to pour. The molten agarose was cooled until you can comfortably hold the flask in your bare hand. The agarose solution was poured into the gel casting tray and the desired comb was inserted into the agarose. When the agarose gel was hardened the running buffer was flooded and the comb was removed slowly. The gel casting tray was transferred into the electrophoresis unit. The chamber was filled with running buffer until the buffer reached 3 mm–5 mm over the surface of the gel. DNA samples was loaded and recycling agarose was run with both TBE and TAE buffer system, at conventional temperature (21°C) and voltage (8V/cm) in the size range of 100–2642 bp at four times.

Results

Six kinds of genomic DNA extracted from different *Fusarium* species and one DNA marker were analyzed on recycling agarose gels with TAE or TBE buffers. TBE buffer is preferred for separation of small DNA (<1 kb) when DNA recovery is not required (Table 8.1). Poor separation of smaller fragments was encountered the in TAE buffer after second recycle (Figure 8.2) While, the smaller DNA fragments were more clearly separated in TBE buffer (Figure 8.3) than TAE buffer on each recycle. Tris-borate buffer (TBE) was advantageous for obtaining a higher resolution of smaller DNA fragments on agarose gels, when compared to the conventional tris-acetate buffer (TAE). The gel resolution of TBE was a better conductive medium than TAE.

Discussion

The reused agarose gels were prepared with the same method as those formed with the original agarose powder, without need for additional steps. Ethidium bromide is dangerous and is a strong mutagent either in buffer or in the gel. To prevent damage to human health and the environment, waste, waste water containing EtBr should only be undertaken after prior decontamination (Lunn and Sansone, 1987). Advantage of reused agarose, recycling also contributes to the reduction of waste discarded in the environment and costs in research labs and services.

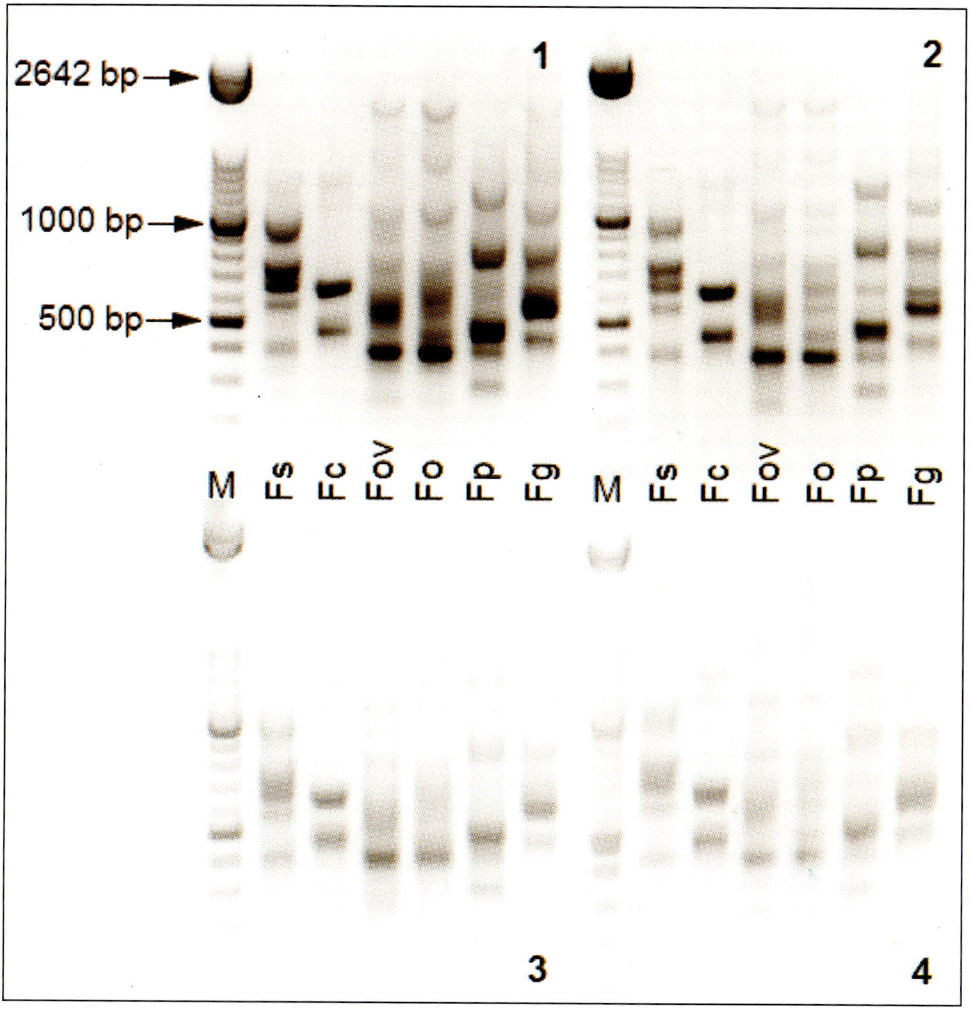

Figure 8.2: Resolution Performance of Four Times Reused Agarose Gel Electrophoresis of Microsatellite-Primed PCR Products from Six *Fusarium* Species Isolates. Separation of DNA in 1.5 per cent agarose gels (Roche Diagonastic GmbH, Mannheim, Germany) in 1X TAE buffer. 20 cm long gels were run at 8 V/cm for ~ 1 hr. Lanes M: 1000–1500 bp DNA marker (Roche), *F. solani* (Fs), *F. chylamdiosporum* (Fc), *F. oxysporum* f. sp. *vasinfectum* (Fov), *F. oxysporum* (Fo) *F. poae* (Fp), and *F. gramanirium* (Fg).

Recycling had disadvantages, tends to become spickly after a while and the background increases with time due to the DNA being left in the gel. Depletes the conductivity of the buffering system quite quickly, making it harder to run an resolve bands nicely on gels.

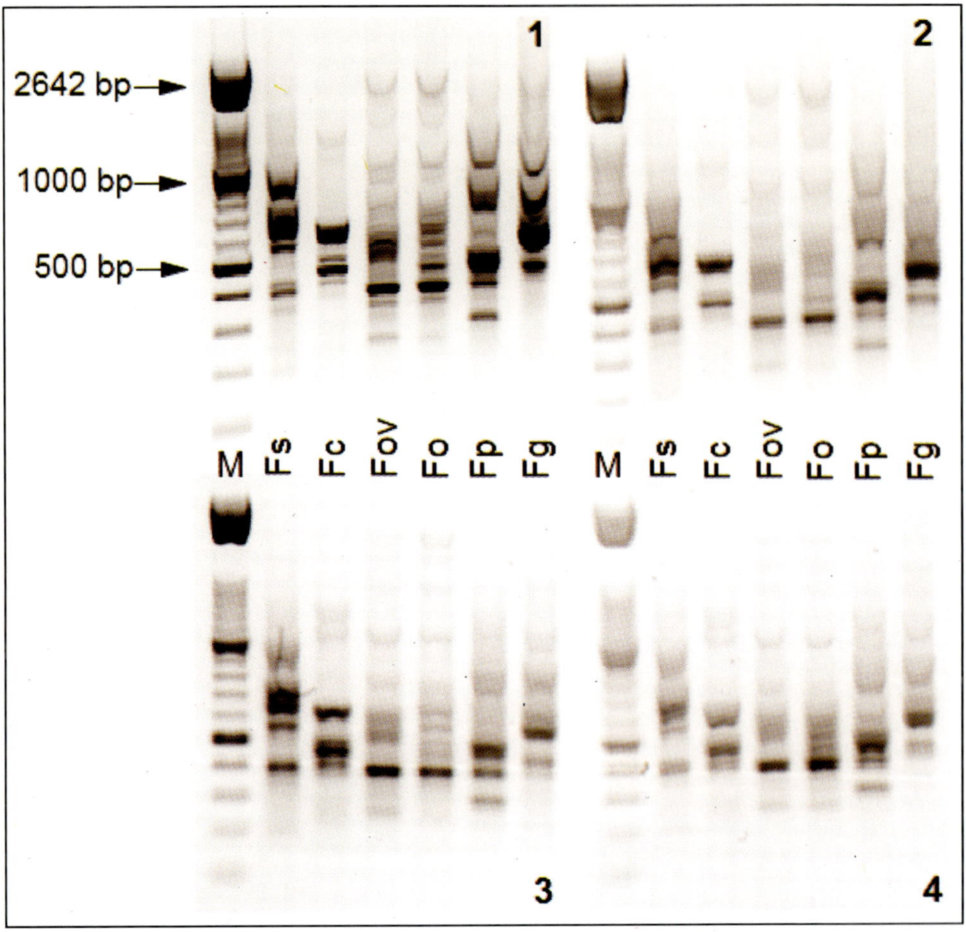

Figure 8.3: Resolution Performance of Four Times Reused Agarose Gel Electrophoresis of Microsatellite-Primed PCR Products from Six *Fusarium* Species Isolates. Separation of DNA in 1.5 per cent agarose gels (Roche Diagonastic GmbH, Mannheim, Germany) in 1X TBE buffer. 20 cm long gels were run at 8 V/cm for ~ 1 hr. Lanes M: 1000–1500 bp DNA marker (Roche), *F. solani* (Fs), *F. chylamdiosporum* (Fc), *F. oxysporum* f. sp. *vasinfectum* (Fov), *F. oxysporum* (Fo) *F. poae* (Fp), and *F. gramanirium* (Fg).

TBE has a more constant and higher buffering capacity. On the contrary, TAE buffering capacity is low, and it tends to become exhausted during successive electrophoresis (Miura *et al.*, 1999). Lower concentrations of boric acid, such as present in 0.5 x TBE and acetate based media, readily permit fragment isolation and cloning from agarose gel slices (Brody and Kern, (2004b).

In conclusion, we can reuse agarose gels if they are not for critical analysis. The recycled agarose was available for immediate use, being storegable at refrigerator for a long time. Conventional Tris media provide good results, producing acceptable resolution of bands and reasonable size separation. These buffers have several additional disadvantages, including high cost, precipitation of stock solutions (specifically TBE), and recipe complexity. Due to the disadvantages of TAE and TBE, others and I hope discover new conductive media in the further study.

Table 8.1: Properties of TAE and TBE Buffer Systems Tested with Recycled Agrose Gel Electrophoresis

Properties	*TAE*	*TBE*
Usage	Use when DNA is to be recovered. Use for electrophoresis of large (>12 kb) DNA.	Use when DNA recovery is not required. Increased resolution of small (<1 kb) DNA.
Buffer capacity	Low buffering capacity–recirculation may be necessary for extended electrophoretic times (>6 hours).	High buffering capacity–no recirculation required for extended run times. Decreased DNA mobility.
Stable	Stable	More stable
Ionic strength	Low ionic strength	High ionic strength
Fast	DNA larger fragments size (>2-kb) migrate faster	DNA smaller fragments (<300-bp) migrate faster
Gel Resolution	Good resolution	Excellent resolution
Buffer preparation	(1X=40 mM Tris base, 40 mM acetic acid, 1 mM EDTA)	(1X=89 mM Tris base, 89 mM boric acid, 2 mM EDTA)

Acknowledgements

This work was supported by the DNA Research Chair, Research Chairs Program, King Saud University. I thank O.E. Amin for excellent technical assistance with PCR analysis.

References

Abd-Elsalam, K. A., Asran-Amal, A. and El-Samawaty, A. (2007). Isolation of high quality DNA from cotton and its fungal pathogens. *J. Pl.Dis. and Prot.* 114: 113-116.

Aaij, C. and Borst, P. (1972). The gel electrophoresis of DNA, *Biochimca Biophysica Acta*. 269: 192-200.

Brody, J.R. and Kern, S.E. (2004b). Sodium boric acid: a Tris-free, cooler conductive medium for DNA electrophoresis. *Biotechniques* 36: 214-216.

Danna, K. and Nathans, D. (1971). Specific cleavage of simian virus 40 DNA by restriction endonuclease of *Hemophilus influenzae, Proceeding Nat. Acad.Sci. USA*. 68: 2913-2917.

Helling, R.B., Goodman, H.M. and Boyer, H.W. (1974). Analysis of endonuclease R *Eco*RI fragments of DNA from lambdoid bacteriophages and other viruses by agarose gel electrophoresis. *J. of Virology*. 14: 1235.

Loewen, P.C. and Khorana, H.G. (1973). Studies on polynucleotides. CXXII. The dodecanucleotide sequence adjoining the C-C-A end of the tyrosine transfer ribonucleic acid gene. *The J. of Biol. Chem.* 248: 3489-3499.

Lunn, G. and Sansone, E.B. (1987). Ethidium bromide: destruction and decontamination of solutions. *Analytical Biochemistry*. 162: 453.

McDonell, M.W., Simon, M.N. and Studier, F.W. (1977). Analysis of restriction fragments of T7 DNA and determination of molecular weights by electrophoresis in neutral and alkaline gels. *J. of Molecular Biol.* 110: 119-146.

Miura, Y, Wake, H. and Kato, T. (1999). TBE, or not TBE; that is the question: Beneficial usage of tris-borate for obtaining a higher resolution of small DNA fragments by agarose gel electrophoresis. *Nagoya Medical Journal*. 43: 1-6.

Palacios, G., Giménez, C. and Garcia, E. (2000). Recycling agarose. *Pl. Molecular Biol. Reptr.* 18: 47-49.

Peacock, A.C. (1968). Dingman, Molecular weight estimation and separation of ribonucleic acid by electrophoresis in agarose–acrylamide composite gels. *Biochem.* 7: 668-674.

Southern, E. (1977). Gel electrophoresis of restriction fragments. *Methods of Enzymology*. 68: 152-176.

Chapter 9

Rapid Detection of *Ganoderma* Disease of Coconut by Immunoassay and PCR

A. Kandan, V. Rajendran,
V. Raguchander and R. Samiyappan*

Department of Plant Pathology, Centre for Plant Protection Studies, Tamil Nadu Agricultural University, Coimbatore – 641 003

ABSTRACT

Tanjore wilt or *Ganoderma* wilt or Basal Stem Rot (BSR) disease caused by *Ganoderma lucidum* (Leys) Karst. is the most destructive disease in coconut plantations throughout South India. This disease can be contained by management practices if the disease is detected in its early stages. For early diagnosis, molecular techniques like immunoassay and nucleic acid hybridization are more specific, rapid and sensitive than the other conventional methods. The internal transcribed spacer (ITS) regions of ribosomal RNA gene (rDNA) have been selected as specific targets for PCR detection of *Ganoderma*. The polyclonal antibodies have been developed against the crude mycelial proteins of *Ganoderma* to detect serologically by applying Indirect Enzyme-linked Immunosorbent Assay (ELISA) and Dot Immunobinding Assay (DIBA). The techniques *viz.*, polymerase chain reaction, ELISA and DIBA are reliable for the early diagnosis of *Ganoderma* under greenhouse and field conditions.

Keywords: *Ganoderma lucidum, Coconut, RAPD, ELISA, DIBA.*

---

* E-mail: raguchander@rediffmail.com.

Introduction

Coconut (*Cocos nucifera)* Linn., is a major plantation as well as oilseed crop in the tropics of the world and has an area, production of 12.5 million hectares, 53.6 billion nuts and in India with an area of 1.92 million hectares and an annual production of 12,141 million nuts and a productivity of 6,345 nuts/ha (All India Final Estimate of Coconut, 2002-03). Coconut is predominantly cultivated in Kerala, Tamil Nadu, Karnataka and Andhra Pradesh. It is affected by more than fifty diseases in different parts of the world. Among which, *Ganoderma* wilt or Basal Stem Rot (BSR) disease caused by *Ganoderma lucidum* (Leys) Karst. is the most destructive disease (Wilson *et al.*, 1987; Bhaskaran *et al.*, 1989; Samiyappan *et al.*, 2006). The disease is also called as Thanjavur wilt, bole rot, *Ganoderma* disease and Anabe roga (Vijayan and Natarajan, 1972; Nambiar and Rethinam, 1986; Bhaskaran *et al.*, 1996; Srinivasulu *et al.*, 2001). The pathogen *G. lucidum* was first recorded on coconut palm in Karnataka by Butler in India (1913). The disease was first noticed in severe form in Thanjavur district of Tamil Nadu after the cyclone of 1952 and 1955, hence it was named as Thanjavur wilt. Now the disease is wide spread in Tamil Nadu occurring in all districts. Infection spreads to neighboring palms by root contact if one palm becomes infected (Turner, 1965). Recent survey revealed that the district wise incidence of BSR in Tamil Nadu ranged from 0.11 to 5.43 per cent (Karthikeyan *et al.*, 2005a).

The incubation period of this disease has been determined to be several years (Turner 1981). Visible disease symptoms appear at a very late stage of infection when more than half of the root tissues have decayed, leaving no chance for the grower to cure the infected palms. Basal stem rot disease of coconut can be contained by management practices if the disease is detected in its early stages. Molecular techniques like immunoassay and nucleic acid hybridization are more specific, rapid and sensitive than the other conventional methods for early diagnosis (Kandan, 2003; Rajendran, 2006; Karthikeyan *et al.*, 2006). A few methods have been reported to be useful to identify diseased palms even before the expression of symptoms, though the methods are non-specific for BSR (Natarajan, *et al.*, 1986; Vijayaraghavan, *et al.*, 1987; Samiyappan *et al.*, 1996). Polymerase chain reaction (PCR) technology has revolutionized the field of plant pathology in diagnosing various plant pathogens (Henson and French 1993). The internal transcribed spacer (ITS) regions of ribosomal RNA gene (rDNA) have been selected as specific targets for PCR detection of Ganoderma (Utomo and Niepold 2000). Secondly, the development of polyclonal antibodies against the crude mycelial proteins of Ganoderma to serologically detect this fungus by applying Indirect Enzyme-linked immunosorbent assay (ELISA) and Dot immunobinding assay (DIBA) techniques is reliable.

Therefore, the objectives of this study were to detect the pathogen at the early stages before the appearance of visual symptoms by applying the enzyme-linked immunosorbent assay (ELISA) and polymerase chain reaction (PCR) technique by using specific primers with ITS 1 region as a target.

Materials and Methods

The *Ganoderma* isolates were collected from coconut, arecanut and other trees from different parts of Tamil Nadu.

Plant Sample Preparation

Root, trunk (3 m from base) and leaf samples were collected from both healthy and BSR infected palms. In addition, infected lesions (oozing portion) from BSR affected plants were also collected. In the case of the pathogenicity test, roots, leaves (bottom, middle and upper) and brackets were collected from pathogen inoculated and uninoculated coconut seedlings. All plant samples collected were washed with distilled water, weighed and ground in 0.1 M phosphate buffer (pH 7.0; 1:2 dilution) in a sterile pestle and mortar at room temperature (30±2°C) and clarified at 12,000 rpm for 10 min at 4°C. The supernatant was stored at –70°C until use for the ELISA and DIBA.

Immunoassay-Based Detection Kit

Indirect ELISA

Antisera have been produced against crude mycelial extract of *Ganoderma* in rabbit. A standard indirect ELISA method as described by Hobbs *et al.* (1987) was used with slight modifications. Microtitre plates (Tarson, India) were coated with 100 µl of samples or 2 h at 37°C and then incubated at 4°C overnight. The plates were emptied and washed three times with Phosphate Buffer Saline-Tween (PBS-T) (3 min each). The primary antibodies diluted in PBS-T (1:1000) containing 2 per cent polypvinylpyrrolidone and 0.2 per cent ovalbumin (PBS-TPO) were added (100 µl per well) separately. After incubation the plates (37°C, 2 h) were washed with PBS-T. Alkaline phosphatase (ALP) conjugated goat anti-rabbit immunoglobulin (Bangalore Genei, India) (1:6000 with PBS-TPO) were added separately (100 µl per well). The plates were incubated for 2 h at 37°C. After washing, 100 µl ALP substrate (1 mg/ml) solution 4-nitrophenyl phosphate (SD fine chemicals, India) dissolved in diethanolamine (Sigma, USA) (pH 9.8) was added after washing in PBS-T. The reaction was terminated by adding 50 µl of 3 M NaOH after incubation for half an hour at room temperature (28±2°C). The colour development was read at absorbance 405 nm with a Microplate reader (Bio Rad Model 3550, USA).

Dot Immunobinding Assay (DIBA)

DIBA assays were performed on a nitrocellulose membrane (NCM) (Sigma, USA) following the method of Hampton *et al.* (1990) with slight modification. Different plant tissue samples and antigen diluted in TBS buffer (0.02 M Tris, 0.5 M NaCl, pH 7.5) were spotted on to the membrane and, after drying, the loaded membrane was incubated in blocking solution (5 per cent spray-dried milk in TBS). After subsequent washings, the membrane was first incubated in Ganoderma antisera (1:1000) and then in enzyme conjugate (1:5000) (Sigma). The substrate solution contained 0.33 mg/ml nitroblue tetrazolium (Sigma) and 0.175 mg/ml 5-bromo 4-chloro, 3-indolyl phosphate (Sigma) in 0.1 M Tris, 0.1 M NaCl, 5 mM $MgCl_2.6H_2O$, pH 9.5 buffer. Immediately after colour development the membranes were washed in distilled water, dried and stored.

DNA Extraction for Cultured Fungi and Plant Samples

Template DNA was extracted from pure cultures of Ganoderma, few saprophytic organisms and from different plant parts of healthy and infected coconut palms and seedlings, obtained from the field and greenhouse as described by Moller *et al.* (1992) with slight modifications.

Molecular Based Detection Kit

Polymerase Chain Reaction

Primers used were derived from the ITS1 region of the rDNA of *G. boninense* (Moncalvo *et al.*, 1995c). A 167 bp PCR product was predicted based on the sequence data (EMBL accession number X78749) and the primers were designed as:

Gan1: 5'–TTG ACT GGG TTG TAG CTG–3'

Gan2: 5'–GCG TTA CAT CGC AAT ACA–3'.

The PCR was carried out with a Mastercycler gradient (Eppendorf, Germany) programmed at: 5 min. preheating at 95°C followed by 48 cycles consisting of denaturation at 94°C for 40 s, annealing at 52°C for 40 s and extension at 72°C for 45s with a final 12 min. extension at 72°C. The PCR products were analysed by electrophoresis on a 1.6 per cent agarose gel, visualized under UV light and photographed and documented with an AlphaImager (Alpha Innotech, California, USA).

RAPD-PCR Analysis

RAPD profiles were produced by the method of Williams *et al.* (1990). Primers and all the reagents were obtained from Bangalore Genei Pvt. Ltd., Bangalore, India. The primers used for molecular analyses are in Table 9.1. Amplification reactions were total volume of 20 µl and standardized as follows for taxonomic analysis of RAPDs: 0.5 to 5 ng of template DNA (in 10 1 of H_2O), 0.6 µM primer, 50 µM each dNTP, 2.5 mM $MgCl_2$, 50 mM KCl, 10 mM Tris-HCl (pH 8.3), and 0.5 U of Taq DNA polymerase. In each amplification reaction, a control sample without DNA was included. Samples were quickly transferred in Mastercycler gradient (Eppendorf, Germany) preheated at 96°C and incubated at this temperature for three min. to

Table 9.1: Primers Used for the RAPD Analysis

Primer Name	Sequence (5'–3')	Primer Type
R1	TGCCGAGCTG	RAPD
R2	AGTCAGCCAC	RAPD
R3	AATCGGGCTG	RAPD
Pgs2	GTTTCGCTCC	RAPD
Pgs3	GTAGACCCGT	RAPD
Pgs4	AAGAGCCCGT	RAPD
$(CAG)_5$	CAGCAGCAGCAGCAG	Micro-satellite
$(GACAC)_3$	GACACGACACGACAC	Micro-satellite
$(TGTC)_4$	TGTCTGTCTGTCTGTC	Micro-satellite
$(GACA)_4$	GACAGACAGACAGACA	Micro-satellite

denature the DNA completely. This was followed by 38 cycles of amplification consisting of 45 sec at 95°C to denature the DNA, 1.5 min. at 40°C to anneal the primers, and two min. at 72°C to extend the annealed primers. A final extension step of 10 min was programmed to ensure complete extension of the amplified products. The amplified fragments were analyzed by electrophoresis of 10 µl of the amplification reaction mixture in 2 per cent agarose gels run in 1X TBE buffer.

RAPD Data Analysis

The band patterns were scored for RAPD and microsatellite primers pair in each *Ganoderma* isolates starting from the small size fragment to large sized one. Presence and absence of each band in each isolate was coded as 1 and 0 respectively. The scores were used to create a data matrix to analyse genetic relationship using the NTSYS-pc program version 2.02 (Exeter Software, New York, USA) described by Rohlf (1990). A dendrogram was constructed based on Jaccard's similarity coefficient (Jaccard, 1908) using the marker data from 35 *Ganoderma* isolates with unweighted pair group method (UPGMA).

Results

Among the different Ganoderma isolates, CRS-1 isolate showed more growth compared to other isolates.

Immunoassay-Based Detection Kit

ELISA

The crude mycelial protein (CMP) antisera showed the values for diseased plant tissues and positive control ranged from 0.445 to 2.65, whereas for healthy root and leaf samples it ranged from 0.172 to 0.256. The highest value was found in positive control followed by artificially inoculated roots of coconut seedling. The same antisera was used to detect the infected field palm samples, which indicates high range for infected roots (0.9) followed by apparently healthy palm (0.84), infected trunk (0.62) and infected leaf (0.36) (Figure 9.1). The antisera developed against the sporophore extract gave positive results to the pathogenicity test especially for roots (0.97) and lower leaf (0.43) of artificially inoculated coconut seedlings. Similar results were obtained for infected field roots (0.798), apparently healthy root (0.648), infected trunk (0.816) and lesions (0.348).

DIBA

Immunoblotting assays have been found to be useful to overcome problems with non-specific interference in ELISA procedures. In the DIBA test, at a 1:10 dilution of antigen, 1:1000 dilution of CMP antisera, 1:5000 dilution of secondary antibody gave clear distinctions in colour development between healthy and diseased samples. In DIBA test, all the four types of antisera were used separately for the CMP antisera the positive control, infected field roots, infected roots from artificially inoculated seedling and infected basal trunk from field palms gave positive reactions whereas leaf and lesions were negative (Figure 9.2).

Molecular-Based Detection Kit

Polymerase Chain Reaction

Primers Gan1 and Gan2 amplified a DNA fragment of the size of 167 bp when additional *Ganoderma* isolates of other host resources were used. Howerver, primers designed for the diagnosis of *Ganoderma* in palms reacted also with other saprophytic fungi. The amplified products of the saprophytic fungi were different in DNA fragment size compared to *Ganoderma* (Figure 9.3).

The amplification of a 167 bp PCR product was achieved in pure *Ganoderma*, infected field roots, field sporophores, artificially inoculated coconut seedling roots and sporophore of infected seedlings. There was no amplification in leaf (lower, middle and upper), trunk and lesion samples (Figure 9.4).

RAPD Analysis

The RAPD marker data were performed by cluster analysis using NTSYS-pc program version 2.02 for all the isolates, to estimate similarity indices and genetic relatedness. The similarity index (SI) values were computed as a ratio of number of similar bands to the total number of bands in pairwise comparison of the isolates (Figure 9.5). The SI values obtained for each pairwise comparison of RAPD bands among the *Ganoderma* isolates. The genetic similarity index calculated from the DNA fingerprint data was used to estimate the genetic relatedness between the isolates. The SI was maximum of 75 per cent between CRS-2 and CRS-3 isolates and least SI 52 per cent was found in CBE-3 isolate to other isolates (Figure 9.6).

Ganoderma spp. are widespread polypore fungi causing basal stem rot or white rot of hardwoods, conifers and palms. In the natural forest, these fungi attack preferably old and declining trees and decay dead wood and stumps but severe diseases have been reported for plantation crops and forest trees (Turner, 1965; Peries, 1974; Bhaskaran *et al.*,1989; Sariah *et al.*, 1994; Lattiffah *et al.*, 2002). The type of decay caused by *Ganoderma* spp. is influenced by several parameters, including the *Ganoderma* spp. and type of wood, and can range from simultaneous decay of all wood components to selective delignification (Adaskaveg *et al.*, 1990). The potential importance of identifying these species have been reviewed (Jong and Birmingham, 1992; Adaskaveg and Gilbertson, 1995). Basal stem rot (BSR) in coconut is a serious disease in India and in severely infected areas, incidence as high as 80 per cent was recorded (Ramadoss, 1991). Based on assessing the severity of the disease, it was planned to detect at the early stage of infection in palms especially on coconut.

Over the past decade, advances in the fields of molecular plant pathology and immunology have greatly increased the accuracy, rapidity and sensitivity of nucleic acid and protein detection (Lamb *et al.*, 1992; Samiyappan *et al.*, 1996; Martin *et al.*, 2000). In the present studies, the work was mainly concentrated on the detection of *Ganoderma* at early stages of infection on palms by applying ELISA, DIBA and PCR technologies. Production of polyclonal antisera against the crude mycelial protein (CMP), extracellular protein (ECP), Monospecific protein (MP) (62 kDa) and sporophore extract (SE) paved the way for detection of pathogen at any form or stage in plant material. Numerous immunoassays have been developed for fungal pathogens

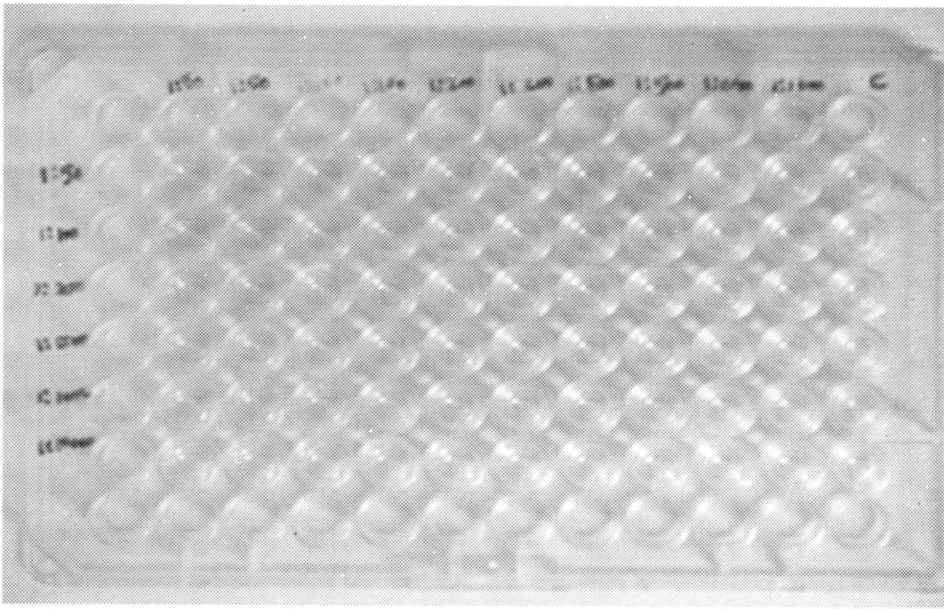

Figure 9.1: Standardization of Titre Value for Antiserum by Indirect ELISA

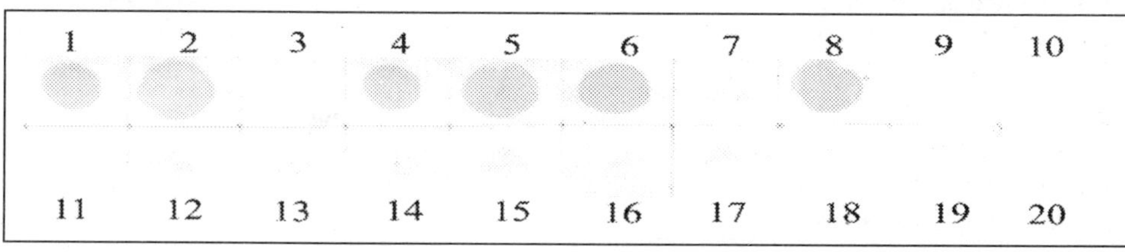

1. Crude protein from MTP isolate	11. Lower leaf from MTP inoculated seedling
2. Crude protein from CRS-1 isolate	12. Middle leaf from MTP inoculated seedling
3. Buffer control	13. Upper leaf from MTP inoculated seedling
4. MTP inoculated seedling roots	14. Lower leaf from CRS-1 inoculated seedling
5. CRS-1 inoculated seedling roots	15. Middle leaf from CRS-1 inoculated seedling
6. Infected field roots	16. Upper leaf from CRS-1 inoculated seedling
7. Healthy roots	17. Uninoculated seedling leaf
8. Infected trunk	18. Leaf from infected tree
9. Lesions	19. Healthy leaf from field
10. Healthy trunk	20. Buffer control

Figure 9.2: Evaluation of Antiserum through DIBA

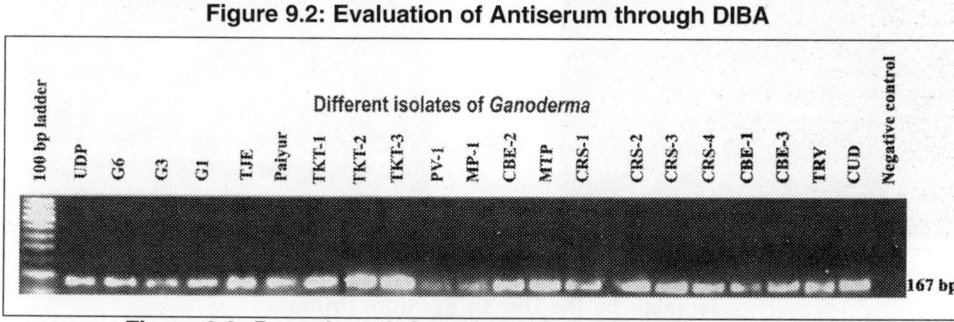

Figure 9.3: Detection of *Ganoderma* from Diseased Palm Roots with Primers Gan1 and Gan2

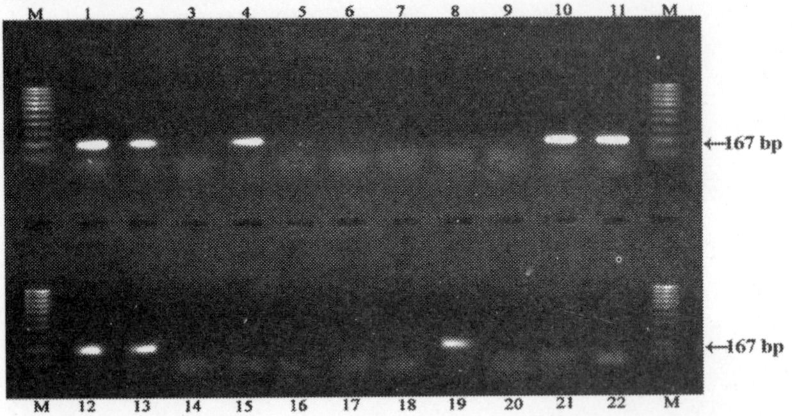

Lane 1. Positive control (MTP isolate)
Lane 2. MTP inoculated seedling roots
Lane 3. Roots from uninoculated seedling
Lane 4. Infected field roots
Lane 5. Lower leaf - MTP inoculated seedling
Lane 6. Middle leaf-MTP inoculated
Lane 7. Upper leaf - MTP inoculated seedling leaf
Lane 8. Uninoculated coconut seedling
Lane 9. Leaf from infected palm
Lane 10. Sporophore (MTP inoculated)
Lane 11. Field sporophore
Lane 12. Positive control (CRS-1 isolate)
Lane 13. CRS-1 inoculated seedling roots
Lane 14. Healthy trunk from field
Lane 15. Lower leaf-CRS-1 inoculated
Lane 16. Middle leaf-CRS-1 inoculated
Lane 17. Upper leaf-CRS-1 inoculated
Lane 18. Healthy leaf
Lane 19. Sporophore-CRS-1 inoculated
Lane 20. Infected basal trunk
Lane 21. Lesions (from field)
Lane 22. Negative control
Lane M. 100 bp DNA ladder

Figure 9.4: Detection of *Ganoderma* from the Artificially Inoculated Coconut Seedlings and Diseased Coconut Palm with Primers Gan1 and Gan2

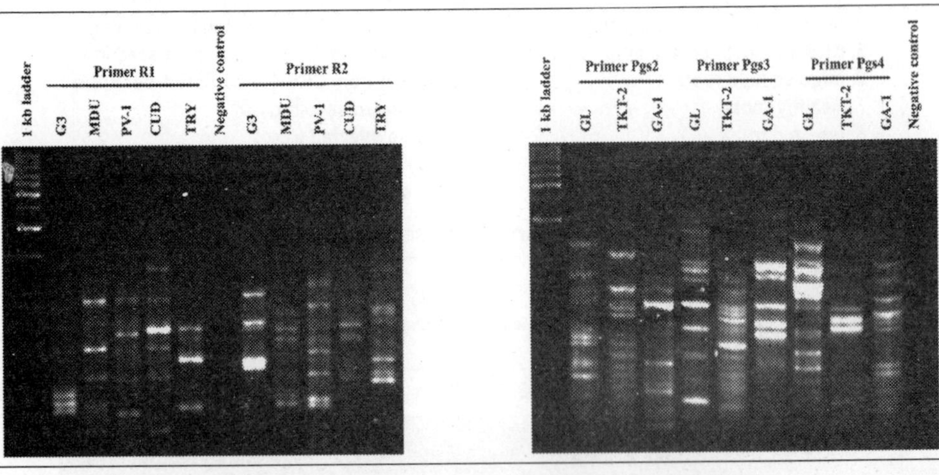

Figure 9.5: RAPD Fingerprinting of *Ganoderma* Isolates by the Use of Random Primers

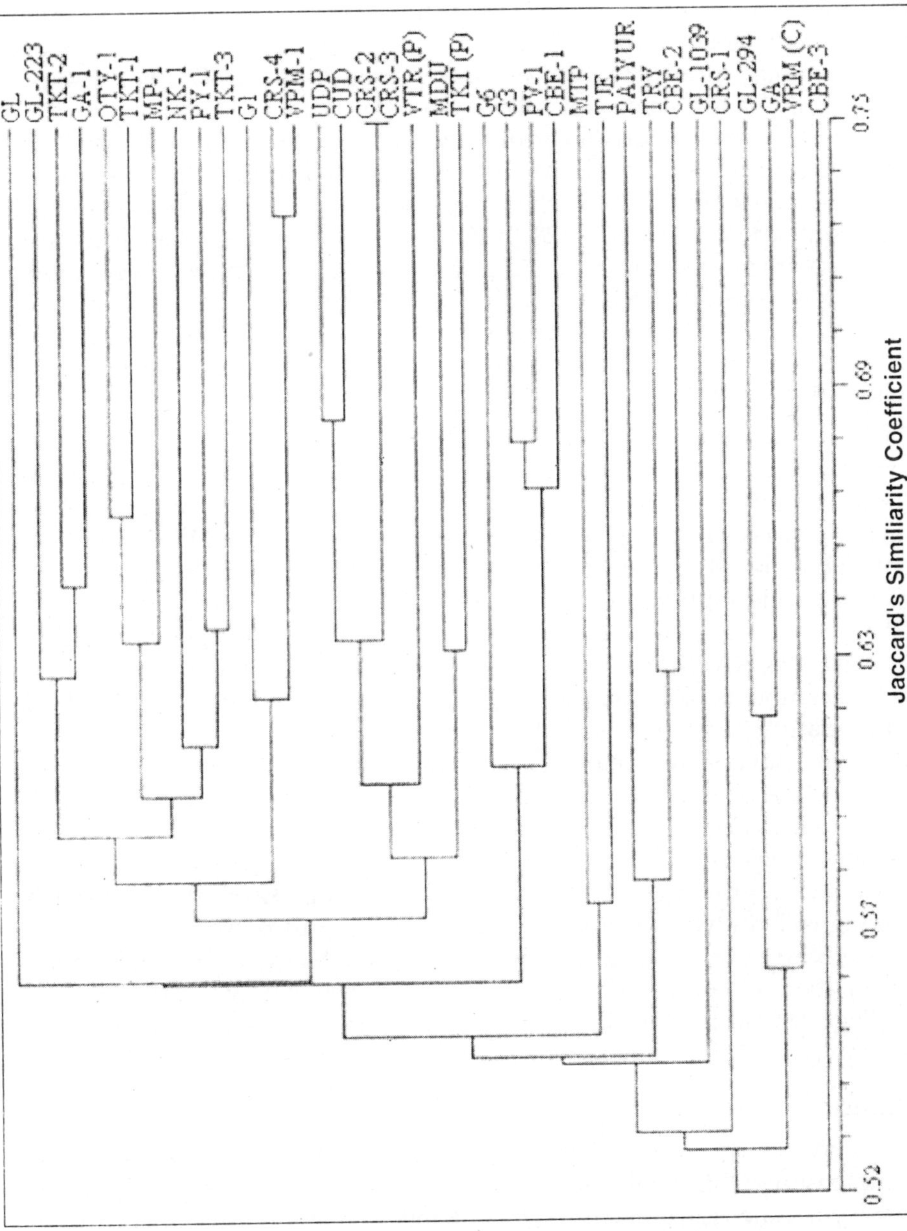

Figure 9.6: Unweighted Pair Group Average Method Dendrogram Based on Similarity Index for *Ganoderma* Isolates

by utilizing polyclonal antisera against whole cells (Kraft and Boge, 1994), crude mycelial extracts (Harrison *et al.*, 1990; Viswanathan *et al.*, 1998), extracellular culture filtrates (Kim *et al.*, 1991; Brill *et al.*, 1994), crude or partially purified soluble proteins (Velichetti *et al.*, 1993; Viswanathan *et al.*, 1998), resulting in varying degrees of specificity for the target fungus. In the present work, we tested the apparently healthy trunk (3 m from base) and lesions from infected field palm which also responded positive reaction against all polyclonal antiserum. Based on testing of all the antiserum, the MP, CMP, ECP and SE antisera ranks in ascending order to detect the *Ganoderma* from both field infected palms and artificially inoculated coconut seedlings. Using different types of antiserum, *Ganoderma* was detected from both infected and apparently healthy palm (palms without expression of any symptoms) by indirect ELISA was supported by Darmono (2000). DIBA is another serological assay to detect the pathogen with more sensitivity than indirect ELISA. In the current investigation, *Ganoderma* was positively detected from infected roots, trunk and lesions, however, negative results were obtained in leaves from artificially inoculated seedlings and infected field palm. This work is in coincidence with earlier work of Mitchell (1988) and Velichetti *et al.* (1993).

The PCR is a relatively new technique (Saiki *et al.*, 1985) that has gained broad acceptance very quickly in many areas of science. For plant pathogenic fungi, specific primers have most often been based on sequences of genes encoding rRNA. These genes have been relatively well studied in fungi (Appels and Honeycutt, 1986), and contain both highly conserved coding regions and variable ITS regions. In the present studies, the application of the two Gan1 and Gan2 primers generated from the ITS1 sequence proved to be useful for the specific detection of plant pathogenic *Ganoderma* isolates. RAPD-PCR is much simpler and faster to carry out, and does not require large amounts of highly purified DNA or the use of radioisotopes (Black, 1993). It is relatively easy means of identifying fungi at the sub-specific level, and several systems have been developed for potential use in pathogen indexing. In the present work, divergent RAPD patterns occurred among isolates from the same region, and even from the same field and host species. The UPGMA cluster analysis grouped the 35 isolates in seven clusters: A to G, with five sub-cluster A_1 to A_5 in cluster A. The slow and moderate growth habits of *Ganoderma* were grouped in cluster A and all the fast growing isolates were grouped in cluster B to G. This present studies are in accordance with several successful works for establishment of phylogenetic relationships between isolates of a morphological species of fungal pathogens such as *Verticillium* spp. (Koike *et al.*, 1996), *Gaeumannomyces graminis* (sacc.) Von Arx and Olivier (Bryan *et al.*, 1999), *F. oxysporum* Schlecht (Paavanen-Huhtala *et al.*, 1999).

Conclusion

The present study showed that, it is possible to detect the *Ganoderma* spp. by ELISA and sensitive DIBA methods in palm tissue, and that sensitivity and specificity of *Ganoderma* detection can be increased by PCR method. Among different plant parts standardized roots alone have been found to be best suited for detection of *Ganoderma* using both molecular and serological kit, but basal stem part is suited only for the serological basis of *Ganoderma* detection. Therefore, both ELISA and DIBA tests may be useful for screening a large number of samples and in case of a positive reaction, PCR assays should be used for confirmation. In future, provision of an immunoassay and molecular based detection kits will help in the detection of infection at the earliest stage of disease development and this will certainly help to adopt suitable management strategies against *Ganoderma* disease in palm crops in advance.

References

Adaskaveg, J. E. and Gilbertson, R. L. (1989). Cultural studies of four North American species in the *Ganoderma lucidum* complex with comparisons to *G. lucidum* and *G. tsugae*. *Mycol. Res.*, 92: 182-191.

Adaskaveg, J. E. and Gilbertson, R. L. (1995). Wood decay caused by *Ganoderma* species in the G. lucidum complex. In: *Ganoderma*–systematics, phytopathology and pharmacology (Eds.) P. K. Buchanan, R. S. Hseu and J. M. Moncalvo, Proceedings of Contributed Symposium 59 A. B. 5th International Mycological Congress, Vancouver, Canada, pp. 79-93.

Adaskaveg, J. E., Gilbertson, R. L. and Blanchette, R. A. (1990). Comparative studies of delignification caused by *Ganoderma* species. *Appl. Environ. Microbiol.* 56: 1932-1943.

All India Final Estimate of Coconut (2002-03). *Directorate of Economics and Statistics*, Ministry of Agriculture, Government of India.

Appels, N. and Honeycutt, R. L. (1986). rDNA: Evolution over a billion years. In: DNA systematics Vol. 2. (Ed.) K. Dutta, CRC Press, Boca Raton, FL, pp. 81-135.

Bhaskaran, R., Rajamannar, M and Kumar, S.N.S. (1996). Basal stem rot disease of coconut. *Technical Bulletin No.30*, Central Plantation Crops Research Institute, Kasaragod, pp.15.

Bhaskaran, R., Rethinam, P. and Nambiar, K. K. N. (1989). Thanjavur wilt of coconut. *J. Plant. Crops.*, 17: 69-79.

Black, W. C. (1993). PCR with arbitrary primers: approach with care. *Insect Mol. Bio.* 2: 1-6.

Brill, L. M., McClary, R. D. and Sinclair, J. D. (1994). Analysis of two ELISA formats and antigen preparations using polyclonal antibodies to Phomopsis longiicolla. *Phytopathology*, 84: 173-179.

Bryan, G. T., Labourdette, E., Melton, R. E., Nicholson, P., Daniels, M. J. and Osburn, A. E. (1999). DNA polymorphic and host range in take-all fungus Gaeumannomyces graminis. *Mycol. Res.* 103: 319-327.

Butler, E. J. (1913). Report of the Imperial Mycologist. *Rept. Agric. Res. Inst. College*, PUSA, 60: 1911-1912.

Darmono, T. W. (2000). Ganoderma in oil palm in Indonesia: Current status and prospective use of antibodies for the detection of infection. In: *Ganoderma* diseases of perennial crops (Eds.) J. Flood, P. D. Bridge and M. Holderness, CAB International, UK, pp. 249-266.

Hampton, R., Ball, E. and De Boer, S. (1990). Serological methods for detection and identification of viral and bacterial plant pathogen, APS Press, pp. 389.

Harrison, J. G., Barker, H., Lowe, R. and Rees, E. A. (1990). Estimation of amounts of Phytophthora infestans mycelium leaf tissue by enzyme-linked immunosorbent assay. *Plant Pathol.* 39: 274-277.

Henson, J. M. and French, R. (1993). The polymerase chain reaction and plant disease diagnosis. *Ann. Rev. Phytopathol.* 31: 81-109.

Hobbs, H. A., Reddy, D. V. R., Rajeswari, R. and Reddy, A. S. (1987). Use of antigen coating and protein A coating ELISA procedures for detection of three peanut viruses. *Plant Dis.* 71: 747-749.

Jaccard, P. (1908). Nouvelles recherches sur la distribution florale. *Bull Sco V and Sci. Nat.*, 44: 223-270.

Kandan, A. (2003). Biotechnological approaches for early detection of *Ganoderma* diseases in Plantation crops. Ph. D. thesis, Tamil Nadu Agrl. Univ Coimbatore. P. 181.

Karthikeyan, G., Karunanithi, K., Rabindran, R., Karthikeyan, A., Natarajan, C. and Aruralraj, S. (2005a). Incidence of basal stem rot (BSR) disease of coconut in coastal and inland districts of Tamil Nadu. *UGC sponsored National Seminar on Emerging Trends in Plant Pathology and their Social Relevance (ETPPSR)*, March 7–8, 2005, Annamalai University, Annamalai Nagar (TN). p. 175.

Karthikeyan, M, Radhika, K., Bhaskaran, R., Mathiyazhagan, S., Samiyappan. and Velazhahan, R. (2007). Pathogenicity confirmation of *Ganoderma* disease of coconut using early diagnosis technique. *J. Phytopathology.* 155: 296–304.

Kim, Y. S., Jellison, J., Goodell, B., Tracy, V. and Chandhoke, V. (1991). The use of ELISA for the detection of white-and brown-rot fungi. *Holzforschung*, 45: 403-406.

Koike, M., Fujita, M., Nagao, H. and Ohshima, S. (1996). Random amplified polymorphic DNA analysis of Japanese isolates of Verticillium dahliae and V. albo-atrum. *Plant Dis.* 80: 1224-1227.

Kraft, J. M. and Boge, W. L. (1994). Development of an antiserum to quantify Aphanomyces euteiches in resistant pea lines. *Plant Dis.* 78: 179-183.

Lamb, C. J., Ryals, J. A., Ward, E. R. and Dixon, R. A. (1992). Emerging strategies for enhancing crop resistance to microbial pathogens. *BioTechnology.* 10: 1436-1445.

Lattifah, Z., Harikrishna, K., Tan, S. G., Tan, S. H., Abdullah, F. and Ho, Y. W. (2002). Restriction analysis and sequencing of the ITS regions and 5.8 S gene of rDNA of *Ganoderma* isolates from infected oil palm and coconut stumps in Malaysia. *Ann. Appl. Biol.* 141: 133-142.

Martin, R. R., James, D. and Levesque, C. A. (2000). Impacts of molecular diagnostic technologies on plant disease management. Ann. Rev. *Phytopathol.* 38: 207-239.

Moller, E. M., Bahnweg, G., Sandermann, H. and Geiger, H. H. (1992). A simple and efficient protocol for isolation of high molecular weight DNA from filamentous fungi, fruit bodies, and infected plant tissues. *Nucleic Acids Res.*, 20 (22): 6115-6116.

Moncalvo, J. M., Wang, H. F. and Hseu, R. S. (1995c). Phylogenetic relationships in *Ganoderma* inferred from the internal transcribed spacers and 25S ribosomal DNA sequences. *Mycologia*, 87: 223-238.

Nambiar, K.K.N. and Rethinam, P. (1986). Thanjavur wilt/*Ganoderma* disease of coconut. *Pamphlet No. 30.* Central Plantation Crops Research Institute, Kasaragod.

Natarajan, S., Bhaskaran, R. and Shanmugam, N. (1986). Preliminary studies to develop techniques for early detection of Thanjavur wilt in coconut. *Indian Coconut J.* 17: 3-6.

Paavanen-Huhtala, S., Hyvonen, J., Bulat, S. A. and Yli-Mattilia, T. (1999). RAPD-PCR, isozyme, rDNA, RFLP and rDNA sequence analysis in identification of Finnish Fusarium oxysporum isolates. *Mycol. Res.* 103: 625-634.

Peries, O. S. (1974). *Ganoderma* basal stem rot of coconut: a new record of the disease in Sri Lanka. *Plant Dis. Rept.* 58: 293-295.

Rajendran, L. (2006). Biotechnological tools and methods for early detection and sustainable management of basal stem rot disease in coconut plantations using microbial consortia. Ph. D. thesis, Tamil Nadu Agrl. Univ., Coimbatore.

Ramadoss, N. (1991). Studies on the epidemiology, pathophysiology and management of thanjavur wilt of coconut. Ph. D. thesis, Tamil Nadu Agrl. Univ. Coimbatore.

Rohlf, F. J. (1990). NTSYS-pc Numerical taxonomy and multivariate analysis system. Exeter software, New York.

Saiki, R. K. Scharf, S., Faloona, F., Mullis, K. B., Horn, G. T., Erlich, H. A. and Arnheim, N. (1985). Enzymatic amplification of β-globin genomic sequences and restriction site analysis for diagnosis of sickle cell anemia. *Science.* 230: 1350-1354.

Samiyappan, R., Bhaskaran, R. and Rethinam, P. (1996). Diagnosis for early detection of *Ganoderma* diseases in perennial crops: approaches and prospects. *J. Plant Dis. Prot.*, 103 (1): 85-93.

Samiyappan, R., Karthikeyan, G., Kandan, A., Raguchander, T. and Rajendran, L. (2006). Basal stem rot disease in coconut–recent developments. *Indian coconut journal*, XXXVII 8: 2-16.

Sariah, M., Hussin, M. Z., Miller, R. N. G. and Holderness, M. (1994). Pathogenicity of *Ganoderma boninese* tested by inoculation of oil palm seedlings. *Plant Pathol.*, 43: 507-510.

Srinivasulu, B., Aruna, K. and Rao, D.V.R. (2001). Biocontrol of *Ganoderma* wilt of coconut palm. National Seminar on changing scenario in production system of Hort. Crops. *South Indian Hort.* 49: 240-241.

Turner, P. D. (1965). Infection of oil palms by *Ganoderma*. *Phytopathology*. 55: 937.

Turner, P. D. (1981). Oil Palm Diseases and Disorders. Oxford Unviersity Press, Oxford. pp. 88-110.

Utomo, C. and Niepold, F. (2000). Development of diagnostic methods for detecting *Ganoderma* infected oilpalms. *J. Phytopathol.* 148: 507-514.

Velichetti, R. K., Lamison, C., Brill, L. M. and Sinclair, J. B. (1993). Immunodetection of Phomopsis species in asymptomatic soybean plants. *Plant Dis*, 77: 70-77.

Vijayan, K. M. and Natarajan, S. (1972). Some observations on the coconut wilt disease of Tamil Nadu. *Coconut Bull.* 2 (12): 2-4.

Vijayaraghavan, H., Ramadoss, N., Ramanathan, T. and Rethinam, P. (1987). Effect of toddy tapping on Thanjavur wilt disease of coconut. *Indian Coconut J.* 17 (12): 3-5.

Viswanathan, R., Samiyappan, R. and Padmanaban, P. (1998). Specific detection of Colletotrichum falcatum in sugarcane by serological techniques. *Sugarcane*, 6: 18-23.

Williams, J. G., Kubelik, A. R., Kenneth, J. L., Rafalski, J. A. and Scott, V. T. (1990). DNA polymorphisms amplified by arbitrary primers are useful as genetic markers. *Nucleic Acids Res.* 18: 6531-6535.

Wilson, K.I., Rajan, K.M., Nair, M.C. and Balakrishnan, S. (1987). *Ganoderma* disease of coconut in Kerala. *International Symposium on Ganoderma wilt diseases on Palms and other Perennial Crops.* Tamil Nadu Agrl. Univ., Coimbatore (Abstr.) pp 4-5.

Chapter 10

Novel Approaches for the Detection of Plant Viruses

S.S. Kang[1] and Abhishek K. Sharma[2]***

[1]Molecular Diagnostic Laboratory for Plant Pathogens, Department of Plant Pathology, Punjab Agricultural University, Ludhiana – 141 004, Punjab India

[2]Department of Vegetable Crops, Punjab Agricultural University, Ludhiana – 141 004, Punjab, India

ABSTRACT

The nature and diversity of plant viruses makes it difficult to use a general method for their detection and characterization. Many crops can thus be infected by unidentified viruses, greatly limiting the possibilities for proper diagnosis and controlling these agents. Detection of harmful viruses in plant material, vector or natural reservoirs is essential to ensure sustainable agriculture. It deals with establishing the presence of a particular target organism within a sample, with special emphasis on symptomless individual. Accurate disease detection requires high level of specificity, sensitivity and speed. Plant virus research has been classified into three eras *i.e.*"Classical Discovery Period" from 1883–1951, "Early Molecular Era" from 1952–1983 and "Recent Period" from 1983 onwards (Zaitlin and Palukaitis, 2000). During early molecular era and recent period, detection techniques available for plant viruses have evolved significantly. With the advancement of highly sensitive detection instruments and nano technology, it is

E-mail: *kang9632@hotmail.com/kang9632@rediffmail.com; **abi_1975@rediffmail.com/pauvirus@gmail.com.

now possible to carry out virus detection from samples camouflaged by a high number of microorganisms, and with low virus titer.

Keywords: *ELISA, DAS-ELISA, DAC-ELISA, PAC-ELISA, DIBA, ISEM, I-CZE, PCR and RT-PCR*

Introduction

Currently, the detection of phytopathogenic viruses is a changing, dynamic and evolving world. The term novel approach is used, because in the past, protocols were established and used routinely for many years, which is not the case today. There is tendency to use polyphasic molecular techniques. Now a day, techniques can be modified or optimized only months after having been developed. Each of these techniques has its own advantages and limitations. Presently, the novel detection techniques can be broadly classified into two categories, serological assay and nucleic acid based assay. In this chapter a brief introduction with protocols, wherever required into each method is given, together with pertinent examples of each.

Serological Assay

Serology is the study of antisera, especially their reaction and properties. In this technique property of a particular type of serum called antiserum (a serum containing antibodies) is used to detect plant pathogens especially plant viruses in unknown samples. The antibodies are produced by purifying a particular plant virus and injecting the purified virus (antigen) into an animal, usually a rabbit. The animal's immune system will then be stimulated to produce antibodies corresponding to the proteins that make up the structure of the virus. After several injections, the blood of the animal is removed (whole serum) and purified to acquire antiserum. This is the fraction containing antibodies that are specific to the viral antigen injected. The antigen/antibody specificity lies in the heart of all detection assays because it allows the antibody to combine to the virus when it is present. As it is the property of the antiserum and its reactions and properties that form the basis of detection, all the techniques for the detection of viruses that use antisera are called serological techniques or immunochemical techniques.

Enzyme Linked Immunosorbent Assay (ELISA)

Among the major developments that have taken place over the past 25 years, it is obvious that the enzyme linked immunosorbent assay (ELISA) is the most significant advent especially in virus detection. Though this is not a new technique but due to its sensitivity and simplicity today it is the most widely used technique, throughout the world.

There are many variants of ELISA *viz.*, DAC (Direct Antigen-Coating), DAS (Double Antibody Sandwich), PAC (Protein A-Coating), DIBA (Dot Immunobinding Assay) but the most common is DAS- ELISA. The technique utilizes the ability of antibodies raised in animal to recognize the antigen (virus) usually the coat protein of the virus of interest. Antibodies (coating) to the solid surface of a well within polystyrene microtitre plate and sap extract from the sample plants are added to the well. If the virus of interest is present in the sample, it will bind to the antibodies fixed on the surface. Any unbound extract is washed-off before a secondary antibody that recognizes the first antibody is added. The secondary antibody (conjugate antibody) is allowed for indirect detection of virus because it has an enzyme (Alkaline phosphatase) linked to it, which acts on substrates to give yellow colour. It is then detected visually by an ELISA reader/spectrophotometer at A_{405nm}.

The detailed protocol of different variants of ELISA are discussed below:

DAS-ELISA

- ☆ Dilute the primary antibody IgG with 0.05M carbonate buffer (coating buffer) at pH 9.6 to obtain a final concentration of 6µg/ml of buffer. Add 200 µl of antibody solution to each well in polystyrene ELISA plates and incubate at 37°C for 4 hours. Empty the wells, wash thrice with 0.15M phosphate buffered saline solution at pH 7.2 containing 0.05 percent Tween 20 (wash buffer PBS-Tween).
- ☆ Add 200 µl samples in well (purified antigens or extracts of infected tissues in PBS-Tween, pH 7.4), incubate at 4°C overnight and wash the wells thrice as described earlier.
- ☆ Add aliquots of 200 µl of enzyme- labeled antibody conjugate dissolved in conjugate buffer (PBS-TPO) to each well; incubate for 4 hours at 37°C and wash the wells.
- ☆ Add enzyme substrate (p-nitrophenyl phosphate) at a concentration of 1 mg/ml in diethanolamine buffer (substrate buffer) pH 9.8 at room temperature (20±5°C). Stop the reaction after 30 minutes by adding 3M NaOH at 50 µl/well.
- ☆ Determine the colour intensity (OD) at 405 nm in an ELISA reader.

DAC-ELISA

In this, direct antigen coating (DAC) technique, plant extracts prepared in a carbonate buffer are applied directly to the wells. In the second step, diluted unfractionated antiserum is added. Ig attached to virus antigen is detected by the addition of enzyme-Ig conjugates. This is the simplest procedure and can be completed within three hours.

- ☆ Add samples @ 200 µl to each well in the ELISA plates; incubate at 37°C for 4 hour and wash the wells thrice with PBS-Tween.
- ☆ Add antiserum at optimum dilution @ 200 µl/well; incubate for 4 hour at 37°C and wash the wells thrice with PBS-Tween.
- ☆ Add enzyme-labelled antirabbit IgG (conjugate antibody) @ 200 µl to each well; incubate for 4 hour at 37°C and wash the wells with PBS-Tween.
- ☆ Follow last two steps of DAS-ELISA.

PAC-ELISA

This protein-A coating (PAC) technique involves four steps. The wells are coated with protein-A (1-10 mg/ml), high dilution of unprocessed antisera are added, test samples are added and finally antisera (cross absorbed with healthy plant components, to minimize non-specific reaction) were added. This is followed by addition of the enzyme substrate for colour reaction.

- ☆ Dissolve protein A (1-10 mg/ml) in carbonate buffer; dispense 200 µl/well in ELISA plate; incubate for 4 hr at 37°C and wash thrice with PBS-Tween.
- ☆ Dispense antiserum (at optimum dilution) at 200 µl/well; incubate for 4 hr at 37°C and wash thrice with PBS-Tween.
- ☆ Dispense 200 µl of samples (purified antigen/extracts of tissues at optimum dilution); incubate at 37°C for 4 hr and wash the wells thrice with PBS-Tween.
- ☆ Dispense enzyme-labeled antirabbit IgG 200 µl/well; incubate for 1 hr at 37°C and wash with PBS-Tween.
- ☆ Follow last two steps of DAS-ELISA.

DIBA/On Site Tissue Printing

Tissue blotting like ELISA, utilizes antibodies raised against viruses sap from the plant tissue is expressed on to blotting paper; nitrocellulose or nylon membrane and the virus is detected by labeled probes often chemiluminescent. The procedure is less labour intensive than ELISA, rapid, sensitive, simple and inexpensive. The detailed protocol is given as under:

- ☆ Dip a nitrocellulose membrane grid (1.0 or 2.5 cm squares) in Tris buffer (0.02 M Tris-CL, 0.5 M NaCl, pH 7.5) (TBS) and dry it on filter paper for 5 minutes.
- ☆ Spot the samples (1µl) prepared in TBS in the center of the grid and dry. Place the grid in blocking solution (3 per cent gelatin, 2 per cent Triton X-100 in TBS) in a Petri dish and agitate for 1 hour.
- ☆ Dip the grid into distilled water; transfer to 50 ml antiserum. (1 mg protein/ml) and 1 per cent gelatin in a Petri dish and agitate for 1 hour.
- ☆ Dip the grid into distilled water; then wash twice by agitation for 10 minutes in TBS containing 0.05 per cent Tween- 20 (TTBS).
- ☆ Dip the grid into distilled water; transfer to horseradish peroxide conjugated 1gG (1/1000 dilution) and 1 per cent gelatin kept in a Petri dish; agitate for 1 hour.
- ☆ Dip the grid into distilled water; wash in TTBS and TBS successfully for 10 minutes.
- ☆ Transfer to substrate solution (dissolve 0.06 g 4-chloro-l-naphthol in 20 ml of 4-C-methyl alcohol, then add 100 ml of TBS and 0.06 ml of 30 per cent hydrogen peroxide); incubate for 10-30 minutes in dark.

Immunosorbent Electron Microscopy (ISEM)

Immunosorbent Electron Microscopy (ISEM), developed by Derrick (1973), is a valuable technique which combines the specificity of serological test with the possibility to visualize the type of viral antigen in the electron microscope. The technique not only helps in the detection of virus particles from mixed infection but also facilitates virus identification.

The method makes use of the trapping of antiserum specific virus particles onto EM grids that have been coated by the specific antiserum. Samples, that are serologically related to the antibodies on the coated grids get concentrated on the grids enabling the detection of a particular virus in samples.

To examine a virus in solution with the electron microscope, the sample is mounted on a grid covered with a thin film of plastic for instance formvar. These grids are copper discs of 3mm diameter, containing a number of apertures (150 to 400 meshes/inch). The virus particles stick to the formvar, the contrast can be improved by shadow casting or negative staining, containing low titre of virus. The antigen antibody binding is visualized after negative staining of the complex.

Potassium phosphotungstate or uranyl acetate is used as negative stains which when added to a viral solution, stain the background. Since these stains contain heavy metal which gets deposited around the virus particle and produce contrast, that's why the screen of the EM displays the virus particle lighter and surrounded by a darker background.

Procedures for Grid Preparation

Leaf Dip Method

In this method a drop of stain is placed on a carbon coated grid. Then an epidermal peel from infected leaf is placed on the drop of stain on grid, after a minute excess fluid from the grid is removed with filter paper and examined under EM.

Leaf Squash and Extraction Method

Small amount of tissues are macerated along with 2-3 drops of stain between two slides. The tissue extract is then examined after loading the extract on grids.

Immuno-Gold Decoration Technique

The ISEM technique has been further modified by incorporating various decoration methods in particular gold conjugated antibodies. The sample is placed on a carbon coated grid and treated with a primary antibody. Prior to primary antibody the nonspecific binding sites are blocked by a blocking agent like BSA(Bovine Serum Albumin). The sample on the grid is reacted with a secondary conjugated with 10 to 15 nm gold particles.

Trapping and Decoration

- ☆ Pipette a drop of crude antisera (approximately 20µl) on parafilm membrane in a moist Petri dish and float a grid with coated side on the drop. Incubate the Petri dish for approximately 30-60 minutes at room temperature (20-22°C).
- ☆ Wash the antiserum-coated grid in phosphate buffer (0.01M, pH 7.8) for 10 min).
- ☆ Pipette a drop (20µl) of extract from infected tissue and float the grid on it. Incubate for 30 minutes to 2 hour.

- ☆ Wash the grid with buffer for 10 min.
- ☆ Pipette a drop (20µl) of 1:50 dilution of antiserum and place the grid on it. Incubate at 37°C for 30-60 minutes.
- ☆ Wash the grid with H_2O followed by staining with 2 per cent aqueous uranyl acetate. Drain and dry the grid before observations under electron microscope.
- ☆ Observe the number of virus particles trapped as compared to the leaf-dip method and their decoration by the antibodies.

Immune Capillary Zone Electrophoresis (I-CZE)/Flow Cytometry

Flow cytometry is a technique for rapid detection of cells or other particles as they pass individually through a sensor in liquid medium. This is an analytical technique first developed by Mikkers *et al.* (1979). The modification of this technique is immune capillary zone electrophoresis, which involves the application of voltage across a hollow capillary filled with uniform electrolyte solution. Separations in CZE are achieved as a result of the unequal rates of migration of different component in a sample under the influence of an externally applied electric field via the combined action of electrophoretic migration flow. The inner walls of capillary are negatively charged under most pH conditions, resulting in buildup positive counter-ions in the buffer solutions adjacent to the capillary wall. When an electric current passes, the layer of positive charge is drawn towards the cathode, resulting in the bulk flow of buffer towards it. Protein molecules are separated based on their net charge-to-mass ratio.

I-CZE is a technique that combines the specificity afforded by serological assay with the sensitive, rapid and fully automated detection capability of CZE. It provides a mean of detecting the antigen-antibody complex in real time. This combination of antibody, which confers specificity, and CZE which provides enhanced sensitivity, speed and automation, offers a new technique for plant virus detection. The application of I-CZE is achieved to detect CymMV and ORSV in crude saps of infected orchids (Eun and Wong, 1999).

Lateral Flow Device/Spot Check Strips

Simple commercial methods for rapid detection are required for testing large numbers of samples by non experienced technicians. For this purpose lateral flow devices has been designed for several plant viruses. The detection specificity is very high along with rapidity. These are based on existing technology similar to pregnancy test kit. The specific antibodies are used and they give results in few minutes. The kit contains a strip coated with specific antibody on it along with a positive control. For preparation of extract small plastic container filled with extraction buffer and beads are provided for extraction by shaking. A drop of samples sap is placed on the strip after extraction, which is allowed to flow laterally and bands are developed. By visualizing these bands the positive and negative samples are decided (Figure 10.1).

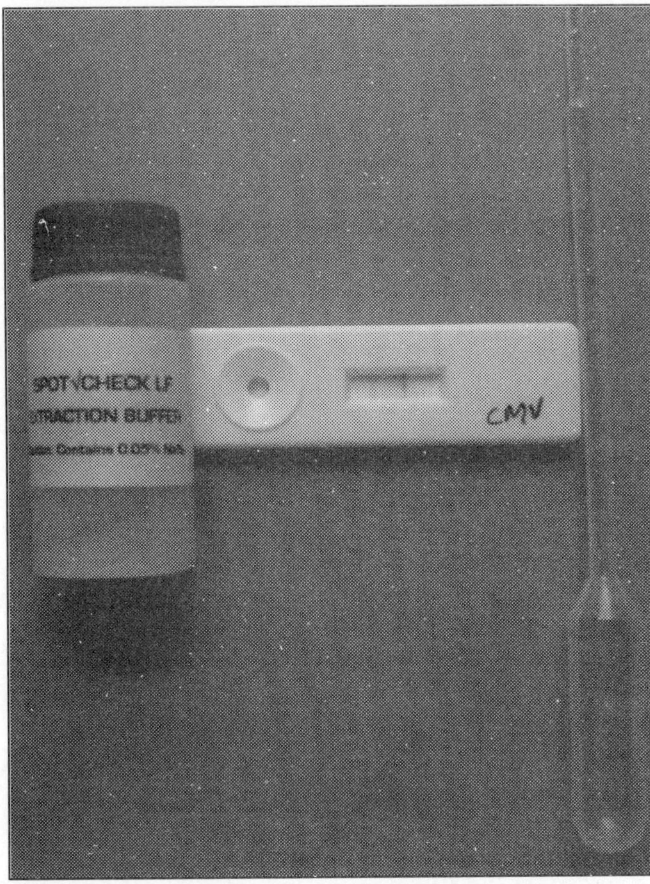

Figure 10.1: Lateral Flow Device/Spot Check Strips

The major constraint of all these methods is the requirement of polyclonal or monoclonal antisera specific for each virus of interest that does not cross react with the plant proteins.

Nucleic Acid Based Assay

With the rapid development of molecular techniques, it is now possible to obtain substantial information about viral nucleic acids. Such nucleic acid based techniques have proved to be very suitable for detection and identification of plant viruses. The critical step in nucleic acid based detection is the quality of nucleic acid to be amplified. Very long, complicated and time consuming protocols developed for nucleic acid extraction in the 1990s, have been replaced by rapid, simple extraction protocols. The commercial kits are available for both RNA and DNA extraction, which are more sensitive, efficient and have been used successfully for different type of plant material. Other parameters that also control the efficiency of nucleic acid based detection methods are polymerase type, buffer composition and stability, purity and composition of dNTPs as well as cycling parameters.

The most common protocols for total nucleic acid, DNA and RNA extractions are given below:

Extraction of Viral Nucleic Acid

If pure viral nucleic acid is necessary for analysis, it is essential to first purify the virus particles from the infected tissue. The procedure used will differ for each virus type. Usually, the tissue is homogenized. The final purification stage is usually density gradient purification in Cesium chloride. Once homogeneous virus particles are obtained, nucleic acid can be separated from viral proteins by conventional techniques. If PCR is to be used to characterize or isolate the viral nucleic acid, it is not necessary to purify the virus. A crude nucleic acid preparation from infected material will contain sufficient viral nucleic acid required for PCR.

Commercial suppliers are now preparing kits with the required reagents and recommended procedures for nucleic acid extraction from tissue. Procedures commonly followed for nucleic acid extraction are mentioned below:

DNA Extraction

- Collect the samples and store them as frozen pellets at –70°C. Alternatively, freshly collected tissue can be quickly frozen in liquid nitrogen.
- Freeze tissue by dropping into liquid nitrogen. Grind to a fine powder in a sterilized mortar that has been pre-cooled with liquid nitrogen. Add fresh liquid nitrogen during grinding if the liquid evaporates completely. Alternatively, use dry ice if liquid nitrogen is not available.
- Transfer the ground, frozen tissue into a Corning tube containing the appropriate volume of lysis buffer *i.e.* CTAB (Cetyl trimethyl ammonium bromide) buffer.
- Make up 500 µg/ml Proteinase K in NTE(100mM NaCl+50mM Tris HCl +20mM EDTA+ 1 per cent SDS adjusted to pH 7.5). Add 2ml proteinase K solution/gm tissue and incubate at 65°C.
- Check digestion periodically, agitate well but gently. The mixture becomes viscous in less than 60 minutes.
- Add 2ml buffered phenol, mix gently, and spin for 5 min at room temperature.
- Transfer aqueous (upper) phase carefully to new tube and repeat the phenol extraction twice.
- Avoid transferring the phenol phase or interface when removing the aqueous phase.
- Add 2.5 volumes of ice cold 100 per cent ethanol to the pooled aqueous phase and mix gently.
- Spool DNA by gently stirring the solution with a sealed glass Pasteur pipette.
- Dissolve DNA in 400 µl TE (10mM EDTA+10 mM Tris HCl at pH 7.8) in microfuge tube.

- Add RNase @ 20 µg/ml and incubate at 37°C for 30 min.
- Add 0.5 ml buffered phenol, mix gently, and spin for 5 min at room temperature.
- Transfer the aqueous phase to a new tube and repeat the phenol extraction.
- Add 2.5 volumes ethanol and mix the solution gently.
- If possible, spool the DNA as described above, or if not, pellet the DNA by spinning.
- Redissolve the DNA in 200 µl TE(10mM EDTA+10 mM Tris HCl at pH 7.8).
- Read A_{260}/A_{280}. The ratio should be about 1.8-2.0.

Precautions

- Don't put the solution on ice or SDS otherwise, it will precipitate.
- Allow the DNA to air dry, but don't allow the DNA to dry completely otherwise it won't re-suspend.

RNA extraction

- Freeze tissue by dropping into liquid nitrogen. Grind to a fine powder in a sterilized mortar that has been pre-cooled with liquid nitrogen. Add fresh liquid nitrogen during grinding if it evaporates completely. Alternatively, use dry ice if liquid nitrogen is not available.
- Transfer powder to a Falcon tube. Add an equal volume of 2X LETS(200mM LiCl+20mM EDTA+20mM Tris HCl+2 per cent SDS adjusted to pH 7.8) (*i.e.*, equal to the original vol. of tissue). Add an equal volume of phenol: chloroform 1:1 (*i.e.*, twice the original volume of tissue).
- Vortex and shake until powder is completely thawed.
- Separate phases in a clinical centrifuge by spinning at 3000 rpm for about 5-10 minutes at room temperature.
- Extract with phenol:chloroform 1:1 for atleast three times until the interface is almost completely clear after spinning.
- Precipitate the total RNA by adding LiCl to 0.2 M and 3 volumes of 100 per cent ETOH, thoroughly mix, and then keep the mixture at –20°C for at least 30 minutes.
- Collect the precipitate by centrifugation in a clinical centrifuge at 3000 rpm for 10-15 minutes at room temperature.
- Rinse the pellet several times with 70 per cent ETOH and re-suspend in DEPC (diethyl pyrocarbonate) treated water or TE.
- Read OD A_{260}/A_{280} nm on an appropriate dilution of the RNA solution; A_{260}/A_{280} nm ratio of 2.0 implies a fairly pure preparation of RNA, if desired.
- Store RNA as an precipitate at –20°C.

Precautions

- The trace amount of DNA contamination can be removed with DNAse.

- ☆ Extract RNA in RNAse are sufficient to destroy RNA.
- ☆ The use of sterile, disposable polypropylene tubes is recommended throughout the procedure.
- ☆ Non-disposable glassware/plastic wares should be treated with 0.1 M NaOH, 1mM EDTA followed by RNAase-free water (0.1 per cent DEPC)

PCR and RT-PCR

The polymerase chain reaction (PCR) technique has developed to detect very small quantities of nucleic acid by amplification of a segment of the DNA situated between two regions of known nucleotide sequence. These segments are generally conserved sequences. Such a segment is flanked by two oligonucleotides which serve as primers for a series of reactions that are catalyzed by an enzyme DNA Polymerase in the presence of buffer and dNTPs. An amplification cycle consists of melting of the double-stranded template DNA at high temperature (denaturation at 94°C), hybridization of the primers with complementary base sequences in the two strands at lower temperature (annealing at 54-60°C depends on GC content of primers) and extension of primers with DNA polymerase (DNA synthesis at 72°C). In this way, at every cycle of high and low temperature, the sequence between the primers will be doubled, so that after n cycles, a 2^n amplification may have been obtained (usually 30-35 cycles of amplification are applied).

To design a primer for particular virus information must be available on the nucleotide sequence of the target virus. The information on nucleotide sequence is available on World Wide Web of *ncbi* (USA) or *embl* (Germany). Now software and online programmes are available to design the primers such as, Primer 3, GeneFisher, PCR Now etc. The selection of primers depends on target sequence. In general, primers should be between 18 to 25 base pair (bp or mer) in length.

PCR is useful in detecting DNA viruses. The technique has been efficiently used in detecting geminivirus even from a single whitefly vector (Deng *et al.*, 1994).

PCR cannot be applied directly to most of the plant viruses as their genome consists of single strand RNA. In such cases, the RNA has to be transcribed first into DNA by reverse transcriptase (RT) enzyme. Therefore the technique involved is known as RT-PCR. The commercial kits for RT-PCR are available which transcribe the RNA into DNA and then coupled with PCR. Different RT-PCR variants have been developed, including immunocapture RT-PCR(IC-RT-PCR), which has been used with plant extracts, or immobilized targets on paper print (PC-RT-PCR) or squash capture (SC-RT-PCR), allowing the detection of minimal quantities of RNA targets from plant material or insect vectors without extract preparation. In these variants, the virus particles are captured by antibodies on a surface and then removed by heating with a non-ionic surfactant such as Triton X-100. The genome of virus is then amplified using RT-PCR. This method is useful in concentrating virus particles from plant samples where virus titre is low or where plant compounds inhibiting PCR are present. Zitikaite and Staniulis, (2006) detected three viruses *viz.*, cucumber mosaic virus (CMV), Tomato ring spot virus (ToRSV) and Tobacco necrosis virus (TNV) from cucumber by RT-PCR.

Nested PCR

When the sensitivity of detection is not good enough, a nested PCR can be helpful. In this method, two rounds of PCR's are carried out, with the first reaction, increasing the amount of template for the second. The method is particularly useful where the virus has very low titre or inhibitors of DNA polymerase are present in the plant extract. Low specificity oligonucleotides, usually degenerate, are used in the first round of amplification. Then an aliquot of the reaction is placed into a fresh tube for a second PCR with primers that anneal within the first amplicon. This at once increases the target molecule and dilutes inhibitors. Earlier the reactions were used to carry out in two different eppendorf tubes, resulting in a high contamination risk. Now this method has been modified and a single compartmentalized eppedorf tube has successfully been used to detect members of *Vitivirus* and *Foveavirus* species in grapes (Dovas and Katis, 2003).

Multiplex PCR

Multiplex PCR allows the simultaneous detection of different RNA and DNA targets in a single reaction. This is useful in plant virology because different RNA viruses frequently infect a single host. Consequently, sensitive detection is needed for the propagation of virus free plant material. The technique has been successfully used in detecting six viruses belonging to four different genera in olive trees (Bertoloini *et al.*, 2001). The only technical difficulty is designing accurate primers for one step RT-PCR amplification in a single closed tube.

Thompson *et al.* (2003) detected four aphid-borne viruses infecting Strawberry (*Fragaria* spp.) *Strawberry crinkle virus* (SCV), *Strawberry mild yellow edge virus* (SMYEV), *Strawberry mottle virus* (SMoV) and *Strawberry vein banding virus* (SVBV) using multiplex reverse transcriptase polymerase chain reaction (RT-PCR).The new variant of multiplex PCR is nested-multiplex PCR which combines the advantages of multiplex RT-PCR with the sensitivity and reliability of nested RT-PCR carried out in a single closed tube. This is a time saving and economically cheap technique as it can be performed in a single reaction, although accurate design of compatible primers is needed. The technique has been used for simultaneous detection of several RNA viruses and bacterial (*Pseudomonas savastanoi* pv *savastanoi)* DNA targets in a single analysis of olive woody plants (Bertolini *et al.*, 2003).

The compartmentalization of single eppendorf tube with a pipette tip allowed multiplex and nested PCR to be combined effectively.

During the first amplification reaction, there is no interference of the external with internal primers because they are physically separated from the initial reaction cocktail. Once the multiplex RT-PCR ends, the internal primers are mixed with the product of first reaction before proceedings to the nested multiplex. Because the concentration of internal primers is very high as compared with that of the external primers (which will also have been consumed by the first amplification), the nested multiplex can be performed with minimal interference.

Real Time PCR/Quantitative PCR/Molecular Beacons

A novel fluorescence-based nucleic acid detection technique was developed by Tyagi and Kramer in 1996. Real-Time chemistries allow the detection of PCR amplification during the early phases of the reaction. Measuring the kinetics of the reaction in the early phases of PCR provides a distinct advantage over traditional PCR detection. Here the probes, which are termed as molecular beacons, are used. Traditional methods use Agarose gel for detection of PCR amplification at the final phase or end-point of the PCR reaction. This facility is available with the authors at Punjab Agricultural University, Ludhiana in the department of plant pathology.

Limitations of End-Point PCR

Agarose gel results are obtained from the end-point of the reaction. End-point detection is very time consuming. Results may not be obtained for days. Results are based on size discrimination, which may not be very precise. As observed later in the section, the end-point varies from sample to sample,while gels may not be able to resolve these variabilities in yield, real-time PCR is sensitive enough to detect these changes. Agarose gel resolution is very poor, about 10 fold. Real-Time PCR can detect as little as a two-fold change.

Some of the limitations with End-Point Detection:

- ✰ Poor Precision
- ✰ Low sensitivity
- ✰ Narrow dynamic range < 2 logs
- ✰ Low resolution
- ✰ Non-automated
- ✰ Size-based discrimination only
- ✰ Results are not expressed as numbers
- ✰ Ethidium bromide for staining is not very quantitative
- ✰ Post PCR processing

As the PCR reaction progresses, the samples begin to amplify in a very precise manner. Amplification occurs exponentially, that is a doubling of product (amplicon) occurs every cycle. This type of amplification occurs in the exponential phase. Exponential amplification occurs because all of the reagents are fresh and available, the kinetics of the reaction push the reaction to favor doubling of amplicon. Real-Time PCR detects the accumulation of amplicon during the reaction. The data is then measured at the exponential phase of the PCR reaction. Traditional PCR methods use Agarose gels or other previous PCR detection methods, which are not as precise. As mentioned earlier, the exponential phase is the optimal point for analyzing data. Real-Time PCR makes quantification of DNA and RNA easier and more precise than previous methods. Molecular beacons are hair-pin shaped oligonucleotides containing a fluorophore on one end and a quenching dye on the opposite end. Under conditions that prevent the oligonucleotide from hybridizing to its complementary target, the fluorescent and quenching dye are proximal to one another, thus preventing fluorescence resonance energy transfer (FRET). Once hybridization occurs, the loop structure is converted to a more rigid conformation causing separation of the fluorophore and quencher leading to fluorescence. For quantitative PCR, molecular beacons bind to the amplified target following each cycle of amplification and the resulting signal is proportional to the amount of template. As with other real time PCR formats, the specific reaction conditions must be optimized for each primer/ probe set to ensure accuracy and precision.

DNA Microarray

DNA microarray or biochips are made of a surface of glass, membrane or different polymer. On these surfaces ssDNA probes are irreversibly fixed as an array of discrete spot of < 150 micron size. Array printed with probes, corresponding to a large number of virus species, can be utilized to simultaneously detect all those viruses within the tissue of an infected plant. Once arrayed, the chip can be exposed to fluorescently labeled DNA/RNA from the sample to be tested. The viral nucleic acids are extracted from the host, reverse-transcribed and amplified where appropriate, then labeled with a fluorescent probe or radioactive probe during the RT reaction. The labeled target molecule is denatured and allowed to hybridize with the arrayed probes. This detection system uses one or several fluorophores, that can be read with laser technology to reveal the targets present in the sample.

Acknowledgement

Authors are highly thankful to the following National Institutes for extending help during the compilation of this chapter. Central Potato Research Institute, Shimla (CPRI), Institute of Himalayan Bio resource Technology, Palampur (IHBT), Indian Institute of Science Bangalore (IISc), Indian Agriculture Research Institute, (IARI), New Delhi, National Botanical Research Institute, Lucknow (NBRI), National Bureau of Plant Genetic Resources, New Delhi (NBPGR) and Punjab Agricultural University, Ludhiana (PAU)

References

Bertolini, E.O., lmos, A., Matinez, M.C., Gorris, M.T. and Cambra, M.(2001). Single step multiplex RT-PCR for simultaneous and colourimetric detection of six RNA viruses in olive trees. *J. Virol. Methods*.96: 33-41.

Bertolini, E.O., lmos, A., Lopez, M.M. and Cambra, M. (2003). Multiplex nested reverse-transcription polymerase chain reaction in a single tube for sensitive and simultaneous detection of four RNA viruses and *Pseudomonas savastanoi* pv *savastanoi* in olive trees. *Phytopathology*. 93: 286-292.

Deng, D. Mcgrath, P. F., Robinson, D. J. and Harrison, B.D. (1994). Detection and differentiation of whitefly-transmitted geminiviruses in plants and vector insects by the polymerase chain reaction with degenerate primers. *Annals App. Biol*.125: 327-336.

Dovas, C.I. and Katis, N. I. (2003). A spot nested RT-PCR method for the simultaneous detection of members of the *Vitivirus* and *Foveavirus* genera in grapevine. *J Virol. Methods*.107: 99-106.

Eun, A.J.C. and Wong, S.M.(1999). Detection of cymbidium mosaic potexvirus and odontoglossum ringspot tobamovirus using immuno-cappilary electrophoresis. *Phytopathology*. 89: 522-528.

Eun, A. J. C. and Wong, S. M. (2000). Molecular Beacons: A new approach to plant virus detection. *Phytopathology*. 90: 269-275.

Lopez, M. M., Bertolini, E., Olmos, A., Caruso, P., Gorris, M.T., Llop, P., Penyalver, R. and Cambra, M. (2003). Innovative tools for detection of plant pathogenic viruses and bacteria. *Int Microbiol*. 6: 233-43.

Mikker, F.E.P., Everrates, F.M. and Verheggen, T. P. E. M. (1979). High performance electrophoresis. *J Chromtogr*. 169: 11-20.

Thompson, J. R., Wetzel, S., Klerks, M. M. , Vaš ková, D., Schoen, C. D., Špak, J. and Jelkmann, W. (2003). Multiplex RT-PCR detection of four aphid-borne strawberry viruses in *Fragaria* spp. in combination with a plant mRNA specific internal control. *J Virol Methods*. 111: 85-93.

Webster, C.G., Wylie, S.J. and Jones, M.G.K. (2004). Diagnosis of plant viral pathogens. *Curr Sci*. 86: 1604-1607.

Zaitlin, M. and Palukaitis, P. (2000). Advances in understanding plant viruses and virus diseases. *Annu. Rev. Phytopathol*. 38: 117-43.

Zitikaite, I. and Staniulis, J. (2006). The use of RT-PCR for detection of viruses infecting cucumber. *Agron. Res*. 4: 471-74.

Chapter 11

Molecular Characterization of Formae Speciales of *Fusarium oxysporum*: A Review

V.C. Khilare[1] and L.V. Gangawane[2]*

[1]Department of Botany, Vasantrao Naik Mahavidyalaya, Aurangabad – 431 003, India

[2]Soil Microbiology and Pesticides Laboratory, Department of Botany, Dr. Babasaheb Ambedkar Marathwada University, Aurangabad – 431 004, India

ABSTRACT

Fusarium oxysporum is significant, ubiquitous complex plant pathogen, which affects to many hosts in many types of habitats. It is an anamorphic species that includes both pathogenic and nonpathogenic strains causes wilt diseases and are grouped into formae speciales. Many formae speciales are comprised of multiple clonal lineages and, in some cases, a pathogenic race is associated with more than one clonal lineage, suggesting independent origins. Genetic diversity studies on formae speciales of *F. oxysporum* are important for relationship between host and pathogen. The host specific forms of *F. oxysporum* are traditionally based on the combination of diagnostic symptoms on the host. Recent studies by RAPDs, RFLPs or DNA fingerprints, gene-genealogy and AFLP-based phylogenies show that the majority of formae speciales in *F. oxysporum* are polyphyletic (unnatural) and do not offer any prospects for the development of molecular diagnostics. In contrast, highly specific PCR primers have been developed for *formae speciales* (or

* E-mail: vikramkhilare@gmail.com

races) that consist of a single clonal lineage, and for monophyletic groups of lineages within a formae specialis. The attempt has made to summarize some of the findings of genetic diversity studies on formae speciales of *F. oxysporum*.

Keywords: *Fusarium oxysporum, RAPD, RFLP, AFLP.*

Introduction

Molecular characterization helps in rapid and reliable detection and identification of potential plant pathogens for taking appropriate and timely disease management measures. For many microbial species of which all strains generally are plant pathogens on a known host range, this has become quite straightforward. However, for some fungal species this is quite a challenge. One of these is *F. oxysporum* which, as a species, has a very broad host range, while individual strains are usually highly host-specific. Moreover, many strains of this fungus are non-pathogenic soil inhabitants. Thus, with regard to effective disease management, identification below the species level is highly desirable. So far, the genetic basis of host specificity in *F. oxysporum* is poorly understood. Furthermore, strains that infect a particular plant species are not necessarily more closely related to each other than to strains that infect other hosts.

The population of fungus is an important aspect for studying the population genetics of fungal plant pathogens. When population has single species becomes easy to identify but amongst population it becomes difficult when it is host specific. This problem is compounded for *F. oxysporum* Schlechtend.:Fr. and other anamorphic fungi, for which even clear delineation of species is lacking. Genetic diversity studies on *F. oxysporum* are beginning to define populations and address the question of the species concept for this important plant-pathogenic fungus. *F. oxysporum* shows species and sub-specific categories which is a common, widespread fungus found in soil. Although the species has been defined by morphology of asexual reproductive structures, considerable variation occurs in these features. The species was placed in the section (*Gruppe*) *Elegans* by Wollenweber and Reinking (1935) along with nine other species in three subsections (*Untergruppen*). Nonetheless the morphological divisions within section *Elegans* are small, and the features themselves highly variable and subject to environmental influence (Nelson, 1991). For this reason, Snyder and Hanson (1940) collapsed section *Elegans* into a single species: *F. oxysporum*. However, these authors also recognized true variants within the species, at least with respect to host plant species specialization. Although many or most isolates of the fungus may be nonpathogenic soil inhabitants (Appel and Gordon, 1994) the concept of form or specialized form (forma specialis) arose to delineate morphologically similar or indistinguishable isolates having the ability to cause disease on different plants.

Formae Speciales

F. oxysporum has received considerable attention from plant pathologists over the past 80 years because of its ability to cause vascular wilt or root rot diseases on a wide range of plants. Despite the broad host range of the species as a whole, host specialization of individual isolates is more circumscribed. Isolates with the same or similar host ranges are assigned to a forma specialis. More than 150 host-specific formae speciales have been described in the *F. oxysporum* complex (Baayen *et al.*, 2000). More often than not, host range is restricted to a few plant species. For example, although many plants may be symptomless carriers of the fungus (Katan, 1971), *F. oxysporum* f. sp. *lycopersici* causes disease only in plants of the tomato genus *Lycopersicon* (Rowe, 1980). However, some formae speciales have broader host ranges, such as *F. oxysporum* f. sp. *radicis-lycopersici*, which at least in greenhouse studies can cause disease on hosts from several other plant families as well as tomato (Rowe, 1980; Menzies, *et al.*, 1990). The categorization of strains by host range may or may not lead to a natural subdivision within the species. The assumption often made is that isolates with shared host range, and thus within the same forma specialis, are more similar genetically than isolates with other host specificities. The evolutionary interpretation resulting from this assumption is that formae speciales are monophyletic and that isolates with a shared host range are likely derived from a single, particularly successful pathogenic genotype. This underlying assumption now can be tested with genetic markers independent of pathogenicity.

Genetic Diversity Studies on Formae Speciales of *F. oxysporum*

Other important problems being addressed by genetic diversity studies are the degree of genetic diversity that exists among subspecific categories and whether there is a correlation between pathogenic phenotype and genotype. Such a correlation might be expected based on the assumptions that (i) formae speciales are monophyletic and (ii) *F. oxysporum* is a completely clonally reproducing microorganism. However, when pathotypes (either forma specialis or race) have been considered, there often has been no clear-cut association between these and genotypes determined by RFLPs, RAPDs, or DNA fingerprints.

Genetic diversity studies have been conducted for a large number of formae speciales and are presented in Table 11.1. Most studies have focused on genetic variation within a single forma specialis and have adopted a phylogenetic approach to analyze diversity. This approach is useful for clonally reproducing organisms in which lineage-specific mutations occur and are maintained without recombination into other genetic backgrounds. The lineages of the pathogen and their ancestry, as a result, can be represented accurately by phylogenetic trees. The examples were listed in Table 11.1 focus on single formae speciales, so the relationship among isolates in different formae speciales usually has not been addressed.

Molecular Characterization: A Review

This approach probably was taken because the researchers often were plant pathologists interested in a single disease or crop. Undoubtedly, this type of analysis also was spurred by early reports that isolates within a forma specialis were highly uniform and could be easily distinguished based on VCG and RFLP data. Few examples of studies on genetic diversity of formae speciales of *F. oxysporum* are mentioned in Table 11.1. Many of these early reports, however, sampled only limited numbers of isolates and few genetic loci (Kistler, 1997). Kim *et al.* (1993) scanned 39 isolates of *F. oxysporum* encompassing five formae speciales causing vascular wilt in cucurbits were examined for genetic similarity RFLP analysis of mitochondrial DNA (mtDNA). Genetic distances generated by UPGMA suggest that *F. o. niveum* was the least diverse forma specialis, while *F. o. cucumerinum* was the most diverse. However, both cluster analysis and parsimony analysis indicated that all of *F. oxysporum* formae speciales in the cucurbits are closely related and, in some cases, isolates of different formae speciales were genetically more similar than isolates of the same forma specialis. Genetic variability within four races of *F. oxysporum* f.sp. *pisi* was assessed by RAPD (Figure 11.1). Races 1, 5 and 6 showed much greater variability within each race than isolate of race 2. Similarities and differences in banding pattern obtained by RAPD could be useful in evolutionary studies of the origins of different races (Grajal-Martin *et al.*, 1993).

Table 11.1: Studies on Genetic Diversity on Formae Speciales of *F. oxysporum*

Formae Specialis	*Method of Analysis*	*References*
Albedinis	RFLP, RAPD	Fernandez and Tantaoui (1994), Fernandez, *et al.* (1995)
Ciceris	RFLP	Perez-Artes, *et al.* (1995)
Conglutinans	FP, RFLP	Bosland and Williams (1987), Hirota, *et al.* (1992), Kistler and Benny (1989)
Cubense	FP, RFLP,RAPD	Koenig, *et al.* (1993), Miao (1990)
Cucumerinum	FP	Namiki, *et al.* (1994)
Cyclaminis	FP, RFLP	Woudt, *et al.* (1995)
Dianthi	FP, RAPD	Manicom and Baayen (1993), Manulis, *et al.* (1994)
Elaeidis	RFLP	Flood, *et al.* (1992)
Gladioli	FP, RAPD	Mes, *et al.* (1994),
Lycopersici	RAPD, RFLP	Elias, *et al.* (1993)
Melonis	RFLP	Jacobson and Gordon (1990)
Niveum	FP, RFLP	Kim, *et al.* (1992),
Pisi	FP, RAPD	Coddington, *et al.* (1987), Grajal-Martín, *et al.* (1993)
Radicis-lycopersici	RFLP	Kistler *et al.* (1987)
Raphani	FP, RFLP	Kistler *et al.* (1987)
Vasinfectum	RAPD, RFLP	Fernandez, *et al.* (1994)

Namiki *et al.* (1994) find out the genetic relatedness of five formae speciales of *F. oxysporum* causing wilts of cucurbit plants was determined by DNA fingerprinting with the moderately repetitive DNA sequences FOLR1 to FOLR4. The fingerprint types detected in each of the formae speciales *cucumerinum, lagenariae, niveum,* and *momordicae* were grouped into a single cluster. However, two different genetic groups occurred in the formae specialis *melonis.* The two groups also differed in pathogenicity: one group caused wilts of muskmelon and oriental melon, while the second was

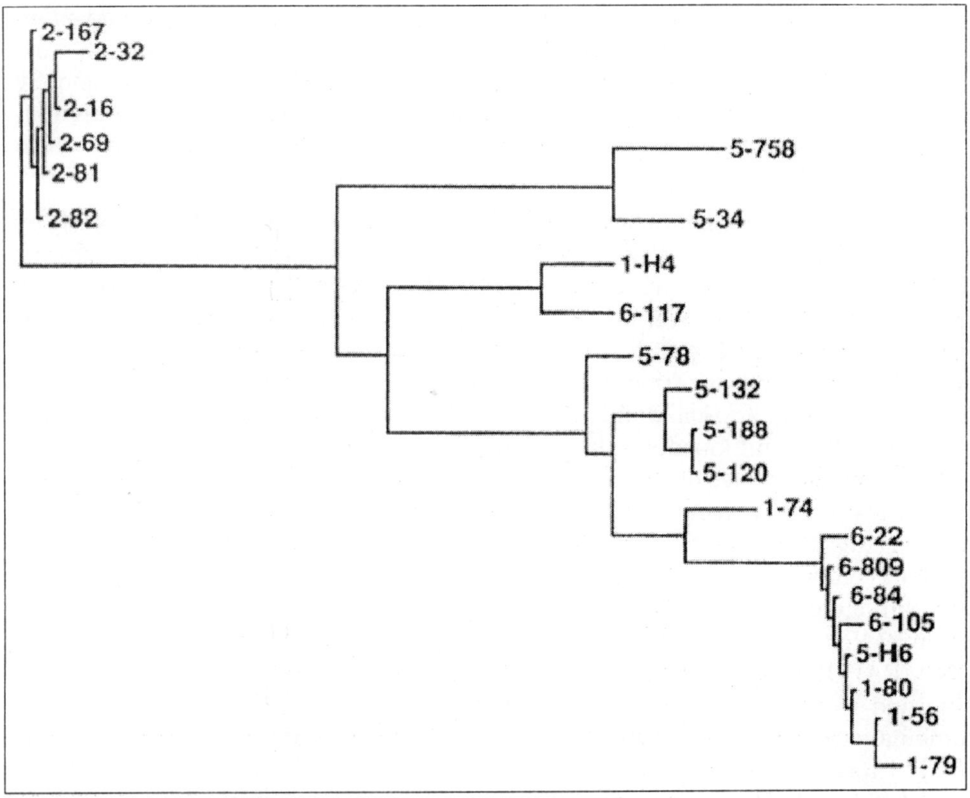

Figure 11.1: Relationship Among *F. oxysporum* f.sp. *pisi* Isolates by RAPD

pathogenic only to muskmelon. The fingerprint types of different formae speciales pathogenic to plants other than cucurbits were distinguishable from one another and from the fingerprints of the cucurbit-infecting strains. The cucurbit-infecting formae speciales are intra-specific variants distinguishable at the DNA level and in their host range. Two distinct clusters were observed which were correlated between two pathotypes *i.e.* yellowing and wilting when amplified for 63 isolates of *F. oxysporum* f.sp. *ciceris* by DNA RAPD technique (Kelly, *et al.*, 1994). Manulis *et al.* (1994) used the random amplified polymorphic DNA (RAPD) to distinguish among pathogenic and nonpathogenic isolates of *F. oxysporum* recovered from carnation. 58 isolates, which were isolated in Israel from different cultivars of carnation, included 42 pathogenic isolates of race 2, one pathogenic isolate of race 4, and 15 nonpathogenic isolates. Compared with other methods of identifying *F. o.* f. sp. *dianthi*, the RAPD procedure is simple, rapid, and reproducible. Assigbetse *et al.* (1994) assessed 46 isolates of *F. oxysporum* f.sp. *vasinfectum* for their genetic diversity. Isolates clustered into three groups corresponding in their pathological reactions. Tantaouti *et al.* (1996) studied 42 isolates of *F. oxysporum* f.sp. *albedinis albedinis* causing Bayoud disease of date palm in Morocco, and 2 isolates from Algeria were also included. All isolates were tested for genetic variability by RFLP and RAPD. Both nuclear and mitochondrial genome were examined. Results showed that no polymorphism was observed either with RAPD or RFLP. Cluster analysis showed that most of the *F. oxysporum* f. sp. *albedinis* isolates were grouped at a small genetic distance. The genetic homogeneity supports the hypothesis that Moroccan populations of *F. oxysporum* f. sp. *albedinis* may belong to a single clonal lineage that originated in Moroccan palm groves, which then reached the Algerian oases.

Woo *et al.* (1996) characterized the isolates of *F. oxysporum* f.sp. *phaseoli* by pathogenicity, vegetative compatibility, restriction fragment length polymorphisms (RFLPs), and random amplified polymorphic DNA (RAPD) analyses. The use of all four methods combined was effective in characterizing and providing insight into the complex relationship of *F. oxysporum* f. sp. *phaseoli*. Schilling *et al.* (1996) were developed PCR amplification of sequence-characterized amplified regions for *F. culmorum, F. graminearum,* and *F. avenaceum* helps in studies of fungal population genetics. Appel and Gordon (1996) showed relationship among pathogenic and nonpathogenic isolates of 15 *F. oxysporum* by IGS of the rDNA and one isolate of *F. sub-glutinans*. The objective of this research was to clarify the origin of virulence within *F. oxysporum*, the relationship between pathogenic and nonpathogenic strains, and the evolution of the different races of *F. oxysporum* f. sp. *melonis*. The nonpathogens and pathogens may share common alleles. Kerenyi *et al.* (1997) studied variability amongst Vegetative compatibility tests and random amplification of polymorphic DNA (RAPD) were used to assess genetic relationships amongst 54 strains of *F. poae* obtained from various geographical regions. The data demonstrate that the combination of traditional and molecular methodologies allows reliable intraspecific subdivisions in an asexual fungus, which is a secondary invader of a wide range of host plants, and so has never been subject to the intense selection pressure of a single host species and lacks pathogenic subgroups. Nelson *et al.* (1997) elicited the genetic complexity amongst 200 isolates of *F. oxysporum* f. sp. *erythroxyli* causing vascular wilt on *Erythroxylum coca* var. *coca* in the coca-growing regions of the Huallaga Valley in Peru. Random amplified polymorphic DNA (RAPD) analysis of isolates of the pathogen was undertaken to elucidate its genetic complexity, as well as to identify a specific DNA fingerprint for the pathogen. The pathogens could be grouped into two subpopulations based on RAPD analysis, and no polymorphism in RAPD pattern was observed among isolates of either subpopulation. Fernandez *et al.* (1997) analyzed microevolution by genetic diversity of the date palm wilt pathogen *F. oxysporum* f. sp. *albedinis* in Algeria. No variation was detected in the mtDNA of a subset of 73 isolates and the RAPD analysis indicated that they were genetically very closely related. The results provide evidence that the Algerian isolates of *F. oxysporum* f. sp. *albedinis* belong to a same clonal lineage and support the hypothesis that they were probably founded by a single virulent clone that originated from the Moroccan oases where the date palm wilt (Bayoud disease) was first detected. Bentley *et al.* (1998) analyzed the genetic variation within a worldwide collection of 208 isolates of *F. oxysporum* f. sp. *cubense*. The genetic diversity and geographic distribution of several of these lineages of *F. oxysporum* f. sp. *cubense* suggests that they have co-evolved with edible bananas and their wild diploid progenitors in Asia. DNA fingerprinting analysis of isolates from the wild pathosystem provides further evidence for the co-evolution hypothesis.

The genetic isolation and limited geographic distribution of four of the lineages of *F. oxysporum* f. sp. *cubense* suggests that the pathogen has also arisen independently, both within and outside of the center of origin of the host (Figure 11.2).

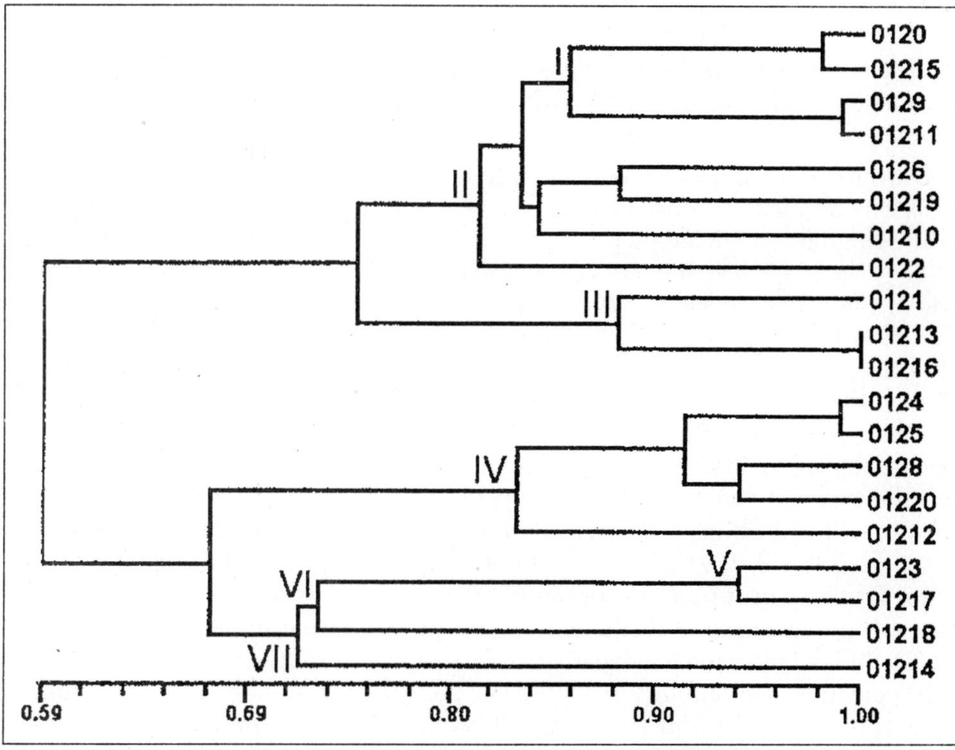

Figure 11.2: Genetic Similarity Between the Reported Vegetative Compatibility Groups (VCGs) of *F. oxysporum* f. sp. *cubense*

Namiki *et al.* (1998) examined pathogenic variation among 41 Japanese strains of *F. oxysporum* f. sp. *melonis*. Genetic variation among strains was analyzed by DNA fingerprinting with four repetitive DNA sequences: FOLR1 to FOLR4. Cluster analysis showed distinct genetic groups correlated with races: the fingerprint types detected in each of races 2 and 1,2y were grouped into a single cluster, and two distinct genetic groups were found in race 0. However, pathogenic variation detected within races 0 and 2 could not be differentiated based on the nuclear markers examined. Mes *et al.* (1999) characterized *F. oxysporum* f. sp. *lycopersici* by random amplified polymorphic DNA (RAPD) analysis to establish the identity and genetic diversity of the isolates. Comparison of RAPD profiles revealed two main groups that coincide with vegetative compatibility groups (VCGs). In addition, several single-member VCGs were identified that could not be grouped in one of the two main RAPD clusters. This suggests that *F. oxysporum* f. sp. *lycopersici* is a polyphyletic taxon. Vakalounakis and Fragkiadakis (1999) studied 106 isolates of *F. oxysporum* obtained from diseased cucumber plants showing typical root and stem rot or Fusarium wilt symptoms were characterized by random amplified polymorphic DNA (RAPD). Of the 106 isolates of *F. oxysporum* from cucumber, 68 were identified by pathogenicity as *F. oxysporum* f. sp. *radicis-cucumerinum*, 32 as *F. oxysporum* f. sp. *cucumerinum*, and 6 were avirulent on cucumber. All 68 isolates of *F. oxysporum* f. sp. *radicis-cucumerinum* belonged to a single RAPD group. A total of 32 isolates of *F. oxysporum* f. sp. *cucumerinum* was assigned to two different RAPD groups. RAPD were effective in distinguishing isolates of *F. oxysporum* f. sp. *radicis-cucumerinum* from those of *F. oxysporum* f. sp. *cucumerinum*. Parsimony and bootstrap analysis of the RAPD data placed each of the two formae speciales into a different phylogenetic (Figure 11.3).

Baayen *et al.* (2000) identified monophyletic and non-monophyletic origin of *F. oxysporum* complex by constructing nuclear and mitochondrial gene genealogies and amplified fragment length polymorphism (AFLP) amongst *F. oxysporum* f. spp. *asparagi, dianthi, gladioli, lilii, lini, opuntiarum, spinaciae*, and *tulipae*. The predictive value of the forma specialis naming system within the *F. oxysporum* complex is questioned. Chakrabarti *et al.* (2001) distinguished races of *F. oxysporum* f.sp. *ciceris* by *Eco*RI restriction pattern of the nuclear ribosomal DNA of four isolates in India indicated that the races could be grouped into three distinct groups; races 1 and 4 representing one group and race 2 and race 3, the other two. Studies were also proposed that out of the four types of races described from India, races 1 and 4 are the same. Jimenez *et al.* (2002) studied *F. oxysporum* f. sp. *ciceris* (Foc), the causal agent of fusarium wilt of chickpea, consists of two pathotypes (yellowing and wilting) and eight races (races 0, 1B/C, 1A and 2–6) of diverse geographical distribution. Foc isolates formed a grouping distinct from other *formae speciales* and nonpathogenic isolates. These results indicate that *F. oxysporum* f. sp. *ciceris* is monophyletic. Gale *et al.* (2003) find out the tomato wilt pathogen *F. oxysporum* f. sp. *lycopersici*, characterized using vegetative compatibility grouping (VCG), nuclear restriction fragment length polymorphism (RFLP), and virulence. All field isolates that could be grouped into VCG belonged to VCG 0033. Population genetic and phylogenetic analyses of 121 isolates indicated that molecular diversity among VCG 0033 isolates was by far the highest. Garcia, *et al.* (2003) carried out experiments to find out genetic relationships among *F.o. melonis, F.o. dianthi, F.o. niveum, F.o. lycopersici, F.o. radis-lycopersici, F.o. lagenaria* and *F.o. luffae* by ITS of rDNA and AFLP analysis. The specialized forms of *F. oxysporum* do not constitute monophyletic lineages because they evolve in a divergent way. The AFLP analysis was used to study the genetic relationships within and between natural populations of five Fusarium spp. The similarity percent of each group oscillated between 87 and 97 per cent. The phenetic dendrogram generated by UPGMA (Figure 11.4) as well as principal coordinate analysis (PCA) grouped all of the Fusarium spp. isolates into five major clusters (Mohmed, 2003).

Jimenez-Gasco and Jimenez-Diaz, (2003) developed the specific SCAR primers and PCR assays to study identification and differentiate isolates of *F. oxysporum* f. sp. *ciceris* and of each of its pathogenic races 0, 1A, 5, and 6. Thirty-four isolates of *F. oxysporum*, obtained cucumber plants. Of these, 23 isolates were identified by pathogenicity as *F. oxysporum* f. sp. *cucumerinum*, and one as *F. oxysporum* f. sp. *radicis-cucumerinum*. The occurrence of *F. oxysporum* f. sp. *radicis-cucumerinum* on cucumber is reported for the first time in China by RAPD method (Vakalounakis *et al.*, 2004).

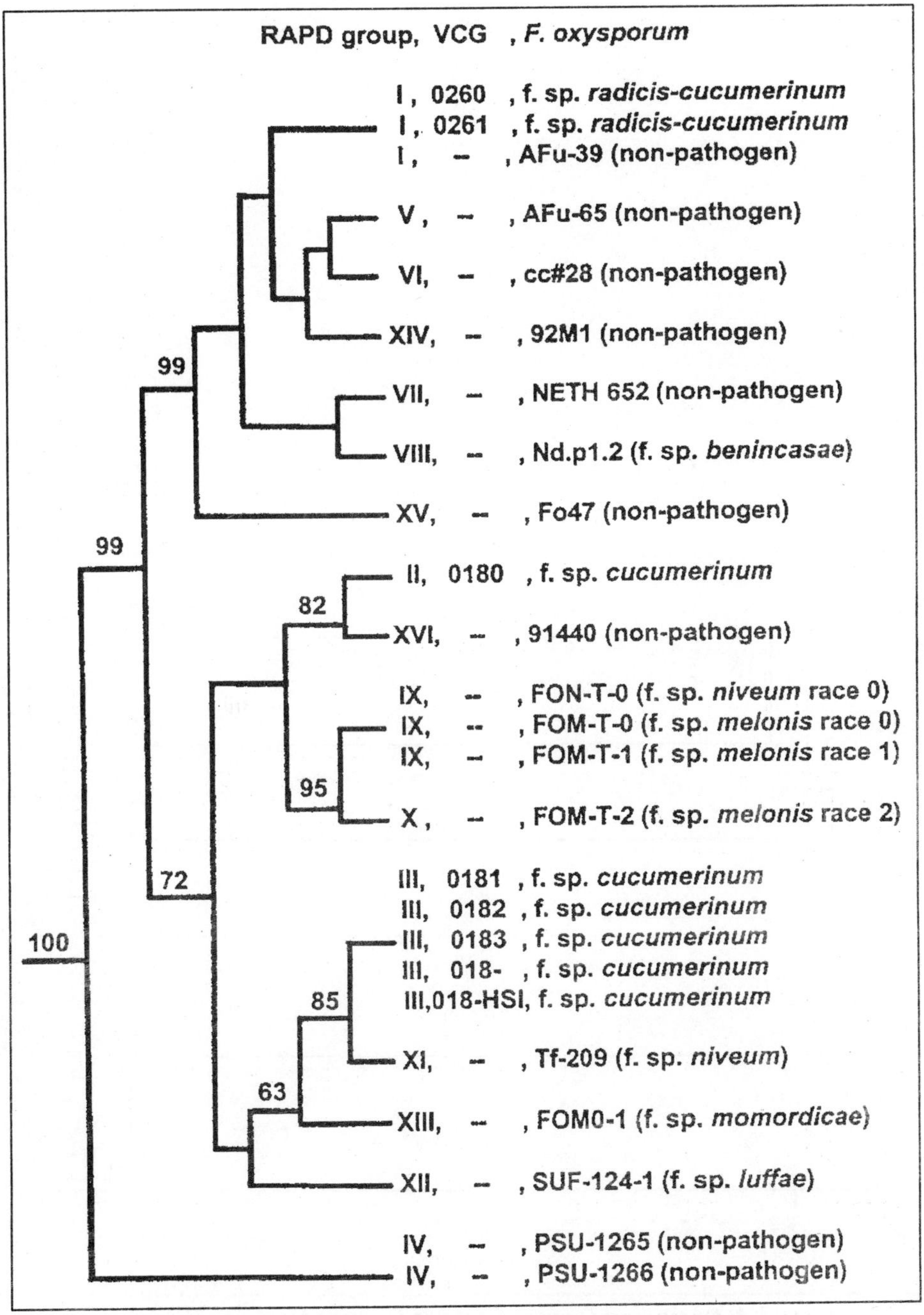

Figure 11.3: Dendrogram Showing Genetic Relationships of *F. oxysporum* Isolates

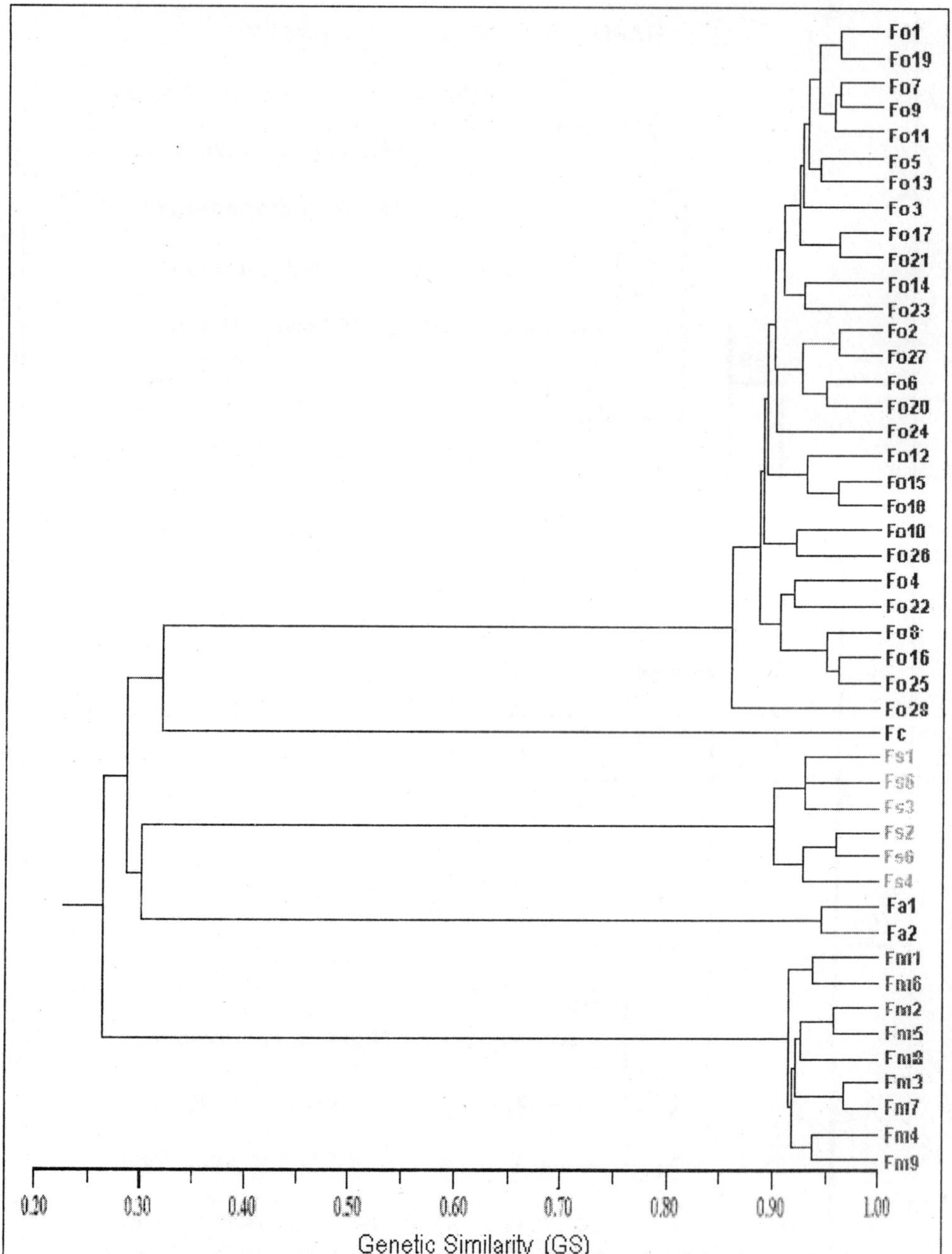

Figure 11.4: Combined Cluster Analysis Derived from AFLP Analysis of 46 *Fusarium* spp. Isolates Using 4 AFLP Primers

Belabid *et al.* (2004) analyzed thirty-two isolates of *F. oxysporum* f. sp. *lentis* in north-west Algeria. There is little genetic variability among a subpopulation of *Fol* as identified by RAPD and AFLP markers and that there is no apparent correlation with geographical origin or aggressiveness of isolates. Also, the data suggest that *Fol* isolates are derived from two genetically distinct clonal lineages. Nagarajan *et al.* (2004) analyzed genetic variation amongst 22 isolates of *F. oxysporum* f. sp. *fragariae* is a fungal pathogen causing strawberry wilt disease by DNA RAPD, RFLPs of IGS region of rDNA were used to identify genetic variation. There was a high level genetic variation among *F. oxysporum* f. sp. *fragariae*. The results of Sharma, *et al.* (2004) indicated a clustering of *Fusarium* wilt resistance genes and that transfer of those genes to improved germplasm can be accomplished relatively easily. Kouabenan *et al.* (2005) studied diversity amongst seventeen isolates of *F. oxysporum* f. sp. *vasinfectum* from the Ivory Coast were characterized using VCG, IGS and MAT idiomorph. The new isolates and reference strains were grouped based upon the three traits. Kim *et al.* (2005) studied thirty isolates of *F. oxysporum* f. sp. *vasinfectum* from California, Australia, China, and the American Type Culture Collection were characterized by partial sequences of translational elongation factor (EF-1α), phosphate permase (PHO), and beta-tubulin (BT) genes, restriction digests of the intergenic spacer (IGS) region of nuclear rDNA, and pathogenicity tests. All isolates belonging to the other lineages caused relatively mild symptoms on both Pima and Upland cultivars. Zhang, *et al.* (2005) developed two species-specific PCR assays for rapid and accurate detection of the pathogenic fungi *Fusarium oxysporum* f. sp. *niveum* and *Mycosphaerella melonis* in diseased plant tissues and soil. Based on differences in internal transcribed spacer (ITS) sequences of *Fusarium* spp. and *Mycosphaerella* spp. Kumar *et al.* (2006) studied *F. oxysporum* f.sp. *cubense* by DNA (RAPD) analysis, and PCR-RFLP analysis of nuclear intergenic spacer rDNA (IGS). UPGMA using genetic distance showed that the isolates belonged to three main groups (Figure 11.5).

Singh *et al.* (2006) showed genetic variation amongst thirty isolates of *F. oxysporum* f. sp. *ciceri* by using RAPD method. There was little genetic variability among the isolates collected from the different locations. At the 0.75 similarity index the isolates divides into three groups (Figure 11.6).

Khilare and Rafi (2009) unpublished work showed genetic variation amongst twenty-four isolates of *F. oxysporum* f. sp. *ciceri* by using RAPD method. There was three major group of a fungus collected from Maharashtra with less polymorphism (Figure 11.7 and Table 11.2).

Table 11.2: Sequence of Primers and Number of Fragments of *F. oxysporum* f. sp *ciceri* Amplified by RAPD Primers

Marker	*Primers*	*Sequences 5'to 3'*	*Amplified Fragments*	*Polymorphic Fragments*	*Per cent Polymorphism*
RAPD	OPA–02	TGCCGAGCTG	135	12	8.88
	OPB–11	GTAGACCCGT	177	11	6.21
	OPE–07	AGATGCAGCC	154	17	11.03
	OPE–11	GAGTCTCAGG	185	05	2.70

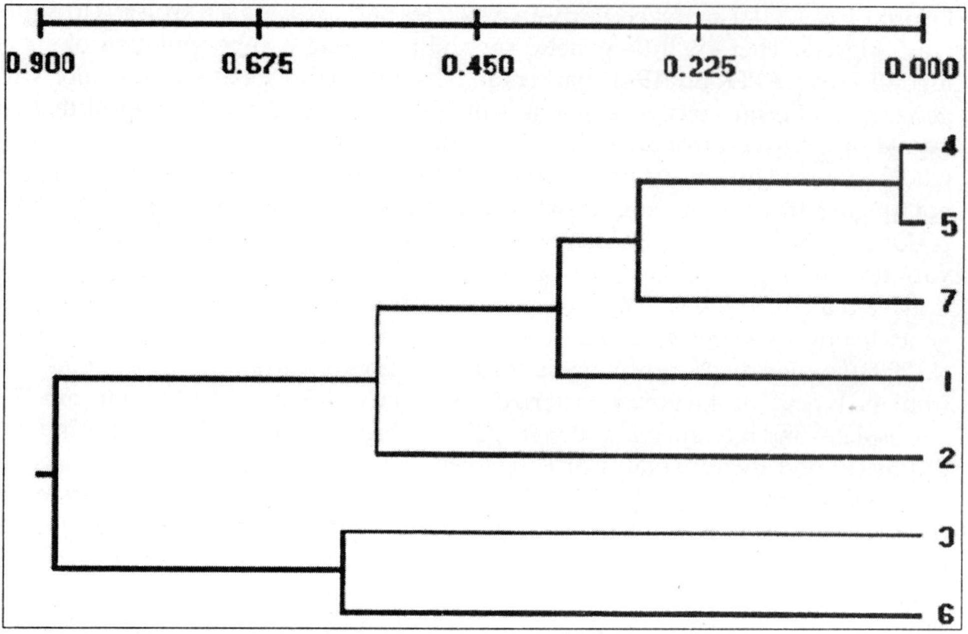

Figure 11.5: Dendrogram of *Foc* Populations Based on UPGMA Cluster Analysis Based on RAPD Analysis

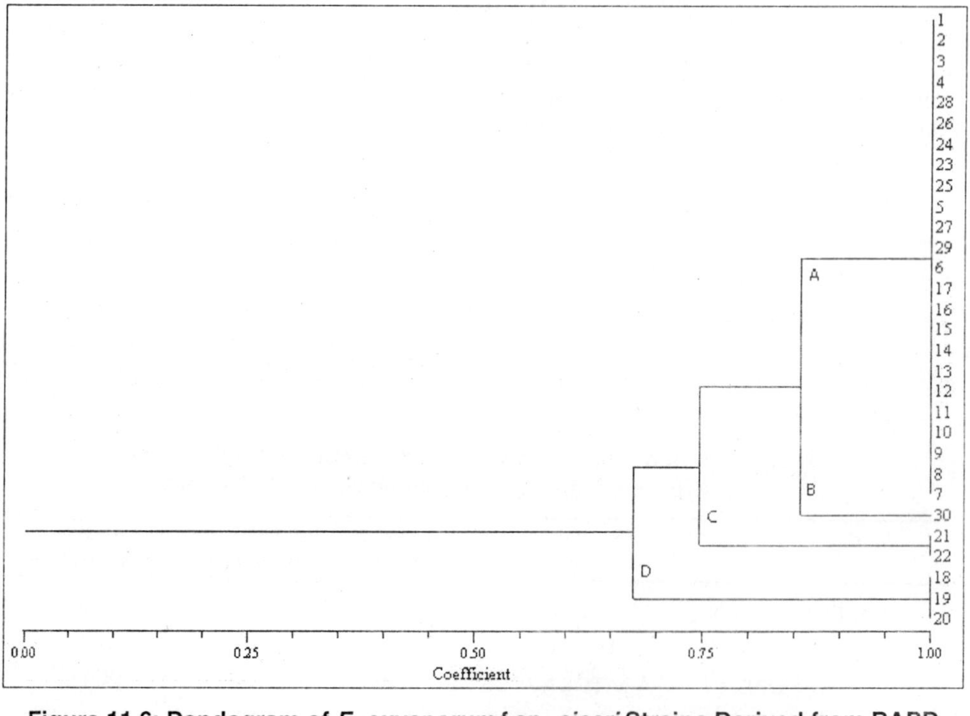

Figure 11.6: Dendogram of *F. oxysporum* f sp. *ciceri* Strains Derived from RAPD Fingerprints Generated by Using Four Different 10-mer Primers

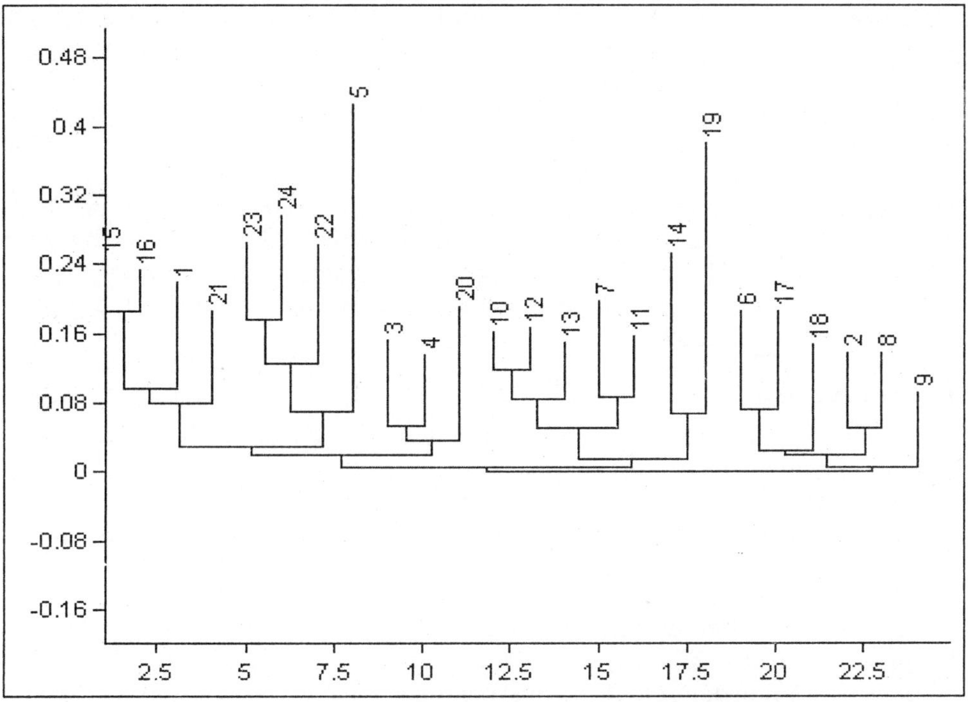

Figure 11.7: Dendogram of *F. oxysporum* f sp. *ciceri* Strains Derived from RAPD Fingerprints Generated by Using Four Different Primers

Genetic diversity in *Fusarium oxysporum* f. sp. *pisi* isolates from three agroclimatically distinct regions indicating that pathogen populations from sub tropical and sub humid regions evolved from three distinct lineages and those from temperate regions from the fourth lineage (Sharma, *et al.*, 2006). Hibar, *et al.* (2006) stated genetic diversity amongst *F. oxysporum* f. sp. *radicis-lycopersici*. This difference of diversity between the two *formae speciales* suggests that *F. oxysporum* f. sp. *radicis-lycopersici* isolates have a foreign origin and may have been accidentally introduced into Tunisia. Genetic variability within 24 isolates representing seven races of *F. oxysporum* f. sp. *ciceris* was assessed by RAPD. UPGMA cluster analysis divided the isolates into seven distinct clusters at 0.55 genetic similarities. The most virulent isolate obtained from wilt sick field of IARI (MB-4C), New Delhi was distinct from others (Honnareddy and Dubey, 2006). Bogale, *et al.* (2006) characterized the isolates of *F. oxysporum* from Ethiopia. For comparative purposes, they also included 18 representatives of *F. oxysporum* formae speciales. They confirmed that pathogenicity of isolates does not necessarily correlate with phylogenetic grouping. The genetic diversity observed among the Ethiopian isolates was low. RFLP for Australian *F. oxysporum* f. sp. *vasinfectum* isolates, which were capable of distinguishing them from other formae speciales of *F. oxysporum*. (Zambounis, *et al.*, 2007). Gullino, (2007) developed rapid and reliable molecular tools for identification of Australian strains of *F. oxysporum* f. sp. *vasinfectum* helps for further research on the disease etiology and epidemiology. Dubey and Singh (2008) analyzed virulence of 64 isolates of *F. oxysporum* f. sp. *ciceris* causing chickpea wilt suggesting the existence of diverse genetic populations of the pathogen at the same location by DNA (RAPD), ISSR, and SSR markers Lievens *et al.* (2008) suggested potential of molecular characterization of *F. oxysporum*. Recently an increasing number of studies have reported the successful development of molecular markers to discriminate *F. oxysporum* strains below the species level. *F. oxysporum* f. sp. *loti* causing severe vascular wilt of trefoil and moderate vascular wilt of pea. It is distinct from other pathogenic *F. oxysporum* and proposed as *F. oxysporum* Schlechtendahl emend. Snyder and Hansen f. sp. *loti* forma specialis *nova* (Wunsch, *et al.*, 2009).

Future Agenda

Accurate identification and early detection of pathogens is the cornerstone of disease management in many crops. Plant pathology principally involves both applied and basic aspects of plant disease management. Many plant pathogens are difficult to identify using morphological criteria, which can be time consuming, challenging and requires extensive knowledge in taxonomy. Molecular detection techniques can generate accurate results rapidly enough to be useful for disease management decisions. For effective management of plant diseases host resistance, biological control and integrated disease management approaches have to be employed for immediate solution. Host pathogen interactions at the molecular level facilitate sustainable production gains in agriculture. Some are challenging issues in molecular studies of formae specials of *Fusarium oxysporum* mentioned below.

1. Studies on 3D structure of R-protein at molecular level in host.
2. Discovery of new resistance genes against *Fusarium oxysporum* and its application in breeding programmes.
3. Analysis of host-pathogen interactions at molecular level.
4. Selection of best gene and its combination in different agro-climatic conditions.
5. Application of 'RNA Silencing' technology helps in developing host resistance.

References

Armstrong, G. M., and Armstrong, J. K. (1981). Formae speciales and races of *Fusarium oxysporum* causing wilt diseases. Pages 391-399 in: *Fusarium*: Diseases, Biology and Taxonomy. P. E. Nelson, T. A. Toussoun, and R. J. Cook, eds. The Pennsylvania State University Press, University Park.

Appel, D. J., and Gordon, T. R. (1994). Local and regional variation in populations of *Fusarium oxysporum* from agricultural field soils. *Phytopathology*. 84: 786-791.

Appel, D.J. and Gordon, T.R. (1996). Relationships among pathogenic and nonpathogenic isolates of *Fusarium oxysporum* based on the partial sequence of the intergenic spacer region of the ribosomal DNA. *MPMI* 9: 125-138.

Assigbetse, K.B. Fernandez, D., Dubois, M.P. and Geiger, J.P. (1994). Differentiation of *Fusarium oxysporum* f.sp. *vasinfectum* races on cotton by random amplified polymorphic DNA (RAPD) analysis. *Phytopathology*.84: 622–626.

Baayen, R. P., O'Donnell, K., Bonants, P. J. M., Cigelnik, E., Kroon, L. P. N. M., Roebroeck, E. J. A., and Waalwijk, C. (2000). Gene genealogies and AFLP analyses in the *Fusarium oxysporum* complex identify monophyletic and nonmonophyletic formae speciales causing wilt and rot disease. *Phytopathology*. 90: 891-900.

Belabid, L., Baum, M., Fortas, Z., Bouznad, Z. and Eujayl, I. (2004). Pathogenic and genetic characterization of Algerian isolates of *Fusarium oxysporum* f. sp. *lentis* by RAPD and AFLP analysis. *African Journal of Biotechnology*, 3 (1) 25-31.

Bentley, S., Pegg, K. G., Moore, N. Y., Davis, R. D., and Buddenhagen, I. W. (1998). Genetic variation among vegetative compatibility groups of *Fusarium oxysporum* f. sp. *cubense* analyzed by DNA fingerprinting. *Phytopathology*. 88: 1283-1293.

Bogale, M., Wingfield, B.D., Wingfield, M.J. and Steenkamp, E.T. (2006). Characterization of *Fusarium oxysporum* isolates from Ethiopia using AFLP, SSR and DNA sequence analyses. *Fungal Diversity*.23: 51-66.

Bosland, P. W. and Williams, P. H. (1987). An evaluation of *Fusarium oxysporum* from crucifers based on pathogenicity, isozyme polymorphism, vegetative compatibility, and geographic origin. *Can. J. Bot*. 65: 2067-2073.

Chakrabarti, A., Mukherjee, P.K., Pramod, D. Sherkhane,P.D., Bhagwat, A.S., Narra B. K. and Murthy, N.B.K. (2001). A simple and rapid molecular method for distinguishing between races of *Fusarium oxysporum* f.sp. *ciceris* from India. *Current Science*, 80 (4), 25 571–575.

Coddington, A., Matthews, P. M., Cullis, C. and Smith, K. H. (1987). Restriction digest patterns of total DNA from different races of *Fusarium oxysporum* f. sp. *pisi*–An improved method for race classification. *J. Phytopathol*. 118: 9-20.

Dubey, S.C. and Shio Raj Singh (2008). Virulence analysis and oligonucleotide fingerprinting to detect diversity among Indian isolates of *Fusarium oxysporum* f. sp. *ciceris* causing chickpea wilt. *Mycopathologica*. 165 (6)389–406.

Elias, K. S., Zamir, D., Lichtman-Pleban, T. and Katan, T. (1993). Population structure of *Fusarium oxysporum* f. sp. *lycopersici*: Restriction fragment length polymorphisms provide genetic evidence that vegetative compatibility group is an indicator of evolutionary origin. *Mol. Plant-Microbe Interact*. 6: 565-572.

Fernandez, D. and Tantaoui, A. (1994). Random amplified polymorphic DNA (RAPD) analysis: A tool for rapid characterization of *Fusarium oxysporum* f. sp. *albedinis* isolates? *Phytopathol. Mediterr*. 33: 223-229.

Fernandez, D., Assigbetse, K., Dubois, M.-P. and Geiger, J.-P. (1994). Molecular characterization of races and vegetative compatibility groups in *Fusarium oxysporum* f. sp. *vasinfectum*. *Appl. Environ. Microbiol*. 60: 4039-4046.

Fernandez, D., Ouinten, M., Tantaoui, A., Lourd, M. and Geiger, J.-P. (1995). Population genetic structure of *Fusarium oxysporum* f. sp. *albedinis*. (Abstr.) *Fungal Genet. Newsl*. 42A: 34.

Fernandez, D., Ouinten, M., Tantaoui, A. and Geiger, J.P. (1997). Molecular records of micro-evolution within the Algerian population of *Fusarium oxysporum* f. sp. *albedinis* during its spread to new oases. *European J. Plant Pathology*. 103 (5) 485–490.

Flood, J., Whitehead, D. S., and Cooper, R. M. (1992). Vegetative compatibility and DNA polymorphisms in *Fusarium oxysporum* f. sp. *elaeidis* and their relationship to isolate virulence and origin. *Physiol. Mol. Plant Pathol*. 41: 201-215.

Grajal-Marin, Simon, M.J. and Muehlbauer, F.J. (1993). Use of random amplified DNA (RAPD) to characterize race 2 of *Fusarium oxysporum* f.sp. *pisi*. *Phytopathology*. 83: 612–614.

Gale, L. R., Katan, T., and Kistler, H. C. (2003). The probable center of origin of *Fusarium oxysporum* f. sp. *lycopersici* VCG 0033. *Plant Dis*. 87: 1433-1438.

Garcia, R.A., Cenis, J.C., Tello, J., Martinez-Zapter, J.M. and Cifuentes, D. (2003). Genetic relationship among seven specialized forms of *Fusarium oxysporum* determined by DNA sequencing of the ITS region and AFLPs. *Spanish J. Agril. Res*. 1 (3): 55–63.

Gulino, Lisa-Maree (2007). *Molecular characterisation and detection of* Fusarium oxysporum *f.sp. vasinfectum* PhD Thesis, School of Integrative Biology, University of Queensland, Australia.

Hibar, K., V. Edel-Herman, Ch. Steinberg, N. Gautheron, M. Daami-Remadi, C. Alabouvette. and M. El Mahjoub (2006). Genetic diversity of *Fusarium oxysporum* populations isolated from tomato plants in Tunisia. *J. Phytopathology*, 155 (3): 136–142.

Hirota, N., Hashiba, T., Yoshida, H., Kikumoto, T. and Ehara, Y. (1992). Detection and properties of plasmid-like DNA in isolates from twentythree formae speciales of *Fusarium oxysporum*. *Ann. Phytopathol. Soc. Jpn*. 58: 386-392.

Honnareddy, N. and Dubey, S. C. (2006). Pathogenic and molecular characterization of Indian isolates of *Fusarium oxysporum* f. sp. *ciceris* causing chickpea wilt. *Current Science*. 91 (5) 661-666.

Jacobson, D. J. and Gordon, T. R. (1990). Variability of mitochondrial DNA as an indicator of relationships between populations of *Fusarium oxysporum* f. sp. *melonis*. *Mycol. Res*. 94: 734-744.

Jimenez-Gasco, M.M., Milgroom, M.G. and Jimenez-Diaz, R.M. (2002). Gene genealogies support *Fusarium oxysporum* f. sp. *ciceris* as a monophyletic group. *Plant Pathology*. 51: 72–77

Jimenez-Gasco, M. M., and Jimenez-Diaz, R. M. (2003). Development of a specific polymerase chain reaction-based assay for the identification of *Fusarium oxysporum* f. sp. *ciceris* and its pathogenic races 0, 1A, 5, and 6. *Phytopathology*. 93: 200-209.

Katan, J. (1971). Symptomless carriers of the tomato Fusarium wilt pathogen. *Phytopathology*. 61: 1213-1217.

Kelly, A., Alcala-Jimenez, A.R., Bainbridge, B.W., Heale, J.B., Perez-Artes, E. Nad. and Jimenez-Diaz, R.M. (1994). Use of genetic fingerprinting and random amplified polymorphic DNA to characterize pathotypes of *Fusarium oxysporum* f.sp. *ciceris* infecting chickpea. *Phytopathology*. 84: 1293–1298.

Kerenyi, Z., Taborhegyi, E., Pomazi, A. and Hornok, L. (1997). Variability amongst strains of *Fusarium poae* assessed by vegetative compatibility and RAPD polymorphism. *Plant Pathology*. 46(6): 882-889.

Khilare, V.C. and Rafi Ahmed (2009). Molecular characterization of *Fusarium oxysporum* f.sp. *ciceri* causing chickpea wilt in Maharashtra. (Unpublished)

Kim, D. H., Martyn, R. D., and Magill, C. W. (1992). Restriction fragment length polymorphism groups and physical map of mitochondrial DNA from *Fusarium oxysporum* f. sp. *niveum*. *Phytopathology*. 82: 346-353.

Kim, D.M., Martyn, R.D. and C. W. Magill, C.W. (1993). Mitochondrial DNA (mtDNA)–Relatedness among Formae Speciales of *Fusarium oxysporum* in the Cucurbitaceae. *Phytopathology*. 83: 91-97

Kim, Y., Hutmacher, R. B. and Davis, R. M. (2005). Characterization of California isolates of *Fusarium oxysporum* f. sp. *vasinfectum*. *Plant Dis*. 89: 366-372.

Kistler, H. C., Bosland, P. W., Benny, U., Leong, S. and Williams, P. H. (1987). Relatedness of strains of *Fusarium oxysporum* from crucifers measured by examination of mitochondrial and ribosomal DNA. *Phytopathology*.77: 1289-1293.

Kistler, H. C. and Benny, U. (1989). The mitochondrial genome of *Fusarium oxysporum*. *Plasmid*. 22: 86-89.

Koenig, R. L., Kistler, H. C. and Ploetz, R. C. (1993). Restriction fragment length polymorphism analysis of *Fusarium oxysporum* f. sp. *cubense*. Page 158 in: 6th Int. Congr. Plant Pathol. National Research Council of Canada, Ottawa, ON.

Kouabenan, A., Keith, K. K., Veronique E., Nadine, G., Dossahoua, T. and Christian, S. (2005). High Genetic Diversity Among Strains of *Fusarium oxysporum* f. sp. *vasinfectum* from Cotton in Ivory Coast (2005). *Phytopathology*. 95: 1391-1396.

Kumar, B.H., Shankar, U.A.C., Kini, R.K., Prakash, H.S. and Shetty, S.H. (2006). Genetic variation in *Fusarium oxysporum* f.sp. *cubense* isolates based on random amplified polymorphic DNA and intergenic spacer.*Archives of Phytopathology and Plant Protection*. 39(2): 151–160.

Lievens, B. Rep, M. and Thomma, B.P. (2008). Recent developments in the molecular discrimination of formae speciales of *Fusarium oxysporum*. *Pest Manag. Sci.* 64 (8): 781–8.

Manicom, B. Q. and Baayen, R. P. (1993). Restriction fragment length polymorphisms in *Fusarium oxysporum* f. sp. *dianthi* and other fusaria from *Dianthus* species. *Plant Pathol.* 42: 851-857.

Manulis, S., Kogan,N., Reuven, M. and Ben-Yephet, Y.(1994). Use of the RAPD Technique for Identification of *Fusarium oxysporum* f. sp. *dianthi* from Carnation. *Phytopathology*. 84: 98-101.

Mes, J. J., Van Doorn, J., Roebroeck, E. J. A., Van Egmond, E., Van Aartrijk, J. and Boonekamp, P. M. (1994). Restriction fragment length polymorphisms, races and vegetative compatibility groups within a worldwide collection of *Fusarium oxysporum* f. sp. *gladioli*. *Plant Pathol.* 43: 362-370.

Mes, J. J., Weststeijn, E. A., Herlaar, F., Lambalk, J. J. M., Wijbrandi, J. Haring, M. A. and Cornelissen, B. J. C. (1999). Biological and molecular characterization of *Fusarium oxysporum* f. sp. *lycopersici* divides race 1 isolates into separate virulence groups. *Phytopathology.* 89: 156-160.

Menzies, J. G., Koch, C. and Seywerd, F. (1990). Additions to the host range of *Fusarium oxysporum* f. sp. *radicis-lycopersici*. *Plant Dis.* 74: 569-572.

Miao, V. P. W. (1990). Using karyotype variability to investigate the origins and relatedness of isolates of *Fusarium oxysporum* f. sp. *cubense*. Pages 55-62 in: Fusarium Wilt of Banana. R. C. Ploetz, ed. The American Phytopathological Society, St. Paul, MN.

Mohmed A. Abdel-Satar, Mohmed. S. Khalil, I. N. Mohmed, Kamel A. Abd-Elsalam. and Joseph A. Verreet. (2003). Molecular phylogeny of Fusarium species by AFLP fingerprint. *African Journal of Biotechnology*. 2 (3): 51–55.

Nagarajan, G., Myeong, H. N., Jeong, Y. S., Sung, J. Y. and Hong, G. K. (2004). Genetic Variation in *Fusarium oxysporum* f. sp. *fragariae* populations based on RAPD and rDNA RFLP analyses. *Plant Pathol. J.* 20 (4): 264-270.

Namiki, F., Shiomi, T., Kayamura, T. and Tsuge, T. (1994). Characterization of the formae speciales of *Fusarium oxysporum* causing wilts of cucurbits by DNA fingerprinting with nuclear repetitive DNA sequences. *Appl. Environ. Microbiol.* 60: 2684-2691.

Namiki, F., Shiomi, T., Nishi, K., Kayamura, T. and Tsuge, T. (1998). Pathogenic and genetic variation in the Japanese strains of *Fusarium oxysporum* f. sp. *melonis*. *Phytopathology*. 88: 804-810.

Nelson, P. E. (1991). History of *Fusarium* systematics. *Phytopathology.* 81: 1045-1051.

Perez-Artes, E., Roncero, M. I. G. and Jimenez-Diaz, R. M. (1995). Restriction fragment length polymorphism analysis of the mitochondrial DNA of *Fusarium oxysporum* f. sp. *ciceris*. *J. Phytopathol.* 143: 105-109.

Rowe, R. C. (1980). Comparative pathogenicity and host ranges of *Fusarium oxysporum* isolates causing crown and root rot of greenhouse and field-grown tomatoes in North America and Japan. *Phytopathology.* 70: 1143-1148.

Schilling, A.G. Moller, E. M. and Geiger, H.H. (1996). Polymerase Chain Reaction-Based Assays for Species-Specific Detection of *Fusarium culmorum*, *F. graminearum*, and *F. avenaceum*. *Phytopathology.* 86: 515-522.

Sharma, P., Sharma, K.D., Sharma, R. and Plaha, P. (2006). Genetic variability in pea wilt pathogen *Fusarium oxysporum* f. sp. *pisi* in north-western Himalayas. *Ind. J. Biotechnology.* 5: 298–302.

Sharma, K.D., Winter, P. and Muehlbauer, F.J. (2004). Molecular mapping of *Fusarium oxysporum* f. sp. *ciceris* race 3 resistance gene in chickpea. *Journal of Theoretical and Applied Genetics.* 108: 1243-1248.

Singh, B.P., Ratul, Saikia, Mukesh, Yadav, Rakesh Singh, V.S. Chauhan and Arora, D.K. (2006). Molecular characterization of *Fusarium oxysporum* f.sp. *ciceri* causing wilt of chickpea.*African Journal of Biotechnology*.5 (6) 497-502.

Snyder, W. C. and Hansen, H. N. (1940). The species concept in *Fusarium*. *Am. J. Bot.* 27: 64-67.

Tantaoui, A., Ouiten, M., Geiger, J.P., Fernandez, D. (1996). Characterization of a single clonal lineage of *Fusarium oxysporum* f. sp. *albedinis* causing Bayoud disease of date palm in Morocco. *Phytopathology*. 86(7) 787-792.

Namiki, F., Shiomi, T., Kayamura, T. and Tsuge, T. (1994). Characterization of the formae speciales of *Fusarium oxysporum* causing wilts of cucurbits by DNA fingerprinting with nuclear repetitive DNA sequences. *Appl Environ Microbiol*. 60(8): 2684-2691.

Nelson, A. J., Elias, K. S., Arevalo G., E., Darlington, L. C. and Bailey, B. A. (1997). Genetic characterization by RAPD analysis of isolates of *Fusarium oxysporum* f. sp. *erythroxyli* associated with an emerging epidemic in Peru. *Phytopathology*. 87: 1220-1225.

Vakalounakis, D. J. and Fragkiadakis, G. A. (1999). Genetic diversity of *Fusarium oxysporum* isolates from cucumber: differentiation by pathogenicity, vegetative compatibility, and RAPD fingerprinting. *Phytopathology*. 89: 161-168.

Vakalounakis, D. J., Wang, Z., Fragkiadakis, G. A., Skaracis, G. N. and Li, D.-B. (2004). Characterization of *Fusarium oxysporum* isolates obtained from cucumber in China by pathogenicity, VCG, and RAPD. *Plant Dis*. 88: 645-649.

Woo, S.L., Zoina, A., Del Sorbo, G., Lorito, M., Nanni, B., Scala, F. and Noviello, C. (1996). Characterization of *Fusarium oxysporum* f. sp. *phaseoli* by Pathogenic Races, VCGs, RFLPs, and RAPD. *Phytopathology*. 86: 966-973.

Wollenweber, H. W. and Reinking, O. A. (1935). Die Fusarien, Ihre Beschreibung, Schadwirkung und Bekampfung. P. Parey, Berlin.

Woudt, L. P., Neuvel, A., Sikkema, A., van Grinsven, M. Q. J. M., de Milliano, W. A. J., Campbell, C. L., and Leslie, J. F. (1995). Genetic variation in *Fusarium oxysporum* from cyclamen. *Phytopathology*. 85: 1348-1355.

Wunsch, M.J., Baker, A.H., Kalb, D.W. and Gary C. (2009). Characterization of *Fusarium oxysporum* f. sp. *loti* Forma Specialis nov., a monophyletic pathogen causing vascular wilt of Birdsfoot Trefoil. *Plant Disease*. 93 (1) 58–66.

Zambounis, A.G., Paplomatas, E. and Tsaftaris, A.S. (2007). Intergenic spacer–RFLP analysis and direct quantification of Australian *Fusarium oxysporum* f. sp. *vasinfectum* isolates from soil and infected cotton tissues. *Plant Disease* 91 (12) 1564–1573.

Zhang, Z. Zhang, J. Wang, Y. and Zheng, X. (2005). Molecular detection of *Fusarium oxysporum* f. sp. *niveum* and *Mycosphaerella melonis* in infected plant tissues and soil. *FEMS Microbiol*, 249 (1): 39–47.

Chapter 12

Molecular and Pathogenic Variability in *Plasmopora halstedii* Causing Downy Mildew of Sunflower

Srikant Kulkarni, Yashoda R. Hegde* and V. Ramesh Kota

Department of Plant Pathology,
University of Agricultural Sciences, Dharwad – 580 005, Karnataka, India

ABSTRACT

Sunflower (*Helianthus annuus* L.), belonging to the family compositae, is one of the important sources of vegetable oils and is popular because of its high oil percentage, quality, short duration, thermophotoin sensitiveness and high multiple seed ratio. In India, Sunflower was introduced around 1969 for commercial cultivation. Sunflower suffers from a number of biotic and abiotic factors. Among diseases, downy mildew caused by *Plasmopara halstedii* is most important one (Kolte, 1985). In India, the sunflower was free from downy mildew till 1984 and was first reported from Maharashtra in 1984. (Mayee and Patil, 1986) Fungus has been distributed by seed trade rapidly. Failure of quarantine regulations resulted in the entry of the pathogen in India.

Keywords: Plasmopora halstedii, RAPD.

---

* E-mail: uasyashoda@gmail.com

Symptoms

Symptoms caused by *Plasmopara halstedii* were evident as seedling damping off, systemic symptoms, local foliar lesions with white downy growth and basal root or stem galls. Damping off of seedlings was observed, whenever there was heavy inoculum and favourable conditions for the disease. Under highly congenial conditions, seedlings were killed even before emergence or immediately after emergence. Reduced plant stand due to death of seedlings is common under such conditions.

Systemic Symptoms

Stunting, veinal chlorosis, were commonly observed. Sunflower plants carrying systemic symptoms were severely stunted. In a susceptible variety, the pathogen colonized the whole plant. Leaves of affected plants were abnormally thick, curled downward and mottled. Downy growth of the fungus consisting of sporangiophores and sporangia was observed on the lower side of the leaves. The corresponding upper surface of the leaves appeared chlorotic yellow. It is observed that the stems and leaves of diseased plants became very brittle and were easily broken. Depending upon the extent of infection, some plants even failed to produce flower heads. Even if flower heads are produced, they seldom produced seeds. Seeds produced from infected plants showed drastic reduction in seed size. When the plants were infected at older stage, they remained symptomless until flowering.

Stunting

Major influence of downy mildew systemic infection was extreme stunting depending on the stage at which infection occurs. The internodes were shortened by 40 per cent.

Veinal chlorosis is very prominent on young seedlings accompanied with whitish downy growth. Systemically infected sunflower plants exhibited typical erect head, with loss of normal bending at flowering. Infected plant show the loss of phototropic responses normally seen in healthy sunflower plant. Flower head and seed: Very small flower head with small seeds and often floral malformation has been noticed (Figure 12.1).

Existance of physiological races is evident from reports of several workers from different parts of world. (Abdullah, 1983; Patil and Mayee, 1990) An attempt was made to document the race flora of pathogen in Karnataka and their molecular variability.

Molecular Variability in *Plasmopara halstedii*

Materials and Methods

Following twelve isolates of *Plasmopara halstedii* were collected for RAPD analysis:

Sl.No.	Locality of Collection of Isolates	Isolate No.
1.	Gadag	Ph 1
2.	Bangalore	Ph 2
3.	Dharwad	Ph 3
4.	Mysore	Ph 4
5.	Narendra	Ph 5
6.	Latur	Ph 6
7.	Hyderabad	Ph 7
8.	Bailhongal	Ph 8
9.	Davanagere	Ph 9
10.	Haveri	Ph 10
11.	Bidar	Ph 11
12.	Parbhani	Ph 12

Collection of Sporangia

Isolates were maintained on susceptible cultivar (morden) by radicle inoculation. Infected leaves were collected and washed with sterile water and dried by keeping them between blotter papers and were placed in humid chambers at 15°C for 6 hours under darkness. Sporangia were collected from freshly sporulated leaf material with camel hair brush or by using an atomizer and made into pellets by centrifuging at 5000 rpm for 5 minutes.

Total Genomic DNA Extraction

Sporangial pellets were taken in microcentrifuge tubes along with 0.2 ml of cold extraction buffer (Triton × 100[2] per cent, SDS 1 per cent, NaCl 100 mM, TrisHCl (8.0) 10 mM, EDTA 1 mM), 0.2 ml phenol chloroform reagent and vortexed at high speed for 10 min. 0.2 ml of TE buffer (Tris HCl pH 8.0 10 mM and EDTA 1 mM) was added again and vortexed for 5 min. Following centrifugation at 10,000 rpm for 5 min at 4°C the supernatant containing DNA was transferred to another tube. The DNA was partially purified by repeated extraction with 0.2 ml of phenol chloroform reagent. It was precipitated by the addition of 3 volumes of cold absolute ethanol and 1/10 volume of 3 M sodium acetate and incubated at –20°C for 3 h. The precipitated DNA was pelletted by centrifuging at 10,000 rpm for 10 min at 4°C. The pellet was washed with cold 70 per cent ethanol, dried under vacuum and dissolved in 100 ml TE buffer. Ten ml of RNAase (10 mg/ml) was added to the DNA solution and stored at –20°C.

Polymerase Chain Reaction (PCR)

A total of 60 random primers, 20 each from OPA set, OPB set and OPF set were used in the present study. Master mix was prepared by mixing 2.50 ml of assay buffer (10X with 15mM $MgCl_2$), 1 ml of dNTPmix (25mM), 1 ml of random Primer (10 mM), 0.20 ml of Taq polymerase and 14.30 ml of double distilled water for each tube and distributed to all tubes at 19 ml/tube and 1 ml of template DNA from the respective

Figure 12.1: Symptoms of Downy Mildew

Damping off

Veinal chlorosis

Contd...

Figure 12.1–Contd...

Stunting

Cottony growth

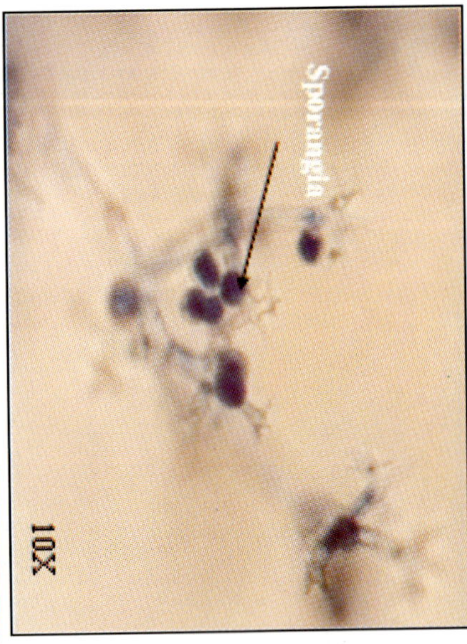

Right angle branching

Erect head

isolates was added making the final volume of 20 ml/reaction. PCR reaction was set in Palm Cycler of Corbett Research, Australia. (1 cycle at 95°C for 5 min; 38 cycles of 95°C for 1 min, 38°C for 1 min and 72°C for 2 min; 1 cycle of 72°C for 10 min). After the completion of the PCR, the products were stored at –4°C until the gel-electrophoresis was done. The PCR product was mixed with 2 ml of loading dye (Bromophenol blue) and was loaded in 1.2 per cent agarose gel of 1x TAE buffer containing ethidium bromide. Gel was run at 70 volts. The gel was photographed by using Alpha DigiDoc Documentation System (Alpha Innotech Corporation.,U.S.A.).

RAPD Analysis

The amplified fragments were scored as 1 for the presence and 0 for the absence of band generating 0 and 1 matrix. The data obtained was analyzed by obtaining the pair wise genetic similarities using Dice similarity coefficient. Clustering was done using the symmetric matrix of similarity coefficient and cluster obtained based on unweighed pair group arithmetic mean (UPGMA) using SHAN module of NTSYSPC version 2.0.

Results and Discussion

Out of 60 primers evaluated, 7 primers (OPA6, OPA9, OPB2, OPB16, OPF7, OPF17 and OPF18) failed to amplify the DNA. Among remaining primers OPA11, OPB3, OPB4, OPB5, OPB14, OPB19, OPF2, OPF8, OPF9, OPF11, OPF12, OPF15 and OPF16 have shown polymorphism. On an average 8.698 bands per primer were amplified and 5.396 bands/primer were polymorphic. Polymorphism per primer ranged from 0 per cent with OPF19 primer to 100 per cent with OPF9 primer. The band profiles obtained with amplified primers are summarized in Table 12.1 and some RAPD gels are presented in Figure 12.2. Amplified primers data on 12 isolates generated a total of 461 amplified fragments out of which 286 (62.03 per cent) were polymorphic.

Isolates formed two main clusters. Among the isolates collected from Karnataka, Haveri (Ph10) and Narendra (Ph5) clustered differently from other isolates of the state. On the contrary, Narendra (Ph5) and Latur (Ph6) though geographically quite apart shared a common cluster. Dharwad (Ph3) and Narendra (Ph5) though geographically very near, clustered separately (Figure 12.3). Existence of physiological races in *P. halstedii* was evident from reports of several workers from different parts of the world (Zimmer, 1972, Yagodkina, 1956). According to Patil and Mayee (1990) the downy mildew isolate in India was European race-1. The low variability among the isolates may be due to homothallic nature of the fungus and narrow base of host genotypes. Inconsistent clustering based on geographical origin may be due to the pathogen being introduced into new areas through seed.

Pathogenic and Morphological Variability in *Plasmopara halstedii*

Material and Methods

Twelve isolates collected from different places were maintained and multiplied on the susceptible cultivar Modern by following radicle inoculation method. To know the race pattern, these isolates were inoculated on 15 international host differentials *viz.*, HA 300, RHA 266A, RHA325, HA61, RHA274R, HA335, HA337, RHA340R,

Table 12.1: Banding Profile of Different Primers for Different Isolates of *Plasmopara halstedii*

Sl.No.	Primer	Total bands	Polymorphic	Per cent Polymorphism	Sl.No.	Primer	Total Bands	Polymorphic	Per cent Polymorphism
1	OPA 1	13	10	76.92	28	OPA11	8	5	62.50
2	OPA 2	13	8	61.50	29	OPA12	8	3	37.50
3	OPA 3	10	4	40.00	30	OPA13	10	7	70.00
4	OPA 4	12	8	66.67	31	OPA 14	10	8	80.00
5	OPA5	13	6	46.15	32	OPA 15	7	4	57.14
6	OPA7	10	6	60.00	33	OPA 17	6	2	33.33
7	OPA8	9	6	66.67	34	OPA 18	7	4	57.14
8	OPA10	7	5	71.42	35	OPA19	10	9	90.00
9	OPA11	11	9	81.82	36	OPA 20	9	7	77.78
10	OPA12	12	9	75.00	37	OPA 1	8	5	62.50
11	OPA 13	12	8	66.67	38	OPA 2	11	9	81.82
12	OPA 14	11	4	36.36	39	OPA3	9	7	77.78
13	OPA 15	9	6	66.67	40	OPA4	6	4	66.66
14	OPA 16	10	6	60.00	41	OPA 5	8	3	37.50
15	OPA17	5	3	60.00	42	OPA 6	5	3	60.00
16	OPA18	13	6	46.15	43	OPA 8	9	8	88.89
17	OPA19	7	2	28.57	44	OPA 9	4	4	100.00
18	OPA20	10	5	50.00	45	OPA 10	5	1	20.00
19	OPA1	5	2	40.00	46	OPA 11	11	10	90.91
20	OPA3	7	6	85.71	47	OPA 12	11	9	81.82
21	OPA 4	10	8	80.00	48	OPA 13	10	6	60.00
22	OPA 5	7	6	85.71	49	OPA 14	6	5	83.33
23	OPA 6	8	2	25.00	50	OPA 15	10	8	80.00
24	OPA 7	7	2	28.57	51	OPA 16	8	7	87.50
25	OPA8	7	3	42.86	52	OPA 19	7	0	0.00
26	OPA9	7	4	57.14	53	OPA 20	7	4	57.14
27	OPA10	5	2	40.00			460	288	

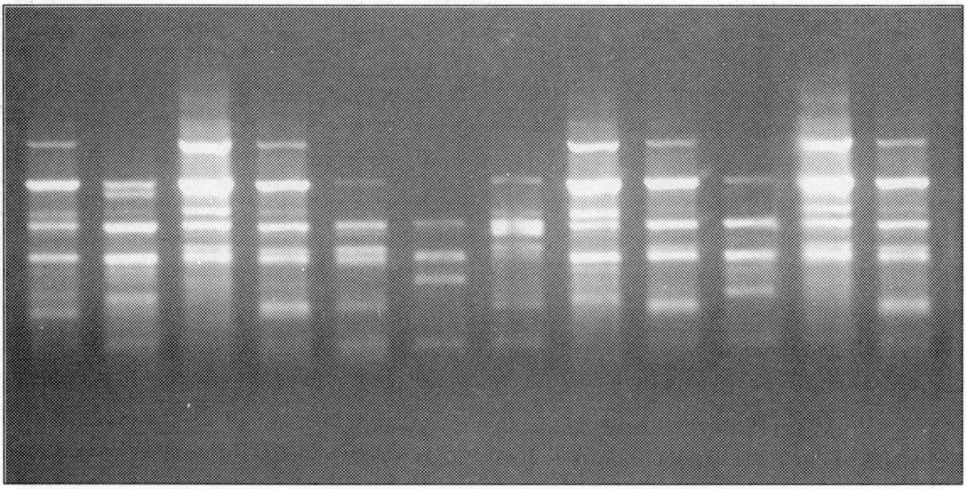

OPB 19

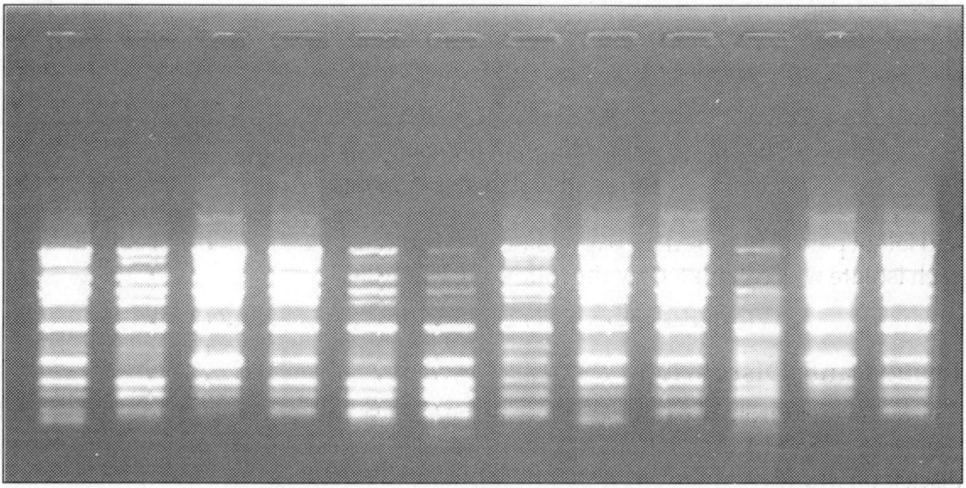

OPA 5

Figure 12.2: RAPD Banding Pattern of Different Isolates of *P. halstedii*

DM2, HIR34, HAR4, HAR5, DM5 and DM6. Seeds of host differential were germinated in rolled paper towel. When radicles were about 3-4 mm size, fresh sporangial suspension prepared in sterile water was sprayed on radicles with the help of atomizer. Inoculated seeds in the roll paper towel were incubated at a temperature of 15ºC for 48 hrs and symptom development was observed.

Virulence Index of Isolates of *P. halstedii*

Virulence index of various isolates was calculated by inoculating them on susceptible cultivar Morden. Observations were recorded on percent disease incidence

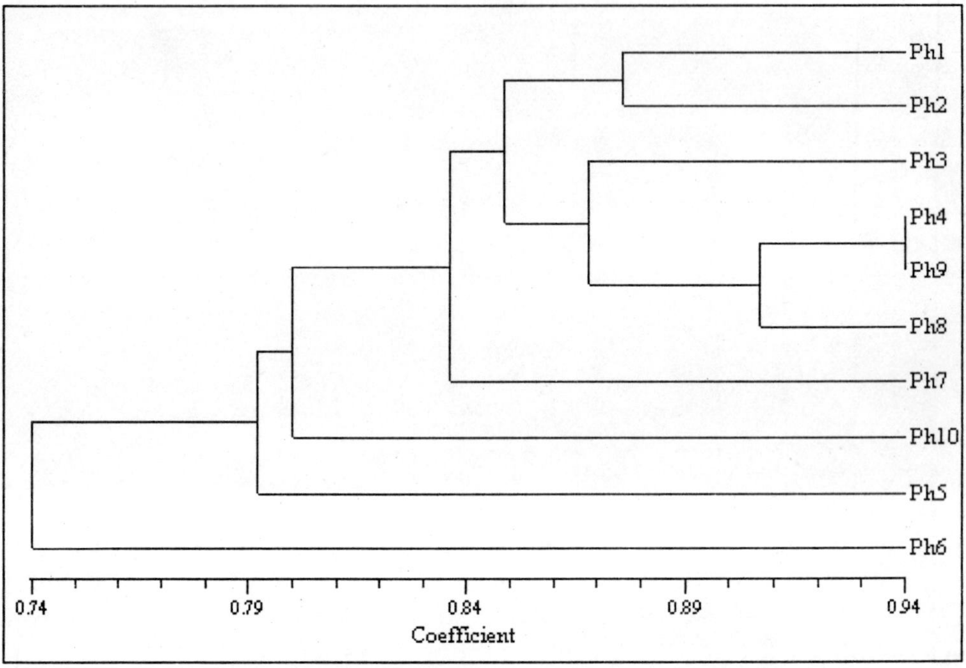

Figure 12.3: Dendrogram Showing the Clustering of Different Isolates of *P. halstedii*

and latent period required for expression of symptoms. Further the virulence index of each isolate was calculated by the following formula.

Virulence Index = Per cent Incidence x 1/Latent Period

Results and Discussion

It is evident from the Table 12.2 that all the isolates collected successfully produced the disease on HA 300, which has no resistance genes. All the isolates produced disease on true leaves only on HA300 and all other differentials were free from infection on true leaves, though sporulation was observed on HA 335 and HA R5 at cotyledon stage. The reaction obtained was similar to race 100 or European race 1 (Table 12.3). Similar reaction of all isolates on differentials indicated these to belong to a single race *i.e.* race 100. This may be due to limited spread of the pathogen, homothallic nature and narrow genetic base in the genotypes grown in India. Similarly Patil and Mayee (1990) reported that downy mildew isolate of Latur was European Race-1.

Morphological Variability

Mean sporangiophore length was highest in Dharwad (*Ph 3*) isolate (620 µm). Length of the sporangiophore ranged between 325 and 770 µm. Sporangiophore width was more in Mysore (*Ph 4*) isolate. Sporangial size varied from 10.50 × 12.50µm to 40.00 × 22.50µm. Mean sporangial size was highest in Mysore (*Ph 4)* and Latur (*Ph 6*) (Table 12.4).

Table 12.2: Reaction of Different Isolates of *Plasmopara halstedii* on Host Differentials of Sunflower Under Study

Sl.No.	Differential	Isolates												Remarks
		Ph 1	Ph 2	Ph 3	Ph 4	Ph 5	Ph 6	Ph 7	Ph 8	Ph 9	Ph 10	Ph 11	Ph 12	
1	HA 300	S	S	S	S	S	S	S	S	S	S	S	S	–
2	RHA 266 R	R	R	R	R	R	R	R	R	R	R	R	R	–
3	RHA 325 R	R	R	R	R	R	R	R	R	R	R	R	R	–
4	HA 61	R	R	R	R	R	R	R	R	R	R	R	R	–
5	RHA 274 R	R	R	R	R	R	R	R	R	R	R	R	R	–
6	HA 335	R	R	R	R	R	R	R	R	R	R	R	R	CLI
7	HA 337	R	R	R	R	R	R	R	R	R	R	R	R	–
8	RHA 340 R		R	R	R	R	R	R	R	R	R	R	R	–
9	DM 2	R	R	R	R	R	R	R	R	R	R	R	R	–
10	HI R 34	R	R	R	R	R	R	R	R	R	R	R	R	–
11	HA R4	R	R	R	R	R	R	R	R	R	R	R	R	–
12	HA R5	R	R	R	R	R	R	R	R	R	R	R	R	CLI
13	DM 4	R	R	R	R	R	R	R	R	R	R	R	R	–
14	DM 5	R	R	R	R	R	R	R	R	R	R	R	R	–
15	DM 6	R	R	R	R	R	R	R	R	R	R	R	R	–

CLI: Cotyledon limited infection considered as resistant as per literature.

Table 12.3: List of International Host Differentials of Sunflower Selected for Race Identification Studies, Genes Present and their Reactions to Different Established Races of *Plasmopara halstedii* in the World (*Gulya *et al.*, 1991)

Sl.No.	*Differential*	*Resistant Genes Present*	*Race reaction* 1	2	3	4	5	6	7
1	HA 300	–	S	S	S	S	S	S	S
2	RHA 266 R	Pl 1	R	S	S	S	S	S	S
3	RHA 325 R	Pl 2, Pl b	R	R	S	S	S	S	R
4	HA 61	Pl 2, 3 +?	R	R	S	S	S	R	R
5	RHA 274 R	Pl 2,9,Plb	R	R	S	S	S	R	R
6	HA 335	Pl 6+?	R	R	R	R	R	R	R
7	HA 337	Pl 7+?	R	R	R	R	R	R	R
8	RHA 340 R	Pl 8+?	R	R	R	R	R	R	R
9	DM 2	Pl 2, 5	R	R	R	S	S	S	S
10	HI R 34	Pl 4+?	R	R	S	S	S	R	R
11	HA R4	Pl f+?	R	R	R	R	R	R	R
12	HA R5	Pl g+?	R	R	R	R	R	R	R
13	DM 4	Pl c+?	R	R	R	R	S	R	S
14	DM 5	Pl d+?	R	R	R	R	S	R	R
15	DM 6	Pl e+?	R	R	R	S	S	R	R

Table 12.4: Morphological Variations Among the Isolates *Plasmopara halstedii*

Sl.No.	*Isolate*	*Sporangiophore Length (mm)*	*Sporangiophore Width (mm)*	*Sporangial Size (mm)*
1.	Gadag (*Ph* 1)	487-770 (602)	5.00-12.50 (9.05)	12.50–27.50 x 12.50–17.50 (21.16 x 18.00)
2.	Bangalore (*Ph* 2)	460-690 (595)	6.50-13.50 (8.95)	14.30-26.30 x 14.15-28.61 (20.65 x 17.97)
3.	Dharwad (*Ph* 3)	575-675 (620)	7.50-10.00 (8.63)	17.50-25.00 x 15.00-20.65 (21.00 x 18.16)
4.	Mysore (*Ph* 4)	525-650 (581)	10.00-15.00 (12.16)	17.50-27.50 x 17.50-22.50 (25.16 x 19.33)
5.	Narendra (*Ph* 5)	325-650 (485)	7.50-10.0 (8.12)	17.50-27.50 x 15.00-20.00 (20.83 x 17.66)
6.	Latur (*Ph* 6)	557-612 (595)	10.00-12.50 (11.05)	22.50-40.00 x 15.00-22.50 (25.17 x 17.67)
7.	Hyderabad (*Ph* 7)	460-695 (580)	6.50-13.50 (8.92)	18.15-28.45 x 15.00-20.00 (22.30 x 17.50)
8.	Bailhongal (*Ph* 8)	325-690 (565)	5.00-13.55 (8.15)	18.35-24.50 x 16.10x21.60 (21.85 x 19.35)

Contd...

Table 12.4–Contd...

Sl.No.	*Isolate*	*Sporangiophore Length (mm)*	*Sporangiophore Width (mm)*	*Sporangial Size (mm)*
9.	Davanagere (*Ph* 9)	470-630 (570)	5.00-12.5 (8.95)	14.65-27.30 x 15.00-21.00 (20.10 x 18.16)
10.	Haveri (*Ph* 10)	430-685 (565)	7.50-10.00 (8.30)	17.85-26.50 x 15.89-21.50 (20.53 x 18.65)
11.	Bidar (*Ph* 11)	470-630 (585)	5.00-13.00 (9.00)	15.00-25.65 x 14.65x21.30 (21.63 x 18.01)
12.	Parbhani (*Ph* 12)	470-630 (565)	10.00-15.00 (11.05)	14.30-27.50 x 15.00-22.50 (20.65 x 17.67)

Figures in parentheses are mean values.

Table 12.5: Virulence Index of Different Isolates of *Plasmopara halstedii*

Sl.No.	*Isolate*	*Virulence Index**
1	Gadag (*Ph* 1)	8.60
2	Bangalore (*Ph* 2)	12.11
3	Dharwad (*Ph* 3)	7.79
4	Mysore (*Ph* 4)	10.17
5	Narendra (*Ph* 5)	7.72
6	Latur (*Ph* 6)	9.81
7	Hyderabad (*Ph* 7)	10.20
8	Bailhongal (*Ph* 8)	7.85
9	Davanagere (*Ph* 9)	8.69
10	Haveri (*Ph* 10)	7.92
11	Bidar (*Ph* 11)	6.26
12	Parbhani (*Ph* 12)	8.90
	SEm ±	0.24
	CD (1 per cent)	0.70

*: Average of three replications.

Virulence Index

Results presented in Table 12.5 indicated that Bangalore (Ph-2) isolate was most virulent of all isolates tested with a virulence index of 12.11 and was followed by Hyderabad isolate (pH 7) with virulence index of 10.20. Bidar isolate was least virulent which recorded the least virulence index of 6.26. Based on the virulence pattern of the isolate, they are grouped into five different groups as follows:

Virulence Group	*Isolates of Plasmopara halstedii*
Group A	*Ph* 2
Group B	*Ph* 4, *Ph* 6, *Ph* 7
Group C	*Ph* 1, *Ph* 9, *Ph* 12
Group D	*Ph* 3, *Ph* 5, *Ph* 8, *Ph* 10
Group E	*Ph* 11

References

Abdullah, M. T. (1984). Sunflower hybrids resistant to Red River of downy mildew; Proceeding of XI International sunflower conference, Mar de Plata Argentina, 89.

Gulya, T.J., Miler J.F., Virnayi, F. and Sackston W.E. (1991). Proposed internationally standardized methods for race identification of *Plasmopora halstedii*. *Helia*. 14: 11-20

Kolte, S. J. (1985). Diseases of annual edible oilseed crops, vol. III sunflower, safflower and niger seed diseases, CRC Press INC., Florida USA, P. 154.

Leppik, E.E. (1964). Mapping the world distribution of seedborne pathogens. *Proc. Int. Seed Test Assoc.* 29: 473-477.

Mayee, C.D. and Patil, M.A. (1986). Downy mildew of sunflower. *Indian Phytopath*, 38: 314.

Patil, M.A. and Mayee, C.D. (1990). Race identity of Indian *Plasmopora halstedii*, cause of downy mildew of sunflower. *Indian Phytopath*. 43: 517-519.

Yagodkina, V.P. (1956). Downy mildew of sunflower in the krasnodar region. *Rev. Appl. Mycol.* 35: 101.

Zimmer, D.E. (1972). Field evaluation of sunflower for downy mildew resistance. *Plant Dis. Rep.*, 56: 478.

Chapter 13

Molecular Variability of Rice Blast Pathogen, *Pyricularia grisea* and Resistant Gene Deployment in Karnataka

S.K. Prashanthi, Srikant Kulkarni, B.S. Meena, Y.R. Hegde and N.G. Hanamaratti*

Department of Plant Pathology, University of Agricultural Sciences, Dharwad – 580 005, Karnataka, India

ABSTRACT

The rice blast pathogen, *Pyricularia grisea* was collected from various agro climatic regions of Karnataka and their molecular diversity was studied through RAPD markers. Primers OPB2, OPB4, OPB9, OPB10, OPB12, OPB14 and OPF3 showed cent percent polymorphism among twelve isolates collected. Cluster analysis delineated the isolates into five different lineages. Resistance status of *Pi*-genes against IB- 41 race of *P. grisea* was studied. Co-39 near isogenic lines with *Pi*-1, *Pi*- 2 and *Pi*- 4b offered good resistance to both leaf and neck blast. Pyramiding of R-genes appears promising for long lasting resistance against blast.

Keywords: Pyricularia grisea, Resistance, RAPD.

* E-mail: prasamhi@rediffmail.com.

Introduction

Rice is the staple food of 70 per cent of the Indian population and being cultivated under varying agro climatic conditions in our country. The crop suffers from several diseases amongst; blast disease is the major constraint to rice production. Around 50 per cent of the production may be lost in a field moderately affected by blast infection. Each year the pathogen destroys rice enough to feed around 60 million people. Rice blast is caused by a filamentous heterothallic ascomycetes fungus *Pyricularia grisea* (Teleomorph = *Magnaporthe grisea*) which can infect more than 50 hosts (Ou,1985).The pathogen infects all aerial parts of the plant causing leaf blast, neck blast, nodal blast and panicle blast (Figure 13.1a,b,c) The destructive nature of the disease is mainly due to the genetic plasticity of this pathogen. The variation in virulence spectrum of *P. grisea* population weakens the stability of resistance and complicates the breeding strategies. Knowledge on the population structure of this pathogen occurring in a particular agro-region is essential to assess the usefulness of resistant gene/s and their effective deployment in different localities. Earlier studies on variability of this pathogen were mainly based on virulence tests on a set of International differentials (Ling and Ou,1969). This is based on phenotypic traits which is highly variable and influenced by environment and hence not a fool-proof method to analyse the variability of the pathogen.

With the advancement of biotechnology in agriculture, different molecular tools *viz.,* MGR-DNA RFLP analysis (Levy *et al.,* 1991), pot 2-based rep PCR analysis (George *et al.*,1998), AFLP and RAPD (Srinivasachary *et al.*, 2002) markers are being used to reveal the genetic diversity of *M. grisea* population. In this context the present research was aimed to study the genetic variability existing among the *P. grisea* isolates from different agroclimatic regions of Karnataka using RAPD (Random amplified polymorphic DNA) markers. In this context, genetic variability existing among the blast isolates from different agroclimatic regions of Karnataka was studied using RAPD marker.

Information on organization and distribution of the blast pathogen population in specific geographic locations helps us to employ the R genes specific to the pathogen population present in that region. In some regions pyramiding of R genes in a cultivar is useful which can confer resistance to entire blast populations by effective complementation. Hence identification of R gene/s against a particular lineage or race group is essential for resistance to be durable. In this view several Near Isogenic Lines (NILs) with Pi gene/s and their pyramids were assessed for blast resistance against IB-41 race of *P.grisea*

Materials and Methods

Studies on Molecular Variability

Collection and Isolation of *P. grisea* Isolates

Twelve isolates of *P. grisea* used in this study were isolated from blast infected leaf, neck and nodal portions of rice plant collected from diverse agroclimatic regions of Karnataka (Table 13.1). The pathogen was isolated by following standard tissue isolation procedure (Tuite, 1969).

Isolation of Genomic DNA

Genomic DNA was isolated by the CTAB/NaCl method. All the monoconidial blast isolates were grown in 30 ml potato dextrose broth contained in 100 ml conical flask for 10 days at 27 ±1ºC. Later fungal mat was separated and freeze dried. 25 mg of lyophilized mycelia was ground in extraction buffer (50 mM Tris HCl pH 8.0; 150 mM Nacl; 100 mM EDTA) using precooled pestle and mortar. The contents were transferred to 1.5 ml eppendorf tube, vortexed until evenly suspended and incubated at 37ºC for 1 hr after adding 50 µl of 10 per cent SDS. Later 75 µl of 5 M NaCl and 65

(a) Leaf blast of rice

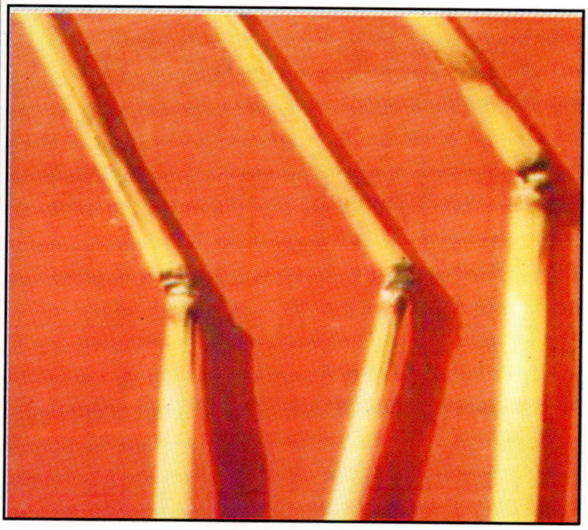

(b) Nodal blast

(c) Neck Blast

Figure 13.1

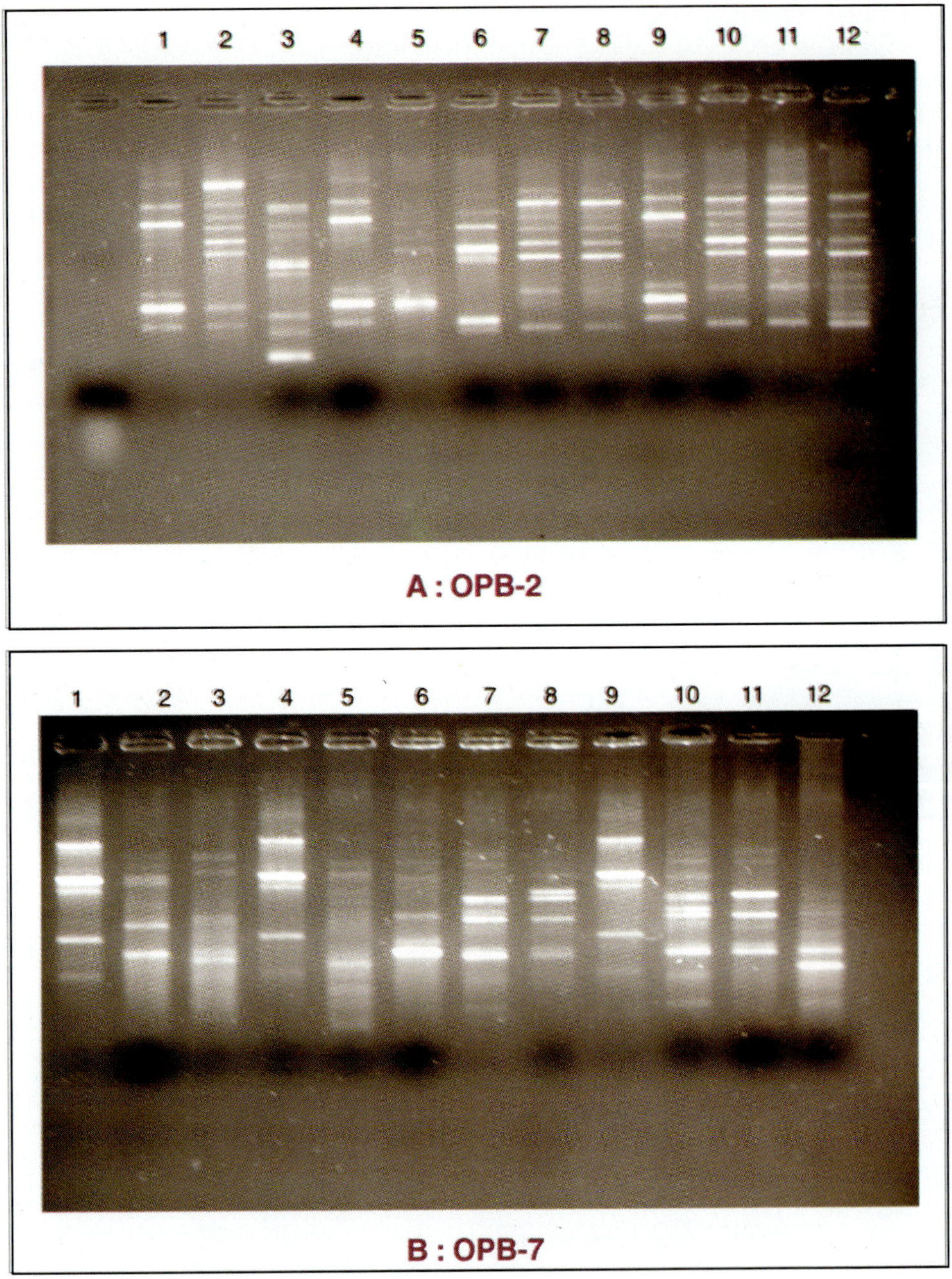

Figure 13.2: Gel Electrophorosis of RAPD Amplification of *Pyricularia grisea* Isolates

µl of CTAB/NaCl solution (10 per cent CTAB in 0.7 M NaCl solution) were added, mixed thoroughly and Incubated at 65ºC for 20 min. The mixture was emulsified with equal volume of chloroform; isoamyl alcohol (24:1) for 5 min by inversion and centrifuged at 10,000 rpm for 12 min. The aqueous phase was transferred to centrifuge tube to which 0.6 V of isopropanol was added and incubated at –20°C for overnight. The tubes were centrifuged at 10,000 rpm for 12 min and supernatant was discarded. The pellet was washed in 70 per cent ETOH and dried completely. DNA was freed from RNases. The pellet was dissolved in 100 ml of $T_{10}E_1$ buffer (Tris 10 mM + 1 mM EDTA) and stored at –20°C.

Table 13.1: Different Isolates of *Pyricularia grisea* Collected from Different Regions of Karnataka

Isolate.no.	*Location*	*Plant Part*	*Variety*	*Zone*
Pg1	Siruguppa	Leaf	BPT-5204	3 (Northern Dry zone)
Pg 2	Bheemarayanagudi	Leaf	BPT-5204	2 (Northern Dry zone)
Pg 3	Raichur	Leaf	BPT-5204	2 (Northern Dry zone)
Pg 4	Sirsi	Leaf	Intan	9 (Hilly zone)
Pg 5	Mundgod	Leaf	Abhilash	9 (Hilly zone)
Pg 6	Shimoga	Leaf	Jyothi	7 (Southern transitional zone)
Pg 7	Mugad	Leaf	HR-12	8 (Northern transition zone)
Pg 8	Haveri	Leaf	Alursanna	8 (Northern transition zone)
Pg 9	Khanapur	Leaf	Dodiga	9 (Northern transition zone)
Pg 10	Mugad	Node	HR-12	8 (Northern transition zone)
Pg 11	Mugad	Neck	HR-12	8 (Northern transition zone)
Pg 12	Sirsi	Neck	Intan	8 (Northern transition zone)

PCR Conditions and Cluster Analysis

RAPD primers OPB and OPF series obtained from Integrated DNA Technologies Sigma Industrial and Lab Equip. Inc., Bangalore,India was used to study the polymorphism among the isolates (Table 13.2). PCR was performed in a 20 μl reaction volume containing 3.0 units Taq DNA polymerase, dNTP mix (2.5 mM each of dCTP, dATP, dTTP, dGTP). 10 X assay buffer (20 mM Tris-HCl, pH 8.4, 50 mM KCl, 15 mM $MgCl_2$; Bangalore Genei Pvt. Ltd., Bangalore, India), 5 pM primer and 25 ng of template DNA. The reaction mixture was vortexed and centrifuged briefly. Amplification was performed in a Thermocycler, Eppendorf, Mastercycler gradient (Eppendorf gradient, 2231, Hamburg, Germany). The PCR amplification for RAPD analysis was performed according to Williams *et al.* (1990) with following modifications; 94°C for 1 min for denaturation followed by primer annealing at 36° for 1 min and primer extension at 72°C for 2 min with total 40 cycles. The initial denaturation of DNA was for 4 min at 94°C and final extension at 72°C for 5 min one cycle each was performed. The PCR products were analysed on 1.2 per cent agarose gel containing ethidium bromide (0.5μg/ml) and electrophoresed at 60 volts for 2 h. The gel was visualized on a UV transilluminator and images were taken by using Gel documentation system. To determine the genetic relationship among isolates, the presence or absence of bands was converted into binary data (1 for presence and 0 for absence of each band).

Table 13.2: Random Primers with Following Sequences were Used in RAPD

Primer	*Sequence*
OPB-2	5 TGA TCC CTG G-3
OPB-3	5 CAT CCC CCT G-3
OPB-4	5 GGA CTG GAG T-3
OPB-5	5 TGC GCC CTT C-3
OPB-6	5 TGC TCT GCC C-3
OPB-7	5 GGT GAC GCA G-3
OPB-8	5 GTC CAC ACG G-3
OPB-9	5 TGG CTG ACT C-3
OPB-10	5 CTG CTG GGA C-3
OPB-11	5 TGA TCC CTG G-3
OPB-12	5 CCT TGA CGC T-3
OPB-13	5 TTC GCT CGC T-3
OPB-14	5 TCC GCT CTG G-3
OPF-1	5 ACG GAT CCT G-3
OPF-2	5 GAG GAT CCC T-3
OPF-3	5 CCA AGC TTC C-3
OPF-4	5 GGT GAT CAG G-3
OPF-16	5 GGA GTA CTG G-3

Analysis of the Profile of the Amplified Fragments

Pair wise genetic similarities (Sij) between isolates were estimated by DICE similarity coefficient. Clustering was done using the symmetric matrix of similarity co-efficient and cluster obtained based on unweighted pair group arithmetic mean (UPGMA) using SHAN model of NTSYS–PC version 2.0 (Rohlf, 1998).The similarity measurements were converted to genetic distance measurements as (1–5M) × 100.

Phenotypic Analysis of Isolines with Pi-genes

The experiment was conducted at Agriculture Research Station- Mugad, University of Agricultural Sciences-Dharwad, situated at an altitude of 697 MSL with 15°-15′ N latitude and 74°-40 E′ longitude. Near isogenic lines each carrying the major genes Pi -1, Pi 2, Pi- 3, Pi-4a and Pi-4b individually and in combinations of two and three genes in the background of susceptible recurrent parent Co-39 were evaluated in Uniform Blast Nursery under field conditions during wet season 2004 and 2005 against leaf and neck blast. In addition, recombinant Inbred lines (RIL) with Pi-5, Pi-7, Pi-12 and wild rice *O. minuta(Pi-9)* were also screened for blast resistance. Each entry was densely sown in two rows of 50 cm length and the susceptible check HR-12 was sown in two rows all around the entries. After every 10 lines one row of HR-12 was sown to ensure uniform and maximum disease pressure. Observations on disease severity and lesion type were recorded twice to know the disease progress in each line with time interval. Leaf blast and Neck blast severity was recorded by adopting 0-9 scale (IRRI, 1996) and the Pi lines were evaluated against race IB-41 prevailed at testing centre (Prashanthi *et al.*, 2005).

Results and discussion

Molecular Polymorphism of *P. grisea* Isolates

Genetic variation was detected among twelve *P. grisea* isolates collected from different rice varieties grown in various agro climatic regions by using RAPD marker. OPB and OPF series primers were used to determine genetic polymorphism and to construct dendrogram. Banding profile of different primers for the isolates of *P. grisea* is given in Table 13.3 and shown in Figure 13.2. Of the 18 primers used for amplification OPB2, OPB4, OPB7, OPB9, OPB10, OPB12, OPB14 and OPF3 showed cent per cent polymorphism. The isolates exhibited overall polymorphism of about 87.8 per cent. Total 135 bands were obtained of which 122 were polymorphic and 13 were monomorphic. The analysis of RAPD data showed the genetic similarity among the isolates. A dendrogram was generated by similarity co-efficient (*Sm*) (Figure 13.3). The genetic similarity co-efficient values for the isolates ranged from 0.34 to 0.86 (Table 13.4). One pair of isolate nodal blast Mugad-Neck blast Mugad were closely related with high similarity co-efficient value (0.86). Cluster analysis of the banding patterns delineates isolates into discrete groups which can be inferred to represent genetically related lineages. UPGMA analysis of total data set of RAPD markers classified the isolates into five DNA fingers print groups.

The first group consisted of two isolates Siruguppa (Pg1) and Shimoga (Pg6) isolates with 57 per cent genetic similarity. The second group comprised five isolates with maximum genetic similarity shared between two pairs of isolates, Mugad neck blast–Nodal blast Mugad (86 per cent): Mugad ; Haveri Leaf blast (82 per cent). The isolates Pg2, Pg5 and Pg3, though they were from diverse agro climatic regions they found genetically related and in a single group (V group). Contrarily, leaf and neck blast isolates collected from the same location (Sirsi) were in different finger print group. Our results indicate the existence of local and geographical polymorphism in *P. grisea*. The diversity and variability of the pathogen originate from the clonal reproduction, mutation, migration, selection, heteroploidy and parasexuality of the pathogen (Ou, 1985). The number of lineages and their composition prevailing in an area depend on the rice varieties grown in that region because varieties with diversified genetic background exert enormous selection pressure on the variability of *P. grisea* (Srinivasachary *et al.*, 2002).

In our study the isolates of *P. grisea* were distributed into five groups/lineages and the variability exhibited by them might be attributed to distinct geographical regions and host genetic diversity. The varieties deployed in the regions are based on crop seasons along with other biotic and geographic factors. Our results are in line with the previous reports (Sharma *et al.*, 2000; Sonia Chadha and GopalKrishna, 2005) who studied the genetic variation among the Indian blast isolates. The present study concludes that, the population of blast pathogen in Karnataka is genetically heterogeneous and the interrelationships among the different isolates can be reliably explained by RAPD marker.

Table 13.3: Banding Profile of Different Primers for Different Isolates of *Pyricularia grisea*

Sl.No.	*Primer*	*Total Bands*	*Polymorphic*	*Per cent Polymorphism*	*Sl.No.*	*Primer*	*Total Bands*	*Polymorphic*	*Per cent Polymorphism*
1	OPB 2	4	4	100	10	OPB 11	11	10	90.90
2	OPB 3	2	1	50	11	OPB 12	13	13	100
3	OPB 4	7	7	100	12	OPB 13	10	8	80.00
4	OPB 5	18	17	94.44	13	OPB 14	4	4	100.00
5	OPB 6	6	4	66.66	14	OPF 1	5	4	80.00
6	OPB 7	6	6	100	15	OPF 2	6	5	83.33
7	OPB 8	12	10	83.33	16	OPF3	7	7	100.00
8	OPB 9	8	8	100	17	OPF 4	7	6	85.71
9	OPB 10	6	6	100	18	OPF 16	3	2	66.66

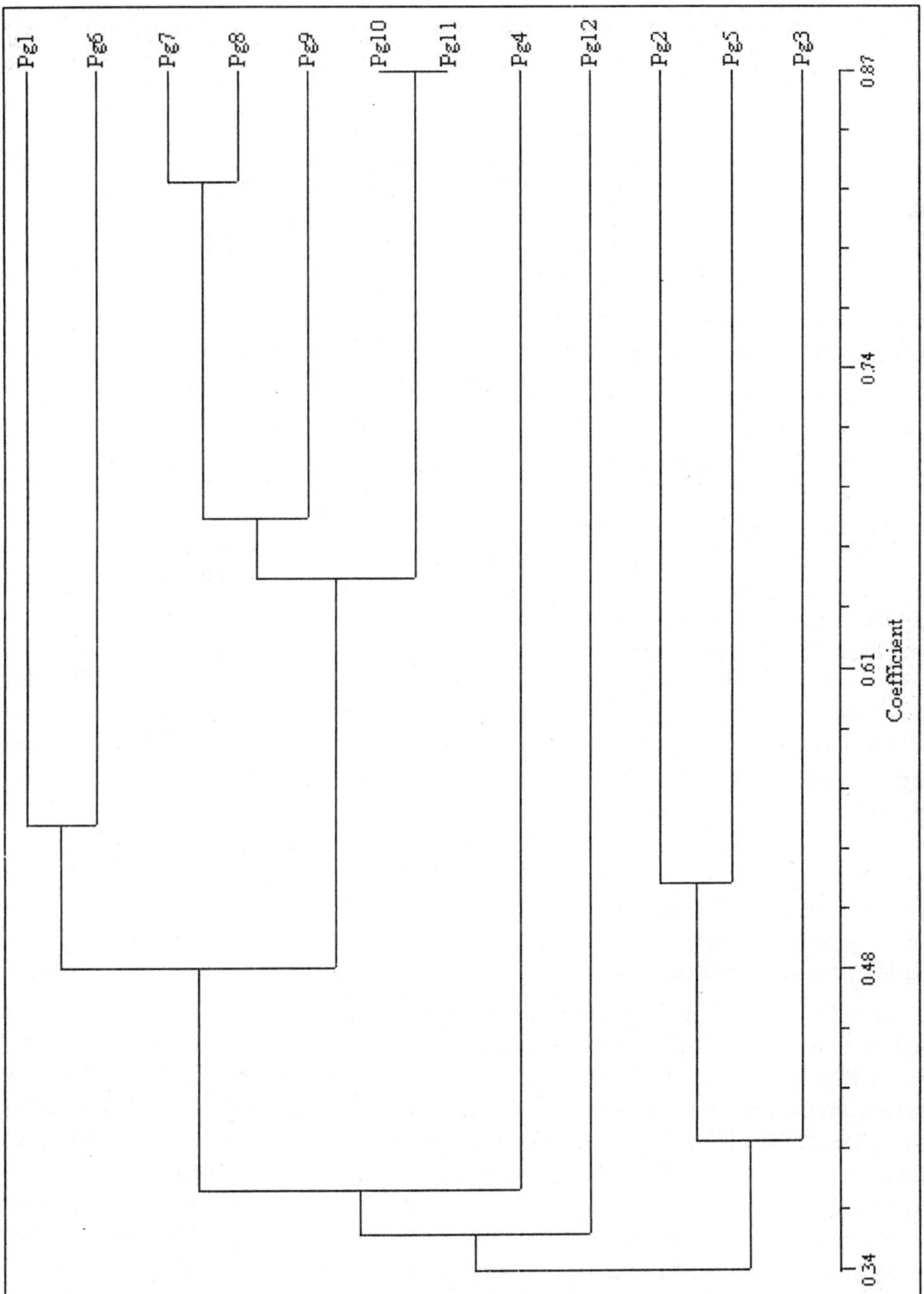

Figure 13.3: Dendrogram Based on RAPD Analysis of 12 Isolates of *Pyricularia grisea*

Table 13.4: Similarity Co-efficient of 12 Isolates of *Pyricularia grisea*

Isolates	*Pg 1*	*Pg 2*	*Pg 3*	*Pg 4*	*Pg 5*	*Pg 6*	*Pg 7*	*Pg 8*	*Pg 9*	*Pg 10*	*Pg 11*	*Pg 12*
Pg 1	1.00											
Pg 2	0.39	1.00										
Pg 3	0.30	0.41	1.00									
Pg 4	0.43	0.6	0.32	1.00								
Pg 5	0.26	0.51	0.38	0.27	1.00							
Pg 6	0.57	0.51	0.34	0.41	0.31	1.00						
Pg 7	0.39	0.41	0.29	0.34	0.33	0.51	1.00					
Pg 8	0.39	0.37	0.29	0.34	0.33	0.51	0.82	1.00				
Pg 9	0.53	0.38	0.29	0.50	0.35	0.49	0.66	0.68	1.00			
Pg 10	0.42	0.42	0.36	0.33	0.32	0.53	0.74	0.63	0.69	1.00		
Pg 11	0.42	0.33	0.34	0.28	0.28	0.48	0.65	0.59	0.56	0.86	1.00	
Pg 12	0.12	0.32	0.35	0.24	0.32	0.34	0.45	0.35	0.39	0.51	0.44	1.00

Resistance Status of Pi-Genes Against Leaf and Neck Blast (IB-41 Race)

The near isogenic lines Co 101 LAC, Co101 A51 and Co 105 TTP with Pi-1, Pi-2 and Pi-4b gene respectively, exhibited high level of resistance to leaf blast and neck blast (Table 13.5). The disease severity (1-2 grade) and lesion type (B) in these was not increased further with time interval. The NILs with Pi-3 and Pi-4a gene were highly susceptible to leafblast (7 grade) as well as neck blast (5 grade).The recurrent parent Co-39 with Pi-a gene was highly susceptible to leaf blast (grade 8) and neck blast (grade 7). The resistance status of RILs having Pi-5 and Pi-12 was very poor with high leaf and neck blast index (grade 8 and lesion type E(D)/E(C)). Pi-7 showed good level of resistance only during leaf stage but it failed to offer resistance to Neck blast. Importance to be given to select major gene/s which has both leaf and neck blast resistance for long lasting resistance. Gene Pi-9 appears promising as it showed resistance to leaf as well as neck blast and there is scope to exploit of this gene for broad spectrum resistance.

Among the two gene combinations Pi-1 + Pi-2 and Pi-2 + Pi-4 were highly resistant to leaf blast and neck blast. The pyramid with three gene combination Pi-1 + Pi-2 + Pi-4 was highly resistant with A lesion type and severity 1 throughout the observation period. In Southern India, the major blast resistance genes Pi-1 and Pi-2 excluded majority of lineages and their combination is also considered useful (Lavanya and Gnanamanickam, 2000). The pyramiding of R-genes offers prolonged resistance through both ordinary gene action and quantitative complementation resulting in durable resistance (Hittalamani *et al.*, 2000). Performance of Pi-1 + Pi-4 combination was not good against IB-41 race and showed leaf blast (severity 5, LT;C) and neckblast (grade 5) infection. This may be attributed to incompatible gene combinations for resistance. Thus indiscriminate combinations of major genes may be ineffective against some populations in the field. Hence, knowledge on population structure of blast pathogen and resistance spectrum of genes against a particular lineage or race group is essential for the success of resistance breeding programme and resistance to be durable.

Table 13.5: Reaction of Rice Lines with Pi-gene/s Against *Pyricularia grisea* Under Upland Ecosystem of Karnataka, India

Sl. No.	*Pi-line*	*Gene*	*2004*			*2005*		
			Leaf Blast		*Neck Blast*	*Leaf Blast*		*Neck Blast*
			LT	*Severity*		*LT*	*Severity*	
1.	C101LAC	Pi-1	B	1	3	B	1	3
2.	C101A51	Pi-2	B	2	3	A	1	3
3.	C104PKT	Pi-3	E(c)	5	5	E(C)	5	5
4.	C101PKT	Pi-4a	E(D)	5	5	D	5	5
5.	C105TTP	Pi-4b	B	1	3	B	1	3
6.	RIL-45	Pi-5	E(C)	8	7	E(C)	8	5
7.	RIL-29	Pi-7	C	2	5	C	3	7
8.	RIL-10	Pi-12	E(D)	8	5	E(C)	6	5
9.	*O.minuta*	Pi-9	B	1	3	C	3	3
10.	BL-122	Pi1+Pi2	A	1	1	A	1	3
11.	BL-142	Pi1+Pi4	E(C)	5	5	E(D)	7	5
12.	BL-245	Pi2+Pi4	C	2	3	B	1	3
13.	A57	Pi1+Pi2+Pi4	A	1	1	B	1	3
	Checks							
14.	Co-39	Pi-a	E(D) and E(C)	8	7	E(C)	7	7
15.	HR-12		E(D)	9	9	E(D)	8	7

LT: Lesion type; A: No symptom/reddish brown speck; B: Minute reddish brown speck; circular spots without ashy center; C: Circular spots (2-3mm) with ashy zone; D: Spindle shaped lesions; E: Large spindle lesions with ashy centre; 3-5mm broad, up to several cm length; E(C) and E (D): Burning of leaves due to coalition of several lesion type C and D.

References

Ou, S.H. (1985). *Rice Diseases* 2nd edition. Commonwealth Mycological Institute. Kew, UK, pp.109-201.

Ling, K.C. and Ou, S.H. (1969). Standardization of the International race numbers of *Piricularia oryzae* cav. *Phytopathology*, 59: 339–342.

Levy, M. Romao, J. Marchetti, M.A. and Hamer, J.E. (1991). DNA fingerprinting with dispersed repeated sequence resolves pathotype diversity in the rice blast fungus. *Plant Cell*. 3: 95–102.

George, M.L.C., Nelson, R.J., Zeigler, R.S. and Leung, H. (1998). Rapid population analysis of *Magnaporthe grisea* by using rep-PCR and Endogenous repetitive DNA sequences. *Phytopathology*. 88: 223–229.

Srinivasachary, S., Shailaja, Hittalmani, Vaishali, M., Shashidhar, H.E. and Girish Kumar K. (2002). Genetic analysis of rice blast fungus of southern Karnataka using DNA markers and reaction of popular rice genotypes. *Current Science*, 42: 25–28.

Tuite, J. (1969). *Plant Pathological Methods, Fungi and Bacteria*. Burges Publ. Co., USA, p. 239.

Williams, J. G. K., Kubelik, A. R., Livik, K.J., Rafalsti, J.A. and Tingey, S.V., (1990). DNA amplified by arbitrary primers are useful genetic markers. *Nucleic acid Research*, 18: 6531–6535.

Rohlf, F.I., (1998). *NTSYS-PC. Numerical taxonomy and multivariate analysis 2.0* Applied Bio-statistics Inc., NewYork.

IRRI, (1996). Standard evaluation system for rice. *International Rice Testing Program*, Philippines.

Prashanthi, S.K., Srikant, Kulkarni and Hanamaratti, N.G., (2005). Virulence pattern of *Magnaporthe grisea*, a rice blast pathogen under rainfed situations of Karnataka. p. 26-27, *National Symposium on Crop Disease Management in Dryland Agiculture*. Jan. 12-14, MAU, Parbhani, Maharastra.

Sharma, T. R., Chauvan, R.S., Singh, B.M., Paul, R., Sagar, V., Sharma, N.R., Rana, M. and Mahajan,Y., (2000). Molecular characterization of genetic variability in *Pyricularia grisea* using RAPD markers. *Indian Phytopathology*, 2: 1211–1213.

Sonia, Chadha and GopalKrishna, T., (2005). Genetic diversity of Indian isolates of rice blast pathogen (*Magnoporthe gresia*) using molecular markers. *Current Science*, 88: 1466-1469.

Lavanya, Babujee and Gnanamanickam, S.S. (2000). Molecular tools for characterization of rice blast pathogen (*Magnaporthe grisea*) population and molecular marker–assisted breeding for disease resistance. *Current Science*, 78: 248-256.

Hittalamani, S., Parco, A., Mew, T.V. and Zeigler, R.S. (2002). Fine mapping and DNA marker assisted pyramiding of three major genes for blast resistance in rice. *Theor. Appl. Genet.*, 100: 1121–1128.

Chapter 14

Diagnosis, Molecular Characterization and Management of Phytoplasma Diseases: A Review

G.P. Rao* and Smriti Mall

Sugarcane Research Station,
Kunraghat, Gorakhpur – 273 008, U.P., India

ABSTRACT

Obligate parasitism is a major barrier in the diagnosis and characterization of phytoplasmas. Efficient detection methods are vital for its management. Diagnostic methods include biological, microscopy, staining, histological, immunological, and molecular approaches in both plant and in insect hosts. Molecular techniques are basic tools in epidemiological studies for identifying vector species, phylogeny and in characterization of strains. In last 15 years application of DNA-based technology has provided an abundance of molecular techniques with high sensitivity that has facilitated the introduction of laboratory test for phytoplasma diseases. PCR assays have provided the most sensitive means available for phytoplasma detections. Methods based on DNA amplification using the PCR, followed by RFLP, and sequencing are now used routinely in phytoplasma detection and characterization, by using the 16S rRNA gene and the

* E-mail: gprao_gor@rediffmail.com.

16/23S spacer region as targets for their phylogenetic and taxonomic significance. Other recent development in the diagnosis of phytoplasma based on nested and real–time PCRs, Q-PCR, NAH assays, HMA assays. Molecular diagnosis and management of phytoplasma is reviewed in this article.

Keywords: *Phytoplasma, Serology, DNA hybridization, NAH, HMA, PCR, cloning, rDNA, RFLP.*

Introduction

Phytoplasma are non-helical mollicutes associated with diseases of several plant species characterized by flower malformation, growth aberrations, yellowing and/or decline, witches'-broom, chlorosis, little leaf, shoot proliferation, multiple meristem and generalized stunting. They were thought to be caused by viruses until discovery by Japanese scientists in 1967 (Doi *et al.*, 1967). They were previously called "mycoplasma like organisms" (MLOs) because of their morphological similarity to the wall-less mollicutes mycoplasmas, known to cause numerous disorders in human and animals. Recent evidence showing that MLOs are distantly related to mycoplasmas led to their designation as "phytoplasma", a name that reflects their primary plant hosts (Sears and Kirkpatrick, 1994). Phytoplasmas are phloem parasites and pleomorphic in shape (Figure 14.1), which colonize almost exclusively the phloem sieve elements of infected plants and are sensitive to the antibiotic tetracycline. The genome phytoplasma genome is the smallest among known cellular plant parasites and pathogens. It is estimated to be between 600 to 1200 kbp, present as one large circular double stranded DNA chromosome.

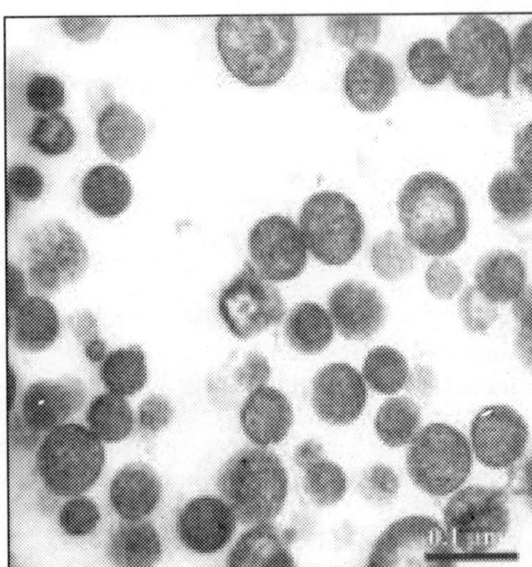

Figure 14.1: Transmission Electron Micrograph of Infected Sieve Tissues of Phloem Cells of Phytoplasma-Infected Plants Showing Pleomorphic Bodies Ranging from 200-800 nm

Phytoplasmas are associated with plant yellows diseases, which were known to occur since the early 1900s, but were then thought to be of viral origin. They cause diseases in several hundreds crop species (Lee *et al.*, 2000), including many important food, vegetable, and fruit crops as well as in ornamental plants and timber and shade trees from temperate to tropical regions (Lee *et al.*, 1997, Jones, 2002, Al-saady and Khan, 2006).

A plant infected by phytoplasmas exhibit variety of symptoms suggests profound disturbances in the normal balance of growth regulators. The common symptoms include virescence/phyllody, sterility of flowers, proliferation of auxiliary buds resulting in witches'-broom behaviour, chlorosis, little leaf, abnormal internodes elongation, shoot proliferation, Leaf and flower malformation and discoloration, multiple meristem, lethal yellowing, enlarged anomalous stipules, yellowing of successively younger leaves and spear necrosis and generalized stunting (Figures 14.2 A-I) (Lee *et al.*, 1997, Al-saady and Khan, 2006).

They present particular problems in pursuit of Koch's Postulates to establish them as the cause of a disease as they cannot be cultured *in vitro*, and are only transmissible by insect, dodder or graft (Davis and Sinclair, 1998; Jones, 2002).

Their high genetic recombination frequency, adaptability and fast evolution rate aggressiveness in colonizing new ecological niches and hosts leds to phytosanitary constraints and considerable economic losses for many countries (Davis and Sinclair, 1998; Bai *et al.*, 2004). The genomic era of mollicutes attracted much attention due to small genomes and their clinical and agricultural impact (Bai *et al.*, 2004). Phytoplasmas are considered to be one of the most molecularly enigmatic genera of plant pathogens because of our lack of knowledge about their host-pathogen-vector interactions (Melarned *et al.*, 2003). Much effort has been directed to develop quick and reliable diagnostic tools and to study their biodiversity, evolution, epidemiology, and to develop sustainable and integrated disease management practices.

Diagnosis of Phytoplasma

Phytoplasmas are wall-less pleomorphic bacteria with varying sizes ranging from 200 to 800 nm, which could survive and multiply only in plant phloem or insect haemolymph. They are strictly host-dependent, but could multiply in insect vectors and also survive in their eggs. Due to the lack of molecular characterization of phytoplasmas, their taxonomy is mainly based on symptoms, host-parasite interaction, host range and insect vector. Microscopic methods including Transmission Electron Microscopy (TEM) and light microscopy have been used to detect phytoplasmas, but most sensitive is the DAPI (DNA-specific-6-diaminido-2-phenylindole) fluorescence microscopy technique (Figures 14.3 A-B). However, these techniques require tissue fixation. Recently bio-imaging methods requiring sensitive, specific and non-toxic fluorescent dyes and the use of confocal or multiphoton microscopy have allowed identification of phytoplasmas in living tissues (Christensen *et al.*, 2004).

The pathogen identification relied for more than two decades on DAPI staining or electron microscopy detection. However, in the last 15 years, the applications of DNA-based technology allowed to preliminary distinguish different molecular clusters inside these prokaryotes. The Phytoplasma Working Team of the International Research Project for Comparative Mycoplasmology (IRPCM) adopted the trivial name 'phytoplasma' to identify the prokaryotes belonging to this group. The '*Candidatus* Phytoplasma' genus has been proposed and adopted in order to start formal classification of these prokaryotes, some of them associated with important or quarantine-subjected plant diseases (IRPCM, 2004). Up to date satisfaction of Koch postulates has not been achieved, but indirect proof, such as phytoplasma and symptoms eliminating after tetracycline treatments, confirmed that they are associated with many plant diseases worldwide.

The phytoplasma genome is very small and its phylogenetic studies propose that the common ancestor for phytoplasmas is *Acholeplasma laiidlawii* in which the triplet coding for tryptophan (trp) is UGG, while in the other prokaryotes, enclosing mycoplasmas and spiroplasmas, trp is coded by UGA. Phytoplasmas are genetically distinguishable from mycoplasmas infecting human and animal by the presence of a spacer region (about 300 bp) between 16S and 23S ribosomal regions, which codes isoleucine tRNA ($tRNA^{Ile}$) and part of the sequences for alanine tRNA ($tRNA^{Ala}$). Sequencing of complete rRNA genes for two phytoplasma strains shows that tRNA coding for valine and asparagine are located downstream from the 5S rRNA gene, and this is a unique feature of phytoplasmas (Ho *et al.*, 2001). First phytoplasma identification and classification was proposed based on specificity of vector transmission, host range, but more recently, on symptom expression of a common host (periwinkle). Experimentally determined plant host range and insect vector species are broader than those observed in nature, and show a considerable amount of overlaps. In case of symptom syndromes induced in infected plants similarities and differences in species of insect transmitting phytoplasmas and in species of plant hosts could reflect genetic differences in pathogens as well as the genetics of the plant host and insect vectors.

Diagnostic methods include biological, histological, immunological, and molecular approaches in both plant and insect hosts. Biological tests are slow and results may be difficult to interpret. Histological detection is based on examination of phloem tissues under a flourense microscope after staining the phytoplasma DNA with the fluoresecent dye DAPI or by electron microscopy. DAPI (4'6 diamidino-2-pheilindole) is a fluorochrome dye that binds to AT-rich dsDNA present in the phytoplasma cells (Seemuller, 1976). Thin sections (20-30um) of young tissues (petioles of young leaves, phloem tissues, of shoots, brnaches and roots) are stained with 1mg/ml aqueous DAPI solution, and subsequently observed under fluorescence microscope. The observation of accumulations of a bluish-white fluorescence in sieve elements indicates the presence of phytoplasmas.

Serology and DNA Hybridization

The introduction of molecular methods has facilitated the significant improvements in phytoplasma extraction from infected plant and insect hosts (Kirkpatrick *et al.*, 1987; Lee and Davis, 1988; Sears *et al.*, 1989; Davis *et al.*, 1997). Polyclonal antisera and monoclonal antibodies have been produced against various phytoplasmas (Kirkpatrick, 1992; Lee and Davis, 1992). The use of serological methods

Figure 14.2(A–I): Phytoplasma Disease symptoms of aster yellows.

A: Gladiolus plants healthy (at left) and affected with multiple sprouts and very weak roots; B: Affected *D. spectabilis* plants with shoot proliferation; C: Bermuda grass with white leaf; D: *Catharanthus* little leaf; E: *Cannabis* witche's–broom; F: Leaf and flower malformation and discoloration of rose; G: *Chrysanthemum* little leaf; H: Apple proliferation-infected apple tree; I: Sugarcane grassyshoot disease with chlorosis and witches broom symptoms.

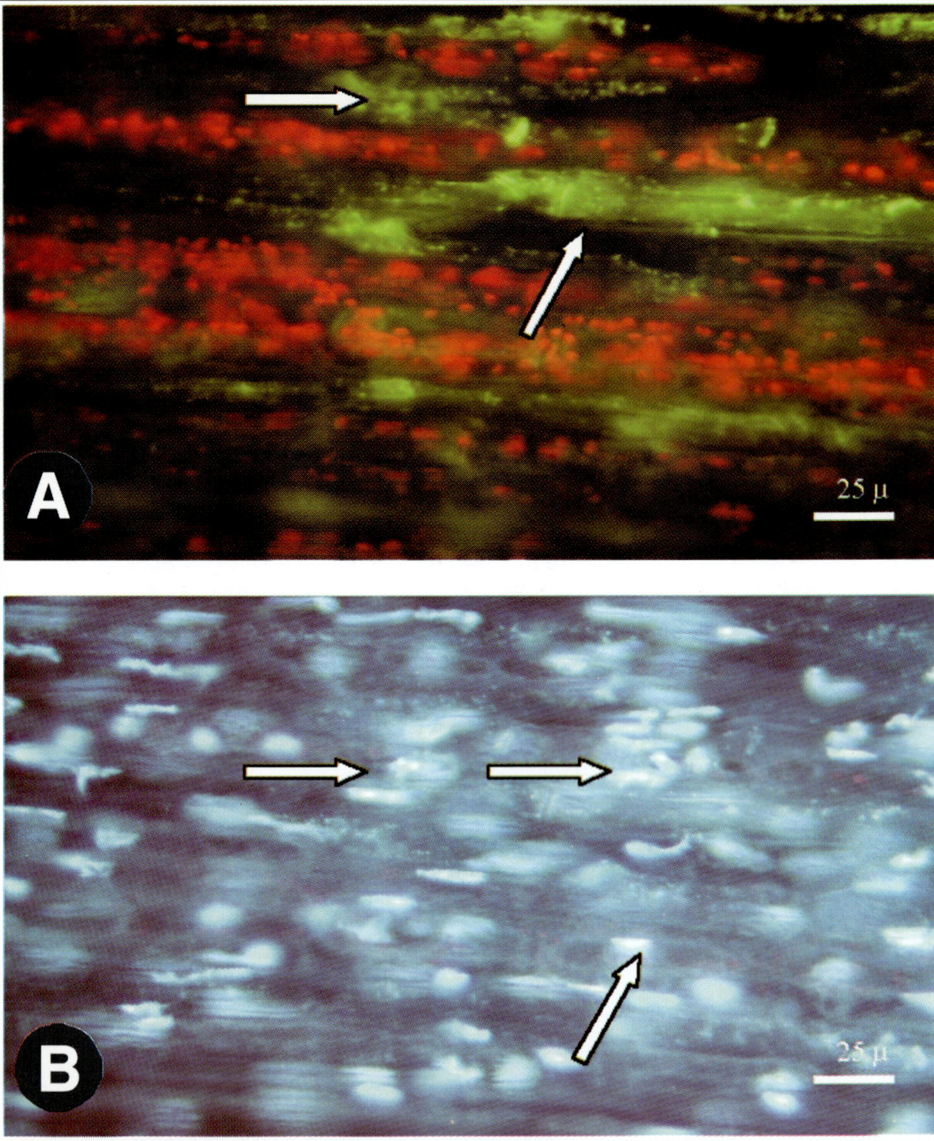

Figure 14.3 (A-B): A: Immunofluorescence staining on longitudinal section of AP-infected stem using AP Mab; Arrows indicate typical green FITC fluorescence; B: Some section described in Figure A. but stained with DAPI; arrows indicated typical DAPI fluorescence.

to detect phytoplasma is mainly limited to the closely related strains that are difficult to distinguish at ribosomal RNA gene level. Evidently, monoclonal antibodies have been found promising for serological detection of phytoplasmas from a wide range of host plants, including woody perennials (Lee *et al.*, 1993a; Guo *et al.*, 1998).

Serological diagnosis of phytoplasma can be employed with polyclonal or monoclonal antibodies. Considering the difficulties on phytoplasma purification from host tissues, due to their fragility, pleomorphic structure and low titre, only a few polyclonal antisera suitable for diagnosis were obtained against phytoplasma infecting herbaceous (Clark *et al.*, 1989). Sous (Chen and Jiang, 1988; Garnier *et al.*,1990; Hsu *et al.*, 1990; Loi *et al.*, 1998) and woody plants (Jiang *et al.*, 1989; Davis and Clark, 1992). Immunofluorescence (IF) is performed directly on plant tissues. *In situ* detection of plant pathogens is well suited for phytoplasmas as they are confined to interior phloem sieve elements of plants. Also, it is possible to combine IF with DAPI staining on the same tissue section, which can improve the clarity of results (Figures 14.3A-B).

Molecular Diagnosis

For many years differentiation and classification of phytoplasmas relied primarily on their biological properties such as symptomatology of affected plants and specificity of plant and insect hosts, these properties are time-consuming, laborious, and sometimes unreliable (Lee *et al.*, 1998; Jones, 2002). The use of molecular probes, phytoplasma-specific cloned DNA and monoclonal antibodies, have made it possible to identify and classify phytoplasmas on the basis of DNA-DNA homology and serological data (Lee *et al.*, 1998; Garcia-Chapa *et al.*, 2004). However, the sensitivity of these assays was insufficient, for example, many phytoplasmas are associated with diseases of woody plants where phytoplasma concentrations are often low (Lee *et al.*, 1998; Seemüller *et al.*, 1998; Jones, 2002).

DNA technology has provided an abundance of molecular techniques with high sensitivity that has facilitated the introduction of laboratory test for phytoplasma diseases (Adam *et al.*, 2001). Molecular techniques are basic tools in epidemiological studies for identifying vector species and phylogeny, and in the characterization of strains. Molecular diagnosis involves a number of steps including sampling the tissues, which is suspected of containing the phytoplasmas, extraction of phytoplasma DNA from the plant tissue, molecular analysis, and interpretation of results. An essential prerequisite for diagnosis is keeping the proper control of phytoplasma strains. Molecular diagnosis is also essential for managing phytoplasma diseases as part of standardized detection schemes used by diagnostic laboratories. Polymerase chain reaction (PCR) assays have provided the most sensitive means available for phytoplasma detections. Currently, phytoplasma detection and characterization are based predominantly on PCR amplification of 16S rRNA genes (rRNAs) (Ahrens and Seemuller, 1992; Lee *et al.*, 1993; Gundersen and Lee, 1996; Okuda, 1997). For phytoplasma detection by PCR, universal primer pairs based on conserved regions of the ribosomal RNA gene operon have been developed (Deng and Hiruki, 1991; Ahrens and Seemuller, 1992; Gundersen, *et al.*, 1994; Smart *et al.*, 1996) which readily amplify rDNA of most, or all phytoplasmas (Table 14.1).

Methods based on DNA amplification using the polymerase chain reaction (PCR), followed by restriction fragment length polymorphism (RFLP), and sequencing are now used routinely in phytoplasma detection and characterization, by using the 16S rRNA gene and the 16/23S spacer region as targets for their phylogenetic and taxonomic significance (Melarned *et al.*, 2003; García-Chapa *et al.*, 2004). Nucleic acid hybridization (NAH) assays are one of the most useful tools for the diagnosis of phytoplasmas, mainly for differentiation of phylogenetically unique species (Harrison *et al.*, 1994; Tymon *et al.*, 1998; Arocha *et al.*, 2004a). Non-radioactive detection methods have largely replaces the use of hazardous radioactive (Tymon *et al.*, 1998; Arocha *et al.*, 2004a). However, a general molecular system has not been standardized for the high throughput screening of phytoplasma-infected crops of economic importance (Arocha *et al.*, 2004a).

Table 14.1: Universal Primers most Commonly Used to Detect the Presence of Phytoplasma Infections

Primer Pair	*Target DNA*	*References*
P1/P7	16S to 5' end of 23S rDNA	Deng and Hiruki, 1991; Smart *et al.*, 1996
R16F1/R0	16SrDNA	Lee *et al.*, 1993
R16F1/R2	16SrDNA	Gundersen and Lee, 1996
758F/1232R	16SrDNA	Gibb *et al.*, 1995
fU3/rU5	16SrDNA	Lorenz *et al.*, 1995
R16F2/m23sr	16SrDNA	Lee *et al.*,1993; Gibb *et al.*, 1999
fP3/P7	Spacer region	Schneider *et al.*, 1995
P4/P7	16S to 5' end of 23S rDNA	Schneider *et al.*, 1995
R16mF2/R1	16SrDNA	Gundersen and Lee, 1996
SN910601/SN910502	16SrDNA	Namba *et al.*, 1993
CYS2Fw/CYS2Rev	16SrDNA	Marzachi and Bosco, 2005
Forward/Reverse	16SrDNA	Christensen *et al.*, 2004

Phytoplasmas have rarely been subjected to quantification. Competitive, nested and real-time PCRs (Christensen *et al.*, 2004), using either phytoplasma-specific or phytoplasma group-specific primers, have provided sensitive and quantifiable data for concentrations of phytoplasma cells in their hosts. High sensitivity, easy handling, and a potential for high throughput, makes quantifiable PCR (Q-PCR) the method of choice for studying the specific interaction between phytoplasmas and their host plants (Christensen *et al.*, 2004). Q-PCR assays have been used to detect or quantify other microorganisms in a range of different environments and have the potential to dramatically improve knowledge of the ecology of microorganisms (Christensen *et al.*, 2004). Microscopical methods have been used to complement biological tests for the identification of phytoplasmas (Davis and Sinclair, 1998). These techniques are not practical for a quick diagnosis due to their high cost and the time consuming nature of sample preparation, occurrence of false negatives, and the fact that they lack of phytoplasma specificity (Jones, 2002). However, a confocal laser scanning microscopy has provided a very important and useful tool for detecting phytoplasmas in sieve tubes. The use of specific vital dyes has allowed physiologically active phytoplasma cells to be discriminated from the total population in a given phloem region. Bioimaging allows the detection of phytoplasmas in living plant tissue and can also identify sieve elements, companion cells, sieve pores and symplasmic contacts (Christensen *et al.*, 2004). The combination of bioimaging and Q-PCR assays could provide the breakthrough to quantify and localize phytoplasmas *in situ*, and to study their functional genomics.

Nucleic Acid Hybridization (NAH)

NAH assays have a number of advantages over PCR since they are easier to manipulate, more practical for analysing a high number of samples, and have the capacity to reduce the number of false positives (Tymon *et al.*, 1998; Arocha *et al.*, 2004a). Non-radioactive methods have sensitivity similar to radioactive ones (Harrison *et al.*, 1994; Kirkpatrick and Smart, 1995; Arocha *et al.*, 2004a) and are safer for human health. Additionally, the detection systems like those using chemiluminescence's increases the functional sensitivity of the assay making this comparable to the radioactive ones (Arocha *et al.*, 2004a). Recently we developed a multi-target non radioactive NAH system for the simultaneous detection of SCYLP and the bacteria causing leaf scald (*Xhantomonas albilineans*) and ratoon stunt (*Leifsonia xyli* subsp *xyli*), three of the most important sugarcane diseases in Cuba (Peralta *et al.*, 2003). The system is based on the use of a probe pool that detects the bacteria intergenic regions and the phytoplasma 16S ribosomal RNA; so, it is a valuable tool for the different phytosanitary, seed certification and breeding programs of the country. rRNA gene sequencing has provided evidence that the, wall-less prokaryotes other than spiroplasmas that colonize plant phloem and insect salivary glands form a large, monophyletic group within the class Mollicutes (IRPCM Phytoplasma/ Spiroplasma Working Team-Phytoplasma Taxonomy Group, 2004). The 16S/23S spacer region (SR) sequence has been also used for phytoplasma comparisons, at only 300 bases in length it is more convenient for differentiating closely-related phytoplasmas than the full-length 16S rRNA sequences of 1200-1800 bases. Additionally, phylogenetic trees constructed from regions flanking the $tRNA^{Ile}$ gene have mirrored those done using full-length 16S rRNA sequences (Kirkpatrick *et al.*, 1994). Similarities between 88 per cent and 94 per cent are required for phytoplasma 16S rDNA to fall into a distinct group, while those between 95 and 98 per cent correspond to subgroups (Lee *et al.*, 1998). Seemüller *et al.* (1998) considered phytoplasmas as belonging to the same group if they showed similarities between their sequences of 97 per cent or more, while values less than 95 per cent would place phytoplasmas in different groups or subgroups.

In 1992, the Phytoplasma Working Team of the International Research Project for Comparative Mycoplasmology (IRPCM) at the 9th Congress of the International Organization of Mycoplasmology, adopted the trivial name 'phytoplasma' to identify prokaryotes that belong to this group (IRPCM Phytoplasma/Spiroplasma Working Team-Phytoplasma Taxonomy Group, 2004). The IRPCM suggested rules for the description of organisms as novel taxa within '*Ca*. Phytoplasma' referred to single, unique 16S rRNA gene sequence (>1200 bp) with < 97.5 per cent similarity to that of any previously described '*Ca*. Phytoplasma' species. In those cases of phytoplasmas that share > 97.5 per cent of their 16S rRNA gene sequence, description of two different species is recommended only when the two Phytoplasmas are transmitted by different vectors; have a different natural plant host or, at least, their behavior is significantly different in the same plant host; or when there is evidence of significant molecular diversity, achieved by either hybridization to cloned DNA probes, serological reaction, or PCR-based assay. Since then, several distinct taxa have been described as 'Candidatus (*Ca.*) Phytoplasma aurantifolia' (Zreik *et al.*, 1995), '*Ca.* Phytoplasma australiense' (Davis *et al.*, 1997), '*Ca.* Phytoplasma australasia' (White *et al.*, 1998), '*Ca.* Phytoplasma fraxini' (Griffiths *et al.*, 1999), '*Ca.* Phytoplasma japonicum' (Sawayanagi *et al.*, 1999), '*Ca.* Phytoplasma brasiliense' (Montano *et al.*, 2001), '*Ca.* Phytoplasma castaneae' (Jung *et al.*, 2002), '*Ca.* Phytoplasma phoenicium' (Verdin *et al.*, 2003), '*Ca.* Phytoplasma ziziphi' and '*Ca.* Phytoplasma oryzae' (Jung *et al.*, 2003a).

Heteroduplex Mobility Assay (HMA)

Heteroduplex mobility assay (HMA), originally developed for the detection and estimation of genetic divergence between human immunodeficiency viruses (Delwart *et al.*, 1993), is a rapid, accurate, and simple screening tool that is not only capable of distinguishing individual strains, but also permitting reliable inferences to be made about phylogenetic relationships among strains of microorganisms. HMA is based on the observation that DNA heteroduplexes formed between sequences have a reduced mobility, proportional to their degree of divergence, in polyacrylamide gels, but not in standard agarose gel. Heteroduplexes are generated by base pairing between complementary single strands derived from the different parental duplex molecules during genetic recombination. Unknown DNA sequences can be compared against themselves or standard reference sequences. The DNA sequences of genetically common or rare variants could, therefore, be determined on a selective rather than random basis. In addition, HMA can be used for tracking specific sequence variants within individual populations and assisting in establishing epidemiological linkages between individual groups (Delwart *et al.*, 1993). The high sensitivity of HMA for distinguishing closely related phytoplasma strains was first demonstrated by its ability to differentiate tomato stolbur phytoplasma isolate (France) from eastern aster yellows phytoplasma isolate (New York), two isolates that were indistinguishably RFLP (Zhong and Hiruki, 1994). The sensitivity of HMA was further demonstrated by its ability to detect DNA substitutions and deletion/insertions derived from the 16S rDNA of aster yellows phytoplasma. HMA was capable of detecting a single base deletion/insertion and differentiating a single base substitution between phytoplasma DNA fragments when a suitable DNA fragment was used as a sequence driver. Thus, HMA is an extremely powerful tool for differentiating closely related phytoplasma strains exploring the molecular epidemiology of phytoplasmas.

HMA has already been applied to identify two types of phytoplasmas associated with clover proliferation and its closely related strains of alfalfa witches'- broom and potato witches'- broom that could not be differentiated previously by DNA-DNA hybridization and RFLP analysis (Lee *et al.*, 1991; Deng and Hiruki, 1991a). Phytoplasma isolates of CP, PWB, and AWB were collected in the field in Alberta from 1990 to 1999 and maintained in a greenhouse. An aster yellows isolate (AY27), a CP isolate, and a PWB isolate maintained in periwinkle (*Catharanthus roseus* (L.)

Don) were used as standard references. DNA fragments that consisted of the entire 16S rRNA gene and the 16S–23S spacer region sequences were initially amplified from various phytoplasma isolates of CP, PWB and AWB using the universal primers P1, (Deng and Hiruki, 1991b) and P7 (Smart *et al.*, 1996). Nested PCR was applied to amplify the 16S rRNA gene using primers R16F2n/R2 and the 16S–23S spacer region using primers P3/P7. HMA is a rapid and powerful means for investigating genetic diversity in closely related phytoplasmas. (Hiruki and Wang, 2008).

Steps in Molecular Diagnosis of Phytoplasma

DNA Extraction

The protocol traditionally used for extracting total DNA from phytoplasma-infected plants is based on the procedure described by Ahrens and Seemuller (1992). Although several other protocols exist for the enrichment of phytoplasma fractions, their yield levels are generally low and time consuming (Jiang *et al.*, 1987; Sears *et al.*,1989; Seddas *et al.*, 1993) or not applicable to most of labs working in this area due to technical difficulties (Garcia-Chapa *et al.*, 2004). Genome sequencing studies largely rely on PFGE purification of phytoplasma chromosomes (Firrao *et al.*, 1996; Liefting and Kirkpatrick, 2003; Garcia-Chapa *et al.*, 2004; Oshima *et al.*, 2004), which is not in the reach of most of the labs and also not completely free from host plant DNA may be applied.

Many different extraction procedures may be applied, depending on the tissue involved and the required sensitivity of the test. Plant contains many secondary metabolites, such as polyphenols and polysaccharides, which may inhibit polymerase. These substances can be significantly increased in phytoplasma-infected plants (Lepka *et al.*, 1999; Musetti *et al.*, 2000; Guthrie *et al.*, 2001; Choi *et al.*, 2004). PCR inhibition, which could be overcome by diluting the DNA extracts, was reported in DNA extracts from mature papya leaves, apricot leaf veins and apple tree roots (Padovan and Gibb, 2001; Heinrich *et al.*, 2001; Brzin *et al.*, 2003). It is therefore vital to remove inhibitory substances during the extraction step for reliable phytoplasma detection. Oxidation of phenols from reactive radicals that covalently bind to DNA, so acre should be taken to avoid air exposure of tissues during sample preparation. Polyvinyl pyrrolidone (PVP soluble) or PVPP (insoluble PVP polymer) to bind phenols and to prevent their oxidation by reducing agents like 2-mercaptoethanol. Co-precipitation with of polysaccharides with DNA can be prevented on the basis of their differential solubility in the presence of CTAB and NaCl. For non-problematic samples, including herbaceous test plants and insects, simpler procedures (Doyle and Doyle, 1990) can be applied and various commercial kits for DNA extraction are available (Green *et al.*, 1999; Constable *et al.*, 2003).

DNA extraction is the first step in characterization of any phytoplasma associated with a plant disease. The best method available so far is the extraction of total DNA from plant tissues employing a phytoplasma enrichment procedure (Ahrens and Seemüller, 1992). In the assays, about 500 mg leaves are cut into pieces and ground in liquid nitrogen. To the powdered sample, 1 ml of pre-warmed CTAB buffer (2 per cent CTAB, 1.4M NaCl, 20mM EDTA, pH 8.0, 100mM Tris HCl, 0.2 per cent mercaptoethanol and 0.5 per cent sodium sulphite) is added (Dellaporta *et al.*, 1983) and incubate at 65°C for 30 min. and then cool to room temperature. Later equal volume of chloroform: isoamylalcohol (24:1) or phenol: chloroform: isoamyl alcohol (25: 24:1) was added. The mixture is centrifuged at 12000 rpm (10 min; 4°C) and the supernatant is carefully pipetted out into a separate tube. To this, 0.8 volume isopropanol or ice-cold ethanol is added. The DNA is collected by spooling or centrifugation. To pellet the DNA, the tube is centrifuged at 12,000 rpm (10 min; 4°C) and it is wash carefully with 76 per cent ethanol twice and air dry. The pellet is resuspended in 20 ml Tris EDTA (TE) buffer.

Recently, Prabhu *et al.* (2008) developed another alternative enrichment methods that result in high phytoplasma: host DNA ratio and user friendly. In their method they tried to take advantage of the filterable ability of phytoplasmas and developed differential filtration method typically used in concentrating microorganisms for detection (Prabu *et al.*, 2008), to achieve the enrichment of the phytoplasma. The soft inner portions of meristematic plant tissue are washed in running tap water for 10 min, treated with commercial detergent (Teepol) for 3 min with a final wash with sterile distilled water. Subsequent steps were carried out in sterile conditions at 4°C. Diced tissue (1: 6 w/v) is homogenized in ice cold 0.2 mol l^{-1} glycine-sodium hydroxide buffer, pH 8.0 containing 0.02 mol l^{-1} magnesium chloride to maintain the membrane integrity of the phytoplasma cells (Razin 1964; Clark *et al.*, 1989; Thomas and Balasundran, 2001). The extract is passed through two layers of sterile cheesecloth followed by filtration through Whatman # 1, 3 and 5 filter papers (Whatman International Limited, England) respectively as phytoplasma particles could easily pass through these filters, whereas the filters retained much of the plant debris. The clear extract is then filtered through 0.45μm Millipore filter (Millipore, USA) due to the pleomorphic morphology of phytoplasma enabled it to pass through the 0.45 μm pores and centrifugat 45,000 g for 45 minutes; the pellet is resuspended in 1 ml of buffer used for homogenization (enriched fraction) (Figure 14.4).

To determine the end point detection level of phytoplasma, 1 μl (50 ng μl^{-1}) DNA samples of enriched fraction infected host tissue using QIAGEN DNA isolation kit are serially diluted with sterile deionized water up to 10^7 and subjected to PCR analysis using phytoplasma 16S rDNA primers, and visualized under UV after 1 per cent (w/v) agarose/EtBr gel electrophoresis. All reactions are performed in triplicates with suitable positive (infected plant DNA) and negative (water; healthy plant DNA) controls. With this protocol, an effective and simple differential filtration technique for resolving the basic difficulty of enriching the phytoplasma with as low as 1 g of tissue sample when compared with the CsCl- bisbenzimide differential centrifugation method and uses inexpensive materials which can be easily accessible to any laboratory. Genomic DNA is also extracted from enriched fractions using commercial DNA isolation kit.

Primers and PCR Amplification

PCR assays using universal primers are most useful for preliminary diagnosis of phytoplasmal diseases. Several universal, and many phytoplasma group-specific primers have been designed for routine detection of phytoplasmas (Doi *et al.*, 1967;

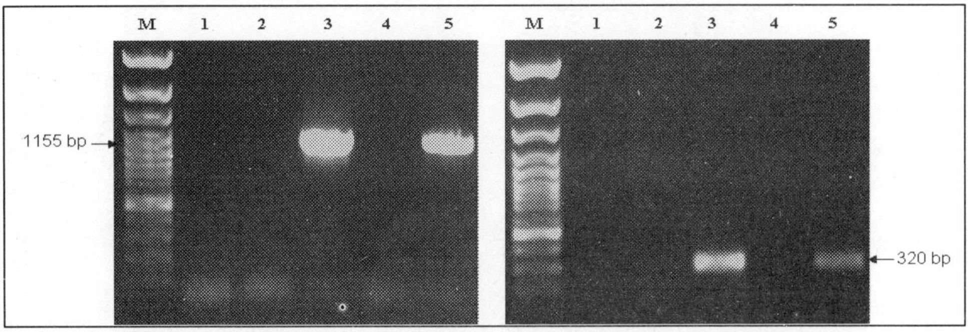

Figure 14.4: PCR Amplification of (a) 16S rDNA (1,155 bp) and (b) 16S/23S rRNA spacer region (320 bp) fragment of SCGS phytoplasma using SCPhy 16SF/SCPhy 16SR and P3/P7 primer pairs. Lane M, 100 bp DNA ladder Mix (MBI Fermentas); Lane 1, Water control; Lane 2, DNA from enriched fraction of Healthy plant (100 ng μl^{-1}); Lane 3, DNA from enriched fraction of infected plant (100 ng μl^{-1}); Lane 4, Healthy sugarcane plant DNA (100 ng μl^{-1}); Lane 5, Infected sugarcane plant DNA (100 ng μl^{-1}). (Source: Prabhu *et al.*, 2008).

Tsai, 1979; Lee and Davis, 1986; Lee *et al.*, 1993b; Gundersen *et al.*, 1994; Seemuller *et al.*, 1994; Davis and Sinclair, 1998; Lee *et al.*, 1998a). Nested-PCR assay increases both sensitivity and specificity, is a valuable technique in the amplification of phytoplasmas from samples in which unusually low titers are present, or substantial inhibitors that may interfere with the PCR efficacy are present (Marwitz, 1990; Lee *et al.*, 1993b). In nested-PCR universal primers are used the preliminary amplification of the DNA fragment and then followed by second amplification using a second group-specific primers. Therefore, nested-PCR enables the detection of dual or multiple phytoplasmas present in the infected tissues in case of mixed infection (Lee *et al.*, 1993b).

After isolation of DNA, the second step for phytoplasma characterization is the use of specific suitable primers. Sometimes single step PCR is sufficient for amplification of phytoplasma DNA but in most of the cases we need second step PCR *i.e.* nested PCR (Figure 14.5). The nested PCR analysis of phytoplasma is generally done by the method used by Schneider *et al.* (1995). Infected plant tissues are fresh frozen in liquid nitrogen and ground to a powder with a pestle and mortar. Extraction buffer (100mM K_2HPO_4, 31 mM KH_2PO_4, 0.3 M sucrose, 44 mM fructose, 0.15 per cent bovine serum albumine fraction, 2 per cent polyvinyl pyrrolidone, 30 mM ascorbic acid, 10 mM EDTA) is added to the powder using 4 ml of buffer/g of tissue. After stirring the mixture for 10 min on ice, the slurry is strained through cheesecloth, and the extract clarified by centrifugation (300g, 10 min) to remove coarse cell debries, and the supernatant subject to further round of centrifugation (20000g, 35 min).

PCR assays using universal primers designed based on conserved sequences of 16S rRNA, tuf and 16S-23S spacer allow the detection of a wide array of unknown phytoplasmas associated with plants and insects (Lee *et al.*, 1993, 1998; Schneider *et al.*, 1995, 1997). PCR with phytoplasma specific oligonucleotide primers derived from rRNA gene sequences make selective amplification of near full-length

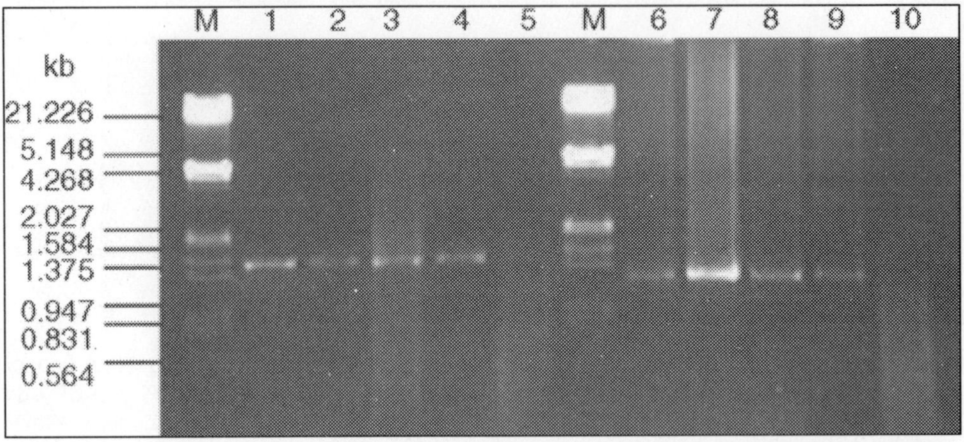

Figure 14.5: Agarose Gel Showing Bands of ~ 1.5 kb Amplified by Direct PCR with P1/P6 Primers from Symptomatic Samples (lanes : 1-4) and ~ 1.2 kb Bands by Nested PCR with R16F2n/R16R2 Primers (lanes : 6-9). Lanes 5 and 10 are PCR and nested PCR products, respectively, from asymptomatic sample. M is Lambda DNA digested with *EcoRI* and *HindIII* (Genei Pvt. Ltd, Bangalore, India) used as marker.

phytoplasma 16S rRNA genes from mixtures with host plant DNA. Several universal, and many phytoplasma group-specific primers have been designed for detection of phytoplasmas (Deng and Hiruki, 1991; Gundersen and Lee, 1996; Khadhair *et al.*, 1998; Lee *et al.*, 1993) (Tables 14.1 and 14.2).

Table 14.2: Group-Specific Primers Used in the Characterization of Phytoplasmas

Primer Pair	*Target DNA*	*References*
fStol/AGY2	16SrDNA	Gibb *et al.*, 1998
R16(X)F1/R16(X)R1	16SrDNA	Lee *et al.*, 1995
F01/r01	16SrDNA	Lorenz *et al.*, 1995
AP5/AP4 ; AP8/AP10	16SrDNA	Jarausch *et al.*, 1994
PA2F/R ; NPA2F/R	16/23SrDNA	Heinrich *et al.*, 2001
LYR1/LYF1	Non–ribosomal	Heinrich *et al.*, 2001a, b
BB88F1/R1	16SrDNA	Gundersen and Lee (1996); Botti and Bertaccini (2003)
fTufu/rTufu	16SrDNA	Schneider *et al.* (1997)

Nested PCR assay, designed to increase both sensitivity and specificity, is performed by using a group-specific primer pair and PCR products amplified with primer pair P1/P6 are used as templates. Nested-PCR is capable of detection of dual or multiple phytoplasmas present in the infected tissues in case of mixed infection.

An alternative method to increase detection sensitivity is Immuno-capture PCR assay, in which the phytoplasma of interest is first selectively captured by specific antibody adsorbed on microtiter plates, and then the phytoplasma DNA is released and amplified using specific or universal primers (Lee and Davis, 1992). RFLP analysis together with the sequencing of 16Sr phytoplasma genes was the first step on this way (Schneider *et al.*, 1995; Gundersen *et al.*, 1996; Chen and Lin, 1997; Berges *et al.*, 1997; Bertaccini *et al.*, 1998) providing the construction of phylogenetic trees of many microorganisms especially in the Mollicutes taxon. Molecular characterisation of the entire phytoplasma genome including its sequencing recently performed will provide, after its full annotation more precise basis for taxonomy, but it will be necessary to do it for several other phytoplasmas in order to achieve comparative genomic analysis that could allow a deeper understanding about physiology of these organisms.

For amplification of phytoplasmal ribosomal DNA (rDNA) by PCR assays, different universal and specific phytoplasma primer pairs are used. The most common primers used for phytoplasma DNA amplication used in Direct PCR and nested PCR specific to 16S rRNA gene of phytoplasma are listed in (Tables 14.1 and 14.2).

Amplifications are performed in a 25µl final volume containing 0.5µM of each primer, 100µM dNTPs,1µ of Taq polymerase,1X reaction buffer and 1 µl of template DNA in a thermocycler. 1 µl of product of the first round of PCR is used in nested - PCR using internal primer pairs which amplify a 1250 bp portion of the 16SrRNA gene of phytoplasmas. The presence/absence of phytoplasmas is demonstrated by electrophoresing the PCR products on 1 per cent agarose gels at 70 volts using 0.5 × TBE running buffer. The gels are stained with ethidium bromide (10mg/ml) and observed on UV transilluminator.

Cloning and Sequencing of PCR Product

After confirmation of phytoplasma DNA amplification, the next step is cloning the amplified product and then sequencing. For sequencing first or second round PCR products (approximately 5 µg) are separated by electrophoresis in agarose gel. Fragments with sizes corresponding to the expected amplified sequences are excised from the gel and eluted using the commercial QIA quick gel extraction kit. DNA fragments are either sequenced directly or cloned prior to sequencing. For cloning, DNA fragments are ligated into plasmid vector *e.g.* pGEM-T and recombinant plasmid used to transform *Escherichia coli* strain DH5a. Selected recombinant clones are then screened for phytoplasmal rDNA inserts by PCR using the primer pair. Plasmid DNA is purified using the NucleoSpin system kit. Sequencing of both strands is performed by ABI's AmpliTaq FS dye terminator cycle sequencing chemistry, based on Sanger's Sequencing method, in an automated ABI Genetic Analyzer. Primers for sequencing PCR products are the same as for PCR amplification whereas the standard primers are used for sequencing the cloned fragments. The sequences are then compared with known phytoplasma sequences in GenBank using the Basic Local Alignment Search Tool (BLAST) software (http://www.ncbi.nlm.nih.gov/blast).

Sequencing and Phylogenetic Analysis of Ribosomal DNA

Molecular identification of the mycoplasmal nature of the phytoplasmas is first obtained by estimating the G+C content of highly enriched phytoplasmal DNA. The very low values obtained (23.0 to 26.2 mol per cent) suggests their close relationship with the culturable *Mollicutes* (Kollar and Seemuller, 1989). The first attempt to sequence the amplified 16S rRNA genes either directly, or after cloning into suitable vectors provide sequences for phylogenetic analyses (Kirkpatrick, 1992; Namba *et al.*, 1993; Gundersen *et al.*, 1994; Seemuller *et al.*, 1994). Later estimation of genome size 600–1150 kb which is within the range for culturable *Mollicutes*, provided more molecular evidence. Phylogenetic relationship of phytoplasmas can be obtained using comparative sequence analysis of the conserved 16S rRNA gene, which was proposed as a universal phylogenetic marker for classifying major groups of prokaryotes. This approach has been used to determine relationships of culturable mollicutes with each other and with walled bacteria.

Lim and Sears (1989) and Kuske and Kirkpatrick (1992) were the first to clone and completely sequence 16S rRNA genes of phytoplasmas. Sequence comparisons indicated that the two aster yellows (AY) phytoplasmas examined were more closely related to culturable mollicutes than to other bacteria. The availability of PCR primers for 16S rRNA gene has resulted in full length or nearly full-length sequences of 16S rRNA genes of 37 representative strains of phytoplasmas (Namba *et al.*, 1993; Gundersen *et al.*, 1994; Seemuller *et al.*, 1994).

Recent analysis (Lim and Sears, 1989; Kuske and Kirkpatrick, 1992) provided further compelling evidence that phytoplasmas constituted a unique, monophyletic clade of Mollicutes more closely related to Acholeplasma than to the true Mycoplasma species. Within the phytoplasma clade, as many as 13 major subclades (groups) represented by the following type strains: (1) stolbur; (2) aster yellows; (3) apple proliferation; (4) coconut lethal yellowing; (5) pigeonpea witches'-broom; (6) X-disease; (7) rice yellow dwarf; (8) elm yellows; (9) ash yellows; (10) sunnhemp phyllody; (11) loofah witches'-broom; (12) clover proliferation and (13) peanut witches'-broom were recognized (Kirkpatrick and Smart, 1995; Schneider *et al.*, 1995). Twenty distinct phytoplasma groups, including eight novel groups absent from previous classifications *viz.*, Australian grapevine yellows (AUSGY), Italian bineweed stolbur (IBS), buckthorn witches'-broom (BWB), Spartium witches'-broom (SpaWB), Italian alfalfa witches'-broom (IAWB), Cirsium phyllody (CirP), Bermuda grass white leaf (BGWL) and Tanzanian lethal decline (TLD) were recently defined (Seemuller *et al.*, 1998). Another reliable phylogenetic marker is the spacer region (SR) separating the 16S from the 23S rRNA gene of phytoplasmas. Phylogenetic analysis of the entire 16-23S SR (Gibb *et al.*, 1998; Kenyon *et al.*, 1998; Khan *et al.*, 2002) or variable regions flanking the $tRNA^{ile}$ gene (Kirkpatrick *et al.*, 1994; Kirkpatrick and Smart, 1995; Schneider *et al.*, 1995) differentiated phytoplasmas into groups concordant with the major groups previously established from analyses of 16S rRNA genes. As a consequence of these studies, the International Committee of Systematic Bacteriology (ICSB) Subcommittee on the Taxonomy of Mollicutes (1993, 1997) has agreed to replace the trivial name MLO by the term phytoplasma. This name will also be under the provisional taxonomic status Candidatus and will become the genus name of the plant pathogenic mycoplasmas.

More than three hundred distinct phytoplasma 16S rDNA genes have been sequenced. However, additional conserved DNA sequence markers can be used as supplemental tools for finer phytoplasma differentiation along with physical map of some phytoplasmas, that are also available. The rp gene sequences reveal more variation than 16S rDNA and the analysis conducted by RFLP or sequencing on tuf and/or SecY genes show clear indications of phytoplasma strains relationships at least with geographical distribution (Schneider *et al.*, 1997; Yu *et al.*, 1998; Marcone *et al.*, 1999; Lauer and Seemuller, 2000; Kakizawa *et al.*, 2001; Myata *et al.*, 2002; Marcone and Seemuller, 2001; Marcone *et al.*, 2000 and Langer and Maixner, 2004; Kakizawa *et al.*, 2004). The use of the 23S rDNA gene was found not to be useful since it appears more or similarly conserved as the 16S. Studies performed comparing similar regions in two phytoplasmas and in phylogenetically closer relatives showed that different organization at this level clearly distinguish phytoplasmas from other phylogenetically related Mollicutes. Hybridization analyses indicated the presence of two sets of 16S rDNA operons, and heterogeneity of these two operons was suggested for some collection maintained as well as from wild strains field collected phytoplasmas (Bertaccini *et al.*, 2005).

The presence of extra chromosomal DNA, similar to plasmid DNA, has been demonstrated in phytoplasmas by using DNA probes; these DNAs ("double stranded covalently closed circle") could be different in different phytoplasma strains, but their role is still unknown in the majority of the cases (Rekab *et al.*, 1999; Nishigawa *et al.*, 2001, 2002). Suggestions have been made about their possible involvement in phytoplasma pathogenicity, but no close correlation between plasmid detected, and symptoms induced by the isolates in periwinkle or in other host plants were found. In some cases a close genetic relationship was detected between part of plasmid sequences and plant viruses containing DNA, such as Gemini viruses.

RFLP Analysis of PCR-Amplified Conserved DNA

Phytoplasmas can be classified based on restriction fragment length polymorphism (RFLP) analysis of PCR products amplified from 16S rDNA sequences. The 16S rRNA gene is universal in prokaryotes and possesses both conserved and variable regions that make it useful for taxonomic studies. Sequence or RFLP analysis of rRNA gene sequences amplified by PCR provided the basis for several comprehensive studies on phytoplasma phylogeny and taxonomy (Lee *et al.*, 1993; Seemüller *et al.*, 1994; Gundersen *et al.*, 1994, 1998). Collective profiles revealed by separate digestion of rDNA products with a selection of frequently cutting endonuclease enzymes varied, according to the particular phytoplasmas that are examined.

In contrast to antibodies and hybridization probes, the 16S rRNA gene is universal in prokaryotes and possesses both conserved and variable regions that make it useful for taxonomic studies. After 16S rDNA became common by PCR amplification, RFLP analysis of PCR-amplified rDNA was used to comprehensively classify the phytoplasmas on a molecular basis for the first time. The use of frequently cutting restriction endonucleases, such as AluI, RsaI and others, identification of approximately 12 major groups and several subgroups of phytoplasma became possible (Lee *et al.*, 1993a). For classification of phytoplasmas on a large scale, RFLP analysis of PCR-amplified 16SrDNA sequences with a number of restriction enzymes was used by (Lee *at al.*, 1993 and Schenider *et al.*, 1993) to differentiate various phytoplasmas by their distinct RFLP patterns Figures 14.6(A-B). This approach in a

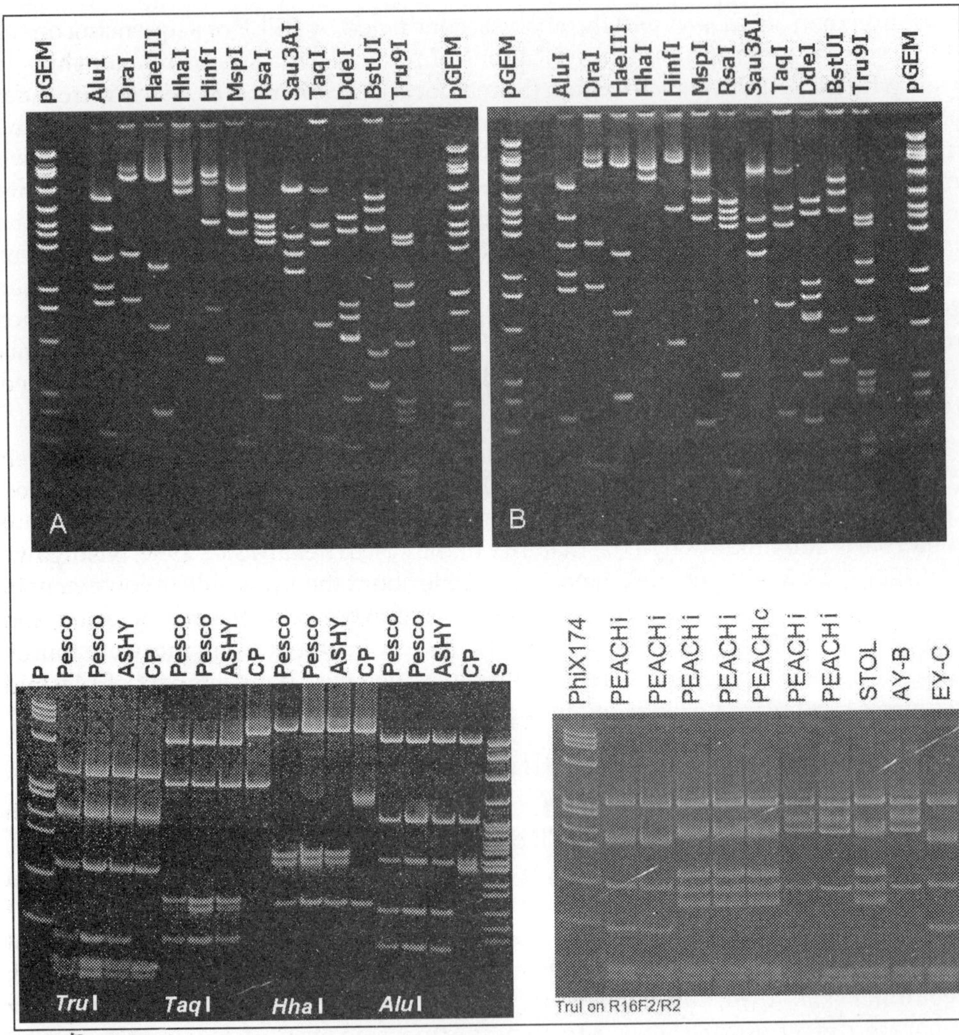

Figure 14.6(A-B): RFLP Analysis of Nested PCR Products Amplified with Pair of Primers. DNA products were digested with different restriction enzymes and separated by electrophoresis.

enabled a simple, reliable and rapid way to identified many phytoplasmas of interest without a need to sequence the gene. At present, RFLP analysis of PCR-amplified rDNA is the routine method for differentiation and classification of phytoplasmas. Classification of phytoplasma by RFLP patterns can be improved by using a large fragment comprising the entire 16S rRNA gene and 16S-23S rDNA spacer region (Schneider *et al.*, 1995). Differentiation can also be improved by increasing the number of restriction enzymes. Lee *et al.* (1993b) used 15 enzymes for the classification of many diverse phytoplasmas. However, from the data available, it appears that classification based only on RFLP analysis of the 16S rRNA gene alone does not reflect the full range of phenotypic diversity. The 16S rRNA gene does not always seem sufficiently variable to allow distinction of phytoplasmas that differ in plant host or vector specificity. More variable are nucleic acid sequences of ribosomal protein genes, which have enabled identification of more RFLP groups than 16S rDNA data. The combined RFLP analysis of both sequences resulted in the most detailed differentiation of the AY and group phytoplasmas so far (Gundersen *et al.*, 1996). Another less conserved sequence, recently examined, is the gene encoding the elongation factor Tu (tuf gene). Due to its variability, a more detailed subdivision of AY group phytoplasmas was obtained than by 16S rDNA analysis (Schneider *et al.*, 1997). RFLP analysis of randomly cloned chromosomal DNA fragments has also been employed, mainly to differentiate closely related phytoplasmas (Jarausch *et al.*, 1994; Daire *et al.*, 1997). The 16SrDNA RFLP classification scheme has been expanded over the last decade. Currently, it comprises 18 groups and more than 50 subgroups and is the most comprehensive classification scheme for phytoplasmas (Lee *et al.*, 2000; Montano *et al.*, 2001; Arocha *et al.*, 2005).

Specific DNA probes used for Southern hybridization in conjunction with restriction fragment length polymorphism (RFLP) analysis, showed genetic relationships among phytoplasmas associated with similar hosts or with symptomatologically similar diseases (Kuske *et al.*, 1991; Kison *et al.*, 1994; Kison *et al.*, 1997). Significantly, major groupings revealed by analysis of probe-generated RFLP patterns were consistent with those defined by monoclonal antibody typing (Lee *et al.*, 1993a) and other molecular methods (Lee *et al.*, 1993b), but differed from previous classification attempts based solely on biological properties.

According to the recommendations of the International Committee of Systematic Bacteriology, subcommittee on the Taxonomy of Mollicutes a new *Candidatus* species may be described when a 16S rDNA sequence (longer than 1200 bp) has less than 97.5 per cent identity with any previously described *Candidatus* species. The naming of new species in the class Mollicutes requires description of species in pure culture, but the DNA homology and phenotypic characteristics used to describe mollicute species are unattainable for uncultured phytoplasmas. Therefore, a provisional classification system using the '*Candidatus*' convention has been adopted for phytoplasmas.

Also two phytoplasmas sharing more than 97.5 per cent of 16S sequence can be designed as separate *Candidatus* species when they meet the following three criteria: (*i*) they are transmitted by different vectors; (*ii*) they have different natural plant host(s); and (*iii*) there is evidence for molecular diversity between the two phytoplasmas (IRCPM, 2004). Tentative classification following 16S ribosomal grouping as parameter is now commonly employed for identification in order to study the phytoplasma-associated plant diseases. (Zreik *et al.*, 1995; Davis *et al.*, 1997; Griffiths *et al.*, 1999; Montano *et al.*, 2001; Jung *et al.*, 2002; Marcone *et al.*, 2003a and Marcone *et al.*, 2003b; Verdin *et al.*, 2003; Jung *et al.*, 2003; Nishigawa *et al.*, 2003; Hiruki and Wang, 2004; IRCPM, 2004; Seemuller and Schneider, 2004;).The other *Candidatus* which have been published with not clear reference to the ribosomal subgroups identified till now are: *Candidatus* Phytoplasma japonicum infecting hydrangea in Japan, *Candidatus* Phytoplasma ramni, *Candidatus* Phytoplasma allocasuarinae infecting those plant species in Germany and in Australia, respectively, *Candidatus* Phytoplasma pini infecting a few Pinns species in Europe, and *Candidatus* Phytoplasma graminis and *Candidatus* Phytoplasma caricae infecting sugarcane, weeds and papaya in Cuba (Sawayanagi *et al.*, 1999; Seemuller and Schneider, 2004; Arocha *et al.*, 2005).

Management

Unfortunately, no effective means of curing phytoplasma diseases are available at present. The disease is complex and its progress is also highly variable. It depends on several factors such as, the pathogen with its different biotypes; the tendency to mutate; the presence and dynamics of the vectors; concentration of host-plants of both phytoplasma and vectors; environmental conditions; agronomic practices. As a consequence no single control fixed control strategy can be adopted. The most important factors to consider before intervening are: disease severity, whether or not to rouge infected plants, rouging strategies, availability of insect vectors, alternative susceptible natural plant reservoirs and economic impact of the disease (Osler and Carraro, 2004).

Control of epidemic outbreak can be carried out theoretically either by controlling the vector or by eliminating the pathogen from the infected plants by antibiotics, mainly tetracycline, or by other chemicals (Dai and Sun, 1995; Dai *et al.*, 1997; Veronesi *et al.*, 2000; Chung and Choi, 2002). Both protection measures were quite ineffective under field conditions, first because it is impossible to eliminate all vectors from environments, second because the use of antibiotics is very expensive, and not allowed in several countries. It is not always effective to prevent the outbreaks by producing clean material or by finding phytoplasma resistant or tolerant varieties, but this could be employed only under restricted and defined environmental conditions (Loi *et al.*, 1995; Parani *et al.*, 1996; Thomas and Mink, 1998 and Kison and Seemuller, 2001). In order to gain information in this fields research is still not very much developed even if some basic knowledge about epidemiology, and physiopathology of phytoplasma associated diseases is available: the knowledge about some phytoplasma membrane-protein as well as about plant gene or plant products possibly involved in pathogenetic mechanisms can improve possibilities to better understand the way to eliminate these dangerous and mostly still unknown pathogens (Sears *et al.*, 1997; Chen and Chen, 1998; Le Gall *et al.*, 1998; Jarausch *et al.*, 1999; Lepka *et al.*, 1999; Blomquist *et al.*, 2001; Jarausch *et al.*, 2001; Khan *et al.*, 2002a; Boti and Bertaccini, 2003; Streten and Gibb, 2003; Choi *et al.*, 2004; Bai *et al.*, 2004 and Garcia-Chapa *et al.*, 2004). Phytoplasmas are normally controlled by the breeding and planting of disease resistance varieties of crops (believed to the most economically viable option) and by the control of the insect vector. Tissue culture can be used to produce clones of phytoplasma infected plants that are healthy. The chances of gaining healthy plants in this manner can be enhanced by the use of cryotherapy, freezing the plant samples in liquid nitrogen, before using them for tissue culture. Work has also been carried out investigating the effectiveness of plant bodies targeted against phytoplasmas. Tetracyclines are bacteriostatic to phytoplasmas that inhibit their growth. However, without continuous use of the antibiotic, disease symptoms will reappear. Thus, tetracycline is not a viable control agent in agriculture, but it is used to protect ornamental coconut trees. So far the elimination of phytoplasmas classified in the kingdom of bacteria, was attempted by the use of antibiotics, such as tetracycline was (Davies and Clark, 1994) rather than by modified temperatures. It has been ling time know that heat therapy can help in eliminating pathogens such as virus and possibly phytoplasmas from plants and mainly fruit trees (Kunkel, 1936; Kassanism, 1954; Nyland, 1962). For the elimination of phytoplasmas at the IAM, the routine method involves a combination of heat therapy and meristem culture (Laimer and Balla 2003), without tetracycline treatment.

Indexing, mass propagation and germplasm conservation are the final steps for the maintenance of elite mother plants. Plantlets obtained from meristem tip culture are propagated *in vitro* as mericlones. As soon as possible, 2–3 plantlets from each mericlone are individually indexed. These tests are repeated several times before deciding, whether a mericlone is virus-free. Indexing can be done by the use of indicator plants, by ELISA tests or PCR. The pathogen-free clones are then propagated by axillary branching and delivered to the nurserymen (Boxus and Druart, 1986).

According to international regulations pathogen-free elite plants are to be kept under conditions avoiding re-infection by air and soil borne vectors, *e.g.* in an insect-proof screen-house. Currently protocols for cryopreservation are being adapted to allow long term storage of this valuable plant material are more adapt to an extensive multiplication of phytoplasmas can help in preventing epidemics of ESFY (Desvignes, 1999). Fruit growers should be very careful in their choice of cultivar and rootstock, and breeders should test the tolerance to this disease before distributing new cultivars (Desvignes, 1999). There is also a certain amount of plants (especially plums), that can be asymptomatic, and represents an important disease reservoir. Asymptomatic plants re-grafted on rootstock, where previously symptomatic scions were observed, showed that 67 per cent of the plants tested were infected by 16SrX-B phytoplasmas within one year. This clearly indicates that the rootstock was colonized by the pathogens that are able to infect the newly grafted scions just within one winter season (Paltrinieri *et al.*, 2004).

Future Prospects

Molecular tools such as monoclonal antibodies, DNA-based probes, and PCR-based sensitive detection procedures have largely replaced traditional procedures based on biological properties, greatly advancing phytoplasmal disease diagnotics and facilitating phytoplasma characterization. It is axiomatic that evolutionary speciation is driven by genes other than those, such as 16SrRNA and ribosomal protein genes, that are commonly used for phylogenetic analysis, classification and delineation and description of '*Candidatus* phytoplasma' species. While conserved genes are useful in phytoplasma strain classification and species delineation, their utility is due to sequence variability that is correlated with, but not determinant of, species evolution. Thus, it is sequence drift in niche–isolated or niche-unique strain populations that results in conserved gene sequence variability. This implies that some closely related species may not be resolved by analysis of conserved genes and that incipient species tend to remain unrecognized. For these reasons and for practicality, it is important at this stage of phytoplasmas taxonomy to link phenotypic or biological characteristic with molecular criteria in definition and description of phytoplasma taxa, especially at species level. Ultimately, as more is learned about factors involved in host-pathogen interactions and nucleotide sequences controlling these interactions, it should be possible to bridge molecular criteria and biological properties through judicious choices of suites of nucleotide sequences applicable for distinguishing and describing phytoplasma species. In our view, use of multigene sequences with varying degrees of variability will eventually afford definitions of phytoplasmas at strain, species or higher level consistent with phylogenies theoretically based on complete genome sequences. This outlook engenders the concept that formal genus phytoplasma taxonomy will become a reality, in spite of inability to isolate these intriguing microbes in artificial cultures, and will one day be based entirely upon gene information. Phytoplasma associated diseases are molecularly distinguishable in most of the cases at the 16Sr DNA level. Therefore, epidemiological studies can be carried out in order to eliminate infected plants to prevent further epidemic spreading. The main limitation to the real application of these procedures that can be very successful in eliminating or reducing the impact of phytoplasma diseases is that agricultural-related problems are not under consideration in many countries worldwide for opposite reasons (over production or not qualified production) people working in this field are not always aware of the risk connected with the trading or the maintenance in field of phytoplasma infected plants. It is presumed that phytoplasma infected natural plants may act as a reservoir and insect vectors feeding on them may carry the destructive pathogen to the agricultural crops. Therefore, the current developed practices in identification and characterization of phytoplasmas discussed in this chapter would be quite helpful in phytoplasmas taxonomy and a sound practical approach could be planned accordingly.

References

Adams, A.N., Davies, D.L. and Kirby, M.J. (2001). Virus and phytoplasma detection in fruit trees. *Outlook on Agriculture*. 30 (1): 45-54.

Ahrens, U. and Seemüller, E. (1992). Detection of DNA of plant pathogenic mycoplasma like organisms by a polymerase chain reaction that amplifies a sequence of the 16S rRNA gene. *Phytopathology*, 82: 828-832.

Anonymous (1993). International Committee 011 Systematic Bacteriology, Subcommittee on the Taxonomy of Mollicutes. Minutes of the Interim meetings, 1 and 2 August, 1992, Ames, Iowa. *Intl. J. Syst. Bacteriol.* 43: 394-397.

Anonymous (1995). International Committee on Systematic Bacteriology Subcommittee on the Taxonomy of Mollicutes. (1995). Minutes of the interim meeting, July 1994, Bordeaux, France. *Int. J. Syst. Bacterial.* 45: 415-417.

Anonymous (1997). International Committee on Systematic Bacteriology, Subcommittee on the Taxonomy of Mollicutes. Minutes of the Interim meetings, 12 and 18 August, 1996, Orlando, Florida, USA. *Intl. J. Syst. Bacterial.*, 47: 911-914.

Anonymous (2004). IRPCM Phytoplasma/Spiroplasma Working Team–Phytoplasma taxonomy group, 'Candidatus Phytoplasma', a taxon for the wall-less, non-helical prokaryotes that colonize plant phloem and insects. *Int. J. Syst. Evol. Microbiol*, 54: 1243-1255.

Al-saady, N. A. and Khan, A. J. (2006). Phytoplasmas that can infectdiverse plant species worldwide. *Physiol. Mol. Biol. Plant.* 12(4): 263-281.

Arocha, Y., Horta, D., Peralta, E. and Jones, P. (2004a). Development of a non radioactive methodology for the generic diagnostic of phytoplasmas in Cuba. *Revista de Protección Vegetal*, 19: 118-22.

Arocha, Y., Horta, D., Pinol, B., Palenzuela, I., Picornell, S., Almeida, R. and Jones, P. (2005). First report of Phytoplasma associated with Bermuda grass white leaf disease in cuba. *Plant Pathology*, 54: 233.

Bai, X., Zhang, J., Holford, I.R. and Hogenhout, S.A. (2004). Comparative genomic identifies genes shared by distantly related insect-transmitted plant pathogenic mollicutes. *FEMS Microbiol. Lettera*, 235: 249-258.

Berges R., Cousin, M.T., Roux, J., Maurer, R. and Seemuller, E. (1997). Detection of phytoplasma infections in declining Populus nigra 'Italica' trees and molecular differentiation of the aster yellows phytoplasmas identified in various Populus species. Europ. *Forest Pathol.*, 27: 33-43.

Bertaccini A., Vorackova, Z., Vibio, M., Franova, J., Navratil, M., Spak, J. and Nebesarova, J. (1998). Comparison of phytoplasmas infecting winter oilseed rape in the Czech Republic with Italian Brassica phytoplasmas and their relationship to the aster yellows group. *Plant Pathology*, 47: 317-324.

Bertaccini, A., Franova, J. Botti, S. and Tabanelli, D. (2005). Molecular characterization of phytoplasmas in lilies with fasciation in the Czech Republic. *FEMS Microbiol. Letters*, 249: 79-85.

Blomquist, C.L., Barbara, D.J., Davies, D.L. Clark, M.F. and Kirkpatrick, B.C. (2001). An immunodominant membrane protein gene from the western X-disease phytoplasma is distinct from those of other phytoplasmas. *Microbiology*. 147: 571-580.

Botti, S. and Bertaccini, A. (2003). Variability and functional role of chromosomal sequences in phytoplasmas of 16SrI-B subgroup (aster yellows and related strains). *Appl. Microbiol.*, 94: 103-110.

Boxus, P.H. and Druart, P.H. (1986). Virus-free trees through tissue culture. In: Biotechnology in Agriculture and Forestry ed. Bajaj Y.P.S. Vol. 1 Trees: 24-30.

Brzin, J., Ermacora, P., Osler, R., Loi, N., Ravnikar, M. and Petrovic, N. (2003). Detection of apple proliferation phytoplasma by ELISA and PCR in growing and dormant apple trees. *Journal of Plant Diseases and Protection*. 110 (5): 476-483.

Chen, M. and Lin, C. (1997). DNA probes and PCR primers for the detection of a phytoplasma associated with peanut witches'-broom. *Europ. PI. Pathoi.* 103: 137-145.

Chen, Y.D. and Chen, T.A. (1998). Expression of engineered antibodies in plants: a possible tool for spiroplasma and phytoplasma disease control. *Phytopathology.* 88: 1367-1371.

Chen,T.A. and Jiang, X.F. (1988). Monoclonal antibodies against the maize bushy stunt agent. *Canadian Journal of Microbiology.* 34(1): 6-11.

Christensen, N., Nicolaisen, M., Hansen, M. and Schulz, A. (2004). Distribution of phytoplasmas in infected plants as revealed by Real-Time PCR and Bioimaging. *MPMI.* 17(11): 1175-1184.

Chung, B. and Choi, G. (2002). Elimination of aster yellows phytoplasma from Dendranthema grandiflorum by application of oxytetracycline as a foliar spray. *Pl. Pathol.* 18: 93-97.

Choi, Y.H., Tapias, B.C., Kim, H.K., Lefeber, A.W.M., Erkelens, C., Verhoeven, J.Th.J., Brzin, J., Zel, J. and Verpoorte, R. (2004). Metabolic discrimination of *Catharanthus roscus* leaves infected by phytoplasma using H-NMR spectroscopy and multivariate data analysis. *Pl. Physiol.* 135: 1-13.

Clark, M.F., Morton, A. and Buss, S.L. (1989). Preparations of mycoplasma immunogens from plants and a comparison of polyclonal and monoclonal antibodies made against primula yellows MLO-associated antigens. *Annals of Applied Biology.*114: 111-124.

Constable, F.E., Gibb, K.S. and Symons, R.H. (2003). Seasonal distribution of phytoplasmas in Australian grapevines. *Plant Pathology.* 52: 267-276.

Dai, Q. and Sun, Z. (1995). Suppressive effects of N-triacontanol on symptoms of mulberry dwarf disease and on the causal phytoplasma. *Pl. Pathol.* 44: 979-981.

Davis, D.L. and Clark, M.F. (1992). Production and characterization of polyclonal and monoclonal antibodies against peach yellow leafroll MLO-associated antigens. *Acta Horticulturae.* 309: 275-283.

Davies, D.L. and Clark, M.A. (1994). Maintenance of mycoplasma-like organisms occurring in Pyrus species by micropropagation and their elimination by tetracycline therapy. *Plant Pathology.* 43: 819-823.

Davis, R.E. and Sinclair,W.A. (1998). Phytoplasma identity and disease etiology. *Phytopathology,* 88: 1372-1376.

Davis, R.E., Dally, EX., Gundersen, D.E., Lee, I.-M. and Habili, N. (1997). "Candidatus phytoplasma australiense", a new phytoplasma taxon associated with Australian grapevine yellows. *Int. J. Syst. Bacteriol.* 47: 262-269.

Dellaporta, S. L., Wood, J. and Hicks, J. B. (1983). A Plant minipreparation: Version II. Plant Molecular Biology Report, 1: 19-21.

Deng, S. and Hiruki, C. (1991). Amplification of 16S rRNA genes from culturable and non-culturable mollicutes. *J Microbiol Methods,* 14: 53-61.

Delwart, E.L., Shpaer, *E.G.,* Louwagie, J., McCutchan, F.E., Grez, M., Rubsamen-Weigmann, H. and Mullins, J.I. (1993). Genetic relationships determined by a DNA heteroduplex mobility assay: analysis of HIV-1 env genes. *Science.* 262: 1257-1261.

Desvignes, J.C. (1999). Virus Diseases of Fruit Trees. CTIFL editor, 113-143.

Doyle, J.J. and Doyle, J.L. (1990). Isolation of plant DNA from fresh tissue. *Focus*, 12: 13-15.

Doi. Y., Teranaka, M., Yora, K. and Asuyama, H. (1967). Mycoplasma or PLT group-like micro-organisms found in the phloem elements of plants infected with mulberry dwarf, potato witches, broom, aster yellow or Paulownia witches' broom. *Ann. Phytopath. Soc. Japan*. 33: 259-266.

Firrao, E., Smart, C.D. and Kirkpatrick, B.C. (1996). Physical map of the Western X-disease phytoplasma chromosome. *Journal of Bacteriology*. 178: 3985-3988.

Firrao, G. (1999). Gemnivirus–related extrachromosomal DNAs of the X-clade phytoplasmas share high sequence similarity. *Microbiology*. 145: 1453–1459.

Garcia-Chapa, M., Batlle, A., Rekab, D., Rosquete, M.R. and Firrao, G. (2004). PCR-mediated whole genome amplification of phytoplasmas. *Journal of Microbiological Methods*. 56: 231-242.

Garneir,M., MartinGros,G., Iskra,M.L., Zreik,L., Gander,J., Fos,A. and Bove,J.M.(1990). Monoclonal antibodies against the MLOs associated with tomato stolbur and clover phyllody. In: Recent Advances in Mycoplasmology, Procedings of the 7th Congress of the International Organization for Mycology Wien 1988, Stanek G., Cassel G.H., Tully J.G., Whitcomb R.F. eds, Fisher Verlag, Stuttgard, New York: 263-269.

Gibb, K.S., Padovan, A.C. and Mogen, B.D. (1995). Studies on sweet potato little-leaf phytoplasma detected in sweet potato and other plant species growing in Northern Australia. *Phytopathology*. 85: 169-174.

Gibb, K. S., Schneider, B. and Padovan, A. C. (1998). Differential detection and relatedness of phytoplasmas in papaya. *Pl. Pathol*. 47: 325-332.

Griffiths, H.M., Sinclair, W.A., Smart, C.D. and Davis, R.E. (1999). The phytoplasma associated with ash yellows and lilac witches' broom: 'Candidatus Phytoplasma fraxini', *Int. J. Sys. Bacteriol*. 49: 1605-1614.

Green, M. J., Thompson, D. A. and MacKenzie, D. J. (1999). Easy and efficient DNA extraction from woody plants for the detection of phytoplasmas by polymerase chain reaction. *Plant Disease*, 83(5): 482-485.

Guthrie, J.N., Walsh, K.B., Scott, P.T. and Rasmussen, T.S. (2001). The phytopathology of Australian papaya dieback: a proposed role for the phytoplasma. *Physiological and Molecular Plant Pathology*. 58: 23-30.

Gundersen, D. E., Lee, I-M., Rehner, S. A., Davis, R. E. and Kingsbury, D. T. (1994). Phylogeny of mycoplasmalike organisms (phytoplasmas): a basis for their classification. *J. Bacteriol*. 176: 5244-5254.

Gundersen, D. E. and Lee, I.-M. (1996). Ultrasensitive detection of phytoplasmas by nested-PCR assays using two universal primer pairs. *Phytopathol Mediterr*. 35: 144-151.

Guo, Y.H., Cheng, Z.M., Walla, J.A. and Zhang, Z. (1998). Diagnosis of X-disease phytoplasma in stone fruits by a monoclonal antibody developed directly from a woody plant. *J. Environ. Hort.* 16: 33-37.

Harrison, N., Richardson, P., Jones, P., Tymon, A., Eden-Green, S. and Mpunami, A. (1994). Comparative investigation of MLOs associated with Caribbean and African Coconut Lethal Decline diseases by DNA hybridisation and PCR assays. *Plant Disease,* 78: 507-511

Heinrich, M., Botti, S., Caprara, L., Arthofer, W., Strommer, S., Hanzer, V., Katinger, H., Bertaccini, A. and Laimer da Câmara Machado, M. (2001). Improved Detection Methods for Fruit Tree Phytoplasmas. *Plant Molecular Biology Reporter.* 19: 169-179.

Hiruki, C. and Wang, K. R. (2004). Clover proliferation phytoplasma: 'Candidatus Phytoplasma trifolii'. *Int. J. Syst. Evol. Microbiol.* 54: 1349-1353.

Hiruki, C. and Wang, K. R. (2008). Clover proliferation phytoplasmas and Their Subgroup Members, pp. 325-351. In: *Characterization, Diagnosis and Management of Phytoplasmas.* (Eds.) Nigel A. Harrison, Govind P. Rao and Carmine Marcone. Plant Pathogens Series-5. Studium Press LLC, U.S.A.

Hsu, H.T., Lee, I.-M., Davis, R.E. and Wang, Y.C. (1990). Immunization for generation of hybridoma antibodies specifically reacting with plants infected with mycoplasmalike organism (MLO) and their use in detection of MLO antigens. *Phytopathology,* 80 (10): 946-950.

Ho, K., Tsai, C. and Chung, T. (2001). Organization of ribosomal RNA genes from a loofah witches' broom phytoplasma. *UNA and Cell Biol.* 20: 115-22.

Jarausch W., Sailiard C., Dosba F. and Bove J.M. (1994). Differentiation of mycoplasmalike organisms (MLOs) in European fruit trees by PCR using specific primers derived from the sequence of a chromosomal fragment of the apple proliferation MLO. *Appl. Environ. Microbiol.* 60: 2916-2923.

Jarausch, W., Lansac, M. and Dosba, F. (1999). Seasonal colonization pattern of European stone fruit yellows phytoplasmas in different Primus species detected by specific PCR. *J. Phytopathol.* 147: 47-54.

Jarausch, W., Jarausch-Wehrheim, B., Danet, J.L., Broquaire, J.M., Dosba, F, Saillard, C. and Gamier, M. (2001). Detection and identification of European stone fruit yellows and other phytoplasmas in wild plants in the surroundings of apricot chlorotic leaf roll-affected orchards in southern France. *Europ. J. PI. Pathol.* 107: 209-217.

Jiang, Y.P. and Chen, T.A. (1987). Purification of Mycoplasma-like organisms from lettuce with aster yellows disease. *Phytopathology.* 77: 949-953.

Jiang,Y.P., Chen, T.A., Chiykowski, L.N. and Sinha, R.C. (1989). Production of monioclonal antibodies to peach eastern X-disease agent and their use in disease detection. *Canadian J. PI. Pathol.* 11: 325-331.

Jones, P. (2002). Phytoplasma: Plant Pathogens pp 126-139. In: Plant Pathologists Pocketbook, Part 12 (eds: Waller M, Lenné JM, Waller SJ), CAB International, Oxford University Press, USA.

Jung, H.-Y., Sawayanagi, T., Kakizawa, T., Nishigawa, H., Miyata, S., Oshima, K., Ugaki, M., Joon-Tak, L. and Namba, S. (2002). 'Candidatus Phytoplasma castaneae', a novel phytoplasma taxon associated with chestnut witches' broom disease. *Int. J. Syst. Evol. Microbiol.* 52: 1543-1549.

Jung, H.-Y., Sawayanagi, T., Wongkaew, P., Kakizawa, S., Nishigawa, H., Wei, W., Oshima, K., Miyata, S., Ugaki, M., Hibi., T. and Namba, S. (2003). 'Candidatus Phytoplasma oryzae', a novel phytoplasma taxon associated with rice yellow dwarf disease. Int. *J. Syst. Evol. Microbiol.* 53: 1925-1929.

Kakizawa, S., Oshima, K., Kuboyama, T., Nishigawa, H., Jung, H., Sawayanagi, T, Tsukikazi, T., Miyata, S., Ugaki, M. and Namba, S. (2001). Cloning and expression analysis of phytoplasma protein translocation genes. *Mol. PI. Microbe Interact.*, 14: 1043-1050.

Kassanis, B. (1954). Heat therapy of virus infected plants. *Ann. Appl. Biol.*, 41(3): 470-474.

Kakizawa, S., Oshima, K., Nishigawa, H., Jung, H.Y., Wei, W., Suzuki, S., Tanaka, M., Miyata, S., Ugaki. M. and Namba, S. (2004). Secretion of immunodominant membrane protein from onion yellows phytoplasma through the Sec protein-translocation system in *Escherichia coli. Microbiology*.150: 135-142.

Kenyon, L., Harrisson, N. A., Ashburner, G. R., Boa, E. and Richardson, P. A. (1998). Detection of a pigeon pea witches'-broom–related phytoplasma in trees of Gliricidia sepium affected by little–leaf disease in Central America.*Pl. Pathol.* 47: 671-680.

Khan, A.J., Botti, S., Al-Subhi, A.M., Gundersen-Rindal, D.E. and Bertaccini, A. (2002). Molecular identification of a new phytoplasma strain associated with alfalfa witches' broom in Oman. *Phytopathology*. 92: 1038-1047.

Kirkpatrick, B. C., Stenger, D. C., Morris, T. J. and A. H. Purcell. (1987). Clonino and detection of DNA from a nonculturable plant pathogenic mycoplasma-like organism. *Science*. 238: 197-200.

Kirkpatrick, B.C. (1992). Mycoplasma-like organisms: plant and invertebrate pathogens. Pages 4050-4067 In: The Prokaryotes, second edition A. Balows, H. G. Truper, M. Dworkin, W. Harder and K. H. Schleifer, eds. Springer-Verlag, New York, USA.

Kirkpatrick, B., Smart, C., Gardner, S., Gao, J., Ahrens, U., Mäurer, R., Schneider, B., Lorenz, K., Seemüller, E., Harrison, N., Namba, S. and Daire, X. (1994). Phylogenetic relationship of plant pathogenic MLOs established by 16/2S rDNA spacer sequences. *IOM Letters*. 3: 228-229.

Kirkpatrick, B.C. and Smart, C.D. (1995). Phytoplasmas: can phylogeny provide the means to understand pathogenicity. Pages 187-212 In: Advances in Botanical Research Vol 21, J. H. Andrews and I. C. Tommerup, eds. Academic Press, NY

Kison, H., Schneider, B. and Seemuller, E. (1994). Restriction fragment length polymorphism within the apple proliferation mycoplasmalike organism. *J. Phytopathol.* 141: 395-401.

Kison, H., Kirkpatrick, B.C. and Seemuller, E. (1997). Genetic comparison of the peach yellow leaf roll agent with European fruit tree phytoplasmas of the apple proliferation group. *Pl Pathol.* 46: 538-544.

Kison, H. and Seemuller, E. (2001). Differences in strain virulence of the European stone fruit yellows phytoplasma and susceptibility of stone fruit trees on various rootstocks to this pathogen. *J. Phytopathol.* 149: 533-541.

Kollar, A. and Seemuller, E. (1989). Base composition of the DNA of Mycoplasma-like organisms associated with various plant diseases. *Journal of Phytopathology.* 127: 177-186.

Kunkel, L.O. (1936). Heat treatments for the cure of yellows and other virus diseases of peach. *Phytopathology.* 26: 809-830.

Kuske, C. R. and Kirkpatrick, B.C. (1992). Phylogenetic relationships between the western aster yellows amycoplasmalike organism and other prokaryotes established by 16S rRNA gene sequence. *Intl. J. Syst. Bacteriol.* 42: 226-233.

Kuske, C. R., Kirkpatrick, B. C., Davis, M. J. and Seemuller, E. (1991). DNA hybridization between western aster yellows mycoplasmalike organism plasmids and extrachromosomal DNA from other plant pathogenic mycoplasmalike organisms. *Mol Plant-Microbe Interact.* 4: 75-80.

Laimer, M. and Balla, I. (2003). Méthodes rapides et fiables pour la détection et l´élimination des Phytoplasmes chez les arbres fruitiers. *Fruit Belge.* 505: 157–161.

Langer, M, and Maixner, M. (2004). Molecular characterisation of grapevine yellows associated phytoplasmas of the stolbur-group based on RFLP-analysis of non-ribosomal DNA. *Vitis.*43: 191-199.

Lauer, U. and Seemuller, E. (2000). Physical map of the chromosome of the apple proliferation phytoplasma. *J. Bacterial.* 182: 1415-1418.

Le Gall, F.I., Bove, J.M. and Gamier, M. (1998). Engineering of a single-chain variable-fragment (scFv) antibody specific for the stolbur phytoplasma (mollicute) and its expression in *Escherichia coli* and tobacco plants. *Appl. Environ. Microbiol.* 64: 4566-4572.

Lee, I-M. and Davis, R.E. (1986) Prospects for in vitro culture of plant-pathogenic mycoplasmalike organisms. *Ann. Rev. Phytopathol.* 24: 339-354.

Lee, I.-M., Davis, R.E. and Hiruki, C. (1991). Genetic iterrelatedness among clover proliferation mycoplasmalike organisms (MLOs) and other MLOs investigated by nucleic acid hybridization and restriction fragment length polymorphism analyses. *Applied and Environmental Microbiology.* 57 (12): 3565-3569.

Lee, I. M. and Davis, R. E. (1992). Mycoplasmas which infect plants and insects. pp 379-390 In: Mycoplasmas: Molecular Biology and Pathogenesis. J. Maniloff, R.

N. McElhaney, L. R. Finch and J. B. Baseman, eds. American Society for Microbiology, Washington, DC, USA.

Lee, I. M., Davis, R. E. and Hsu, H. T. (1993a). Differentiation of strains in the aster yellows mycoplasmas like organism strain cluster by serological assay with monocolonal antibodies. *Plant Disease.*77: 815-817.

Lee, I. M., Hammond, R. W., Davis, R. E. and Gundersen, D. E. (1993b). Universal amplification and analysis of pathogen 16S rDNA for classification and identification of mycoplasmas like organisms. *Phytopathology.* 83: 834-842.

Lee, I.-M., Pastore, M., Vibio, M., Danielli, A., Attathorn, S., Davis, R.E. and Bertaceini, A (1997). Detection and characterization of a phytoplasma associated with annual blue grass (Poa annua) white leaf disease in Southern Italy. *European Journal of plant Pathology.* 103: 251-254.

Lee, I. M., Gundersen-Rindal, D. E., Davis, R. E. and Bartoszyk, M. (1998). Revised classification scheme of phytoplasmas based on RFLP analyses of 16S rRNA and ribosomal protein gene sequences. *Intl. J. Syst. Bacteriol.* 48: 1153-1169.

Lee, I-M. and Davis, R.E. (1988). Detection and investigation of genetic relatedness among aster yellows and other mycoplasmalike organisms by using cloned DNA and RNA probes. *Mol. Plant-Microbe Interne.* 1: 303-310.

Lee, I. M., Gundersen-Rindal, D. E., Davis, R. E. and Bartoszyk, M. (1998). Revised classification scheme of phytoplasmas based on RFLP analyses of 16S rRNA and ribosomal protein gene sequences. *Intl. J. Syst. Bacteriol.* 48: 1153-1169.

Lee, I. M., Davis, R.E. and Gundersen-Rindal, D.E. (2000). Phytoplasma: phytopathogenic mollicutes. *Annual Review of Microbiology.* 54: 221-255.

Lepka, P., Atitt, M., Moll, E. and Seemuller, E. (1999). Effect of phytoplasmal infection on concentration and translocation of carbohydrates and amino acids in periwinkle and tobacco. *Physiol. Mol. Plant Pathol.* 55: 59-68.

Liefting, L.W. and Kirkpatrick, B.C. (2003). Cosmid cloning and sample sequencing of the genome of the uncultivable mollicute, western X-disease phytoplasma, using DNA purified by pulse-field gel electrophoresis. *FEMS Microbiology Letters.*221: 203-211.

Lim, P. O. and Sears, B. B. (1989). 16S rRNA sequence indicates that plant-pathogenic mycoplasmas like organisms are evolutionarily distinct from animal mycoplasmas. *J. Bacteriol.* 171: 5901-5906.

Loi, N., Carraro, L., Musetti, R., Firrao, G. and Osler, R. (1995). Apple proliferation epidemics detected in scab-resistant apple trees. *J. Phytopathol.* 143: 581-584.

Loi, N., Ermacora,P.,Carraro, L. and Osler, R. (1998). Monoclonal antibodies for the detection of Tagetes witches-broom agent. *Joural of Plant Pathology.* 80(2): 171-174.

Loi, N., Ermacora, P., Carraro, L.,Osler, R. and Chen, T.A. (2002b). Production of monoclonal antibodies against apple proliferation phytoplasma and their use in serological detection. *European Journal of Plant Pathology,* 108(1): 81-86.

Lorenz, K.-H., Schneider, B., Ahrens, U. and Seemüller, E. (1995). Detection of the apple proliferation and pear decline phytoplasmas by PCR amplification of ribosomal and nonribosomal DNA. *Phytopathology.* 85: 771-776.

Marcone, C., Neimark, H. Ragozzino, A., Lauer, U. and Seemuller, E. (1999). Chromosome sizes of phytoplasmas composing major phylogenetic groups and subgroups. *Phytopathology.* 89: 805-810.

Marcone C., Lee I. M., Davis R. E., Ragozzino A. and Seemüller, E. (2000). Classification of aster yellows-group phytoplasmas based on combined analyses of ribosomal RNA and tuf gene sequences. *Int. J. Syst. Evol. Microbiol.* 50: 1703-1713.

Marcone, C., Gibb, K. S., Streten, C. and Schneider, B. (2003a). Candidatus Phytoplasma spartii, Candidatus Phytoplasma rhamni and Candidatus Phytoplasma allocasuarinae, respectively associated with spartium witches-broom, buckthorn witches-broom and allocasuarina yellows diseases.*Int. J. Syst. Evol. Microbiol.* 54: 1025–1029.

Marcone, C., Schneider, B. and Seemüller, E. (2003b). 'Candidatus Phytoplasma cynodontis ', the phytoplasma associated with Bermuda grass white leaf disease. *Int. J. Syst. Evol. Microbiol.* 54: 1077–1082.

Marwitz, R. (1990). Diversity of yellows disease agents in plant infections. *Zentralblatt fur Baktenologie, Suppl.* 20: 431-434.

Marzachì, C. and Bosco, D. (2005). Relative quantification of phytoplasma in their plant and insect hosts: a Real Time PCR based method to quantify CY (16Sr I) phytoplasma in infected daisy and leafhopper vector. *Molecular Biotechnology,* 30: 117-127.

Melarned, S., Tanne, E., Ben-Haim, R., Edelbaum, O., Yogev, D. and Sela, I. (2003). Identification and characterization of phytoplasmal genes, employing a novel method of isolating phytoplasma genomic DNA. *Journal of Bacteriology.* 185(22): 6513-6521.

Musetti, R., Favali, M.A. and Pressacco, L. (2000). Histopathology and polyphenol content in plants infected by phytoplasmas. *Cytobios.* 102(401): 133-147.

Montano, H. G., Davis, R. E., Dally, E. L., Hogenhout, S., Pimentel, P.P. and Brioso, P.S.T. (2001). 'Candidatus Phytoplasma brasiliense', a new phytoplasma taxon associated with hibiscus witches' broom disease. *Int. J. Syst. Evol. Microbiol.* 51: 1109-1118.

Myata, S., Furuki, K., Oshima, K., Sawayanagi, T., Nishigawa, H. Kakizawa, S., Jung, H.Y., Ugaki, M. and Namba, S. (2002). Complete nucleotide sequence of the SlO-spc operon of phytoplasma: gene organization and genetic code resemble those of Bacillus subtilis. *DNA Cell Biol.* 21: 527-534.

Namba, S., Katao, S., I wanami, S., Oyaizu, H., Shiozawa, H. and Tsuchizaki, T. (1993). Detection and differentiation of plant pathogenic MLOs using PCR. *Phytopathol.* 83: 786-781.

Nishigawa, H., Miyata, S., Oshima, K., Sawayanagi, T., Komoto, A., Kuboyama, T., Matsuda, L, Tsuchizaki, T. and Namba, S. (2001). In planta expression of a protein encoded by the extrachromosomal DNA of a phytoplasma and related to geminivirus replication proteins. *Microbiology*. 147: 507-513.

Nishigawa, H., Oshima, K., Kakizawa, S., Jung, H.Y., Kuboyama, T., Miyata, S., Ugaki, M. and Namba, S. (2002). Evidence of intermolecular recombination between extrachromosomal DNAs in phytoplasma: a trigger for the biological diversity of phytoplasma. *Microbiology*. 148: 1389-1396.

Nishigawa, H., Wey, W., Oshima, K., Miyata, S., Ugaki, M., Hibi, T. and Namba, S. (2003). 'Candidatus Phytoplasma oryzae', a novel phytoplasma taxon associated with rice yellow dwarf disease. *Int. J. Syst. Evol. Microbiol.* 53: 1925–1929.

Nyland, G. (1962). Thermotherapy of virus infected fruit trees. Proceedings of the 5th European Symposiuum on Fruit Tree Virus Diseases, Bologna pp. 156-160.

Okuda, S.,Prince, J.P.,Davis,R.E.,Dally,E.L., Lee,I.M., Morgan,B. and Kato,S. (1997).Two groups of phytoplasmas from Japan distinguish on the basis of amplication and restriction analysis of 16SrDNA. *Plant Disease*. 81: 301-305.

Osler, R. and Carraro, L. (2004). Gli scopazzi del melo. *Informatore Fitopatologico* 5: 3-6.

Oshima,K., Kakizawa, S., Nishigawa, H., Jung,H.Y., Wei,W., Suzuki,S., Arashida,R., Nakata,D., Miyata,S., Ugaki, M. and Namba, S. (2004). Reductive evolution suggested from the complete genome sequence of a plant-pathogenic phytoplasma. *Nature Genetics*. 36: 27-29.

Padovan, A.C. and Gibb, K.S. (2001). Epidemiology of Phytoplasma Diseases in Papaya in Northern Australia. *Journal of Phytopathology*, 149: 649-658.

Parani, M., Singh, K.X., Rangasamy, S.R.S. and Ramalingam, R.S. (1996). A study on mechanism of phyllody disease resistance in Ses\amum alatum Thonn. *Current Sci*. 70: 86-89.

Paltrinieri, S., Lugli, A., Monari, W. and Bertaccini, A. (2004). Three years of molecular monitoring of phytoplasma spreading in a plum growing area in Italy. *Acta Horticulturae*. 657: 501-506.

Peralta, E., Arocha, Y., Iglesia, A., Diaz, M., Álvarez, E., López, M., Pino, O., Miranda, I., Milián, J., Carvajal, O., Matos, M. and Curbelo, I. (2003). Completamiento de la caracterización de los patógenos causantes de las enfermedades más importantes de la caña de azúcar en Cuba y desarrollo de métodos de mayor eficacia para su control. Informe Final de Proyecto Academia Nacional de Ciencias de Cuba, 70p.

Prabu, G.R., Kawar, P.G., Dixit, G.B. and Theertha,P.D. (2008). First report of a phytoplasma and virus pathogens associated with sugarcane grassy shoot disease in India. Sugar Cane International (In press).

Razin, S. (1964). Factors influencing osmotic fragility of mycoplasmas. *Journal of General Microbiology*. 36: 451-459.

Rekab, D., Carraro, L., Schneider, B., Seemüller, E., Chen, J., Chang, C.J., Locci, R. and Firrao, G. (1999). Gemnivirus–related extrachromosomal DNAs of the X-clade phytoplasmas share high sequence similarity. *Microbiology*. 145: 1453–1459.

Sawayanagi, T., Horikoshi, N., Kanehira, T., Shmohara, M., Bertaccini, A., Cousin, M.T., Hiruki, C. and Namba, S. (1999). 'Candidatus Phytoplasma japonicum, a new phytoplasma taxon associated with Japanese Hydrangea phyllody. *Int. J. Syst. Bacterial.* 49: 1275-1285.

Schneider, B., Seemüller, E., Smart, C. D. and Kirkpatrick, B. C. (1995). Phylogenetic classification of plant pathogenic mycoplasma-like organisms or phytoplasmas, pp. 369-380. In: Molecular and Diagnostic Procedures in Mycoplasmology, Vol. I. Molecular Characterization (Razin S., Tully J. G., Eds).-Academic Press Inc., San Diego, California, USA.

Schneider, B., Marcone, C., Kampmann, M., Ragozzino, A., Lederer, W., Cousin, M.T. and Seemüller, E. (1997). Characterization and classification of phytoplasmas from wild and cultivated plants by RFLP and sequence analysis of ribosomal DNA. *European Journal of Plant Pathology.* 103: 675-686.

Sears, B.B., Lim, P.O., Holland, N., Kirkpatrick, B.C. and Klomparens, K.L. (1989). Isolation and characterization of of DNA from a mycoplasma-like organism. *Mol. Plant-Microbe Interne.* 2: 175-180.

Sears, B.B., Klomparens, K.L., Wood, J.I. and Schewe, G. (1997). Effect of altered levels of oxygen and carbon dioxide on phytoplasma abundance in Oenothera leaf tip cultures. *Physiol. Mol. PL Pathoi.* 50: 275-287.

Sears, B.B. and Kirkpatnck, B.C. (1994). Unveiling the evolutionary relationships of plant-pathogenic mycoplasmalike organisms. *American Soc. Microbiol. News.* 60: 307-312.

Seemuller, E. (1976). Investigations to demonstrate mycoplasmalike organisms in diseased plants by fluorescence microscopy. *Acta Horticulturae.* 67: 109–112.

Seemuller, E., Schneider, B., Mauree, R., Ahrens, U., Daire, X., Kison, H.,Lorenz, K.H., Firrao, G., Avinent, L., Sears, B.B. and Stackebranett, E. (1994). Phylogenetic classification of phytopathogenic mollicutes by sequence analysis of 16SrDNA. *Int. J. Syst. Bacteriol.* 44: 440-446.

Seemuller, E., Morcone, C., Lauer, U., Ragozzino, A. and Goschl, M. (1998). Current status of molecular classification of the Phytoplasmas. *Journal of Plant Pathology.* 80: 3-26.

Seemuller, E. and Schneider, B. (2004). Taxonomic description of 'Candidatus Phytoplasma mali' sp. nov., 'Candidatus Phytoplasma pyri' sp. nov. and 'Candidatus Phytoplasma prunorum' sp. nov., the causal agents of apple proliferation, pear decline and European stone fruit yellows, respectively. *Int. J. Syst. Evol. Microbiol.* 54: 1217-1226.

Seddas, A., Meignoz, R., Daire, X., Boudon-Padieu, E. and Goschl, M. (1993). Purification of grapevine flavescence doree MLO (mycoplasma-like organism) by immunoaffinity. *Current Microbiology.* 27: 229-236.

Smart, C.D., Schneider, B., Blomquist, C.L., Guerra, L.J., Harrison, N.A., Ahrens, U., Lorenz, K.H., Seemuller, E. and Kirkpatrick, B.C. (1996). Phytoplasma specific

PCR primers based upon sequences of the 16-23SrRNA spacer region.*Applied and Environmental Microbiology*, 62: 2988-2993.

Streten, C. and Gibb, K.S. (2003). Identification of genes in the tomato big bud phytoplasma and comparison to those in sweet potato little leaf-V4 phytoplasma. *Microbiology*,149: 1797-1805.

Thomas, P.E. and Mink, G.I. (1998). Tomato hybrids with nonspecific immunity to viral and mycoplasma pathogens of potato and tomato. *Hortscience*. 33: 764-765.

Thomas, S. and Balasundran, M. (2001). Purification of sandal spike phytoplasma for the production of polyclonal antibody. *Current Science*. 80: 1489-1494.

Tsai, J. H. (1979). Vector transmission of mycoplasmal agents of plant diseases. Pages 265-307. In: The Mycoplasmas III, Plant and Insect Mycoplasmas, R. F. Whitcomb and J. G. Tully (eds.). Academic Press, New York, USA.

Tymon, A., Jones, P. and Harrison, N. (1998). Phylogenetic relationships of coconut phytoplasmas and the development of specific oligonucleotide PCR primers. *Annals of Applied Biology*, 132: 437-452.

Verdin, E., Salar, P., Danet, J. L. Choueiri, E., Jreijiri, F., Zammar, S. E1., Gelie, B., Bove, J. M. and Garnier, M. (2003). Candidatus Phytoplasma phoeniceum, a new phytoplasma associated with an emerging lethal disease of almond trees in Lebanon and Iran. *Int. J. Syst. Evol. Microbiol.* 53: 833–838.

Veronesi, F., Bertaccini, A., Parente, A., Mastronicola, M. and Pastore, M. (2000). PCR indexing of phytoplasma-infected micropropagated periwinkle treated with PAP-II, a ribosome inactivating protein from Phytolacca ainericana leaves. *Acta Hort.*, 530: 113-119.

White, D.T., Blackhall, L.L., Scott, P.T. and Waldh, K.B. (1998). Phylogenetic positions of phytoplasmas associated with dieback, yellow crinkle and mosaic diseases of papaya and their proposed inclusion in 'Candidatus Phytoplasma australiense' and a new taxon, 'Candidatus Phytoplasma australasia'. *International Journal of Systematic and Evolutionary Microbiology*. 48: 941-951.

Yu, Y., Yeh, K. and Lin, C. (1998). An antigenic protein gene of a phytoplasma associated with sweet potato witches' broom. *Microbiology*. 144: 1257-1262.

Zreik, L., Carle, P., Bové, J. and Garnier, M. (1995). Characterization of the mycoplasmalike organism associated with witches´ broom of lime and proposition of a Candidatus taxon for the organism "Candidatus phytoplasma aurantifolia". *Int. J. Systematic and Bacteriol.* 453: 449-53.

Zhong, Q. and Hiruki, C. (1994). Genetic differentiation of phytoplasma isolates determined by a DNA heteroduplex mobility assay. *Proc. of the Japan Acad.* 70B: 127-131.

Chapter 15

Biochemical and Molecular Variability in *Pythium aphanidermatum* Causing Rhizome Rot of Ginger

Shalini D. Sagar* and Shrikant Kulkarni

Department of Plant Pathology,
University of Agricultural Sciences, Dharwad – 580 005, Karnataka, India

ABSTRACT

In the biochemical studies, *i.e.*, peroxidase, polyphenoloxidase and catalase, polymorphic banding pattern was observed, *i.e.*, some isolates producing one/two extra bands indicating variation among the isolates or may be these are virulent races. Cluster analysis of these enzymes results was inconsistent, because in some cluster comprised of the isolates with farer geographical origin. Molecular variability among 12 isolates of *Pythium aphanidermatum* was studied by using PCR based RAPD method. RAPD analysis of using 40 random primers, resulted in total of 313 amplicon levels. Out of these, total of 245 polymorphic bands levels were observed. The cumulative analysis of similarity values placed 12 isolates in 2 clusters. In the present investigations the results obtained showed the possibility of using RAPD technique to distinguish variability among the isolates of *P. aphanidermatum*. The information could then be used to determine specific primers that would allow the identification of the fungus directly from plant material or in the soil.

Keywords: *Pythium aphanidermatum, Rhizome rot, Ginger, Variability, Isozyme, Molecular, RAPD*

---

* E-mail: shalupat@gmail.com

Introduction

India is considered as *'the land of spices'* and enjoys from time immemorial a unique position in the production and export of ginger. These crops are cultivated for their underground rhizomes, which are used in many ways. Its medicinal value is increasingly being recognized nowadays. Ginger originated in south-east asia, probably in India (Burkill, 1966, Purseglove *et al.* 1981). The name itself supports this view. The sanskrit name 'singabera' has given rise to greek *'zingiberi'* and later the generic name *zingiber*.

Ginger is used as flavoring agent, a preservative, used in pickling and ginger oil in soft drinks. Among the major constraints for growing ginger is the rhizome rot. Even though important foliar diseases do exist in these crops, rhizome rot is very important in view of severe crop losses. It occurs in several parts of India wherever these crops are grown. The term rhizome rot is loosely used for all the diseases affecting the rhizome irrespective of pathogens involved, since the ultimate result is the partial or total loss of rhizome. The pathogens involved decide the nature of damage and also symptoms expression. The major disease identified is the soft rot resulting in wet rot caused by *Pythium aphanidermatum*. Studies on physiological specialization of predominant pathogen like *Pythium aphanidermatum* is important in order to screen the germplasm for resistance against the virulent race. PCR based technique like Random Amplification Polymorphic DNA helps to distinguish variability among the isolates of the organism. This information helps to determine specific primers that would allow identification of the fungus directly from plant material or in the soil. Hence the present investigation on the biochemical and molecular variability of *Pythium aphanidermatum* was undertaken.

Materials and Methods

Twelve isolates of *Pythium aphanidermatum* were collected from different places and designated as *Pa1* (Basavakalyan), *Pa2* (Humnabad), *Pa3* (Chickmagalur), *Pa4* (Hassan), *Pa5* (Hanagal), *Pa6* (Hirekerur), *Pa7* (Kodagu), *Pa8* (Mysore), *Pa9* (Sorab), *Pa10* (Sagar), *Pa11* (Sirsi) and *Pa12* (Siddapur).

Biochemical Variability

The possible existence of qualitative variation among pathogenic isolates of *P. aphanidermatum* was assessed by adopting the vertical Poly Acrylamide Gel Electrophoresis (PAGE). Peroxidase, Poly Phenol Oxidase and Catalase, isozyme studies were undertaken as described here under.

Extraction and Staining of Peroxides

Peroxidase isozyme was extracted for individual isolate. For this purpose isolates were grown on 30 ml V-8 broth, contained in 150 ml flasks for 10 days at temperature 28±1°C. Peroxidase extracted from above fungal mat by finely homogenizing it in two ml chilled extraction buffer consisting of 0.1 M tris, 17 per cent sucrose, 0.1 per cent ascorbic acid, 0.17 per cent cysteine hydrochloride and pH 8.0 in precooled mortar and pestle (Farkas and Stahmann, 1966). The resulting homogenate mixture was centrifuged at 5000 rpm for 10 min. The supernatant was decanted and used as enzyme source. Electrophoresis was performed according to Davis (1964).

In order to stain peroxidase isozyme, the gel was incubated in Na- Acetate buffer (0.2M pH 5.6), Benzidine and H_2O_2 solution till brown bands appeared.

Extraction and Staining of Polyphenol oxidase : The polyphenoloxidase was extracted by homogenizing the material in two ml of chilled 0.2 M sodium acetate buffer at 5.6 pH (Park *et al.*, 1980). In order to stain the polyphenoloxidase isozyme, the gel after the run was incubated in 0.03 M catechol containing 0.05 per cent P-phenylene diamine in phosphate citrate buffer (pH 6.0) for one hour.

Extraction and Staining of Catalase

Catalase isozyme extraction and the electrophoretic technique was essentially the same as in case of peroxidase isozyme studies except the staining procedure. The isozymes of catalase were localized by incubating the gels in H_2O_2 (0.01 per cent) solution for 5-10 minutes depending upon activity level. Staining solution consisting of Ferric chloride 500 mg, potassium ferriccyanide (500 mg) and water 50 ml were added to the gels and shaken gently until achromatic bands were developed. The catalase activity was revealed as achromatic zones at green background.

Analysis and interpretation of isozyme banding patterns : The banding patterns or the zymograms, so obtained were analysed based on procedure given for identifying the putative loci as described by Wendel and Weeden (1989). The measurement of band positions was made by using the following formula:

$$Rm = \frac{\text{Distance moved by the isozyme}}{\text{Distance moved by the tracking dye}}$$

The genetic similarity co-efficient was estimated using NTSYS PC-2.0 Software programme (Rohlf, 1998).

Molecular Variability

Random Amplified Polymorphic DNA (RAPD) analysis was used to detect the variations among the isolates of *P. aphanidermatum*. Standard protocols were used for the isolation of DNA and RAPD analysis.

Total Genomic DNA Extraction and Purification

2-3g of fungal mat grown V-8 juice broth was taken and homogenized using pestle and mortar in 4 ml of 2 per cent SDS for 5 minutes. To the solution 6 ml of lysis buffer was added. The suspension in pestle and mortar was extracted with equal volume of phenol: Chloroform: isoamylalcohol (1:1W/V) in centrifugation tube and centrifuged at 10, 000 rpm for 10 minutes. Supernatant was taken in fresh centrifuge tube. To this 1/10th volume of 3M Sodium acetate and 0.54 volume of isopropanal were added. Mixed by gentle inversion and kept for 30 minutes at –20°C. Centrifuged @ 10,00rpm for 10 minutes at 4°C for the necessary of DNA pellet and was washed with 70 per cent ethanol, air dried and resuspended in 500il of $T_{10}E_{1.}$ This DNA obtained was further quantified by agarose gel electrophoresis.

Random Primers and PCR Amplification

A total of 40 primers of 10 mer (Kits OPB and OPF) were used. The RAPD-PCR amplifications were carried out in 20 µl containing 15-25 ng of genomic DNA (Williams *et al.*, 1990). The reaction buffer consisted of 10 x assay buffer with 15 mM $MgCl_2$ (2.00 µl), 0.17 µl of Taq DNA polymerase (6.0U μl^{-1}), 1.0 ìl of dNTPs mix (2.5 mM each), 1.0 µl of RAPD primers (5PM/ìl), 1.0 ìl of template DNA (25 ng/µl) and sterile distilled water (14.83 µl). The PCR amplications were performed using Thermal Cycler programmed for initial denaturation of 94°C for 4 min followed by 40 cycles at 94°C for 1 min, 36°C for 1 min, 72°C for 2 min and a final cycle at 72°C for 5 min. All amplified DNA products were resolved by electrophoresis on agarose gel (1.2 per cent) in TAE (1X) buffer, stained with ethidium bromide and photographed.

Data Analysis

Gel photographs were scored for the presence (1) or absence (0) of bands of various molecular sizes. Binary matrices were analyzed to obtain Jaccard's coefficients among the isolates using NTSYS-pc (version 2.0). Jaccard's coefficients were clustered to generate dendrograms using the SAHN clustering program, selecting the unweighted pair-group method with arithmetic average (UPGMA) algorithm in NTSYS-pc (Rohlf, 1998).

Results and Discussion

Differences were noticed in the peroxidase enzyme activity of different isolates of *P. aphanidermatum*. The variation within these isozyme bands of different isolates usually involved in the position of faint minor bands or the comparative thickness or density of bands (Table 15.1). Band number 1, 2 and 3 were wide, prominent and dense. These three bands were common to all 12 isolates, whereas band number 4 with Rm values of 0.40 was seen only in the isolates *viz., Pa4, Pa5, Pa6, Pa9* and *Pa10.* The isolates *viz., Pa1, Pa2, Pa11* and *Pa12* produced an extra band with Rm values of 0.50. In case of Polyphenol oxidase band number 1 was wide, prominent and dense band present in all isolates with Rm value of 0.10 (Table 15.1). In case of isozyme catalase band number 1 was present in all the isolates with Rm value of 0.04 (Table 15.1).

Further, the dendrogram constructed from the pooled data clearly showed two major clusters at similarity coefficient of 0.63 (Figure 15.1). First cluster consists of *Pa1*(Basavakalyan) and *Pa2*(Humnabad) isolate belonging to Bidar district. In second two cluster two subgroups were obtained consisting of different isolates there was variation among the isolates because some isolates produced an extra band with respect to different enzymes. In all the three enzymes, *i.e.*, peroxidase, polyphenoloxidase and catalase, polymorphic banding pattern was observed, *i.e.*, some isolates producing one/two extra bands indicating variation among the isolates or may be these are virulent races. It is clear from the cluster analysis that, the isolates from nearer geographical locations were grouped in one cluster and also included isolates from distinct locations. The reasons may be due to the pathogen being introduced into new areas through rhizomes. The variation may be attributed due to intraspecific variation and difference in geographical location. These findings are in

Table 15.1: Relative Mobility (Rm) Values of Peroxidase, Polyphenol Oxidase and Catalase of Twelve Ginger Isolates of *Pythium aphanidermatum*

Isolate	*Peroxidase/Rm Values*						*Polyphenol Oxidase/Rm Values*					*Catalase/Rm Values*				
	No. of Bands	*0.05*	*0.25*	*0.30*	*0.40*	*0.50*	*No. of Bands*	*0.10*	*0.30*	*0.50*	*0.60*	*No. of Bands*	*0.04*	*0.20*	*0.30*	*0.40*
Pa1	4	+	+	+	–	+	3	+	+	–	+	1	+	–	–	–
Pa2	4	+	+	+	–	+	3	+	+	–	+	1	+	–	–	–
Pa3	3	+	+	–	+	–	3	+	–	+	+	3	+	+	–	+
Pa4	4	+	+	–	+	–	2	+	–	+	–	3	+	+	+	–
Pa5	4	+	+	+	+	–	3	+	+	+	–	3	+	+	+	–
Pa6	4	+	+	+	+	–	3	+	+	+	–	3	+	+	+	–
Pa7	3	+	+	+	+	–	3	+	+	+	–	2	+	+	–	–
Pa8	3	+	+	+	+	–	3	+	+	+	–	2	+	+	–	–
Pa10	4	+	+	+	+	–	3	+	+	+	–	3	+	+	–	+
Pa10	4	+	+	+	+	–	3	+	+	+	–	3	+	+	–	+
Pa11	4	+	+	–	+	+	2	+	–	+	–	3	+	+	–	+
Pa12	4	+	+	–	+	+	2	+	–	+	–	3	+	+	–	+

+: Present; –: Absent.

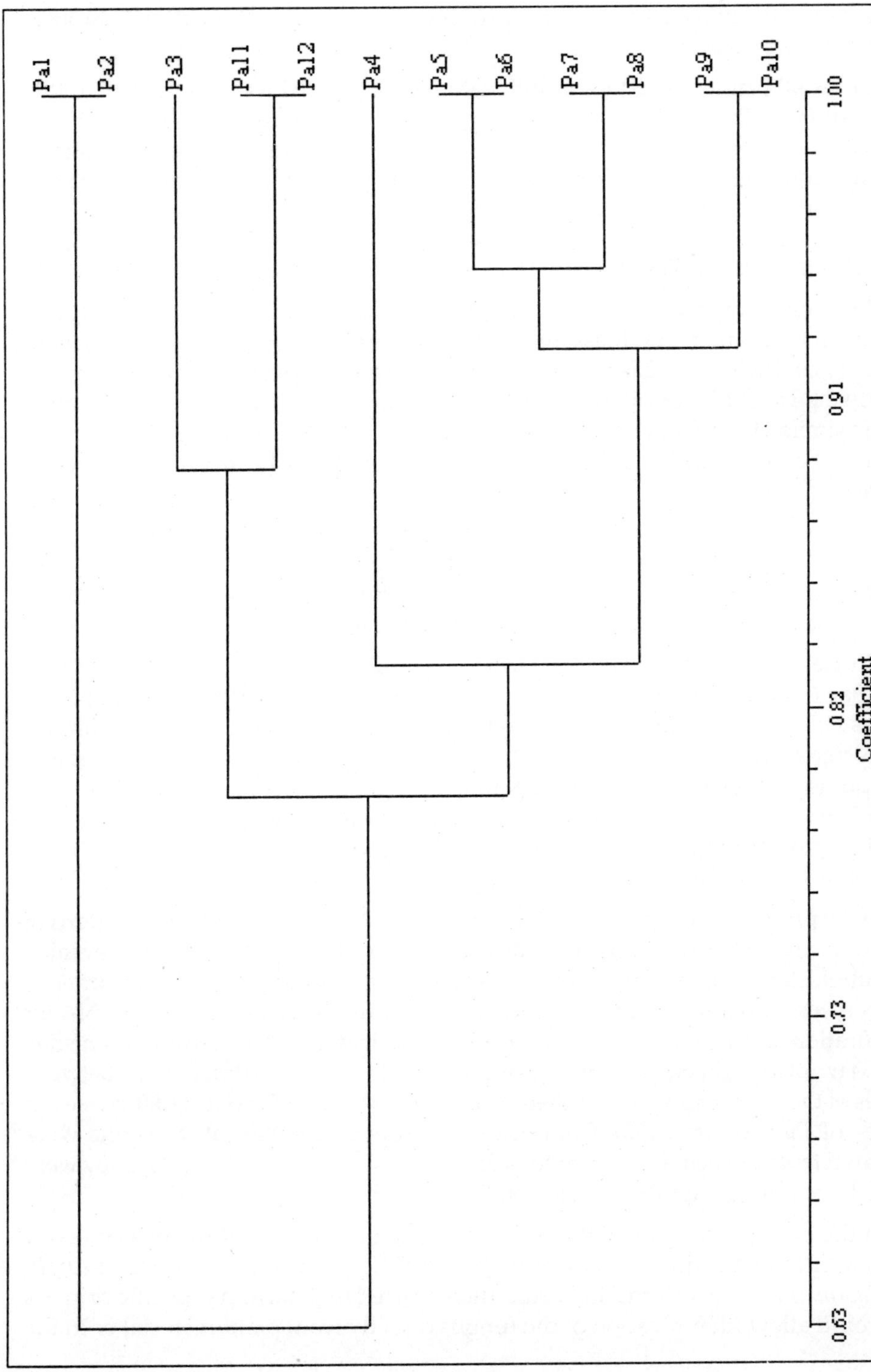

Figure 15.1: Dendrogram Based on Peroxidase, Polyphenol Oxidase and Catalase Pattern of Twelve Ginger of *Pythium aphanidermatum*

accordance with reports by Chen *et al.* (1992), who reported that isozyme analysis for 204 isolates of *Pythium* species from different geographical locations could be differentiated by banding pattern.

Of the 40 primers used for amplification OPB1, OPB2, OPB14, OPB17, OPF6, OPF11 and OPF 20 showed 100 per cent polymorphism (Figure 15.2). Three primers showed monomorphism and two primers did not show any amplification. A total of 313 amplicon levels resulted from 38 primers and were available for analysis (Table 15.2). Out of these, total of 245 polymorphic bands levels were observed. On an average there were 7.82 amplicon levels per primer of which 6.12 were polymorphic, indicating there is a molecular variability among the isolates of *P. aphanidermatum*.

Based on simple matching coefficient a genetic similarity matrix was constructed to assess the genetic relatedness among the isolates of *P. aphanidermatum*. Genetic similarity coefficient of 12 isolates of *P. aphanidermatum* based on RAPD analysis is given in the Table 15.3. Similarity coefficient ranged from 0.55 to 0.93. The maximum genetic similarity of 93 per cent was between *Pa5* and *Pa6*. There was 89 per cent similarity between *Pa11* and *Pa12* and 85 per cent similarity between the isolate *Pa1* and *Pa2*.

Further, the dendrogram constructed from the pooled data clearly showed two major clusters A and B at similarity coefficient of 0.61 (Figure 15.3). Cluster A was classified upto sub-sub cluster A11 and Cluster B was classified upto two minor clusters comprising *Pa1* and *Pa2* isolates belonging to Bidar district.

Cluster A was subgrouped into A1 and A2. A2 comprising only one isolate *i.e. Pa10* (Sagar isolate). A1 was subgrouped into A3 and A4. A4 comprising of two isolates *Pa5* and *Pa6* (Hanagal and Hirekerur isolate) belonging to Haveri district. Sub cluster A3 comprising of A5 and A6. A6 comprising of *Pa11* and *Pa12* (Sirsi and Siddapur isolate) belonging to Uttara Kannada district. Sub cluster A5 was grouped into A7 and A8. A7 comprising of only one isolate Pa3 (Chickmagalur isolate). A8 classfied into A9 and A11. A11 comprising of only *Pa4* (Hassan). A9 comprising of two isolate *Pa7* and *Pa8* (Kodagu and Mysore).

In the present investigation, the results revealed that, geographical locations of isolates were closely related. So the results obtained from the cluster analysis revealed that, sub cluster group composed of isolates belonging to same geographical locations with very less variability. The suitability of random amplified polymorphic DNA for identification of *Pythium aphanidermatum* was investigated by Herrero and Klemsdal, (1998). Two of the primers gave fingerprints that could be used to differentiate between isolates of the *Pythium* species studied. Matsumoto *et al.* (2000) collected forty-seven isolates of *Pythium irregulare* from different hosts and geographic origins were compared from molecular, morphological and physiological viewpoints. They were divided into four groups (I-IV) based on RAPD analysis.

In the present study also, the results obtained here showed the possibility of using RAPD technique to distinguish variability among the isolates of *P. aphanidermatum*. The information could then be used to determine specific primers that would allow identification of the fungus directly from plant material or in the soil.

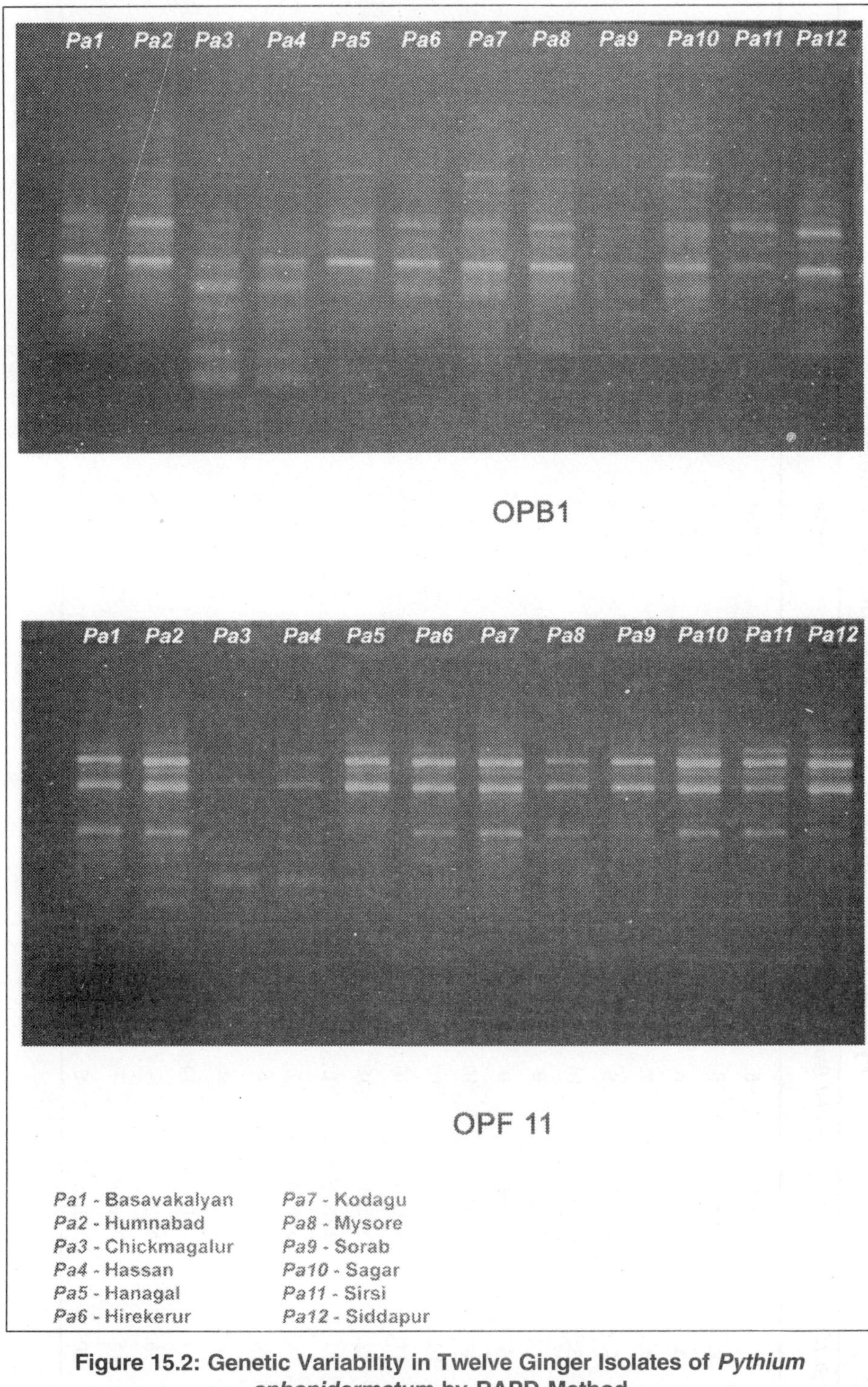

Figure 15.2: Genetic Variability in Twelve Ginger Isolates of *Pythium aphanidermatum* by RAPD Method

Table 15.2: Banding Profile of Different Primers for Different Isolates of *Pythium aphanidermatum* a Causal Agent of Rhizome Rot of Ginger

Sl. No.	Primer	Total Bands	Polymorphic Bands	Per cent Polymorphism	Sl No	Primer	Total Bands	Polymorphic Bands	Per cent Polymorphism
1.	OPB1	5	5	100.00	21.	OPF1	11	10	90.91
2.	OPB2	5	5	100.00	22.	OPF2	7	5	71.43
3.	OPB3	6	4	66.67	23.	OPF3	4	2	50.00
4.	OPB4	7	5	71.43	24.	OPF4	4	–	0.00
5.	OPB5	9	7	77.78	25.	OPF5	8	7	87.50
6.	OPB6	14	11	78.57	26.	OPF6	6	6	100.00
7.	OPB7	9	5	55.56	27.	OPF7	4	–	0.00
8.	OPB8	9	7	77.78	28.	OPF8	5	4	80.00
9.	OPB9	10	7	70.00	29.	OPF9	–	–	–
10.	OPB10	10	6	60.00	30.	OPF10	14	11	78.57
11.	OPB11	10	7	70.00	31.	OPF11	12	12	100.00
12.	OPB12	12	11	91.67	32.	OPF12	11	10	90.91
13.	OPB13	8	6	75.00	33.	OPF13	12	12	100.00
14.	OPB14	7	7	100.00	34.	OPF14	11	10	90.91
15.	OPB15	8	7	87.50	35.	OPF15	8	5	62.50
16.	OPB16	4	3	75.00	36.	OPF16	13	11	84.62
17.	OPB17	14	14	100.00	37.	OPF17	11	8	72.73
18.	OPB18	–	–	–	38.	OPF18	6	5	83.33
19.	OPB19	7	4	57.14	39.	OPF19	11	8	72.73
20.	OPB20	6	–	0.00	40.	OPF20	6	6	100.00

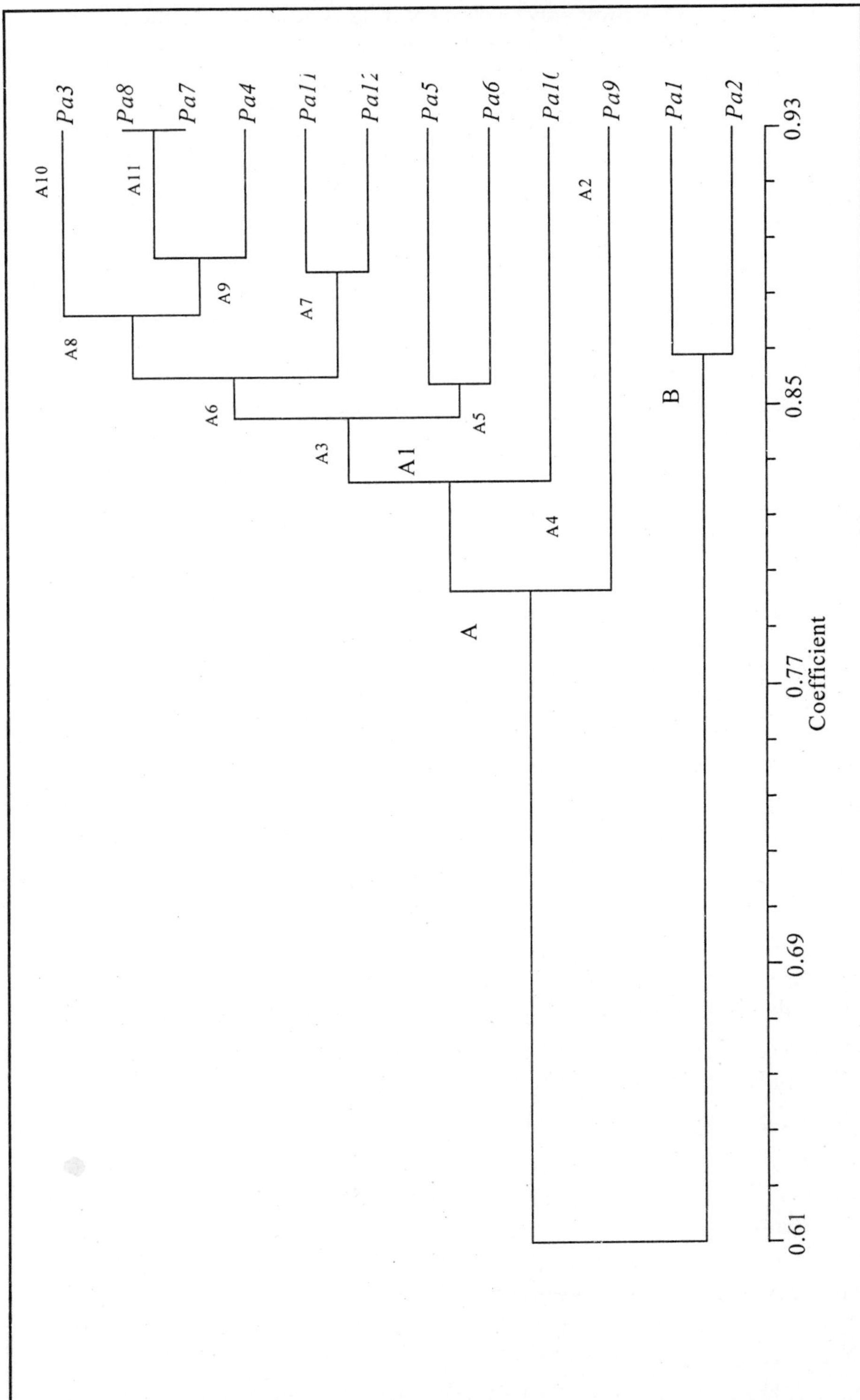

Figure 15.3: Dendrogram Based on RAPD Analysis of Twelve Isolates of *Pythium aphanidermatum*

Table 15.3: Similarity Coefficient of 12 Ginger Isolates of *Pythium aphanidermatum* Obtained by RAPD Analysis

	Pa1	*Pa2*	*Pa3*	*Pa4*	*Pa5*	*Pa6*	*Pa7*	*Pa8*	*Pa9*	*Pa10*	*Pa11*	*Pa12*
	1.00											
Pa2	0.85	1.00										
Pa3	0.64	0.58	1.00									
Pa4	0.62	0.59	0.86	1.00								
Pa5	0.87	0.83	0.62	0.66	1.00							
Pa6	0.87	0.85	0.62	0.63	0.93	1.00						
Pa7	0.84	0.81	0.68	0.68	0.84	0.86	1.00					
Pa8	0.88	0.84	0.62	0.63	0.88	0.90	0.88	1.00				
Pa9	0.84	0.80	0.62	0.61	0.84	0.85	0.85	0.87	1.00			
Pa10	0.76	0.77	0.55	0.58	0.79	0.82	0.78	0.81	0.82	1.00		
Pa11	0.84	0.81	0.58	0.58	0.87	0.89	0.83	0.88	0.83	0.84	1.00	
Pa12	0.83	0.82	0.60	0.58	0.81	0.86	0.83	0.87	0.83	0.78	0.89	1.00

References

Chen, W., Schneider, R. W. and Hoy, J. W. (1992). Taxonomic and phylogenetic analysis of ten *Pythium* species based on isozyme polymorphisms. *Phytopathology*. 82: 1234–44.

Burkill, I.H. (1966). *A Dictionary of the Economic products of the Malaysia*, Ministry of Agriculture and Co-operatives, Peninsula, Kuala Lumpur.

Davis. B.J. (1964). Disc electrophoresis, method and application to human serum proteins. *Annals of New York Academic Sci.* 121: 104.

Farkas, G.L. and Stahmann, M.A. (1966). On the nature of changes in peroxidase isozyme in bean leaves infected by southern bean mosaic virus. *Phytopath.* 56: 669-671.

Herrero, M.L. and Klemsdal, S.S. (1998). Identification of *Pythium aphanidermatum* using the RAPD technique. *Mycol. Res.* 102: 136-140.

Matsumota, C., Kageyama, K., Suga, H. and Hyakumachi, M. (2000). Intraspecific DNA polymorphism of *Pythium irregulare*. *Mycol. Res.* 104: 1333-1341.

Park, Y.K., Sato, H.H., Almeida, T.D. and Moretti, R.H. (1980). Poly Phenol Oxidase of mango variety, Haden. *J. of Food Science*. 45: 1619-1621.

Purseglove, J. W., Brown, E. G., Green, C.L. and Robbins, S.R. J. (1981). *Spices* Vol. 2. Longman Inc, New York. USA, p.2.

Rohlf, F. J. (1998). *ntsys-pc Numerical Taxonomy and Multivaraite Analysis Version* 2.0. Applied Biostatics Inc., New York.

Wendel, J.F. and Weeden, N.F. (1989). Visualization and interpretation of plant isozymes. In: *Isozymes in Plant Biology* (Eds. Soltis, D.E. and Soltis, P. S.). Dioseovides Press, USA pp. 5-46.

Williams, J.G.K., Kubelik, A.R., Livak, K.J., Rafalski, J.A. and Tingey, S.V. (1990). DNA polymorphism amplified by arbitrary primers are useful as genetic markers. *Nucleic Acid Res.* 12: 6531-6535.

Chapter 16

Molecular Based Techniques for Characterization, Detection and Diagnosis of Plant Viruses, Viroids and Phytoplasmas

Satya Prakash* and S.J. Singh

Virology Group, Agharkar Research Institute,
G.G. Agarkar Road, Pune – 411 004, India

ABSTRACT

Control of plant pathogenic viruses, viroids and phytoplasmas is difficult and hence preventive measures are essential to minimize the losses they cause in various crops. In this context, rapid and accurate methods for detection and diagnosis of these pathogens are required to apply. Plant viruses are generally detected and identified by particle morphology under electron microscope, host range and the serological assays. Electron microscopy is most convenient approach of direct detection of viruses but it is generally not used for routine diagnostic purpose. Moreover, the negative results in electron microscopy does not necessarily mean the absence of viral pathogens as it is quite likely the tissue used for electron microscopy may not have virus particles. Host range studies or biological indexing though, useful but it is basically a time consuming procedure and requires a well-equipped glass house and long-term maintenance of test host. Serological techniques such as ELISA and its variants are used in most cases for detection of viruses and are sensitive for most viruses where titre of antibodies is higher enough for ELISA testing. Cross-reactivity of antisera raised against viruses from different groups has frequently been used for detection and

* E-mail: satya.dwivedi2006@gmail.com

establishment of taxonomic relationships. However, nucleic acid sequence data are accumulating rapidly and allow more accurate relationship to be established between the individual members of virus groups than serological methods do. Invention of PCR and nucleic acid hybridization techniques revolutionized the detection and diagnosis of virus, viroid and phytoplasma infections in the plants. They are around 100-1000 fold more sensitive than serological assays such as ELISA. Furthermore, PCR has been greatly improved by the introduction of the second generation PCR, known as the Real time PCR where closed-tube fluorescence detection and quantification during PCR amplification (in real time) is possible eliminating the need for laborious post-PCR sample processing steps which greatly reduces the risk of carryover contamination. Using Real Time PCR, it is possible not only to detect the presence or absence of the target pathogen, but it is also possible to quantify the amount present in the sample allowing the quantitative assessment of the number of the pathogen in the sample. PCR and nucleic acid hybridization techniques can be applied for detection diagnosis and characterization of many plant viruses, viroids and phytoplasmas occurring in low concentration in plant tissues and also against viruses that are poor immunogen. The nucleic acid hybridization is particularly effective in the detection of viroids, which lack coat protein. The DNA Microarray technology, originally designed to study gene expression and generate single nucleotide polymorphism (SNP) profiles, is currently a new and emerging pathogen diagnostic technology, which in theory, offers a platform for unlimited multiplexing capability. The current trend in protocols for the detection of plant pathogens is to combine conventional, serological and molecular techniques in integrated approaches.

Keywords: *Molecular techniques, Virus, Viroid, Phytoplasma.*

Introduction

Diseases caused by viruses, viroids and phytoplasmas are a constant and major problem for crop production worldwide. The method of diagnosis, detection and identification of these agents in plants play a vital role. Conventional diagnosis of plant viruses requires bioassay (an indicator plant, determination of host range, symptomatology), virus particle morphology (size and shape) and vector relations. A single diagnostic test or assay may provide adequate information on the identity of a virus but a combination of methods is generally needed which are specific, sensitive and inexpensive (Naidu and Hughes, 2003). However, progress in molecular biology, biochemistry and immunology has led to the development of many new, accurate, rapid and less labour-intensive methods of virus, viroid and phytoplasma detection. Technologies for the molecular detection of plants pathogens have already undergone two major break throughs well over the past three decades. The first was the advent of antibody-based detection, in particular monoclonal antibodies and enzyme-linked immunosorbent assay (Kohler and Milstein, 1975; Clark and Adams, 1977). There are various immuno- and molecular-diagnostic techniques presently available in the field of virology, which are of two kinds: (1) Protein based techniques-include precipitation/agglutination tests, enzymes linked immunosorbent assay (ELISA), Immunosorbent electron microscopy (ISEM), fluorescent antibody test, dot immunoblotting assay (DTBIA) and (2) Viral nucleic acid based techniques- dot blot hybridization/slot blot hybridization, polymerase chain reaction (PCR), nucleic acid hybridization with radioactive and nonradioactive-labeled probes, DNA/RNA probes. Appropriate screening procedures have been conducted in order to certify any plant free of certain pathogen using ELISA, PCR, DNA probes (Akinjogunla *et al.*, 2008).

Recent advances in molecular biology and biotechnology are being applied to the development of rapid, specific and sensitive tools for the detection of plant virus, viroid and phytoplasma infections in plants (Martelli *et al.*, 1993; Minafra *et al.*, 1997; Maria *et al.*, 2003). Immunodiagnostic assays have reached the point of practical application in agriculture and the DNA probe technology is established in research laboratories. Immuno-assays are being routinely used to detect viruses in vegetatively propagated material and seeds for the purpose of quarantine, certification and maintenance. Use of immuno-reagents such as monoclonal antibodies in ELISA systems has the potential for enhanced precision and sophistication of detection.

When a virus is prone to antigenic variation or occurs without coat protein (Harrison and Robinson, 1978) and in viroids, immunological detection systems are unsuitable. Indirect methods for comparing viral nucleic acids are routine and have been used to detect viroids (Owens and Diener, 1984). The nucleic acid-based detection systems, which make use of, cloned DNA or DNA probes in a dot-blot or related assays have the potential to detect even single nucleotide differences. The ability to make DNA copies (cDNA) of a part or the whole plant viral RNA genome has revealed many new possibilities. The nucleotide sequence of the DNA copy can be determined, though it is a time-taking process, but it can be considered as a diagnostic procedure in certain special cases. Basically, four approaches can be used to detect and diagnose the viral nucleic acids: (1) Type, molecular size of the viral associated nucleic acids, (2) Cleavage pattern of viral DNA or cDNA, (3) Hybridization between nucleic acids and (4) Amplification studies of desired viral nucleic acid (Polymerase chain reaction).

Morris and Dodds (1979) developed a method for the isolation and detection of dsRNA from virus-infected plants. The properties of dsRNAs associated with RNA viral infections have been used for diagnosis (Dodds, 1993). These dsRNAs are very resistant to enzymatic degradation and not normally present in healthy plants. Simplification, coupled with improved equipment for nucleic acid analysis, has made the technique more practical and attractive to plant virus diagnosis.

The current trend in protocols for the detection of plant pathogens is to combine conventional, serological and molecular techniques in integrated approaches. The use of integrated approaches for detection is advised, especially when the targets are plant quarantine pathogens (Lopez *et al.*, 2008). This chapter presents up-to-date information on various molecular methods that have been in practice for detection, diagnosis and characterization of plant viruses, viroids and phytoplasmas throughout the world.

Molecular Detection Methods

Molecular hybridization or Nucleic acid hybridization

A viral particle is composed of nucleic acids [ribonucleic acid (RNA) or deoxyribonucleic acid (DNA)] and a capsid made up of several dozen to a thousand copies of coat protein subunit. In some cases, the virus possesses an envelope composed of viral proteins integrated in membranes deriving from the host cell. Serological techniques detect the virus by specific recognition of the coat protein by specific antibodies developed in animals against this protein. Molecular hybridization techniques detect viral nucleic acids by specific recognition of their nucleotide sequence.

Nucleic acids are long, linear polymers of nucleotide molecules. Each nucleotide is in turn composed of several elements: a nitrogen-containing base linked to a phosphate group and a sugar molecule (ribose for RNA and deoxyribose for DNA). DNA contains four different bases: adenine (A), guanine (G), cytosine (C) and thymine (T). In the case of RNA, thymine is replaced by uracil (U), the three other bases being the same.

DNA is usually found in a double-stranded configuration, *i.e.* two chains of DNA associate through specific base pairing (A pairs with T and C pairs with G). Base pairing is extremely specific and creates non-covalent hydrogen bonds that unite the molecules associated in this way. RNA is most commonly found in a single stranded configuration but, like DNA, it possesses the capacity to form double-stranded structures through A-U and G-C pairing.

The specific pairing of the bases composing nucleic acids constitutes the basis for the formation of hybrids (double-stranded structure) between complementary molecules and thus for the use of molecular hybridization as a diagnostic technique.

Nucleic acid molecules differ from one another in the order and sequence of alignment of their nucleotides (= nucleotide sequence). Two molecules of complementary sequences will form double-stranded hybrids under suitable conditions. For example, TCGGCGTAT will pair with AGCCGCATA to make a DNA-DNA hybrid.

A probe used for virus detection in molecular hybridization experiments is a single-stranded nucleic acid molecule prepared from a viral nucleic acid, with a nucleotide sequence complementary to that of the target viral RNA molecule. Thus a DNA probe with the sequence TCGGCGTAT will specifically detect RNA and DNA molecules with the respective sequences AGCCGCAUA and AGCCGCATA. An RNA probe with the same specificity would be UCGGCGUAU.

Nucleic acid hybridization is a powerful technique for the diagnosis of many plant viruses which are not easily detected by serological techniques. It is particularly effective in the detection of viruses occurring in low concentration in plant tissues and also against viruses that are poor immunogens or viroids which lack coat protein.

Molecular hybridization analysis also referred to as the 'nucleic acid hybridization 'or' spot hybridization' or 'dot blot' technique, has been shown to be a highly sensitive and specific procedure for identifying RNA or DNA viruses (Abu-Samah and Randles, 1983; Gould and Symons, 1983; Maule *et al.*, 1983) and plant viroids (Palukaitis and Symons, 1978; Owens and Diener, 1981). This technique has been extensively used for detection of grapevine viruses, viroids and phytoplasmas

(Bretout *et al.*, 1988; Semansic, 1993; Barba *et al.*, 1995). Basis of molecular hybridization of viral nucleic acids was discussed by Hull (1993).

The principle of hybridization analysis for indexing of viroids and viruses is relatively simple. The first requirement is to prepare highly radio active complementary DNA (cDNA) to the viroid or viral nucleic acid using ^{32}P or ^{3}H isotopes. The cDNA is prepared enzymatically on the viral or viroid nucleic acid so that its sequence is exactly complementary. When this ^{32}P or ^{3}H cDNA is incubated with a nucleic acid extract of a test plant under appropriate conditions, it hybridizes to any viral or viroid sequences present to produce a double strand ^{32}P cDNA-RNA hybrid in case of viroid and RNA viruses or ^{32}P cDNA-DNA hybrid in case of DNA viruses. The extent of hybrid formation is a measure of the concentration of viral or viroid sequences in the plant extract (Symons, 1984).

Nucleic acid specific hybridization techniques are being used for testing plant viruses and it involves the use of labeled complementary DNA (cDNA) or RNA (cRNA) prepared from purified viral nucleic acid as a probe or a recombinant clone of such viral nucleic acid as a probe. These labeled probes are used to detect the presence of plant viral nucleic acid by forming hybrid with them. Presently nucleic acid hybridization includes the formation of DNA-DNA, DNA-RNA, RNA-RNA complexes (Singh and Dhar, 1998). To detect the plant viruses and viroides three types of "Molecular hybridization" tests are being used, (*a*) Solution hybridization test (*b*) Filter hybridization test or Dot-blot hybridization or Nucleic acid spot hybridization (NASH) and (*c*) *In-situ* hybridization test.

Two procedures are available for the molecular hybridization analysis of viroid or virus infections, similar in principle differ in the way the ^{32}P cDNA: RNA hybrids are detected.

In-situ hybridization with oligonucleotide probes to specifically localize phytoplasmas in plant tissue in EM has been developed by Bonfiglioli *et al.* (1997) for the detection of phytoplasma diseases of grapevine.

Liquid Hybridization Assay

In this method, nucleic acid hybridization takes place in solution. Most of basic studies were performed using both the target and probe DNAs in solution hence it is called as liquid–liquid hybridization. There are also some data available for RNA: RNA and DNA: DNA interaction in solution. Presently most of plant pathogens diagnosis was involved by mixed phase hybridization with the target immobilized on a solid matrix, the theory developed for liquid-liquid hybridization is very relevant (Hull, 2002). Liquid-liquid hybridization has been used to examine the relationship between plant viruses. The main principal of the technique is to hybridize the target nucleic acid with the probe that has been radioactively labeled, than to digest away and unhybridized ss probe, precipitate the ds hybrid on a filter membrane and count the radioactivity. It has been extensively used for indexing avocado sunblotch viroid (ASBV) disease of avacados (Palukaitis *et al.*, 1981; Allen and Dale, 1981; Allen *et al.*, 1981), coconut cadang cadang viroid (CCCV) in extracts of coconut and other palms (Randles and Palukaitis, 1979; Randles *et al.*, 1980; Boccardo *et al.*, 1981) and also to a very limited extent for detection of chrysanthemum stunt viroid (CSV) in extracts of infected plants (Palukaitis and Symons, 1979). This method has also been used to assess relationships between tombus viruses and a sensitive non-radioactive procedure has been developed for detections, BaMV and its associated satellite RNA in meristem-tip cultured plants (Hsu *et al.*, 2000).

Dot-Blot Hybridization Assay or Nucleic Acid Spot Hybridization (NASH)

NASH is one of the widely used techniques to detect the plant viruses and viroids. In this technique the target nucleic acid is immbolized on a membrane (Nitrocellulose or Nylone) and then hybridized to a labeled specific nucleic acid probe. Crude leaf or seed extracts are applied to the membrane and labeled with specific probe of DNA or RNA. Serological methods (ELISA) gives little information on the virus detection but nucleic acid hybridization gives more genetic information on the viral or viroids nucleic acids. Due to lack of coat protein in the viroids, the nucleic and hybridization techniques are the only detection methods to characterize the genome. The identifying probes were made into two types: (*a*) Radio labeled probes and (*b*) Non-radio labeled probes.

This method is common for indexing of viroids, viruses and phytoplasmas and a popular method on large scale testing (Martelli *et al.*, 1993; Chen *et al.*, 1994; Del Serrone *et al.*, 1995). The basic procedures have been described by Thomas (1980) and Owens and Diener (1981), which are extensions and modifications of earlier techniques, which are developed for use in general recombinant DNA technology. The major difference between liquid hybridization and dot-blot assay is that in the latter test nucleic acid is immobilized on a cellulose nitrate sheet for hybridization rather than being in solution.

The preparation of samples in this procedure is quite simple. The crude seed/leaf extract samples are simply spotted onto nitro-cellulose filters and dried. The filters are subsequently hybridized with radio-labelled DNA probes which are copies of the viral genome. After hybridization and washing to remove the non-hybridised probe, the presence of virus is detected by autoradiography (Baulcombe *et al.*, 1984). A modification of the dot-blot assay, squash-blotting has been used to detect maize streak virus (MSV) (Boulton and Markhan, 1986). Complementary DNA (cDNA) probes require the use of radioactive label (^{32}P), which is not amenable for common use, and very limited success has been achieved in their application. Prehybridization and the hybridization steps are carried out in a specialized manner (Dijkstra and de Jager, 1998).

This technique has been used by Gallitelli *et al.* (1985) and Saldarelli *et al.* (1994) for the detection of grapevine virus A (GVA) and grapevine leafroll associated virus 3 (GLRaV-3), respectively in grapevines. Salazar *et al.* (1983) used this technique for the detection of PSTV in mixtures of potato seed extracts equivalent to one infected seed among healthy ones. Ahlawat (2007) mentioned usefulness of nucleic acid hybridization in detection of citrus viruses (citrus ring spot virus and citrus yellow mosaic virus) and viroid infections. This technique is being used in routine testing of seeds at International Potato Centre, Peru. Bijaisoradat and Kuhn (1988) used it to detect the presence of PMV and PStV in peanut seeds. Both the viruses have been detected readily in one mg of infected seed tissue and when extracts from seeds have been diluted to 1/62,500 with water. One part of the infected seed can be reliably detected when mixed with 99 parts of healthy seeds. This sensitivity will be 8-10 times greater than that achieved by the use of ELISA. Besides sensitivity, rapidity and simplicity, this test is highly specific. Hsu *et al.* (2000) developed dot-blot hybridization by using non-radioactive method to detect the BaMV and its sat RNA.

Types of Probes

Radioactive Labeled Probes

Several improvements have been made to the application of complementary DNA or RNA probes to detect viral pathogens. Mostly ^{32}P labeled nucleotides were used to label these nucleic acids radioactively, with specific activities up to 10^{10} cpm μg^{-1}, which can detect 1 pg of target sequences. Nucleic acid hybridization between DNA and RNA was done in liquid but presently done by blotting the test samples on membrane followed by hybridization with a radio labeled (^{32}P) viral nucleic acid that serves as probe. This technique was more sensitive than ELISA for the detection of potato virus X (PVX) and potato leaf roll virus (PLRV) in potato. Varveri *et al.* (1988) used ^{32}P labeled transcribed RNA probe for dot-hybridization detection of plum pox virus. The disadvantages of the radio labeled probes are, they are radioactive bio-hazardous.

Non-Radioactive Labeled Probes

Due to bio-hazardous nature of Radio labeled probes, the non-radio labeled probes are preferred to do the hybridization studies for the viral nucleic acids. The most used non-radioactive labeled compounds are Biotin Vitamin (H) and hapten digoxigenin (DIG). These two compounds are detected by indirect way, using labeled, specific-binding protein, either streptavidin (Biotin) or anti-digoxigenin serum and later visualized by a chemiluminescent or colorimetric reaction. The biotin-streptavidin system is very sensitive and efficient but we can get positive or background occurrence. Genome of the various viroids (246 to 375 bases long) is being used for hybridization studies. Non-radioactive labeled hybridization has been successfully used for detection of PSTVd in potato (Borkhardt *et al.*, 1994) and GFLV in grapevines (Gemmrich *et al.*, 1993). An advantage of non-radio labeled probes are- Stability of probes, no radioactivity, inexpensive, rapid detection and disadvantages are- Possible stearic hindrance during hybridization, because of hapten presence. Schematically the technique can be divided into five steps:

Probe Labeling

This is achieved by incorporation of a labeled (radioactive or biotinylated) nucleotide triphosphate precursor during in vitro reactions (nick translation for DNA probes or transcription for RNA probes).

Sample Preparation

The nucleic acids present in the plant extract are immobilized on a nitrocellulose membrane by baking it for two hours at 80°C under vacuum.

Hybridization

The labeled probe will form double-stranded structures under suitable conditions with the target nucleic acid immobilized on the membrane.

Washing(s)

Non-hybridized probe molecules are removed by successive washings of the membrane under stringent conditions.

Hybrid Detection

For radioactive probes, this is achieved by contact of an X-ray film with the membrane (autoradiography), usually for 24 hours. In the case of biotinylated probes, three additional steps are required:

1. Incubation in the presence of streptavidin, which reacts specifically with the biotin molecules, fixed on the probe.
2. Incubation in the presence of a biotinylated enzyme, which will be trapped by the streptavidin already retained on the membrane
3. An enzymatic reaction that results in the formation of a coloured product at the fixing point of the complex probe biotin-streptavidin-biotinylated enzyme.

Experimental Protocol

Probe Labeling

DNA Probes

Purified recombinant plasmid DNA is labeled (by incorporation of either ^{32}P sCTP, biotinylated dUTP or dCTP) by the technique of nick translation, using one of the several commercially available kits (*e.g.* BRL, Amersham).

RNA Probes

After linearization of the purified recombinant plasmid downstream of the viral cDNA with a suitable restriction endonuclease, labeled RNA is produced by *in- vitro* transcription using one of several commercially available kits (*e.g.* Promega, Biotec, Boehringer). Either ^{32}P or biotin-labeled CTP is usually incorporated.

Sample Preparation

Many different plant samples can be used, consisting of leaves, stems, tubers, barks or fruits. There is no standard protocol; each protocol should be optimized for a given host/virus combination.

Sample Grinding

One gram of plant sample is ground in 4 ml of grinding buffer using a pestle and mortar (or other apparatus such as an electric press or Polytron homogenizer when available). It is extremely important to use a buffer that will optimize the signal-tonoise ratio. The extract is then clarified by centrifugation for 10 minutes at 10,000 rpm.

The samples can, if necessary, be deproteinized by including one volume of a 1:1 mixture of water-saturated phenol and chloroform during the grinding. This step is optional for the use of radioactive probes but necessary when using biotinylated probes.

Sample Denaturation

The nucleic acids contained in the supernatant are then denatured if necessary, to ensure good binding of the nitrocellulose and availability of the sequences for hybridization. This step is important for the detection of viroids but of no utility for most viruses. In a small microcentrifuge tube, 50 µl of sample are added to 50 µl of formaldehyde denaturation buffer. The mixture is then incubated for 60 minutes at 60°C (the length of this incubation should be reduced for viruses). At this point, samples are ready for spotting on the membrane. They can also be stored for up to several months at –20°C. It is found that concentration of the nucleic acids present in the extract by ethanol precipitation is detrimental since it usually increases the non-specific background reactions; it is therefore not recommended.

Nitrocellulose Membrane Preparation

Soaking of the membrane in a high-salt solution is required for proper binding of nucleic acids in the samples. The membrane is first soaked for 2 minutes in pure distilled water and then equilibrated for 10 minutes in 20X SSC buffer.

Sample Application and Fixation

Next, 20 µl of sample are applied to the nitrocellulose membrane using a BRL "Hybri-dot" apparatus. Alternatively, 3 to 5 µl of sample can be applied directly (using a micropipette) to nitrocellulose that has been air dried after soaking in 20X SSC. The membrane is then dried at room temperature and baked for a further 2 hours at 80°C under vacuum to ensure stable binding of the nucleic acids to the nitrocellulose membrane. This can conveniently be achieved by using an electrophoresis slab gel drier or a vacuum oven. At this point, the membranes can be directly processed or sealed in a plastic bag and stored (at 4°C or 20°C) for up to several months.

Hybridization Reaction

Pre-hybridization

In order to prevent nonspecific binding of the probe to the membranes, they are pre-incubated in the hybridization mixture (pre-hybridization). The membranes are sealed in a plastic bag in the presence of 1 ml of hybridization buffer for each 10 cm^2 of membrane, taking care to avoid trapping any air bubbles. The bag is then incubated for 2 to 4 hours in a water bath at 42°C.

Probe Denaturation

This step is included to remove any secondary structure of the probe and is especially important for DNA probes which are essentially double-stranded after the labeling reaction. A suitable quantity of probe is placed in a small disposable tube and incubated for 10 minutes (DNA probe) or 3 minutes (RNA probe) at 100°C in a water bath and then quickly chilled by placing the tube in an ice-bucket.

Hybridization

The pre-hybridization buffer is discarded and replaced by the hybridization buffer to which the denatured probe has been added. Use approximately 1 ml of buffer containing radioactive probe of 1 to 2 x 10^6 cpm per ml or 200 mg per ml of biotinylated probe per 15 cm^2 of membrane (See techniques for determination of probe-specific activity, below). The plastic bag is then resealed and incubated in a water-bath overnight at 50°C. Schematic representation of the hybridization of a probe to nucleic acids immobilized on a nitrocellulose membrane is given in Figure 16.1.

Washing

After hybridization is completed, the membrane is removed from the plastic bag and washed in a small plastic tray. After washing, the nitrocellulose membranes should be air-dried at room temperature.

DNA Probes

Wash at room temperature for 5 minutes in three changes of 'washing buffer A', and then proceed with two 15-minutes washes at 50°C in 'washing buffer B'.

RNA Probes

Carry out four 20-minute washes at 60°C in 'washing buffer C'.

Hybrid Detection

Radioactive Probes

An X-ray film (Kodak XAR or equivalent) is exposed to the membrane for 24 hours at –70°C using intensifying screens. After autoradiography, the film is developed using Kodak LX 24 developer and Ilford Hypam fixer. Within the limits of linearity of the response of the film, the intensity of the spots is proportional to the concentration of viral RNA present on the membrane. No non-specific signal should be obtained with healthy plant controls. Autoradiograph of hybridization is given in Figure 16.2.

Biotinylated Probes

Several commercially available kits can be used for the detection of biotinylated probes (*e.g.* BRL). The composition of the buffers is given above.

- ☆ The membranes are first soaked for 1 minute at room temperature in buffer 1, then for 20 minutes at 42°C in buffer 2 to saturate the protein-fixing sites on the membrane. They are then dried and baked for 10 to 20 minutes at 80°C under vacuum.
- ☆ Following the treatment, the membranes are rehydrated for 10 minutes in buffer 2 and then incubated on a Petri dish in the streptavidin solution: 6 µl of a 1-mg per ml streptavidin solution diluted in 3 ml of buffer 1. Incubate for 10 minutes at room temperature, shaking occasionally.
- ☆ The membranes are then washed well with at least three changes of buffer for 3 minutes each time. Incubate on a Petri dish with 3 ml of buffer 1 containing 3 µl of a solution of biotinylated polymers of alkaline phosphatase (polyAP) at 1 mg per ml. Incubate for 10 minutes at room temperature with occasional shaking.
- ☆ Wash abundantly with two changes of buffer 1 and then with two changes of buffer 3. The developing solution should be prepared at the last moment in the following way: add 33 µl of the nitro-blue tetrazolium solution to 7.5

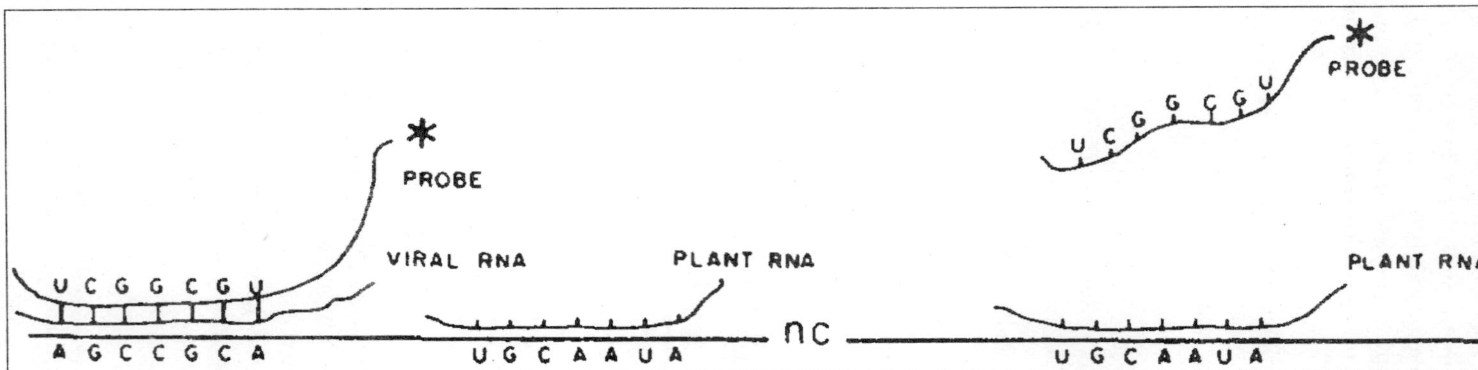

Figure 16.1: Schematic Representation of the Hybridization of a Probe to Nucleic Acids Immobilized on a Nitrocellulose Membrane (left) Infected Sample Containing Normal Plant Cell RNAs and Viral RNA (The target sequence to which the RNA probe hybridizes). Because of this hybridization the labeled probe will be retained on the membrane (right) Healthy sample containing only normal plant cell RNAs. The probe cannot hybridize with any sequence; it will not be retained on the membrane and will be eliminated upon washing.

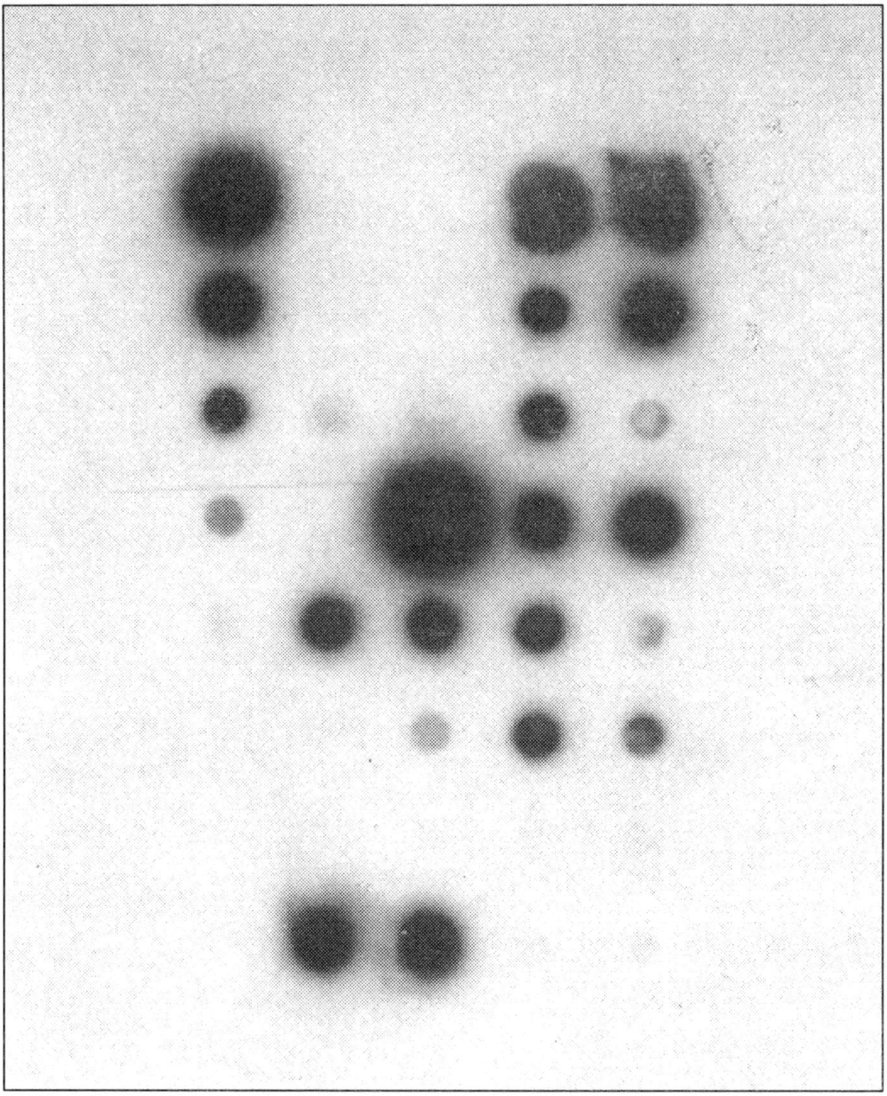

Figure 16.2: Autoradiograph Showing Dot-Blot Hybridization

ml of buffer 3. Mix thoroughly, and then add 25 µl of the 5-bromo-4-chloro-3-indolyl phosphate (BCIP) solution mix. Incubate the membrane in this solution in a sealed plastic bag protected from light.

☆ Maximum colour development is usually achieved within 4 hours. To stop the development, simply wash the membrane in 20 mM Tris-HCI pH 7.5, 5 mM EDTA. The dried membranes can then be stored for several months in the dark to preserve the colour.

Determination of the Probe Specific Activity

Following the labeling reaction, the radioactive DNA probe (1 µg is precipitated with ethanol), freed from the unincorporated labeled nucleotides by several 70 percent ethanol washes, dried and finally taken up in 100 µl of sterilized distilled water. Then 2 µl of the probe are mixed with 3 ml of a 10 percent trichloroacetic acid (TCA) solution along with 10 µl of 3 mg per ml calf thymus DNA used as a carrier. The mixture is left for 30 minutes at 0°C and then filtered through a Whatman GF/C fibreglass filter. The filter is rinsed with 20 ml of a 5 percent TCA solution and then with 5 ml of ethanol before being dried. The radioactivity retained on the filter is then determined by liquid scintillation counting. The specific activity, in cpm per µg, is given by cpm x 100/2.

Besides determining the specific activity of the probe, this technique can also help to calculate how much of the probe should be added to the hybridization reaction. Radioactive RNA probes can be counted in the same way.

For biotinylated probes, the result of the labeling reaction can be estimated by spotting dilutions of the probe on a membrane and comparing with a standard of known activity provided in the labeling kit.

Buffers

Grinding Buffer for Viroids (GPS)

- ☆ 200 mN glycine
- ☆ 100 mM Na_2HPO_4
- ☆ 600 mM NaCI
- ☆ 1 per cent SDS (sodium dodecyl sulphate)
- ☆ Adjust pH to 9.5, autoclave 20 minutes at 120°C, then add:
- ☆ 0.1 per cent DIECA (sodium diethyldithiocarbamate)
- ☆ 0.1 mM DTT (dithiothreitol)

Grinding Buffer for Viruses

- ☆ 50 mM sodium citrate, pH × 0.3
- ☆ 2 per cent PVP (polyvinylpyrrolidone) Autoclave 20 minutes at 120°C, then add:

  1 mM EDTA

  20 mM DIECA

Phenol/Chloroform Mixture

- ☆ 1 volume water-saturated phenol
- ☆ 1 volume chloroform/pentanol (24/1)

20X SSC Buffer

- ☆ 3 M NaCI
- ☆ 0.3 M sodium citrate Adjust pH to 7.0.

Formaldehyde Denaturation Buffer

- ☆ 5X SSC
- ☆ 25 mN Na_2HPO_4
- ☆ 5X Denhart [0.1 percent each of bovine serurn albumin (BSA), Ficolland PVP 360] 50 per cent deionized formamide 200 mg per ml of denatured calf thymus DNA.

DNA Probes Hybridization Buffer

- ☆ 4 volumes DNA probes prehybridization buffer 1 volume 50 per cent dextran sulphate.

RNA Probes Pre-Hybridization and Hybridization Buffer

- ☆ 50 per cent formamide
- ☆ 50 mM phosphate buffer pH 6.5
- ☆ 5X SSC
- ☆ 0.1 per cent SDS
- ☆ 1 mM EDTA
- ☆ 0.05 per cent Ficoll
- ☆ 0.05 per cent PVP 360
- ☆ 200 mg per ml of denatured salmon
- ☆ Sperm DNA

Washing Buffers

Washing Buffes A

- ☆ 2X SSC 0.1 per cent SDS

Washing Buffer B

- ☆ 0.2X SSC 0.1 per cent SDS

Washing Buffer C

- ☆ 0.1 X SSC 0.1 per cent SDS

Development Buffers for Biotinylated Probes

Buffer 1

- ☆ 100 mM Tris-HCI pH 7.5
- ☆ 100 mM NaCI
- ☆ 2 mM $MgCI_2$
- ☆ 0.05 per cent Triton X 100

Buffer 2

- ☆ Buffer 1 plus 3 per cent BSA

Buffer 3

- ☆ 100 mM Tris-HCI pH 9.5
- ☆ 100 mM NaCI

Double-Stranded RNA (dsRNA)

Almost 90 per cent of plant viruses have RNA genomes that can be single stranded. During replication of ssRNA viruses in plant cells, high molecular weight double stranded RNA (dsRNA) is produced as an intermediate product called as Replication form (RF) or Replicative Intermediate (RI), consistently present when plant is virus infected. Most of the dsRNAs are associated with plant RNA viruses. dsRNAs associated with RNA viral infections have been used for diagnosis (Dodds, 1993). Some viruses have dsRNA genomes during replication, others produce dsRNA. Detection of high molecular weight dsRNA has been used to study plant diseases of suspected viral etiology for which no virus-like particles could be identified (Dodds and Bar-Joseph, 1983; Jordan *et al.*, 1983; Morris and Dodds, 1979).

Double-stranded RNA (dsRNA), an indicator for the presence of viruses in plants, can be isolated rapidly by simple methods from small tissue weights of seedlings grown out of infected seed, using relatively simple laboratory equipment. The number, size, intensity and complexity of dsRNA detected by gel electrophoresis are distinctive and consistent for different virus groups and also aid in the diagnosis of a virus to a group or even a strain. Detection of dsRNA of appropriate size may provide insight into the cause of diseases of unknown aetiology (*i.e.* little cherry, lettuce big vein, black currant reversion) and for which no reliable assay is available (Dodds *et al.*, 1984). Additionally, it was expected that detection of dsRNA molecules would improve awareness of latent infections in apparently healthy plants because any suspected dsRNA species could be labeled directly or after cloning. DNA copies can be multiplied in bacteria, thereby facilitating their routine use in ELISA-based systems or in nucleic acid hybridization assays (Dodds, 1986). The method for extraction of dsRNA from plant tissues, purification and analysis is described by Valverde (2008) and Tzanetakisa and Robert (2008). Plant virus infections so far tested have been in Tobamo, Cucumo, Potex, Carla, Poty and Closterovirus groups which contain disease specific dsRNA, molecules of the size expected for the molecular weight of the genomic ssRNA. Bar-Joseph *et al.* (1983) showed TMV and CMV could readily be diagnosed in singly or double infected plants by dsRNA analysis. Use of dsRNA for the detection and diagnosis of grapevine viruses is discussed by Boscia (1993).

This technique has several advantages over other methods for virus diagnosis. The dsRNA technique overcomes these problems: (1) The technique is simple and relatively inexpensive, (2) dsRNA can be obtained regardless of the host or the RNA virus, (3) Results are obtained in a relatively short time (8-12 hrs), (4) Interfering host components and instability of the virus or viral RNA are two main problems encountered by plant virologists while using traditional methods, (5) The technique detects mixed infection, which often go undetected with other methods and result in inadequate diagnoses, (6) Unlike most other diagnostic techniques, dsRNA analysis is nonspecific and can be used to distinguish not only different viruses but also strains of the same virus and satellite RNAs (Valverde and Dodds,1986; Valverde *et al.*, 1986) and, (7) The purified dsRNA could then be used as a reagent for inoculation, probe preparation or molecular cloning (Dodds, 1986; Jordan and Dodds, 1983; Rosner *et al.*, 1983).

The dsRNA analysis has some limitations as only RNA viruses can be detected. The knowledge about the number and size of viral RNAs of different viral groups is required. Some plants contain cryptic viruses and/or cellular dsRNAs that yield ds RNAs similar in size to those associated with ssRNA viruses. Certain viral groups, such as the luteoviruses and most potyviruses, yield very low quantities of dsRNA, making the method impractical for their routine diagnosis (Valverde *et al.*, 1990).

Gel Electrophoresis

Electrophoresis is the migration of charged particles like proteins, nucleic acids or polysaccharides to either cathode or anode under the influence of an electric field. It is usually carried out in gels formed in tubes, slabs or on a flat bed. In electrophoretic units, the gel is mounted between two buffer chambers containing separate electrodes so that the only electrical connection between the two chambers is through the gel. The sample may be run in agarose or polyacrylamide or agarose-acrylamide composite gel. At the end of run, the gels are stained and used for scanning or visual recording of results.

The major useful systems for viruses and viroids are: (1) Agarose gel electrophoresis and (2) Polyacrylamide gel electrophoresis (PAGE).

Agarose Gel Electrophoresis

The standard method used to separate, identify and purify nucleic acid (RNA or DNA) fragments is electrophoresis through agarose gels. This technique is simple and rapid to perform, and capable of resolving mixtures of nucleic acid fragments that cannot be separated adequately by other methods, such as density gradient centrifugation. Furthermore, the location of nucleic acids within the gel can be determined directly by staining with low concentration of fluorescent or ethidium bromide dye and subsequent examination of the gel in ultraviolet light.

Several different designs of apparatus have been used. Currently, agarose gel electrophoresis is being carried out with horizontal slab gels. Several different electrophoresis buffers are available containing tris-acetate, tris-borate or tris-phosphate, and among them tris-acetate is the most commonly used buffer.

Polyacrylamide Gel Electrophoresis

Polyacrylamide gels are used to analyze and prepare fragments of DNA/RNA. They may be casted in a variety of polyacrylamide concentrations, ranging from 3.5 per cent to 20 per cent, depending on the size of the fragments of interest.

Polyacrylamide gels are poured between two glass plates that are held apart by spacers. In this arrangement, most of the acrylamide solution is shielded from exposure to the air, so that inhibition of polymerization by oxygen is confined to a narrow layer at the top of the gel.

Polyacrylamide gels can range in length from 10 cm to 100 cm, depending on the separation required and are invariably run in the vertical position (Maniatis *et al.*, 1982; Sambrook, 2002).

The other major usefulness of PAGE system is determination of the molecular weights of the polypeptides or nucleic acids, detection of cryptic virus, dsRNAs in virus or viroid infected seed material.

Return-polyacrylamide Gel Electrophoresis (R-PAGE)

Polyacrylamide gel electrophoresis (PAGE) has played a pivotal role in the diagnosis of diseases caused by viroids which have low molecular weight RNA (Singh, 1992). A modification of PAGE specially designed for the rapid and sensitive detection of circular RNAs, termed as return PAGE (R-PAGE), and has further facilitated viroid detection in various ways (Schumacher *et al.*, 1986; Singh *et al.*, 1988 and Singh *et al.*, 1993).

In R-PAGE assay, nucleic acids are subjected to two electrophoretic runs, one under non-denaturing and the other under denaturing conditions. Because of the denaturation, circular RNAs lose their double stranded configuration, become single-stranded covalently closed circular forms, migrate much more slowly in the second electrophoresis and thus are well separated from non-circular molecules. The circular RNAs from the lowest bands, on the electrophorograms, rendering them easily distinguishable from non-infected plant extracts. The procedure takes one day and requires simple laboratory equipment and non-radioactive chemicals. Viroid concentrations as low as 15-20 Pg can be detected reliably by R-PAGE (Singh, 1989). For example, it detects viroids in one infected potato leaf disc mixed with 500 healthy leaf discs and standard nucleic acid extracts diluted to 1 : 1024 times.

R-PAGE has been successfully utilized to detect viroid from dormant potato tubers or from infected true potato seeds (TPS) singly or mixed with 100 healthy TPS (Singh *et al.*, 1988). This technique has also been widely used to detect viroid infections (*i.e.* citrus exocortis, hop stunt and citrus yellow corky vein) in citrus (Ahlawat, 2007).

It has been used to identify several viroids from various crop plants in developing countries. A unique feature of the R-PAGE has been the separation of viroid strains on the basis of their mobility on the gel, which has greatly aided the studies on cross-protection (Singh, 1989). R-PAGE can be used to assess the PSTV content of various seed lots before planting either from seeds or from *in vitro* seedlings, when germplasm is valuable and available in small quantities (Singh *et al.*, 1988 and 1992).

Sodium Dodecyl Sulfate-Polyacrylamide Gel Electrophoresis (SDS-PAGE)

It is a technique used in biochemistry, genetics and molecular biology to separate proteins according to their electrophoretic mobility (a function of length of polypeptide chain or molecular weight as well as higher order protein folding, posttranslational modifications and other factors). In virology, SDS-PAGE is used for the determination of molecular weight of the viral capsid protein (Ravi *et al.*, 1997; Thomas *et al.*, 1997; Dujovny *et al.*, 1998).

cDNA Probes

cDNA technology utilizes the specific recognition between the viral RNA and its complementary DNA (cDNA) (Watson *et al.*, 1983). As majority of the plant viruses have single stranded RNA (ssRNA) as the genetic material, by using the reverse transcriptase, ssRNA can be copied into cDNA which then by a recombinant DNA technology can be cloned in a suitable cloning vehicle. In this way, cDNA, specific to each virus can be mass produced.

DNA or RNA viruses can be detected in seeds by using cDNA probes which are labeled with radioactive markers (^{32}P or ^{3}H) or non-radioactive markers such as biotin. Sensitivity of detection may be related to the probe size and larger probes give more sensitive assays. It is possible to make diagnostic cDNA probes to regions of the certain viral genome where the extent of sequence similarity has been shown to distinguish the viruses and strains (Hull and Davies, 1983).

Restriction Fragment Length Polymorphism (RFLP)

Every organism has a set of DNA patterns that separate it from others, which form its genetic "finger print". RFLP offers simple and quick DNA analysis techniques in which organism may be differentiated by analysis of patterns derived from cleavage of their DNA. RFLP is a difference in homologous DNA sequences that can be detected by the presence of fragments of different lengths after digestion of DNA samples with specific restriction endonucleases.

RFLP is used in combination with PCR to identify differences between viruses based on the presence or absence of restriction enzyme-recognition sites. After PCR amplification, the amplicon is digested with a restriction enzyme(s) and the fragment sizes analysed by gel electrophoresis.

RFLP is a method that can be used to differentiate isolates of viruses without the expenses of cloning and sequencing. Its effectiveness relies on polymorphism within restriction enzyme-recognition sites. RFLP was used to show that only members of subgroup 2 of cucumber mosaic virus were present in Western Australian lupin crops (Wylie *et al.*, 1993; Webster *et al.*, 2004).

When PCR products are cleaved with selected restriction endonucleases and the DNA fragments are subsequently separated by gel electrophoresis, characteristic banding patterns are obtained. These restriction enzyme profiles can be used to differentiate virus species following PCR amplification of DNA fragments with universal, group-specific primers, as demonstrated for Luteo-, Gemini- and Tospoviruses (Robertson *et al.*, 1991; Rojas *et al.*, 1993; Dewey *et al.*, 1996). PCR-RFLP also enables classification of virus isolates into subgroups or pathotypes and may allow differentiation of serologically indistinguishable isolates (Dietgen, 2002). Pathotyping tobamoviruses by PCR-RFLP appears to be a simple alternative to bioassays on a set of differential *Capsicum* genotypes. One RFLP group derived from the CP gene PCR product of CTV isolates appears to represent "mild" strains and could be used for rapid presumptive discrimination of mild and severe strains (Gillings *et al.*, 1993). Two primer pairs specific for SCMV or sorghum mosaic virus were used in PCR-RFLP for rapid discrimination between strains of the two viruses without the need to interpret symptoms on differential hosts (Yang and Mirkov, 1997). Genetic distances of PCR- RFLP of the PVY CP gene differentiated isolates into three strain groups that correlated with biological strains (Blanco-Urgoiti *et al.*, 1996).

Polymerase Chain Reaction (PCR)

The advent of Polymerase Chain Reaction (PCR) by Kary B. Mullis in the mid-1980s (Mullis *et al.*, 1986) revolutionized molecular biology as we know it. PCR is a fairly standard procedure now, and its use is extremely wide-ranging. Its advantages (speed, sensitivity, specificity) are far more important than its drawbacks (risk of contamination, sensitivity to inhibitors, complexity, cost), and several modifications to solve these problems have been performed with success. At its most basic application, PCR can amplify a small amount of template DNA (or RNA) into large quantities in a few hours. This is performed by mixing the DNA with primers on either side of the DNA (forward and reverse), Taq polymerase (of the species Thermus aquaticus, a thermophile bacterium whose polymerase is able to withstand extremely high temperatures), free nucleotides (dNTPs for DNA, NTPs for RNA), and buffer. The temperature is then alternated between hot and cold to denature and reanneal the DNA, with the polymerase adding new complementary strands each time. In addition to the basic use of PCR, specially designed primers can be made to ligate two different pieces of DNA together or add a restriction site, in addition to many other creative uses. Clearly, PCR is a procedure that is an integral addition to the molecular biologist's toolbox, and the method has been continually improved upon over the years (Purves *et al.*, 2001).

Polymerase chain reaction is a method for oligonucleotide primer directed enzymatic amplification of a specific DNA sequence of interest. This technique is capable of amplifying of a sequence 105 to 106 fold from nanogram amounts of template DNA within a large background of irrelevant sequences (e. g. from total genomic DNA). A prerequisite for amplifying a sequence using PCR is to have known, unique sequences flanking the segment of DNA to be amplified so that specific oligonucleotides can be obtained. It is not necessary to know anything about the intervening sequence between the primers. The PCR product is amplified from the DNA template using a heat-stable DNA polymerase from *Thermus aquaticus* bacterium (*Taq* DNA polymerase) and using an automated thermal cycler to put the reaction through 30 or more cycles of denaturing, annealing of primers, and polymerization. After amplification by PCR, the products are separated by agarose gel electrophoresis and are directly visualized after staining with ethidium bromide.

Before understanding the exact mechanism of how PCR works in amplification of target DNA, let us get into the kinetics of how a DNA molecule replicates in *in-vivo*. In the first step of replication, DNA double helix strand is unwound into two individual strands by the enzyme helicase at the site of replication initiation. The initial primer required for the DNA replication initiation is synthesized by the enzyme primase. Then the DNA polymerase synthesizes the DNA strands by utilizing the dNTP's present in the cytoplasm and the synthesis is carried out which leads to the formation of okazaki fragments in a replication fork and the synthesis is completed by filling the fragments in the replication fork.

The basis of invention of PCR is to overcome difficulties in amplification of the desired copy of DNA. Old procedures included cloning and replication of clone for multiplying the copy of desired DNA. The supplements of the DNA replication are provided artificially to replicate a desired copy of DNA in PCR reaction. The primary most problem faced by molecular biologists in *in-vitro* amplification of DNA is the stability of the DNA polymerases. Most of the DNA polymerases are not stable at higher temperatures like 90-95°C where two strands of DNA get separated by heat denaturation. Studies have shown that certain hot spring living bacteria like *Thermus aquaticus* have polymerases that are stable at higher temperatures. They are used in PCR as a replacement to conventional DNA polymerases. In the PCR reaction, other requirements are supplemented artificially like *Taq* DNA buffer (Which provides buffered environment for necessary action of *Taq* DNA polymerase), $MgCl_2$ (essential for activity of *Taq* DNA polymerase and also aids by binding of dNTPs (dATP, dGTP, dCTP and dTTP) for polymerization, Oligos (short stretch of DNA sequences that are complementary to the target DNA, popularly known as primers that will decide the range of amplification), template DNA from which the amplification of the specific target DNA is achieved, deionised water (to make up the concentration so the components). The PCR reaction is set in polycarbonate tubes or strips or plates designed for the purpose depending on the requirement process. The kinetics of PCR are chiefly dependent on the concentrations of the components like $MgCl_2$ concentration, dNTP's concentration, amount of enzyme used, concentration of primers, amount of the DNA templates added into the reaction mixture. All the components are added in the reaction mixture. All the components are mixed in a PCR tube, the further reactions are carried out on thermal cycler machines specially designed for the PCR.

PCR reaction involves three chief steps that include Denaturation, Annealing and Extension (Joan and Roy, 1993). Denaturation step is carried out to separate double helix strands, generally done at approximately 90-95°C followed by annealing where the temperature of the reaction is lowered to ensure the binding of Oligos (Primers) to the target. Generally these temperatures are lowered such that the annealing temperature is below the Tm of the primers (Temperature of Dissociation of Primer Duplexes). The final step is known as Extension where the temperature of the reaction is adjusted to approximately 72°C where most of the *Taq* DNA polymerases have their maximum activity extending the primer by polymerization of dNTPs towards the 3'OH end thus forming the new copy of the target DNA completely from the template DNA. Additionally, initial denaturation is done for longer time to ensure proper and clear denaturation of template DNA molecules. Also final extension is done to fill the protruded ends of the incomplete fragments. During each cycle, one copy of template DNA is doubled so the total number of copies of the target obtained after the n cycles will be 2^n. Generally, these repetitions are done from 25 to 40 cycles depending upon the amount of DNA copies.

PCR is used routinely in many laboratories; but there are still a considerable number of problems in getting good and reproducible result. Plant tissues have a lot of PCR inhibitors; tissues like potato, banana, papaya and betelvine have inhibiting substances. Hence it is important to use a good protocol for purification of the nucleic acid, which can be used in the enzymatic reaction.

It is important to standardize the optimum concentration for every PCR study. For long PCR, special enzyme mixes are available. For standard detection to plant viruses it is better to optimize the concentrations of all components of the mix.

Taq DNA Polymerase Concentration

Taq DNA polymerase conc. can be between 1 and 2.5 units per 100 µl reaction. Enzyme requirements may vary with respect to individual size of target templates or primers. Non-specific background products may occur if the enzyme concentration is too high, and if too low, desired product may not be sufficiently amplified.

dNTP's

One of the two mM each dNTP can be used as working concentration. The stability of the dNTPs during repeated cycles of PCR is such that approximately 50 per cent remains as dNTP in the mix after 50 cycles. The four dNTPs should be used at equivalent concentrations to minimize misincorporation errors. Low dNTPs concentrations minimize mispriming at non-target sites and reduce the likelihood of extending misincorporated nucleotides. One has to decide on the lowest dNTP concentration appropriate for the length and composition of the target sequence.

Concentration of Mg^{2+}

Taq DNA polymerase requires free Mg^{2+} on top of that bound by template DNA, primers and dNTPs. It is necessary to optimize the magnesium ion concentration. The magnesium concentration may affect primer annealing, strand dissociation temperatures of template and PCR product, product specificity, formation of primer-dimmer artifacts, and enzyme activity and fidelity. The optimum range of magnesium in PCRs is 10-25 mM over the total dNTP concentration.

Buffer and Monovalent Cations

A recommended buffer for PCR is 10 to 100 mM Tris-HCl (pH 8.3 to 9.0) at 20°C. To facilitate primer annealing up to 500 mM KCl can be included in the reaction mixture.

Primer Annealing

The temperature and length of time required for primer annealing depend upon the base composition, length and concentration of the amplification primers. An applicable annealing temperature is 5°C below the melting temperature of the primers. Because *Taq* polymerase is active over a range of temperatures, primer extension will occur at low temperature also, including in the annealing step. The range of enzyme activity varies by two orders of magnitude between 20°C and 85°C. Annealing temperature in the range of 45 to 65°C generally yield the best results. At typical primer concentrations (0.2 to 0.5 µM) annealing will required only a few seconds. Stringent annealing temperatures especially during the first fewer cycles will help to increase specificity.

Primer Extension

Extension time to be given for the primers depends upon the length and concentration of the target sequence and temperature. Primer extensions are traditionally performed at 72°C because this temperature was found to be optimal for extending primers. The rate of nucleotide incorporation at 72°C varies from 35 to 100 nucleotides second^{-1} depending upon the buffer, pH, salt concentration, and the nature of the DNA template. The polymerization rate of *Taq* polymerase is 2000 nucleotides/minute at the optimal temperature (72–78°C). An extension time of one minute at 72°C is considered sufficient for the products up to 2kb in length. However, longer extension times may be helpful in early cycles if the substrate concentration is very low, and at late cycles when product concentration exceeds enzyme concentration. For the last cycle PCR, many investigators use an extension time that is three times longer than in previous cycles, to allow completions of all amplified products. However, in reality it is not correct.

Denaturation Step

Complete denaturation of the target template is necessary for obtaining good amount PCR product. The higher the proportion of G+C, the higher is the temperature required to separate the strands of template dsDNA. If the template is longer, more time is required to separate the two strands. By reducing the temperature in the PCR cycle, the denatured template DNA will anneal and will be in native form.

Typical denaturation conditions are 95°C for 15 seconds. However, higher temperature may be appropriate, especially for G+C rich targets. It only takes a few seconds to denature DNA at its strand-separation temperature (T_{ss}); however, there may be a lag time involving in reaching T_{ss} inside the reaction tube. Incomplete denaturation allows the DNA strands to "snap back" and thus, reduces product yield. In contrast, denaturation steps that are too high and/or too long lead to unnecessary loss of enzyme activity. The half-life of *Taq* DNA polymerase activity is > 2 hours, 40 minutes and 5 minutes at 92.5, 95 and 97.5°C, respectively.

Number of Cycles

The cycles for amplification depend on the number of template DNA copies present in the start of the cycle. Number of cycles should aim to increase the amplification products whereas nonspecific products should be barely detectable. Too many cycles increase the amount and complexity of nonspecific background products. The number of cycles can be chosen based on the table given below:

Number of Initial Template Molecules	*Number of Cycles*
$<5 \times 10^4$	25 to 30
1.0×10^3–5×10^4	30 to 35
100–1000	35 to 40
10–100	40 to 45

Primer Concentration

Optimal concentration of primers is between 0.1 and 0.5 µm. Mispriming and accumulation of nonspecific product increase the probability of generating a template-independent artifact termed a primer-dimmer are more common with high concentration of primers in the mix. Nonspecific products and primer dimer artifacts acts as substrate for PCR and compete with the desired product for polymerase, dNTPs and primers, resulting in a lower yield of the desired product.

Primer Designing

Primers are key component of specific detection of plant viruses. They are short, artificial DNA strands usually 18 to 25 base pair long nucleotides that are complementary to the beginning and end of the DNA fragment to be amplified. There are a variety of computer programs available to design primers. Try to test several sets of primers, and select one, which is more specific and reliable and conserved. There are no definite rules to guarantee the success of a primer pair. Because some primer pairs are 100 to 1000 fold more sensitive than others for elusive reasons, several sets of primers should be tried out. Design the primers by visual inspection of the nucleic acid sequence. Then, if available, use a primer design program.

- ☆ The primers should be 18-30 nucleotides long, with a Tm of 55°C or above. Tm is calculated by adding 2°C for each A or T and 4°C for G and C.
- ☆ Before primers can be selected, information must be available on the nucleotide sequence of the target virus. Consult nucleotide databases.
- ☆ The target nucleic acid region should be conserved one and preferably be between 200 and 1000 base pairs in size, if less, the product can be more easily confused with primer-dimers, and, if larger, the product my not be efficiently amplified.
- ☆ Primers forming hairpin loops, or pairs having complementarity at 3′ ends should be avoided.
- ☆ Do not use oligo (dT) on its own: Have this anchored, by using a mix of 3 oligo (dT) primers with A, C, or G at 3′ end.
- ☆ Mismatching of T bases in target sequence or primer occurs at a higher rate than for other bases, so it is best to avoid a T at the 3′ end of the primer for discriminating similar DNAs. In most cases, try to ensure that the 3′ end has a GC clamp. Avoid runs of three or more of the same base.

Preparation of Nucleic Acid Template

Preparation of DNA template for direct PCR or RNA template for cDNA synthesis is an extremely important parameter for PCR or RT-PCR amplification. This part is often the most labour intensive stage of detection assays, and therefore may contribute significantly to the over all economy of the process. The most widely used method for template preparation is to isolate total DNA or RNA from virus infected tissue. This generally involves grinding of plant tissues in an extraction buffer (some times by using liquid nitrogen to reduce the plant material to a powder first) and then to perform one or more rounds of phenol and/or chloroform extraction to remove plant proteins and contaminating substances. The nucleic acids are then precipitated by ethanol or isopropanol and resuspended in sterile water and used for PCR amplification. A large variety of protocols have now been published for the preparation of these purified nucleic acid preparations (Candresse *et al.*, 1998). Commercial nucleic acid extraction kits are now often used but it involves high cost for template preparation. Alternative to complex procedures of nucleic acid isolation or costly commercial nucleic acid extraction kits, cost effective simple approaches have been developed. The simplest approach is to prepare crud plant homogenates in sterile water or simple buffers and then dilute the extract to dilute out the inhibitors while retaining sufficient quantities of viral nucleic acids to allow successful amplification. It ranged from 10 fold dilution (*Cucumber mosaic virus* in cucumber, De Blas *et al.*, 1994) to 10^6-10^9 (*Cymbidium mosaic virus* in *Cymbidium*, Lim *et al.*, 1993). The other alternative is to immobilize nucleic acid targets by placing crude extract on membranes and then nucleic acid can be released in sterile water or a buffer. The eluted nucleic acid can be directly used for PCR/RT-PCR (Singh *et al.*, 2004, 2006; Osman and Rowhani, 2006). Immunocapture of virus particles from crude extract is another method in which virus is first trapped by specific antibodies coated on the wall of PCR tube and then nucleic acid of virus is released for cDNA synthesis and subsequent PCR. This method is commonly referred as immuno-capture PCR (IC-PCR). Similar to IC-PCR, there is a print capture PCR (PC-PCR) which allows virus detection from infected plants without grinding the sample (Olmos *et al.*, 1996).

Isolation of Total DNA (Based on Banana Bunchy Top Virus)

For PCR experiments the DNA need to be isolated from the sample, which is suspected to have viral infection. If the sample is infected with DNA viruses the method given below can be used for PCR analysis.

Materials Required

Autoclaved- pestle and mortar, measuring cylinder, centrifuge tubes, eppendorf tips and tubes, spatula, doubled distilled water; tissue roll; sprit; 70 per cent alcohol; SDS; 100 per cent alcohol; chloroform; isoamyl alcohol, and isopropanol.

Buffers and Reagents

CTAB Buffer

- ✰ 2 per cent (w/v) hexadecytrimethyl bromide (CTAB)
- ✰ 100 mM Tris HCl pH 8.0
- ✰ 1.4 M NaCl
- ✰ 20 mM EDTA
- ✰ 4 per cent PVP
- ✰ 0.1 per cent Ascorbic acid
- ✰ 0.1 per cent DIECA
- ✰ 2 per cent (w/v) mercaptoethanol- add immediately prior to use

Solution II

- ✰ 100 mM Tris HCl pH 8.0
- ✰ 1.4 M NaCl
- ✰ 50 mM EDTA
- ✰ 2 per cent (w/v) mercaptoethanol- add immediately prior to use

20 per cent SDS

5 M Potassium Acetate (pH 5.2)

Nuclease Free Water

Protocol

1. Grind 200 mg of leaf tissue in liquid nitrogen.
2. Add ground tissue to 100 µl of CTAB buffer, 650 µl of solution II and 50 µl of 20 per cent SDS.
3. Follow vigorous mixing.
4. Incubate at 65°C for 20 min, with occasional mixing by inversion.
5. Add 250 µl of 5 M potassium acetate.
6. Centrifuge at 13000 rpm for 20 min at 4°C
7. Take out aqueous phase, add 750 µl of ice cold isopropanol and mix by inversion.
8. Store at –20°C for 1 hr.
9. Centrifuge at 13000 rpm for 20 min at 4°C
10. Discard the supernatant
11. Dissolve pellet in 70 per cent alcohol. Blot dry on paper towel.
12. Dissolve the DNA in 25 µl of nuclease free water.

PCR Amplification

Materials Required

Disposable globes, Ice Bucket with ice flakes, Microfuges, Microwave oven or heating mantle, Mini coolers, Parafilm membrane, PCR apparatus, PCR tubes, Eppendorf tips, Vortexing machine, etc.

Protocol

☆ The below given solutions are added to each tube in the following order. Use always a fresh pipette tip for each solution.

Sl.No.	*Reagents*	*Volume (µl)*
1	10 X PCR buffer with $MgCl_2$	5.0
2	10 mM dNTP Mix	4.0
3	*Taq* Polymerase (5 units/µl)	0.5
4	Distilled water	34.5
5	Forward primer	1.0
6	Reverse primer	1.0
7	Template	4.0
	Total	50.0

- ☆ Vortex the mixture well.
- ☆ Centrifuge the tubes for 10-20 sec to drain the walls.
- ☆ Paraffin oil/mineral oil should be over layered to prevent evaporation during repeated cycles of heating and cooling. If PCR apparatus has heated lid, this step is not necessary.
- ☆ Place the tubes in the PCR apparatus and set the programme in thermocycler (PCR machine).
- ☆ The amplification conditions are as follows:

  PCR Programme

  94°C for 4 min

  94°C for 1 min- 30 cycles

  53°C for 1 min- 30 cycles

  72°C for 2 min- 30 cycles

  72°C for 10 min

After PCR, if necessary, keep the reaction mixtures at 4°C overnight. After the PCR, the amplicons can be resolved in 1.5 per cent agarose gel electrophoresis.

Analysis of PCR Products by Agarose Gel Electrophoresis

The amplified PCR product of viral DNA/viral RNA has to be resolved in agarose gel electrophoresis. The sample along with appropriate DNA ladder need to run for comparison and measuring the size of the amplicon.

Materials Required

Electrophoresis apparatus with power pack, Cellotape, Heating mantle, Balance, Pipettes, Parafilm, UV transilluminator, Image analyzer, Protective specks/goggles or full safety mask that effectively block UV light.

Buffer and Reagents

0.5 X Tris Borate buffer, Agarose, Ethidium bromide, Gel loading dye.

Protocol

1. Prepare horizontal agarose gels by sealing the short ends of a plastic tray supplied with the electrophoretic apparatus with tape. Place the tray in a perfectly horizontal position.
2. Dissolve 1.5 g agarose in 100 ml of buffer or 0.5 X TBE using a microwave oven or a water-bath or heating mantle.
3. When the agarose is dissolved, thoroughly mix the agarose solution cooled down to approx. 60°C with 2.5 µl of stock ethidium bromide.
4. Place the comb 0.5-1.0 mm above the bottom of the tray and pour the warm agarose solution into the mould so as to get a gel thickness of 3-5 mm.
5. Allow the gel to set in 10-15 min.

6. As soon as the gel is completely set, carefully remove the comb and tape (or casting gates), and mount the gel in the electrophoresis tank.
7. Cover the gel with a thin layer (approx. 1mm) of 0.5 X TBE.
8. Prepare the samples to be analysed by pipetting 3 µl gel- loading buffer and 10 µl PCR product on parafilm membrane. Mix well.
9. Slowly pipette each mixture into the slots of the submerged gel.
10. Close the lid of the tank and attach the electrical leads.
11. Electrophoresis at 80 V and maximum amperage. Watch the bromophenol blue front migrate from the wells into the rest of the gel.
12. When the bromophenol blue front has migrated 5 cm, turn off the electric current. Remove the leads and lid from the tank.
13. Place the gel under UV light to generate fluorescence emitted by ethidium bromide-DNA complexes.
14. Take the photograph of the gel in an image analyzer or use a thermal printer.

Variants of PCR and their Applications

PCR is a potent tool in molecular biology since its advent. It was modified variably to serve many purposes. These modified types are termed as variants of PCR *viz.*, RT-PCR, Nested PCR, Multiplex PCR, LAMP, Direct binding PCR, IC-PCR, IC-RT-PCR, CF-PCR, PCR-RFLP, RT-PCR-RFLP, IC-RT-PCR-RFLP, PCR-ELISA, Co-operational PCR, Generic spot nested RT-PCR, Real-time PCR, Magnetic capture hybridization RT-PCR (MCH-RT-PCR), One step multiplex PCR, and Single plex and multiplex real time PCR, PCR heteroduplex mobility assay (Bariani *et al.*, 1994; Brandt and Himmler, 1995; Chevalier *et al.*, 1995; Gorsane *et al.*, 1999; Notomi *et al.*, 2000; Walsh *et al.*, 2001; Angelini *et al.*, 2003; Dovas and Kaitis, 2003; Harper *et al.*, 2003; Deloni *et al.*, 2003; Little, 2005; Faggioli and La Starza, 2006; Fillipin, 2006; Mnari-Hattab and Ezzaier, 2006).

PCR has been in use for detection of diseases caused by virus, viroid and phytoplasma since its invention in mid-1980's (Ahrens and Seemuller, 1992; Brandt and Himmler, 1993; Brandt *et al.*, 1993; Nolasco *et al.*, 1993; Martelli *et al.*, 1993; Rowhani *et al.*, 1993; Walter *et al.*, 1994; Levy *et al.*, 1994; Bertaccini *et al.*, 1994, 1996; Minafra *et al.*, 1995; Ahlawat, 2007).

Reverse Transcription-Polymerase Chain Reaction (RT-PCR)

RT-PCR is the "gold standard" molecular method used for the detection of plant viruses due to its high sensitivity and specificity. As the majority of them are RNA viruses, an initial step of reverse transcription that converts single strand RNA to cDNA, is necessary for PCR based molecular amplification (Olmos *et al.*, 2005) involving an enzyme known as reverse transcriptase. In RT-PCR, PCR is done in two steps: (1) Reverse transcription and (2) PCR. Reverse transcription is a process where cDNA is synthesized to an RNA template involving an enzyme known as Reverse transcriptase. This process also requires an important component for cDNA synthesis i.e a primer possibly an OligodT for the viruses with poly-A tail or a random primer or a reverse primer of the PCR itself. In this mechanism, the RNA is initially denatured, along with the primer and snap chilled to unwind the super secondary structures present in RNA and also to aid the primer annealing. Then the other components like dNTPs, RT buffer and RNAse inhibitors are added and the reverse transcription is carried out at a specific temperature, generally 42°C for most of the RT's and a final step of termination is held by heating the reaction mix to higher temperatures to inactivate RT. The cDNA can be used as a template in the next step PCR whose details are discussed earlier. In case of viruses where a Poly A tail is not present, either a reverse primer or PCR or a Random primer (probably a Hexamer) is used for cDNA synthesis. Reports of usage of terminal dinucleotidyl transferases to generate poly A tails for cDNA synthesis is also there in case of certain types of Ilar and Bromoviruses. This technique has been used for detection and identification of several viruses belonging to various genera (Henry *et al.*, 1995; Choi *et al.*, 1999; Saade *et al.*, 2000; Gillaspie *et al.*, 2001; Dietzgen, 2002; Raj Verma *et al.*, 2006; EL-Kewey *et al.*, 2007; Tobias *et al.*, 2008).

Total RNA Isolation (Total RNA Isolation from ZYMV Infected Bottle Gourd)

Procedure

Total RNA can be isolated from 100 mg of virus infected sample using RNeasy Mini Kit (Qiagen, USA) as per manufacturer's protocol.

Protocol

- ☆ Grind 100 mg of leaf tissue in liquid nitrogen.
- ☆ Add ground tissue to 450 µl of RLT buffer containing 10 µl mercaptoethanol.
- ☆ Collect the above in a collection tube and incubate at 56°C for 1-3 min.
- ☆ Centrifuge for 2 min at 10,000 rpm.
- ☆ Take out aqueous phase, add 225 µl of ethanol (96- 100 per cent), mix well and keep it on ice.
- ☆ Apply the above onto an RNeasy mini column placed in 2 ml collection tube.
- ☆ Centrifuge for 15 sec at 10000 rpm and discard the flow through.
- ☆ Add 700 µl RW1 buffer and centrifuge for 15 sec at 10000 rpm.
- ☆ Discard the flow through.
- ☆ Transfer RNeasy column into new 2 ml collection tube and add 500 µl RPE buffer.
- ☆ Centrifuge for 15 s at 10000 rpm. Discard flow through.
- ☆ Add 500 µl RPE buffer and centrifuge for 2 min at 10,000 rpm.
- ☆ To elute the RNA, transfer the above into a new 1.5 ml collection tube.
- ☆ Add 30-50 µl of RNase free H_2O, wait for 1 min and centrifuge for 1 min at 10000 rpm.
- ☆ Take elute in a separate tube and place in on ice.
- ☆ Incubate for 5 min at 65°C.
- ☆ Use total RNA for RT-PCR

Total RNA can be stored in small aliquots at –80ºC for periods of up to 2.5 years (stock sample) and at –20ºC for periods of up to 8 months (working sample). Avoid thawing and re-freezing too many times that may lead RNA degradation.

RT-PCR

Procedure

Reverse transcription and PCR are done separately. For first strand cDNA synthesis 5µl of elution from RNeasy column is denatured for 10 min at 65ºC and quenched on ice. RT is carried out using Omniscript RT kit (Qiagen) in 20 µl reaction mixture using OligodT primer (1 µM) and Omniscript RT (4 units) at 42ºC for 1h. PCR amplification is carried out using farward and reverse primers. For 50 µl reaction mixture, 1 µM of primer, 200 µM each of dNTPs, 0.05 unit/µl of *Taq* DNA polymerase, 1 X reaction buffer, 1.5 mM of $AgCl_2$ and 5 µl of cDNA template are required. Samples can be amplified for 30 cycle using any brand of thermocycler. Each cycle consists of danaturation at 94ºC (30s), primer annealing at 58ºC (30s), and extension at 72ºC (45s), and a final extension of 10 min at 72ºC. The PCR products are electrophoresed in 1 per cent agarose gels in Tris-acetate EDTA followed by staining with ethidium bromide (Sambrook and Russel, 2001) and can be viewed in an UV transilluminator (Figure 16.3).

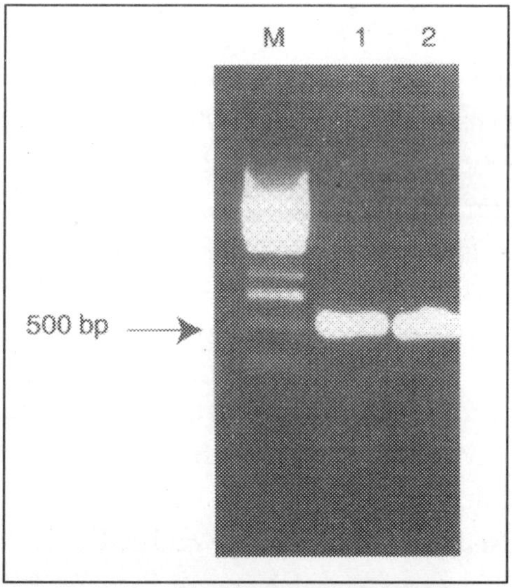

Figure 16.3: RT-PCR Amplification of C-Terminal CP Region and 3' UTR of ZYMV from Leaves of Bottle Gourd. Lane M: 1 kb DNA ladder, Lane 1 and 2 amplified PCR product from diseased leaves of bottle gourd.

Precautions to be Taken for RT-PCR Work

In RT-PCR, sometimes false positives or no amplifications are encountered. As RNase are omnipresent in the open environment, degradation of RNA is very quick and while attempting to detect RNA viruses, considerable care must be taken to limit exposure of extracted RNA to RNases. Singh (1998) suggests having separate rooms for preparation of samples, cDNA synthesis, amplification and visualization of amplified products. Separate dedicated micropipettes and sterile, disposable and filtered tips. Laboratory disciplines such as wearing disposable gloves and minimum disruption of work while preparing PCR mix. Decontamination of all the reagents and equipments with exposing them to UV light is necessary. If mineral oil has to be over layered, mix 8-hydroxyquinoline with the oil prior to layering. Tools and containers resistant to heat can be sterilized at 180°C for two hr., heat autoclavable containers and solutions with 0.1 per cent (v/v) diethyl pyrocarbonate (DEPC) overnight in fume hood, followed by autoclave them. Disposable plastic-wares such as gloves, tubes and tips, are usually RNase-free when supplied by manufacturers, and hence treatment may not necessary, but they should handled by wearing clean gloves.

Duplex and Multiplex PCR

This variant of PCR is targeted for detection of multiple viral or viroid pathogens in a single reaction. Multiplex primers are designed specifically for different viruses but with same annealing temperatures and different sized product amplicons. Care is also taken that no secondary structure formations are there between these set of primers. When a PCR reaction is carried out from the templates having a mixture of viruses (Mixed infections) only the viruses that are present are amplified. When these products are analyzed, basing upon the sized fragment obtained, the viruses are identified. This is done in case of many viruses for example six citrus pathogens (Roy *et al.*, 2005). For the six citrus pathogens that include CMBV, CTV, CLRV, ICRSV, CVV and citrus exocortis viroid, multiplex primers were designed and used successfully. Multiplex immunocapture PCR with colourimetric detection of banana viruses was developed by Sharman *et al.* (2000). There are several examples of simultaneous detection of several targets and the amplification by multiplex PCR of two or three plant viruses (Minafra and Hadidi, 1994; Nemchinov *et al.*, 1995; Singh *et al.*, 1996; Russo *et al.*, 1999; Grieco and Gallitelli, 1999; Jacobi *et al.*, 1998; Saade *et al.*, 2000; Sharman *et al.*, 2000). Multiplex PCR has also been used to detect citrus yellow mosaic virus (CYMBV) and greening bacterium simultaneously in citrus (Ahlawat, 2007). Nevertheless, there are still only a few examples in which more than three plant viruses were amplified in a single PCR-based assay (Bariano *et al.*, 1994; Nasuth *et al.*, 2000; Nie and Singh, 2000; Okuda and Hanada, 2001), probably due to the technical difficulties of designing a reaction involving many compatible primers. One of them is the simultaneous detection of the six major characterised viruses described in olive trees, which belong to four different genera: Cucumovirus (CMV), Nepovirus [CLRV, SLRSV and Arabis mosaic virus (ArMV)], Necrovirus [Olive latent virus-1 (OLV-1)] and Oleavirus (Olive latent virus-2) (Bertolini *et al.*, 2001). This includes accurate design of six primer pairs for one-step RT-PCR amplification in a single closed tube and specific probes, enabling the detection of all major viruses described in olive trees, which are problematic for RNA extraction. This technique has also been used in detection of five seed-borne legume viruses (Bariana *et al.*, 1994). Sometimes RNA and DNA viruses are also detected simultaneously. Multiplex PCR is useful in plant pathology because different RNA viruses frequently infect a single host and consequently sensitive detection is needed for the propagation of pathogen-free plant material.

Nested PCR

It is a variant of PCR designed to get amplification of desired fragment specifically. In this technique, the product of the first few cycles (using a first primer set) is used as a template for the second set of nested primers that are designed to amplify the desired fragment. It is designed to increase the specificity of detection and also to enhance the copies of viral nucleic acid in case of low virus titers. This technique was successfully used in Vitivirus and Foveavirus species detections in grape vines (Dovas and Katis, 2003 a, b) and for beet necrotic yellow vein virus detection (Cambra *et al.*, 2006; Samuitiene *et al.*, 2006).

Multiplex Bested RT-PCR

This newly developed method combines the advantages of multiplex RT-PCR with the sensitivity and reliability of nested RT-PCR carried out in a single closed tube. It enables the simultaneous detection of several viral RNA and bacterial DNA targets in a single analysis, performed with woody plants. It also saves time and reagent costs because it can be performed in a single reaction, although accurate design of compatible primers is needed. The compartmentalization of a single Eppendorf tube with a pipette tip (Olmos *et al.*, 1999; Olmos *et al.*, 2003) allowed multiplex-PCR and nested PCR to be combined effectively. During the first amplification reaction there is no interference of the external with internal primers because they are physically separated from the initial reaction cocktail. Once the multiplex RT-PCR ends, the internal primers are mixed with the products of the first reaction before proceeding to the nested multiplex. Because the concentration of internal primers is very high compared with that of the external primers (which will also have been consumed by the first amplification), the nested multiplex can be performed with minimal interference. Consequently, sensitivity is increased at least 100-fold over that of multiplex RT-PCR for the detection of viruses. A multiplex nested RT-PCR in a single closed tube has been developed for simultaneous and sensitive detection of the viruses CMV, CLRV, SLRSV, and ArMV from olive plants (Bertolini *et al.*, 2003) using 20 compatible primers in a compartmentalized tube.

Direct Binding-PCR (DB-PCR)

It is a method similar to the IC-PCR, but it applies direct binding of the virus particles from the crude plant sap or seed extract to the PCR tube, washing of the unbound particle and debris, lysis of viral particles in a medium and PCR detection of the target. Although this technique is simple and affordable, rate of success and level of detection is lower than that of IC-PCR for many of the virus hosts with heavy polyphenolics. DB-PCR detection of banana bunchy top virus could not be useful, but it could give consistent result with respect to BBrMV and CMV. This technique has also been used to detect the episomal form of BSV. This assay is cost effective and less time consuming.

Competitive Fluorescence PCR (CF-PCR)

It is a modified method of PCR which could be used to differentiate between the different strains of virus where 5′ fluorescent dye labeled Oligos are used for amplification. Upon obtaining the amplicon, the dye fluoresces only in a double stranded hybrid. This technique is used chiefly to differentiate the viruses with divergence in 3′ end nucleotide sequences. This was chiefly used to differentiate the multiple strains of the Potato virus Y (Walsh *et al.*, 2001; Webster *et al.*, 2004).

Co-operational PCR

This is a new PCR concept of high sensitivity for the amplification of viral RNA from plant material has recently been described (Olmos *et al.*, 2002). The method has been patented as Co-PCR (Spanish patent P20002613; 31 October 2000). The Co-PCR (co-operational amplification) technique can be performed easily in a simple reaction based on the simultaneous action of four or three primers. The reaction process consists of the simultaneous reverse transcription of two different fragments from the same target, one internal to the other, the production of four amplicons by the combination of the two pairs of primers, one pair external to the other, and the cooperational action of amplicons for the production of the largest fragment. The Co-PCR technique has been used successfully, both in metal block and capillary air thermal cyclers, for the detection of plant RNA viruses [Cherry leaf roll virus (CLRV), strawberry latent ringspot virus (SLRSV), Cucumber mosaic virus (CMV), Plum pox virus (PPV) and Citrus tristeza virus CTV]. Coupled with colorimetric detection, the sensitivity observed is at least 100 times higher than that achieved with RT-PCR and is similar to that of nested RT-PCR. Co-PCR usually produces the largest amplicon, in contrast to nested- PCR, which requires two sequential reactions and obtains the smallest fragment. Metal block and capillary air thermal cyclers have been employed for the detection of some plant RNA viruses from different genera, and to a bacterium, but by using only three primers (Caruso *et al.*, 2003), which shows the possibilities of this new approach. The low amount of reagents (ten times less than in conventional PCR) probably increases susceptibility to inhibitors, requiring prior RNA extraction for sensitive virus detection.

LAMP (Loop Mediated Isothermal Amplification)

It is a modified, efficient and specific method for the amplification of the DNA templates. The reaction is performed with two sets of primers containing two specific regions of the target sequence for amplification (Notomi *et al.*, 2000). The process starts with annealing of the first primer set for synthesis of the complementary strand that forms a loop out structure for further amplification by second primer set. The amplification is carried out at isothermal conditions (65°C for 1 hour) with Bst DNA polymerase that has strand displacement activity. Target amplification is detected by measuring the turbidity of the reaction mixture. The amplification takes place at 60-65°C for 60 min. The amplification products are stem-loop DNA structures with several inverted repeats of the target and cauliflower-like structures with multiple loops, yielding >500 !g/ml. Although it was initially developed for DNA it can be adapted to amplify RNA (RT-LAMP) (Fukuta *et al.*, 2003). This was used efficiently for detection of TSWV from chrysanthemum (Fukuta *et al.*, 2004) by combining LAMP with IC-RT steps. Wang *et al.* (2008) and Mumford *et al.* (2006) used LAMP for diagnosis of plant viruses.

Immunocapture PCR (IC-PCR)

IC-PCR is a variant of PCR that combines the capture of the virus particles by antibodies coated in a PCR tube. The bound antigen (virus particle) is then lysed into a buffer that will release the total viral nucleic acid content from the virion into the medium and PCR is carried out by addition of component to the medium. This is a reliable technique to detect the virus in materials to be indexed. This technique has been widely used for many plant and animal viruses successfully. False positive can be there as sometimes traces of host DNA present in the tubes even after washing. Optimization of test is required for extraction and other parameters in IC-PCR. This method is specially useful in concentrating virus particles from plant species where virus titre is low, or where compounds that inhibit PCR are present, for example plum pox tree sap containing PPV (Wetzel *et al.*, 1992) and sugarcane streak mosaic virus (SCSMV) (Hema *et al.*, 2003). A comparison of IC-PCR to other detection methods including IBIA, ELISA and Dot-blot immunoassay for the detection of Florida hibiscus virus (FHV) found it the most sensitive of the methods tested, able to detect 500 pg/ml of virus or 16 to 32 fold more, than DAS-ELISA (Kamenova and Adkins, 2004). Sharman *et al.* (2000) also found IC-PCR up to 625 times more sensitive than ELISA for multiplexed detection of four viruses of banana. It has been used to detect episomal banana streak virus (BSV) parts of whose genome is present within the banana genome, and therefore increasing the chance of false positive from standard PCR test (Harper *et al.*, 1999). IC-RT-PCR has also been develoved for the detection of RNA viruses (Mnari-Hattab *et al.*, 2006).

Hot Start PCR/Touchdown PCR

Hot Start PCR is a modified PCR in which *Taq* polymerase is added after the initial denaturation of the DNA to increase the specificity of amplification. In case of Touch Down PCR, the annealing temperature for the primer is higher in the first cycle and will be reduced in the following cycles step by step to the estimated annealing temperature to avoid amplification of unspecific products.

Touchdown PCR aims to reduce nonspecific background by gradually lowering the annealing temperature as PCR cycling progresses. The annealing temperature at the initial cycles is usually a few degrees (3-5°C) above the T_m of the primers used, while at the later cycles, it is a few degrees (3-5°C) below the primer T_m. The higher temperatures give greater specificity for primer binding, and the lower temperatures permit more efficient amplification from the specific products formed during the initial cycles (Don *et al.*, 1991).

In situ PCR

A PCR conducted in a section of tissue inside a cell. This technique can be utilized to locate the viral genome in the tissues or the cell. Here the reaction takes place in the glass slides. The amplification is confirmed based on colour reaction. This technique can also be used for *in situ* hybridization with radioactive probes.

Real Time or Quantitative PCR

It is a modified method of PCR where quantification can be done while the amplification is under process. This is also called "quantitative PCR" because it allows quantifying the increase in the amount of DNA as it is amplified. RT-PCR is performed on a real time machine that has a function of both thermal cycling and also quantitative Fluorimetry that will quantify the amount of DNA as it is amplified during the process. In addition the sensitivity and specificity, this technique has certain advantages over PCR; it reduces the risk of cross-contamination, obviates post PCR manipulations, provides higher throughput, and enables quantification of virus load in a given sample. Real-time reaction monitoring with specific instruments and fluorescent probes combines amplification, detection and quantification in a single step and has been applied to the detection of several pathogens (Weller *et al.*, 2000). However, this technology requires expensive and special equipment and reagents compared with conventional PCR technology. Deloni *et al.* (2003) developed real time PCR assay for rapid detection of episomal banana streak virus.

Rapid-cycle real-time PCR methods may revolutionize the manner in which plant pathogens are identified and diseases are diagnosed. As the genomics age progresses and more and more DNA sequence data become available, highly specific primers and fluorescent probe sequences can be designed to yield target amplicons to unique regions of a pathogen's genome. Rapid real-time PCR diagnosis can result in appropriate control measures and (or) eradication procedures more quickly and accurately than traditional methods of pathogen isolation. Disease losses are minimized and control costs reduced (Schaad and Frederick, 2002).

Several different types of real-time PCR systems are available having their own advantages and disadvantages. These are as follows:

TaqMan probes

In the TaqMan system, an oligonucleotide probe sequence of approximately 25–30 nucleotides in length is labeled at the 5′ end (Holland *et al.*, 1991) with a fluorochrome (Lee *et al.*, 1993), usually 6-carboxyfluorescein (6-FAM) and a quencher fluorochrome, usually 6-carboxytetramethyl-rhodamine (TAMRA), at the 3′ end. The TaqMan probe is degraded by the 5–3µ exonuclease activity of the *Taq* polymerase as it extends the primer during each PCR amplification cycle and the fluorescent chromophore is released. The amount of fluorescence is monitored during each amplification cycle and is proportional to the amount of PCR product generated.

Eclipse Probes

The eclipse probes are also similar in function to TaqMan probes in quantification of the amplicon but differ in a single aspect that they are not cleaved during the polymerization process. The probes use MGB technology with FAM on the 5′ end and TAMRA, an eclipse quencher molecule on 3′ end. Here the Fluorescence Resonance Energy Transfer (FRET) is arrested between the reporter dye and the eclipse quencher when the probe hybridizes to the target amplicon after amplification is carried out.

Molecular Beacons

Molecular beacons are fluorescent oligonucleotide probes that are designed to include stem-loop folding and also use FRET to detect and quantify the PCR amplicons *via* a fluorophore coupled to the 5' end and a quench attached to the 3' end of an oligonucleotide. They are complementarity nucleotide sequences that are complementary to the target amplicon. A fluorescent chromophore is attached at the 5′ end of the probe and a quencher molecule is attached at the 3′ end. A stem structure is formed by annealing of the complementary arm sequences that are added on both sides of the probe sequence. When a stem structure is formed, the fluorophore transfers energy to the quencher, and no fluorescence is emitted. However, when the probe hybridizes to the target amplicon during PCR amplification, the fluorophore and quencher become separated from each other and fluorescence can be detected (Didenko, 2001; Cockerill and Smith, 2002).

They slightly differ from TaqMan probes as they are designed to remain intact during the amplification reaction, and must rebind to target in every cycle for signal measurement. Molecular Beacons form a stem-loop structure when free in solution. So the fluorophore and quencher molecule come nearer that prevents the probe from fluorescing. When a Molecular Beacon is denatured and hybridized to a target, the fluorescent dye and quencher is separated, FRET does not occur, and the fluorescent dye emits light upon irradiation. Molecular Beacons, like TaqMan probes, can be used for multiplex assays by using spectrally separated fluor/quench moieties on each probe. As TaqMan probes, Molecular Beacons can also be expensive to synthesize, with a separate probe required for each target.

Scorpion Probes

With Scorpion probes, sequence-specific priming and PCR product detection is achieved using a single oligonucleotide. The Scorpion probe maintains a stem-loop configuration in the unhybridized state. The fluorophore is attached to the 5' end and is quenched by a moiety coupled to the 3' end. The 3' portion of the stem also contains sequence that is complementary to the extension product of the primer. This sequence is linked to the 5' end of a specific primer via a non-amplifiable monomer. After extension of the Scorpion primer, the specific probe sequence is able to bind to its complement within the extended amplicon thus opening up the hairpin loop. This prevents the fluorescence from being quenched and a signal is observed. The advantage of this configuration is that the unimolecular nature of the primer-probe, scorpion probe allows for faster hybridisation kinetics. This may be useful for high-volume screening assays.

SYBR Green Dyes

SYBR Green provides the simplest and most economical source for detecting and quantifying PCR products in real-time PCR reactions. SYBR Green binds double-stranded DNA, and upon excitation emits fluoresence. Thus, as a PCR product accumulates, fluorescence increases. The advantages of SYBR Green are that it is inexpensive, easy to use, and sensitive. The disadvantage is that SYBR Green will bind to any double-stranded DNA in the reaction, including primer-dimers and other non-specific reaction products, which results in a wrong quantification of the target concentration. For single PCR product reactions with well designed primers, SYBR Green can work extremely well, with spurious non-specific background only showing up in very late cycles.

The 'Amplifluor Universal Detection System', developed by Serologicals Corporation (Norcross, Ga.), also uses the paired fluorophore-quencher hairpin structure. Invitrogen recently developed a new class of real-time probes called LUX (light upon extension) fluorogenic primers. Like hairpin probes, LUX primers adopt a stem-loop structure in solution and, like Scorpion probes, are intended for use as PCR primers. A recent comparison of TaqMan and molecular beacon chemistries found them to be more or less equivalent in sensitivity, accuracy, and reproducibility (Gloffke, 2003). The fluorescent probes are expensive, but that price pales in comparison to the thermal cycler itself. These instruments offer a variety of options and support a range of chemistries, but they are all expensive, and not every laboratory can afford to upgrade to real-time quantitative capabilities. A less expensive alternative are the portable rapid cycling real-time PCR platforms such as the RAPID (Idaho Technologies) and Smart Cycler (Cepheid, Charleston, USA) that can be used for rapid on-site diagnosis. However, only few data are available to compare the sensitivity level reached by their protocols with other currently available molecular protocols.

The same Real time PCR principle can be used to detect, quantify the presence and titre of viruses in plant samples as well as the seed samples in case of seed borne plant viruses. Realtime immunocapture RT-PCR was used for detection of Pepino Mosaic virus in tomato seed by Ling *et al.* (2005), whereas Multiplex real-time fluorescent reverse transcription polymerase chain reaction was developed for detection of Potato mop top virus and Tobacco rattle virus (Mumford *et al.*, 2000), *Pepino mosaic virus* (Ling *et al.*, 2007) and *Cherry leaf roll virus* (Jalkanen *et al.*, 2007).

PCR-RFLP

This method is a variant of PCR which combines amplification of a specific PCR target and distinguishing the products by the principle of RFLP (Restriction Fragment Length Polymorphism) that involves use of rare enzymes cutting the obtained amplicon and basing upon the number of fragments obtained in analysis, the virus strains could be distinguished. It can be used specifically to differentiate between the various strains of same virus. A similar modified method known as AFLP (Amplified length polymorphism) is also known that differentiates the strains of a virus based on the length of the PCR amplicon obtained. PCR-RFLP was applied for distinguishing the members of sub group 2 of *Cucumber Mosaic virus* present in Western Australian Lupian crops (Wylie *et al.*, 1993), between the strains of Rice tungro bacilliform virus belonging to Tungrovirus genus (Joshi *et al.*, 2003) and also for distinguishing between the types of Cassava mosaic disease causing viruses (Patil, *et al.*, 2005).

RT-PCR-RFLP has also been developed to characterize the RNA plant viruses. This technique has been used for the identification and characterization of pepper veinal mottle virus by Gorsane *et al.* (1999). Mnari-Hattab *et al.* (2006) developed IC-RT-PCR-RFLP and used pathotype identification of PMMoV.

PCR-ELISA

This is modified method of PCR, where PCR products are directly labeled during amplification by incorporation of DIG-11Dutp and hybridisation is carried out with the specific biotinylated capture oligo nucleotide that is complementary to PCR amplicon obtained, the hybrid is immobilized on streptavidin coated microtiter wells and the process of ELISA is carried out to detect the DIG labeled molecule by anti DIG IgG peroxidase conjugate and its corresponding substrate. This principle was applied for detection of various plant viruses like PVX (Soliman *et al.*, 2000), for apple viruses (Menzel *et al.*, 2003) and (Nolasco *et al.*, 1993).

Array Technologies

Array technology has revolutionized the world of viral diagnosis because of its efficiency in screening a large volume of field samples in a single array plate or reaction. Basic principle of array technology combines the binding of DNA on to a solid support such as Membrane filter or array plate and followed by Hybridization technology with a specific probe that will detect the target DNA. This technology was first invented and applied for gene expression studies, later they have been used in variant pathogen diagnosis. There are mainly two types of arrays: (1) Macroarrays and (2) Microarrays based on the volume of the sample and the droplet size used for the analysis.

Macro-arrays

Macro-array is a technique where the amount of sample used is higher than Micro-arrays and the droplet size is more than 200 µm space. Principle involves simple blotting of Oligoprobes of virus specific sequences either by Dot-Blot or Slot-Blot method followed by Nucleic acid hybridization with specific sample in which the virus is to be detected. This technique was first applied in plant virology for detection of various RNA viruses and also for detection of eleven potato viruses and a potato infecting viroid (Agindotan *et al.*, 2007, 2008).

Agindotan and Perry (2007) compared relative sensitivity of the macroarray with that of three commonly used assays for plant RNA viruses: DAS-ELISA, conventional RT-PCR, and real-time RT-PCR. In testing for PLRV in *S. tuberosum*, a 10-fold dilution series was prepared of an infected plant RNA (or sap) extract in an uninfected plant RNA (or sap) extract. The end point of detection in the macroarray was 10^{-2}, comparable with results obtained with the DAS-ELISA. Conventional RT-PCR was three orders of magnitude more sensitive, with an end point of detection of 10^{-2}. These levels of detection were observed in two independent experiments using plants infected by either isolate PLRV#4 or PLRV-WI. In real-time RT-PCR analyses of the infected plant RNA extracts, the endpoint of detection was 10^{-7} in two independent experiments, a further improvement of two orders of magnitude. In order to estimate the viral copy number being detected in these assays, a 10-fold dilution series of a PLRV RNA transcript was employed in real-time RT-PCR experiments and the data were used to prepare a standard curve. These quantitative PCR measurements showed that the minimum copy number of PLRV template detected was 2×10^2 for real-time RT-PCR and 2×10^4 for the conventional RT-PCR. Furthermore, in the macroarray experiments, PLRV was detected only if there were at least 2×10^6 copies of the viral RNA per 4 μg of total RNA. The quantitative PCR measurements also were employed to estimate the copy number of viral RNA in sample preparations before and after the standard amplification procedure for the macroarray. In the starting aliquot of 4 μg of total RNA, there were an estimated 2×10^9 copies of viral RNA. Following amplification, the copy number increased 100 to 1,000-fold to between 2×10^{11} and 2×10^{12} (Agindotan and Perry, 2007). Membrane-based macroarrays provide a relatively inexpensive technology with the potential to detect hundreds of pathogens in a single assay. For the simultaneous detection of a large number of pathogens, it is necessary to obtain sufficient nucleic acids for labeling, and any amplification reactions need to be performed using unbiased, pathogen-nonspecific primers. A nonradioactive macroarray system is described to test for plant RNA viruses using 70-mer oligonucleotide probes immobilized on nylon membranes. Starting with a total plant RNA extract, complementary DNA (cDNA) and second-strand syntheses were carried out using an anchor primer sequence with random pentamers coupled at the 3μ end. Subsequent synthesis by polymerase chain reaction using the anchor primer alone resulted in a relatively unbiased amplification of plant and viral RNAs. These cDNAs were chemically labeled and the product used as a target in hybridization analyses. The system was validated using RNA extracts from plants infected with *Cucumber mosaic virus*, *Potato virus Y*, and *Potato leaf roll virus* (PLRV). Despite the relative excess of host-derived nonviral sequences, viral RNAs were amplified between 100- and 1,000-fold and were detected in single and mixed infections. The macroarray sensitivity was comparable to that of double-antibody sandwich enzyme-linked immunosorbent assay, with PLRV being detected in sap dilutions of 1:100 (Agindotan and Perry, 2007).

Micro-arrays

Micro-array is a technique where the amount of sample used is much lesser than Macroarrays and the droplet size is less than 200 μm space. The invention of Microarrays was chiefly for the detection of differential gene expression patterns in cells and was applied in case of plant viruses by Boonham *et al.* (2003) and Bystricka *et al.* (2005). Hadidi *et al.* (2004) and Bianco *et al.* (2003) emphasized the importance of Microarrays in detection of plant viruses and phytoplasmas, respectively.

Microarray technology or the DNA chips are used generally for identification of differential gene expression patterns of an organism or tissue. Chiefly used in functional genomics for understanding the different types of tissues, developmental stages, disease conditions. The principle is mainly based on Nucleic acid hybridization and fluorescent dyes for labeling the hybridization probes and hence termed as a process of *in-situ* hybridization.

DNA microarrays or biochips are made of a surface on which are linked multiple capture probes, each one being specific for a DNA or RNA sequence of the targets. Their purpose is the detection of numerous sequences in a single assay. Various supports are currently in use for the elaboration of microarrays, including glass, nylon and different polymers. Up to 30,000 DNA probes (gene sequences) can be arrayed onto a single chip. The probes arrayed can be PCR products amplified to high concentrations or relatively short (30–50 bp) oligonucleotide probes.

A Microarray is a slide, to which single-stranded DNA molecules are attached at fixed locations (Figure 16.4). Generally this slide is either made of glass or Nitrocellulose that can bind biomolecules covalently. Each spot will require 0.25-1 nl solution and diameter on the glass will be 100-150 μm separated by a space of 200-250 μm. Since the samples are spotted onto the microtitre plates in the micro volumes, in the form of array, so the technology is defined as Microarray technology which is a highly developed technology.

High density nucleic acid samples are isolated and purified which can be used as samples or else some times cDNA derived from reverse transcription of expressed mRNA or RNA itself some times (termed as RNA Microarrays). In some cases oligonucleotides are synthesized directly onto the microarray plates or the glass slides (known as *Insilico* synthesis). To print these samples, the scientist use high throughput robotic systems. These Nucleic acid samples are then immobilized on to the substrate. The next step includes hybridization of probes. Probes are prepared from many sources, some from cell extracted nucleic acids, some are synthesized oligonucleotides, some are from PCR amplified products, some include the different cloned inserts, etc. Generally, in order to identify the expression patterns in genes, mRNA or cDNA derived from mRNA are used as probes. In case of DNA microarrays, DNA samples or fragments are used as probes. The probes are isolated and purified nucleic acids and they are labeled by two types of fluorescent dyes one is CY3 that is a green channel excitation fluorophore and the other is CY5 that is a red channel excitation fluorophore. The labeled probes are hybridized to the microarray plate in a hybridizer. After hybridization, the un-hybridized probe will be washed out with appropriate buffers. This micro array plate is ready for scanning. Many microarray scanners are available now-a-days which apply briefly two methods: (1) Sequential scanning and (2) Simultaneous scanning. The image obtained by scanning of the microarray plate in a scanner is subjected to analysis by using softwares like Scanalyze, etc., for the expression pattern.

Similar principles could be applied in case of plant viruses where the Microarray slide could be printed with oligo's that are synthesized specific to the plant viruses. At a time, two samples labeled with two different fluorophores could be used for hybridization and the presence of specific virus or its genotype could be identified easily.

A prototype microarray system was produced and shown to be able to detect, if not all, the grape viruses. The chip accuracy was validated for several viruses using viral genomic libraries and well known infected grapevines (Engel and Valenzuela, 2005).

The potential of microarray technology in the detection and diagnosis of plant diseases is very high, due to the multiplex capabilities of the system. Moreover, it can be coupled with other systems, *i.e.* to perform nucleic-acid extraction on the chip (Liu

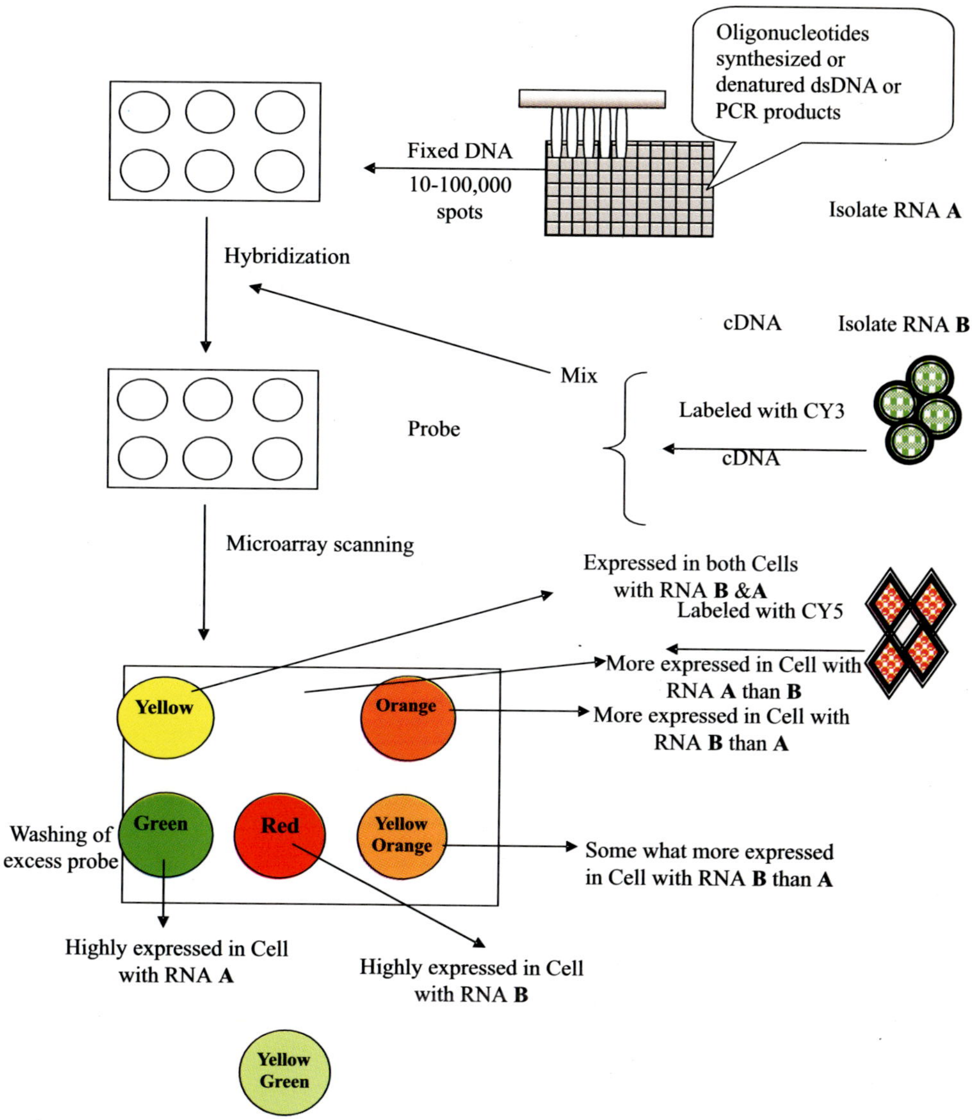

Figure 16.4: Explaining Methodology of Microarrays

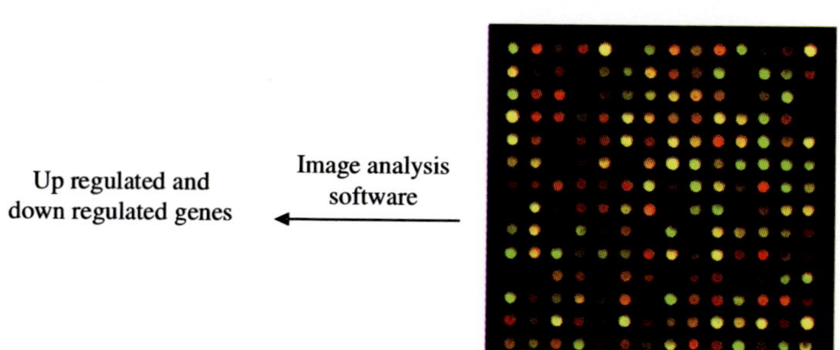

Gene Chip Expression Probe Array

et al., 2007), achieve PCR reactions and their detection on the same device (van Doorn *et al.*, 2007) or even mix all the systems in one (Lee *et al.*, 2006), providing the possibility of automation that can be of great importance and utility. This possibility, with the coupling with previous steps of the analyses (extraction, PCR, detection) promises a wider use in future protocols (Bonants *et al.*, 2005; Lee *et al.*, 2006; Boonham *et al.*, 2007; van Doorn *et al.*, 2007; Liu *et al.*, 2007). Furthermore, new developments, like the labeling of total bacterial RNA (François *et al.*, 2003), the direct detection of DNA or RNA without previous PCR amplification (Call *et al.*, 2003), or multiplex detection based on padlock probe technology (pUMA) (Bonants *et al.*, 2005), may make this technique simpler.

Table 16.1: Comparison of Sensitivity, Specificity, Feasibility, Rapidness and Cost of Different Techniques in Detection of Plant Pathogenic Bacteria and Viruses (Lopez *et al.*, 2008)

Technique	*Sensitivity*[a]	*Specificity*[b]	*Feasibility*[c]	*Rapidness*	*Cost*
Molecular hybridisation	+[d]	++++	++	+	+++
Fish	++	++	+++	+	++
Conventional PCR	+++	++++	+++	+++	+++
Nested PCR in a single tube	++++	++++	+++	++	+++
Cooperational-PCR[e]	++++	++++	+++	+++	+++
Multiplex PCR	+++	++++	+++	+++	+++++
Multiplex nested PCR	++++	++++	++	+++	+++++
Real-time PCR[f]	+++++	+++++	++++	+++++	+++
NASBA[g]	+++++	++++	++++	++++	++
LAMP	++++	++++	+++	++++	++
Microarrays	+	+++++	+	++	+

a: Sensitivity: probability of detecting true positives; b: Specificity: probability of detecting true negatives; c: Feasibility: practicability in routine analysis, execution and interpretation; d: The number of + symbols indicates how methods rate regarding each considered criterion, from acceptable (+) to optimum (+++++); e: Coupled with hybridisation and colorimetric detection; f: Using TaqMan probes; g: Using Molecular Beacons probes.

Molecular Markers

Many plant viruses are recognized as emerging or re-emerging or new emerging viruses which have to be studied rapidly and thoroughly for their etiology, ecology and epidemiology. Emergence of new strains or new viruses is a major thrust area of research. Brown's development of molecular marker to track the movement of viruses in the plant system. Attempts were made in this regard to locate genes conferring resistance to certain seed borne viruses. For example, Timmerman *et al.* (1993) determined the location of Sbm-1 on the *Pisum sativum* genetic map by linkage analysis with eight synthetic molecular markers. Analysis of the progeny of two crosses confirmed that Sbm-1 is on chromosome six. The inclusion of Fed-1 (encoding ferrodoxin-1) and Prx-3 (encoding peroxidase) among the markers facilitated the comparison of this map with the classical genetic map of pea. The Sbm-1 gene is most closely linked to RFLP markerGS 185 as a marker for Sbm-1 in breeding programmes. The GS 185 hybridization pattern and virus resistance phenotypes were compared in a collection of breeding lines and cultivars. Quarantines must be strengthened by molecular marker methods to detect the seed-borne viruses and to prevent new introductions of viruses into a country.

Comparison of sensitivity, specificity, feasibility, rapidness and cost of different molecular techniques are summarized in Table 16.1.

References

Abu-Samah, N. and Randles, J. W. (1983). A comparison of Australian bean yellow mosaic virus isolates using molecular hybridisation analysis. *Ann. App. Biol.* 103, 97-107.

Agindotan, B.O. and Perry, K.L. (2007). Macroarray detection of plant RNA viruses using randomly primed and amplified complementary DNAs from infected plants. *Phytopath.* 97: 119-127.

Agindotan, B.O., Sheil, P.J. and Berger, P.H. (2007). Simultaneous detection of potato virus, PLRV, PVA, PVX and PVY from dormant potato tubers by TaqMan (®) real time RT-PCR. *J. Virol. Methods* 142: 1-9.

Ahlawat, Y.S. (2007). Citrus decline: Problems and preventions. *Indian Phytopath.* 60: 1-12.

Ahrens, U. and Seemüller, E. (1992). Detection of DNA of plant pathogenic mycoplasmalike organisms by a polymerase chain reaction that amplifies a sequence of the 16S rRNA gene. *Phytopath.* 82: 828-832.

Akinjogunla, O. J., Taiwo, M. A. and Kareem, K. T. (2008). Immunological and molecular diagnostic methods for detection of viruses infecting cowpea (*Vigna unquiculata*). *African J. Biotech.* 7: 2099-2103.

Allen, R.N. and Dale, J.L. (1981). Appication of rapid biochemical methods for detecting avocado sunblotch disease. *Ann. Appl. Biol.* 98: 451-461.

Allen, R.N., Palukaitis, P. and Symons, R.H. (1981). Purified avocado sunblotch viroid causes disease in avocado seedlings. *Aust. Pl. Pathol.* 10: 31-32.

Angelini, E., Negrisolo, E., Clair, D., Borgo, M. and Boudon-Padieu, E. (2003). Phylogenetic relationships among Flavescence doree isolates and related phytoplasmas determined by Heteroduplex Mobility Assay and sequences of ribosomal and non-ribosomal DNA. *Plant Path.* 52: 663-672.

Barba, M., Del Serrone, P., Minucci, C., Boccardo, G. and Conti, M. (1995). Diagnosi molecolare di fitoplasmi della vite (Molecular diagnosis of grapevine phytoplasmas). *Petria* 5: 299-300.

Bar-Joseph, M., Rasner, A., Moscovitz, M. and Hull, R. (1983). A simple procedure for the extraction of double–stranded RNA form viral infected plants. *J. Virol. Methods* 6: 1-18.

Bariana, H.S., Shannon, A.L., Chu, P.W.G., Waterhouse, P.M. (1994). Detection of five seedborne legume viruses in one sensitive multiplex polymerase chain reaction test. *Phytopath.* 84: 1201–1205.

Baulcombe, D., Flavell, R.B., Boulton, R.E. and Jellis, G. J. (1984). The sensitivity and specificity of a rapid nucleic acid and hybridization method for the detection of potato virus X in crud sap samples. *Plant Pathol.* 33: 361-370.

Bertaccini, A., Boscia, D., Faoro, F. and Minafra, A. (1994). Metodi di diagnosi delle malattie da virus, viroidi e micoplasmi della vite (Methods for the diagnosis of grapevine virus, viroid and phytoplasma diseases), p. 281-294. In Atti Giornate Fitopatologiche 1994, Montesilvano Lido (Pescara), 9-12 maggio 1994, Vol.2. Cooperativa Libraria Universitaria Editrice Bologna (CLUEB), Bologna, Italy.

Bertaccini, A., Murari, E., Vibio, M., Danielli, A., Davis, R.E., Borgo, M., Consolaro, R. and Sancassani, G.P. (1996). Identificazione molecolare dei fitoplasmi (Molecular identification of phytoplasmas). *L'Informatore Agrario* 52: 55-59.

Bertolini, E., Olmos, A., Martý'nez, M.C., Gorris, M.T. and Cambra, M. (2001). Single-step multiplex RT-PCR for simultaneous and colourimetric detection of six RNA viruses in olive trees. *J. Virol. Methods* 96: 33–41.

Bertolini, E., Olmos, A., Lo´Pez, M.M. and Cambra, M. (2003). Multiplex nested reverse-transcription polymerase chain reaction in a single tube for sensitive and simultaneous detection of four RNA viruses and Pseudomonas savastanoi pv. savastanoi in olive trees. *Phytopath.* 93: 286–292.

Bianco, P.A., Frosini, A., Casati, P. and De Bellis, G. (2003). Identification of phytoplasmas infecting grapevine by ligase reaction and universal array. *In: Extended Abstracts 14th Meeting ICVG*, Locorotondo, 2003, p 55.

Bijaisoradat, M. and Kuhn, C.W. (1988). Detection of two viruses in peanut seeds by complementary DNA hybridization tests. *Plant Dis.* 72: 956-959.

Boccardo, G., Beaver, R.G., Randles, J.W. and Imperial, J.S. (1981). Tinangaja and bristle top, coconut diseases of uncertain etiology in Guam, and their relationship to cadang-cadang disease of coconut in the Phillipines. *Phytopath.* 71: 1104-1107.

Bonants, P.J.M., Schoen, C.D., Szemes, M., Speksnijder, A., Klerks, M.M., van den Boogert, P.H.J.F., Waalwijk, C., van der Wolf, J.M. and Zijlstra, C. (2005). From single to multiplex detection of plant pathogens: pUMA, a new concept of multiplex detection using microarrays. *Phytopathol. Polonica* 35: 29-47.

Bonfiglioli, R., Lherminier, J., Daire, X. and Boudon-Padieu, E. (1997). The use of *in-situ* hybridization with Oligonucleotide probes to specifically localize phytoplasmas in plant tissues in electron microscopy, p. 81-82. In: O.A. Sequeira de; J.C. Sequeira, and M.T. Santos (ed.), Extended abstracts 12 Meeting ICVG, Lisbon, Portugal, 29 Sept.-2 October 1997. Dept. Plant Pathology, Estacao Agronomica Nacional, Oeiras, Portugal.

Boonham, N., Tomlinson, J. and Mumford, R. (2007). Microarrays for rapid identification of plant viruses. *Annu. Rev. Phyto.* 45: 307-328.

Boonham, N., Walsh, K., Smith, P., Madagan, K., Graham, I. and Barker, I. (2003). Detection of potato viruses using microarray technology: towards a generic method for plant viral disease diagnosis. *J. Virol. Methods* 108: 181-187.

Borkhardt, B., Vongsasitorn, D. and Albrechtsen, S. E. (1994). Chemiluminescent detection of potato spindle tuber viroid in true potato seed using a digoxigenin labeled DNA probe. *Potato Research* 37: 249-255.

Boscia, D. (1993). Isolation and analysis of double stranded RNAs, p. 217-218. In: G.P. Martelli (ed.), Graft transmissible diseases of grapevines. Handbook for detection and diagnosis. FAO, Rome.

Boulton, M.I. and Markham, P.G. (1986). The use of squash blotting to detect plant pathogens in insect vectors. In Jones RAC, Torrance, L. (eds.), Developments and applications in Virus Testing, Association of Applied Biology, Wellasbourne, Warwick, England, pp. 55-69.

Brandt, S. and Himmler, G. (1993). Detection of grapevine fanleaf virus (GFLV) from woody material by using immunocapture polymerase chain reaction, p. 150. In: P. Gugerli (ed.), Extended abstracts 11th Meeting ICVG, Montreux, Switzerland, 6-9 September 1993. Federal Agricultural Research Station of Changins, CH-1260 Nyon, Switzerland.

Brandt, S and Himmler, G. (1995). Detection of Nepoviruses in ligneous grapevine material by using IC/PCR. *Vitis* 34: 127-128.

Brandt, S., Himmler, G. and Katinger, H. (1993). Anwendung der Immunocapture Polymerase Chain Reaction (IC/PCR) für den Nachweis von Rebenviren aus holzigem Material (Use of immunocapture polymerase chain reaction (IC/PCR) for detection of grapevine viruses in woody material). *Mitt. Klosterneuburg* 43: 143-147.

Bretout, C., Candresse, T., Gall Le, O., Brault, V., Ravelonandro, M. and Dunez, J. (1988). Virus and RNA specific molecular hybridization probes for two nepoviruses. *Acta Hort.* 235: 231-238.

Bystricka, D., Lenz, O., Mraz, I., Piherova, L., Kmoch, S. and M. Sip (2005). Oligonucleotide-based microarray: A new improvement in microarray detection of plant viruses. *J. Virol Methods* 128: 176-182.

Call, D.R., Borucki, M.K., and Loge, F.J. (2003). Detection of bacterial pathogens in environmental samples using DNA microarrays. *J. Microbiol. Methods* 53: 235-243.

Candresse, T., Hammond, R.W. and Hadidi, A. (1998). Detection and identification of plant viruses and viroids using polymerase chain reaction (PCR). In Plant Virus Disease Control. Eds. A. Hadidi, R.K. Khetarpal and H. Koganezavwa. APS Press, Minnesota, USA, 399-416.

Caruso, P., Bertolini, E., Cambra, M. and Lo´ pez, M.M. (2003). A new and co-operational polymerase chain reaction (Co-PCR) for rapid detection of Ralstonia solanacearum in water. *J Microbiol Methods* (in press)

Chen, K.H., Credi, R., Loi, N., Maixner, M. and Chen, T.A. (1994). Identification and grouping of mycoplasmalike organisms associated with grapevine yellows and clover phyllody diseases based on immunological and molecular analyses. *Appl. Environ. Microbiol.* 60: 1905-1913.

Chevalier, S., Greif, C., Clauzel, J. M., Walter, B. and Fritsch, C. (1995). Use of an immunocapture-polymerase chain reaction procedure for the detection of grapevine virus A Kober stem grooving infected grapevines. *J. Phytopath.* 143: 369-373.

Choi, S.K., Choi, J.K., Park, W.M. and Ryu, K.H. (1999) RT-PCR detection and identification of three species of cucumoviruses with a genus-specific single pair of primers. *J Virol Methods* 83: 67-73.

Clark, M.F. and Adams, A.N. (1977). Characteristics of the microplate method of enzyme-linked immunosorbent assay for the detection of plant viruses. *J. Gen. Virol.* 34: 475-483.

Cockerill, F.R., III, and Smith, T.F. (2002). Rapid-cycle real-time PCR: a revolution for clinical microbiology. *Am. Soc. Microbiol. News* 68: 77–83.

De Blas, C., Borja, M.J., Saiz, M. and Romero, J. (1994). Broad spectrum of cucumber mosaic virus (CMV) using the polymerase chain reaction. *J. Phytopath.* 141: 323-329.

Del Serrone, P., Minucci, C. and Barba, M. (1995). Diffusione del Giallume Fitoplasmale della vite in impianti laziali (Diffusion of grapevine phytoplasma yellows in vineyards of Latium). *Riv. Vitic. Enol.* 48: 11-16.

Delanoy, M., Salmon, M., Kummert, J., Frison, E. and Lepoivre, P. (2003). Development of real time PCR for the rapid detection of episomal Banana streak virus (BSV). *Plant Dis.* 87: 33-38.

Dewey, R.A., Semorile, L.C. and Grau, O. (1996). Detection of *Tospovirus* species by Rt-PCR of the N-gene and restriction enzyme digestion of the products. *J. Virol. Methods* 56: 19-26.

Didenko, V.D. (2001). DNA probes using fluorescence resonance energy transfer (FRET): designs and applications. *BioTechniques* 31: 1106–1120.

Dietzgen, R. G. (2002). *Application of PCR in plant virology.* In: Plant Viruses as Molecular Pathogens (Editors Jawaid A. Khan, Dijkstra, J.). p. 471-491.

Dijkstra, J. and de Jager, C.P. (1998). *Practical Plant Pathology: Protocols and Exercises.* Springer-Verlag, Berlin.

Dodds, J.A. (1993). dsRNA in diagnosis. In: REF Matthews (ed.) Diagnosis of Plant Virus Diseases, pp. 273-294. CRS Press, Boca Raton, FL.

Dodds, J.A. and Bar-Joseph, M. (1983). Double-stranded RNA from plants infected with Closteroviruses. *Phytopath.* 73: 419-423.

Dodds, J.A., Morris, T.J. and Jordan, R.L. (1984). Plant viral double-stranded RNA. *Annu. Rev. Phytopath.* 22: 151-168.

Don, R.H., Cox, P.T., Wainwright, B.J., Baker, K. and Mattick, J.S. (1991). "Touchdown' PCR to circumvent spurious priming during gene amplification". *Nucl Acids Res 19: 4008.*

Dovas, C.I. and Katis, N.I. (2003 a). A spot nested RT-PCR method for the simultaneous detection of members of the *Vitivirus* and *Foveavirus* genera in grapevine. *J. Virol. Methods* 107: 99-106.

Dovas C. I. and Katis, N. I. (2003 b). A spot multiplex nested RT-PCR for the simultaneous and generic detection of viruses involved in the etiology of grapevine leafroll and rugose wood of grapevine. *J. Virol. Methods* 109: 217-226.

Dujovny, G., Usugi, T., Shohara, K. and Lenardon, S. (1998). Characterization of a potyvirus infecting sunflower in Argentina. *Plant Dis.* 82: 470-474.

EL-Kewey, S.A., Sidaros, S.A., Abdelkader, H. S., Emeran, A. A. and EL-Sharkawy, M. (2007). Molecular detection of *Broad bean stain* Comovirus (*BBSV*) and *Cowpea aphid borne mosaic* Potyvirus (*CABMV*) in Faba Bean and Cowpea Plants. *J. Appl. Sci. Res.* 3: 2013-2025.

Engel, E. and Valenzuela, P.D.T. (2005). Simultaneous detection of grapevine viruses using Microarrays. *In: XLVIII Meeting of the Biological Society of Chile,* October, 2005, Pucon, Chile.

Faggioli, F. and La Starza, S. (2006). One-step multiplex RT-PCR for simultaneous detection of eight grapevine viruses and its application in a sanitary selection programme. *In: Proc. 15th Meeting ICVG,* 3-7 April, Stellenbosch, South Africa, pp. 120-121.

Filippin, L., Angelini, E., Bianchi, G., Morasutti, C. and Borgo, M. (2006). Singleplex and multiplex real time PCR for the detection of phytoplasmas associated with grapevine yellows. *In: Proc. 15th Meeting ICVG,* 3-7 April, 2006, Stellenbosch, South Africa.

Francois, P., Bento, M., Vaudaux, P., and Schrenzel, J. (2003). Comparison of fluorescence and resonance light scattering for highly sensitive microarray detection of bacterial pathogens. *J. Microbiol. Methods* 55: 755-762.

Fukuta, S., Lida, T., Mizukami, Y., Ishida, A., Ueda, J., Kanbe, M. and Ishimoto, Y. (2003). Detection of *Japanese yam mosaic virus* by RT-LAMP. *Arch. Virol.* 148: 1713-1720.

Fukuta, S., Ohishi, K., Yoshida, K., Mizukami, Y., Ishida, A. and Kanbe, M. (2004). Development of immunocapture reverse transcription loop mediated isothermal amplification for the detection of tomato spotted wilt virus from chrysanthemum. *J. Virol. Methods* 121: 49-55.

Gallitelli, D., Savino, V. and Martelli G. P. (1985). The use of a spot hybridization method for the detection of grapevine virus A in the sap of grapevine. *Phytopath. Medit.*, 24: 221-224.

Gemmrich, A. R., Link, G. and Seidel, M. (1993). Detction of grapevine fanleaf virus (GFLV) in infected grapevines by non-radioactive nucleic acid hybridization. *Vitis* 32: 237-242.

Gillaspie, A.G.J., Pio-Ribeiro, G., Andrade, G.P. and Pappu, H.R. (2001). RT-PCR detection of seedborne Cowpea aphid-borne mosaic virus in peanut. *Plant Dis.* 85: 1181-1182.

Gillings, M., Broadbent, P., Indsto, J. and Lee, R. (1993). Characterization of isolates and strains of citrus tristeza closterovirus using restriction analysis of the coat protein gene amplified by the polymerase chain reaction. *J. Virol. Methods* 44: 305-317.

Gorsane, F., Fakhfakh, H., Tourneur, C., Makni, M. and Marrakchi, M. (1999). Some biological and molecular properties of pepper veinal mottle virus isolates occurring in Tunisia. *Plant Mol. Biol. Prot.* 17: 149-158.

Gould, A.R. and Symons, R.H. (1983). A molecular biological approach to relationships among viruses. *Annu. Rev. Phytopath.* 21: 179-199.

Gloffke, W. (2003). Quantitative PCR up date. *Scientist* 21: 41–43.

Grieco F. and Gallitelli D. (1999). Multiplex reverse transcriptase polymerase chain reaction applied to virus detection in globe artichoke. *J. Phytopathol.* 147: 183–185.

Hadidi, A., Czosnek, H. and Barba, M. (2004). DNA microarrays and their potential applications for the detection of plant viruses, viroids, phytoplasmas. *J. Plant Pathol.* 86: 97-104.

Hema, M., Kirthi, N., Shreenivasulu, P. and Savithri, H.S. (2003). Development of recombinant coat protein antibody based IC-RT-PCR for detection and discrimination sugarcane streak mosaic virus isolates from Southern India. *Arch. Virol.* 148: 1185-1193.

Harper, G., Dahal, G., Thottapilly, G. and Hull, R. (1999). Detection of episomal banana streak badnavirus by IC-PCR. *J. Virol. Methods* 79: 1-8.

Harrison, B.D. and Robinson, D.J. (1978). The tobamoviruses. In: Advances in Virus Research, Vol. 23, pp. 25-27. Eds. M.A. Lanfer, F.B. Bang, K. Maramorosch and K.M. Smith. Academic Press, New York.

Holland, P.M., Abramson, R.D., Watson, R. and Gelfand, D.H. (1991). Detection of specific polymerase chain reaction product by utilizing the 52 to 32 exonuclease activity of *Thermus aquaticus* DNA polymerase. *Proc. Natl. Acad. Sci. U.S.A.* 88: 7276–7280.

Hsu, Y.-H., Annamalai, P., Lin, C.S., Chen, Y.Y., Chang, W.C. and Lin, N.S. (2000). A sensitive method for detecting bamboo mosaic virus (BaMV) and establishment of BsaMV-free meristem-tip cultures. *Pl. Pathol.* 49: 101-107.

Hull, R. (1993). Nucleic acid hybridization procedures. In: Diagnosis of plant virus diseases, REF Matthews, ed. (Florida, USA: CRC Press), pp. 253-271.

Hull, R. (2002). Matthews' Plant Virology, 4th Edition. pp. 1001. London: Academic Press.

Hull, R. and Davies, J. W. (1993). Genetic engineering with plant viruses and their potential as vectors. *Adv. Virus Res.* 28: 1-33.

Jacobi, V., Bachand, G.D., Hamelin, R.C. and Castello, J.D. (1998). Development of a multiplex immunocapture RT-PCR assay for detection and differentiation of tomato and tobacco mosaic tobamoviruses. *J. Virol. Methods* 74: 167–178.

Joan, M.H. and Roy, F. (1993). The polymerase chain reaction and plant disease diagnosis. *Annu. Rev. Phytopathol.* 31: 81-109.

Jordan, R.L. and Dodds, J.A. (1983). Hybridization of 5′ end-labeled RNA to plant viral RNA in agarose and acrylamide gels. *Pl. Mol. Biol. Reptr.* 1: 31-37.

Jordan, R.L., Dodds, J.A. and Ohr, H.D. (1983). Evidence for virus like agents in avocado. *Phytopath.* 73: 1130-1135.

Joshi, R., Kumar, V and Dasgupta, I. (2003). Detection of molecular variability in rice tungro bacilliform viruses from India using polymerase chain reaction–restriction fragment length polymorphism. *J. Virol Methods* 109: 89-93.

Kamenova, I. and Adkins, S. (2004). Comparison of detection methods for a novel Tobamovirus isolated from Florida Hibiscus. *Plant Dis.* 88: 34-40.

Kohler, G. and Milstein, C. (1975). Continuous culture of fused cells secreting antibody of predefined specificity. *Nature* (Lond.) 256: 495-497.

Lee, L.G., Connell, C.R. and Bloch, W. (1993). Allelic discrimination by nick-translation PCR with fluorogenic probes. *Nucleic Acids Res.* 21: 3761–3766.

Lee, JG., Cheong, K.H., Huh, N., Kim, S., Choi, J.W. and Ko, C. (2006). Microchip-based one step DNA extraction and real-time PCR in one chamber for rapid pathogen identification. *Lab Chip.* 6: 886-895.

Levy, L., Lee, I.M. and Hadidi, A. (1994). Simple and rapid preparation of infected plant tissue extracts for PCR amplification of virus, viroid, and MLO nucleic acids. *J. Virol. Methods* 49: 295-304.

Lim, S.T., Wong, S.M., Yeong, C.Y., Lee, S.C. and Goh, C.J. (1993). Rapid detection of cymbidium mosaic virus by polymerase chain reaction (PCR). *J. Virol. Methods* 41: 37-46.

Little, A. (2005). Complete sequence, improved detection and functional analysis of Grapevine leafroll associated virus 1 (GLRaV-1). *Ph. D. Thesis, University of Adelaide*, Australia, 93 pp.

Liu, Y., Cady, N.C. and Batt, C.A. (2007). A plastic microchip for nucleic acid purification. *Biomed Microdevices* 9: 769-776.

Lopez, M.M., Bertolini, E., Olmos, A., Caruso, P., Gorris, M.T., L Pop, P., Penyalver, R. and Cambra, M. (2003). Innovative tools for detection of plant pathogenic viruses and bacteria. *Int. Microbiol.* 6: 233-243.

Lopez, M., Liop, P., Olmos, A., Marco-Noales, E., Cambra, M. and Bertolini, E. (2008). Are molecular tools solving the challenges posed by detection of plant pathogenic bacteria and viruses?. *M. Biol.* (Curr. Issues) 11: 13-46.

Martelli, G.P., Avgelis, A.D., Boscia, D., Cambra, M., Candresse, T., Caudwell, A., Dunez, J., Garnsey, S.M., Golino, D.A., Lee, R.F., Lehoczky, J., Macquaire, G., Namba, S., Roistacher, C.N., Rumbos, I.C., Savino, V., Semancik, J.S. and Walter,

B. (1993). *Graft transmissible diseases of grapevines*. Handbook for detection and diagnosis. p. 1-263. In G. P. Martelli (ed.),. FAO Publication Division, Rome.

Maule, A.J., Hull, R. and Donson, J. (1983). The application of spot hybridization to the detection of DNA and RNA viruses in plant tissues. *J. Virol. Methods* 6: 215-224.

Menzel, W., Zahn, V. and Maiss, E. (2003) Multiplex RT-PCR-ELISA compared with bioassay for the detection of four apple viruses. *J. Virol. Methods* 110: 153-157.

Minafra, A., Grieco, F., Gallitelli, D. and Martelli, G.P. (1995). Improved PCR procedures for multiple identification of some artichoke and grapevine viruses. *Bulletin OEPP/EPPO Bulletin* 25: 283-287.

Minafra, A., Greif, C. and Romero, J. (1997). Molecular tools for the detection of grapevine viruses, p. 157-170. In B. Walter (ed.), Sanitary selection of the grapevine. Protocols for detection of viruses and virus-like diseases (Les Colloques no 86). INRA Editions, Paris.

Minafra, A. and Hadidi, A. (1994) Sensitive detection of grapevine virus A, B or leafroll-associated virus III from viruliferous mealybugs and infected tissue. *J. Virol. Methods* 47: 175–187.

Mnari-Hattab, M. and Ezzaier, K. (2006). Biological, serological and molecular characterization of pepper mild mottle virus (PMMoV) in Tunisia. *Tunisian J. Pl. Prot.* 1: 1-12.

Morris, T.J. and Dodds, J.A. (1979). Isolation and analysis of double stranded RNA from virus infected plant and fungal tissue. *Phytopath.* 69: 854-858.

Mullis, K.B., Faloona, F., Sharf, S., Saiki, R., Horn, G. and Elrich, H. (1986). Specific enzymatic amplification of DNA *in-vitro*. Polymerase chain reaction. *Cold Spring Harbor Symp. Quant. Biol.* 51: 263-273.

Mumford, R.A., Barker, I., Walsh, K. and Boonham, N. (2000). Detection of potato mop-top virus and tobacco rattle virus using a multiplex real time fluorescent reverse-transcription polymerase chain reaction assay. *Phytopath.* 90: 448-453.

Mumford, R.A., Boonham, N., Tomlinson, J. and Barker, I. (2006). Advances in molecular phytodiagnosis-new soutions for old problems. *Eur. J. Plant Pathol.* 116: 1-19.

Naidu, R.A. and Hughes, J.D.A. (2003). Methods for the detection of plant viral diseases in plant virology in sub-Saharan Africa, Proceedings of plant virology, IITA, Ibadan, Nigeria. Eds. Hughes JDA, Odu B, pp. 233-260.

Nassuth, A., Pollari, E., Helmeczy, K., Stewart, S. and Kofalvi, S.A. (2000). Improved RNA extraction and one-tube RT-PCR assay for simultaneous detection of control plant RNA plus several viruses in plant extracts. *J. Virol. Methods* 90: 37–49.

Nemchinov, L., Hadidi, A., Foster, J.J., Candresse, T. and Verderevskaya, T. (1995). Sensitive detection of apple chlorotic leaf spot virus from infected apple or peach tissue using RT-PCR, IC-RT-PCR, or multiplex IC-RT-PCR. *Acta Hortic.* 386: 51–62.

Nie, X. and Singh, R.P. (2000) Detection of multiple potato viruses using an oligo(dT) as a common cDNA primer in multiplex RT-PCR. *J. Virol. Methods* 86: 179–185.

Nolasco, G., de Blas, C., Torres, V. and Ponz, F. (1993). A method combining immunocapture and PCR amplification in a microtitre plate for the detection of plant viruses and subviral pathogens. *J. Virol. Methods* 45: 201-218.

Notomi, T., Okayama, H., Masubuchi, H., Yonekawa, T., Watanabe, K., Amino, N. and Hase, T. (2000). Loop-mediated isothermal amplification of DNA. *Nucleic Acid Res*. 28: E63.

Okuda, M. and Hanada, K. (2001). RT-PCR for detecting five distinct Tospovirus species using degenerate primers and dsRNA template. *J. Virol. Methods* 96: 149–156.

Olmos, A., Bertolini, E. and Cambra, M. (2002). Simultaneous and cooperational amplification (Co-PCR): a new concept for detection of plant viruses. *J. Virol. Methods* 106: 51–59.

Olmos, A., Bertolini, E., Gil, M. and Cambra, M. (2005). Real-time assay for quantitative detection of non-persistently transmitted plum pox virus RNA targets in single aphids. *J. Virol. Methods* 128: 151-155.

Olmos, A., Cambra, M., Esteban, O., Gorris, M.T. and Terrada, E. (1999). New device and method for capture, reverse transcription and nested PCR in a single closed-tube. *Nucleic Acids Res*. 27: 1564–1565.

Olmos, A., Dasí, M. A., Candresse, T. and Cambra, M. (1996). Print-capture PCR: a simple and highly sensitive method for the detection of Plum pox virus (PPV) in plant tissues. *Nucleic Acids Res.* 24: 2192–2193.

Olmos, A., Esteban, O., Bertolini, E. and Cambra, M. (2003). Nested RT-PCR in a single closed tube. In: Bartlett JMS, Stirling D (eds) Methods in molecular biology, 2nd edn, vol 226. *PCR Protocols*. Humana Press, Ottowa, pp. 156–161.

Osman, F. and Rowhani, A. (2006). Application of a spotting sample preparation technique for the detection of pathogen in woody plants by RT-PCR and real-time PCR (Taqman). *J. Virol. Methods* 104: 41-54.

Owens, R.A. and Diener, T.O. (1981). Sensitive and rapid diagnosis of potato spindle tuber viroid disease by nucleic acid spot hybridization. *Science* 213: 670-672.

Owens, R.A. and Diener, T.O. (1984). Spot hybridization for detection of viroids and viruses. *Methods Virol*. 7: 173-187.

Palukaitis, P., Rakowski, A.G., Alexander, D. Mce and Symons, R.H. (1981). Rapid indexing of the sunbblotch disease of avacados using a complementary DNA probe to avacado sunblotch viroid. *Ann. Appl. Biol.* 98: 439-449.

Palukaitis, P. and Symons, R. H. (1978). Synthesis and characterization of a complementary DNA probe for chrysanthemum stunt viroid. *FEBS Letters* 92: 268–272.

Palukaitis, P. and Symons, R. H. (1979). Hybridization analysis of chrysanthemum stunt viroid with complementary DNA and the quantitation of viroid RNA sequences in extracts of infected plants. *Virology* 98: 238–245.

Patil, B.L., Rajasubramaniam, S., Bagchi, C. and Dasgupta, I. (2005). Both Indian cassava mosaic virus and Sri Lankan cassava mosaic virus are found in India and exhibit high variability as assessed by PCR-RFLP. *Arch. Virol.* 150: 389-397.

Purves, *et al.* (2001). *Life: The Science of Biology*. Sixth Edition. Freeman and Company: USA;. pg. 214-217.

Raj Verma, Baranwal, V.K., Satya Prakash, Tomer, S.P.S., Jitender Singh, Pant, R.P. and Ahlawat, Y.S. (2006). Biological and molecular characterization of zucchini yellow mosaic virus (ZYMV) from naturally infected bottle gourd. *Indian J. Virol.* 17: 96-101.

Randles, J.W., Boccardo,G. and Imperial, J.S. (1980) *Phytopath.* 70, 185-189.

Randles, J. W. and Palukaitis, P. (1979). *In vitro* synthesis and characterization of DNA complementary to cadang-cadang associated RNA: *J. Gen. Virol.* 43: 649-662.

Ravi, K.S., Joseph, J., Nagaraju, N., Reddy, H.R. and Savithri, H.S. (1997). Characterization of a pepper vein banding virus from chilli pepper in India. *Plant Dis.* 81: 673-676.

Robertson, N.L., French, R. and Gray, S.M. (1991). Use of group-specific primers and the polymerase chain reaction for the detection and identification of luteoviruses. *J. Gen. Virol.* 72: 1473-1477.

Rojas, M.R., Gilbertson, R.L., Russel, D.R. and Maxwell, D.P. (1993). Use of degenerate primers in the polymerase chain reaction to detect whitefly transmitted geminiviruses. *Plant Dis.* 77: 340-347.

Rosner, A., Bar-Joseph, M., Moscovitz, M and Mevarech, M. (1983). Diagnosis of specific viral RNA sequences in plant extracts by hybridization with polynucleotide kinase-mediated, ^{32}p-labeled, double stranded RNA probe. *Phytopath.* 73: 699-702.

Rowhani, A., Chay, C., Golino, D.A. and Falk, B.W. (1993). Development of polymerase chain reaction technique for the detection of grapevine virus in grapevine tissue. *Phytopath.* 83: 749-753.

Roy, A., Fayad, A., Barthe, G. and Brlansky, R.H. (2005). A multiplex polymerase chain reaction method for reliable, sensitive and simultaneous detection of multiple viruses in citrus trees. *J. Virol. Methods* 133: 130-136.

Russo, P., Miller, L., Singh, R.P. and Slack, S.A. (1999). Comparison of PLRV and PVY detection in potato seed samples tested by Florida winter field inspection and RT-PCR. *Am. J. Potato Res.* 76: 313–316.

Saade, M., Aparicio, F., Sanchez-Navarro, J.A., Herranz, M.C., Di-Terlizzi, A.M.B. and Pallas, V. (2000). Simultaneous detection of the three ilarviruses affecting stone fruit trees by nonisotopic molecular hybridization and multiplex reverse-transcription polymerase chain reaction. *Phytopath.* 90: 1330–1336.

Salazar, L.F., Owens, R.A., Smith, D.R. and Diener, T.O. (1983). Detection of potato spindle tuber viroid by nucleic acid spot hybridization: evaluation with tuber sprouts and potato seed. *American Potato Journal* 60: 587-597.

Saldarelli, P., Minafra, A., Martelli, G. P. and Walter B. (1994). Detection of grapevine leafroll-associated closterovirus III by molecular hybridization. *Plant Path.* 43: 91-96.

Sambrook, J. and Russell, D.W. (2001). Molecular cloning. *A Laboratory Manual.* Cold Spring Harbor Laboratory Press, New York.

Samuitiene, M. and Navalinskiene, M. (2006). Molecular detection and characterization of phytoplasma infecting *Celosia argentea* L. plants in Lithuania. *Agronomy-Research* 4: 345-348.

Schaad, N.W. and Frederick, R.D. (2002). Real-time PCR and its application for rapid plant disease diagnostics. *Can. J. Plant Pathol.* 24: 250-258.

Schumacher, J., Meyer, N., Riesner, D. and Weidemann, H.L. (1986). Diagnostic procedure for detection of viroids and viruses with circular RNAs by 'return-gel electrophoresis'. *J. Phytopath.* 115: 332-343.

Semancik, J.S. (1993). Detection and identification of viroids, p. 199-215. In G.P. Martelli (ed.), Graft–transmissible diseases of grapevines. Handbook for detection and diagnosis. FAO, Rome.

Sharman, M., Thomas, J. and Dietzgen, R.G. (2000). Development of a multiplex immunocapture PCR with colourimetric detection for viruses of banana. *J. Virol. Methods* 89: 75–88.

Singh, H., Le Bowitz, J.H., Baldwin, A.S. and Sharp, P.A. (1988). Molecular cloning of an enhancer binding protein: Isolation by screening of an expression library with a recognition site DNA. *Cell* 52: 415.

Singh, R.P. (1989). Molecular hybridization with complementary DNA for plant viruses and viroid detection. In: Agnihotri, V.P., Singh, N., Chambe, H.S., Singh, U.S., Dwivedi (eds.), Perspectives in Phytopatholgy, Today and Tomorrow's Printers and Publishers, New Delhi, India, pp. 51-60.

Singh, R.P., Boucher, A., Somerville, T.H. and Dhar, A.K. (1993). Selection of a monoclonal antibody to detect PVY^N and its use in ELISA and DIBA assays. *Can. J. Plant Pathol.* 15: 293-300.

Singh, R.P., Kurz, J. and Boiteau, G. (1996). Detection of stylet-borne and circulative potato viruses in aphids by duplex reverse transcription polymerase chain reaction. *J. Virol. Methods* 59: 189–196.

Singh, R.P., Dilworth, A.D., Singh, M. and Mclaren, D.I. (2004). Evaluation of a simple membrane based nucleic acid preparation protocol for RT-PCR detection of potato viruses from aphid and plant tissues. *J. Virol. Methods* 121: 163-170.

Singh, R.P., Dilworth, A.D., Singh, M. and Babcock, M. (2006). An alkaline sodium simplifies nucleic acid preparation for RT-PCR and infectivity assay of viroids from crude sap and spotted membrane. *J. Virol. Methods* 132: 204-211.

Soliman, A.M., Shalaby, A. A., Barsoum, B. N., Mohamed, G. G., Nakhla, M. K., Mazyad, H. M. and Maxwell, D. P. (2000). Molecular characterization and RT-PCR-ELISA detection of a potato virus X (PVX) isolate from Egypt. *Ann. Agric. Sci.* (Sp. Issue) 4: 1791-1804.

Symons, R. H. (1984). Diagnostic approaches for the rapid and specific detection of plant viruses and viroids. In *Plant-Microbe Interactions: Molecular and Genetic Perspectives*, ed. T. Kosuge and E.W. Nester, vol. 1, pp. 93–124 (Macmillan Publishing Co., NewYork).

Thomas, J.E., Geering, A.D.W., Gambley, C.F., Kessling, A.F. and White, M. (1997). Purification, properties, and diagnosis of banana bract mosaic potyvirus and its distinction from abaca mosaic potyvirus. *Phytopath.* 87: 698-705.

Thomas, P. (1980). Hybridization of denatured DNA and small RNA fragments transferred to nitrocellulose. *Proc. of the National Academy of Sciences*, USA 77: 5201-5205.

Tobias, I., Szabo, B., Salanki, K., Saria, L., Kuhlmann, H. and Palkovics, L. (2008). Seedborne transmission of Zucchini yellow mosaic virus and Cucumber mosaic virus in Styrian Hulless group of Cucurbita pepo. 1Cucurbitaceae 2008, Proceedings of the IXth EUCARPIA meeting on genetics and breeding of Cucurbitaceae (Pitrat M, ed), INRA, Avignon (France), May 21-24th, 2008, p. 189-197.

Tzanetakis, *I.E.* and Robert R.M. (2008). A new method for extraction of double-stranded RNA from plants. *J. Virol. Methods* 149: 167-170.

van Doorn, R., Szemes, M., Bonants, P., Kowalchuk, G.A., Salles, J.F., Ortenberg, E. and Schoen, C.D. (2007). Quantitative multiplex detection of plant pathogens using a novel ligation probe-based system coupled with universal, high-throughput realtime PCR on Open Arrays. *BMC Genomics*. 14: 276.

Valverde, R. A. (2008). DsRNA as a detection and diagnostic tool for plant viruses. Pages: 49-60 in Techniques in diagnosis of plant viruses. Studium Press, LLC.

Valverde, R.A. and Dodds, J.A. (1986). Evidence for a satellite RNA associated naturally with the U5 strain and experimentally with the U1 strain of tobacco mosaic virus. *J. Gen. Virol.* 67: 1875-1884.

Valverde, R.A., Dodds, J.A. and Heick, J.A. (1986). Double stranded fibonucleic acid from plants infected with viruses having elongated particles and undivided genomes. *Phytopath.* 76: 459-465.

Varveri, C., Candresse, T., Cugusi, M., Ravelonandro, M. and Dunez, J. (1988). Use of ^{32}p-labeled transcribed RNA probe for dot hybridization detection of plum pox virus. *Phytopath.* 78: 1280-1283.

Walsh, K., North, J., Barker, I. and Boonham, N. (2001). Detection of different strains of Potato virusYand their mixed infections using competitive fluorescent RT-PCR. *J. Virol. Methods* 91: 167–173.

Walter, B., Greif, C. and Martelli, G. P. (1994). Recent progresses in the detection of viruses and phytoplasmas of the grapevine: application to sanitary selection, p. 141-144. In VIth International Symposium on Grape Breeding, Yalta, Crimea, Ukraine, 4-10 September 1994. Office International de la Vigne et du Vin (OIV), Paris, France.

Webster, C.G., Wylie, S.J. and Jones, M.G.K. (2004). Diagnosis of plant viral pathogens. *Curr. Sci.* 86: 1604-1607.

Weller, S.A., Elphinstone, J.G., Smith, N.C., Boonham, N. and Stead, D.E. (2000). Detection of Ralstonia solanacearum strains with a quantitative multiplex, real-time, fluorogenic PCR (Taqman) assay. *Appl. Environ. Microbiol.* 66: 2853–2858.

Wetzel, T., Candresse, T., Macquaire, G., Ravelonandro, M. and Dunez, J. (1992). A highly sensitive immunocapture polymerase chain reaction method for plum pox potyvirus detection. *J. Virol. Methods* 39: 27-37.

Wylie, S.J., Wilson, C.R., Jones, R.A.C. and Jones, M.G.K. (1993). A polymerase chain reaction assay for cucumber mosaic virus in lupin seeds. *Aust. J. Agric. Res.* 44: 41-51.

Yang, Z.N. and Mirkov, T.E. (1997). Sequence and relationships of sugarcane mosaic and sorghum mosaic virus strains and development of RT-PCR-based RFLPs for strain discrimination. *Phytopath.* 87: 932-939.

Chapter 17

Characterization of *Trichoderma* spp.

Pratibha Sharma*, A. Muthu Kumar and Monica Henry

Division of Plant Pathology,
IARI, Pusa Campus, New Delhi – 110 01, India

ABSTRACT

Identification based on morphological characters consent a relatively simple method for classification of *Trichoderma* as genus, but the species perceptions are complex to construe and there is considerable confusion over the application of specific names. Bisby (1939) could identify only one species *T.viride* after examining several isolates and collection of *Trichoderma* without finding any reliable characters to distinguish them, he concluded *Trichoderma* as a monotypic genus. It was only few years ago the factual character of *Trichoderma* has been recognized. Pioneers in *Trichoderma* like Rifai (1969) and Bissett (1991a-c) observed certain cultural characters that could be used for identification and description of these species *viz,* tuft or cushion of hyphae on natural substrate composed of conidiophores, spores and some sterile hyphae, conidiophores indefinite, branched or unbranched hyphae bearing phialides laterally or terminally, phialides oust by heads rarely short or in chains, spore hyaline or brightly coloured, one celled. Nonetheless, class affinities of the genus for the development of species are still very slow. Rifai (1969) classified the *Trichoderma* into nine species aggregates, further it was elaborated by Bissett (1984, 1991a-c, 1992), covering thirty-five species, their classification reflected the importance of microscopic characters for delimiting the *Trichoderma* species.

Keywords: Trichoderma, Characterization, Morphological, Molecular, Biological control, Plant growth promotion.

---

* E-mail: psharma032003@yahoo.co.in

Occurrence and Distribution

Trichoderma are widely distributed all over the world and they are cosmopolitan in nature (Domsch *et al.*, 1980). Their occurrence is found in most of the soils and other natural habitats, especially in those containing or consisting of organic matter. Individual species aggregates may be restricted in their geographic distribution (Daneilson and Davey, 1973a).

The first report on the occurrence of species of *Trichoderma* in soil was made by Oudemans and Koning (1902), who obtained in culture an isolate which the former author named in honour of his junior author as *Trichoderma koningii Oud.* With increasing interest in fungi living in soil, more and more soil isolates of *Trichoderma* were reported from various parts of the world and it soon became evident that this genus was one of the main components of the soil mycoflora. *Trichoderma* seems to be a secondary colonizer, as indicated by its frequent isolation from well-decomposed organic matter. *Trichoderma* is also found on root surfaces of various plants, on decaying bark, especially when it is damaged by other fungi and on sclerotia or other propagules of various fungi (Weindling *et al.*, 1934).

Danielson and Davey (1973b) reported certain strains of *T. hamatum* and *T. pseudokoningii* are adapted to conditions of excessive soil moisture and that of *T. viride* and *T. harzianum* is most commonly found in warm climatic regions, where as *T. hamatum* and *T. koningii* are widely distributed in areas of diverse climatic conditions. Widden and Scattolin (1988) reported the variation in competitiveness of *Trichoderma* species to colonize spruce litter under a range of temperature conditions.

Widden and Aribtol (1980) studied the seasonal effect on the distribution of *Trichoderma* species. Temperature, soil moisture and competition from different *Trichoderma* play a significant role in its distribution. *T. polysporum* was most abundant in the fall and winter, *T. viride* in the spring and fall and *T. koningii* was most abundant in the summer month. The iron content of the soil may also be an important determinant of the microsite preference of *Trichoderma*. A study made by Huang *et al.* (1983), showed that soil *Pseudomonas* limited the efficacy and establishment of *T. hamatum* in soil and rhizosphere containing little available iron.

Trichoderma species may be sensitive to some environmental pollution, as is indicated by the low rate of recovery of *T. viride* from coniferous forests that had been subjected to alkaline dust for a period of 25 years, the high pH (6.6) of the humus layer was blamed (Fritze and Baath, 1993). Though some studies have revealed the capacity of this microbe to work under abiotic stresses also.

Morphological Characterization

The literature revealed that the genus *Trichoderma* was introduced into mycological literature by Persoon in 1794 to accommodate four species of fungi namely *Trichoderma viride* Pers. Ex S. F. Gray, *Xylohypha nigrescens* (Pers. Ex Fr.) Mason, *Sporotrichum aureum* Pers. Ex Fr. and *Trichothecium roseum* (Pers.) Link ex S. F. Gray. These species now are commonly considered to be unrelated to one another.

After the first report given by Oudemans and Koning (1902) about the occurrence of *Trichoderma* in soil, several reports were reported from various parts of the world and it soon became evident that this genus, *Trichoderma* was one of the main components among the soil microorganisms. In 1916, Waksman described six strains of *Trichoderma*, which he distinguished chiefly on the macroscopic characters of the colonies and based on phialides size.

Abbott (1926) studied seven isolates of *Trichoderma,* which he considered to represent four species. He tabulated the characters, which he used in distinguishing these species, such as the difference in the macroscopic appearance *viz.*, floccose against tufted colonies, greenish, and yellowish or whitish to grayish colouration, the appearance of the colonies on different media etc. The tabulation made by Abbott was used by Gilman (1957) to construct a key to species *Trichoderma,* which was largely based on macroscopic characters of the colonies in the form of 'A Manual of Soil Fungi'.

Niethammer (1937), worked on the macroscopic characters of *T. lignorum* and the report was in contradiction to the key given by Gilman and Abbott. Thus there was no agreement on the correct application of a name to certain isolates and the delimitation of each species was obscure. Bisby (1939) examined a large number of isolates and collections of *Trichoderma* without finding any reliable characters to distinguish them; therefore he concluded that *Trichoderma* was a monotypic genus. Most of the mycologists accepted the conclusion given by Bisby and hence after this any culture with green spored was named as *T. viride* without any discrimination. While Gutter (1957) observed there was some difference in response of the isolates of *T. viride* to light. Similarly Moubashe (1963) reported the isolates of *T. viride* showed variation in their tolerance to duramycin contradicting monotypic nature of this genus. Thus there was confusion and no clarity in the species concept of *Trichoderma.*

Rifai (1969) and Bissett (1984, 1991a, 1991b, 1991c) who have done elaborate work in characterization and differentiation of *Trichoderma* species based on morphological characters emphasized the difficulties inherent in defining morphological species of *Trichoderma*. Characters useful for characterization and identification in other Hyphomycete genera frequently are not as useful in differentiating *Trichoderma* species, usually because of the narrow range of variation of the simplified morphology in *Trichoderma*, or because descriptive terms to describe variation in colour or pattern are not sufficiently precise to define differences between species. Nonetheless, careful morphological observations often are sufficient for identification of species and strains of *Trichoderma*, at least to the extent that taxa have been adequately differentiated morphologically and described in the existing literature. Identifications based on morphological characters involving both macroscopic and microscopic characters remain the primary method for identification and verification of species in *Trichoderma*. Rifai (1969) distinguished nine species aggregates based on microscopic characters. Samuels (1996) also provided detailed observations and commented on the utility of morphological characters involving microscopic features to define species in *Trichoderma*. However the importance of microscopic characters in delimiting the genus *Trichoderma* was emphasized by Harz as early as in 1871.

Similarly several studies on the biosystematics and morphology of the genus, and detailed studies on the relationship between the anamorph (*Trichoderma*) and the teleomorph (*Hypocrea*) have been made based on morphological characters (Cook and Baker, 1983; Doi, 1972; Domsch *et al.*, 1980).

Colonies

Colony characters can be distinctive and characteristic of a species. However, colony appearance is difficult to describe with sufficient precision to be very useful for identification. Gilman and Abbott (1927) generated a key for the species of *Trichoderma*, which was largely based on macroscopic characters of the colonies of the colony, based, on which they identified *T. nigrovirens* Goggard (1913) with *T. lignorum*, because the descriptions of the colonies of the two species were similar.

Gutter (1957) described in majority of the species the conidial areas were formed in distinct and characteristic ring-like zones which had been shown to be mediated by light. When the colonies get old the zonation often becomes less obvious because new conidiophores may be formed outside the primary conidial areas. In isolates with floccose colonies zonation can be detected at very young stages only.

The colour of the colonies was considered as one of the important macroscopic character in the identification of *Trichoderma*. Rifai (1965) reported probably the pH of media had some effect in colour of the colonies. More over the presence of sterile hyphal elongations over the tufts of conidiophores of *T. hamatum* made colonies appear characteristically whitish or grayish-green. Bissett (1984) revised the genus *Trichoderma*. I. Section *Longibrachiatum* sect., by differentiating it from other species of *Trichoderma* by the growth of the colony and the colour produced.

Mycelium

The characterization study of nine species of *Trichoderma* done by Rifai (1969) showed that the mycelium of most of the species was composed of hyaline, septate, much branched and smooth-walled hyphae.

Lejeune *et al.* (1995) studied the microscopic morphology of *Trichoderma reesei* QM 9414, growing in submerged culture, by image analysis. The morphology was characterized by the total hyphal length, the total number of tips, the number of actively growing tips, and the length of the main hyphae. To describe the growth of a single mycelium a simple model was set-up with some features like, saturation type kinetics for the tip extension of the individual branches within the mycelium; and random branching with a frequency function, which was proportional to the total hyphal length. The model was used to simulate a population of mycelia, where spore germination was described with a log-normal distribution. From the simulation of the population, the average properties of the mycelia, *e.g.*, the average total hyphal length, were calculated, and by fitting the model to experimental data the model parameters were estimated. Finally, the distribution function with respect to the mycelia properties *i.e.* number of tips and total hyphal length was calculated, and it corresponded well with the experimental determination of the distribution function.

Conidiophores

The pattern of conidiophore branching and aggregation of conidiophores into fascicles and pustules were useful for identification of strains of *Trichoderma* to sections and species aggregates. Compact pustules were characteristic of many species in section Pachybasium, although also of numerous strains in other sections. Conidiophore branching can be regularly verticillate or more irregular. Branches can be broad and straight or relatively narrow and flexuous. The conidiophore apex in some species in section Pachybasium characteristically ends in a sterile elongation which may be straight, undulate or coiled. Sparsely branched conidiophores were studied by Bissett (1991b) for the variability in strains of the *Trichoderma*, based on which he could characterize and differentiate them.

Rifai (1969) found the conidiophores of *Trichoderma* were conical or pyramidal in outline and length of the side branches of the main branch of each conidiophore increased regularly and proportionately with the distance from the apex of the latter. In general, it was stated that each side branch was a miniature of its bearer, because it may also put out further smaller side branches and so on. He compared the conidiophores of *T. hamatum* and *T. polysporum* and predicted the main branches of the conidiophores were relatively long but they had short and thick side branches. In most isolates of these two species aggregates the apex of each main branch was typically terminated by a long, whip like, sterile hyphal elongation. In a number of isolates seen, however, these sterile hyphal elongations may be completely reduced and terminated by a phialide, or they become fertile as in the other species of *Trichoderma*, so that superficially they may resemble the conidiophores of the genus *Botryosporium* Corda. He also found the species *T. longibrachiatum* Rifai had long main branches of conidiophores with short side branches, but unlike *T. hamatum* and *T. polysporum* the structure of its conidiophores was much simpler because they produced very few side branches which in turn was very simply constructed. The conidiophores of *T. viride, T. koningii* and its related species generally were dendroid like those of some species of the genus *Verticillium*.

Phialides

It has been found that with in certain limits the structure of conidiophores and the size, shape and disposition of phialides of *Trichoderma* are quite and reliable characters for delimiting the species as well as for determining its relationships. The importance of these characters for taxonomic purposes however should not be over-emphasized, but in any attempt to establish a natural and workable system of classification of *Trichoderma* in the future, a higher evaluation should be accorded to them rather than to the other dubious characters such as the macroscopic appearance of the colony (Rifai and Webster, 1966).

The general appearance of the phialides was given by Rifai (1969). Most of the phialides were flask or nine-pin-shaped, slightly narrower at the base than above the middle and from there attenuated towards the apex into a narrow conical or subcylindrical neck. They stood at a wide angles and bend slightly at the middle towards the apex of the bearer, hence from side view they gave a horn shape appearance. The phialides were either arranged in irregular whorls of upto five around the end of the penultimate cells of the branches, or they arose pleurogenously along the branch either singly and alternate or in opposite pairs. In most of the cases the terminal phialides were slightly longer than those that arose beneath them.

Phialide shape is characteristic of the section; phialides were characteristically short and plump in section Pachybasium, whereas in section Longibrachiatum they were elongate and lageniform to nearly cylindrical. Terminal phialides in most species tend to be more elongate and narrower and frequently more or less subulate. Subterminal cells of the conidiophore may produce conidia through a short lateral neck, thus intercalary phialides or what Gams (1971) called aphanophialides; these were rather commonly seen in *Trichoderma* sect. *longibrachiatum*.

Webster (1966) reported the phialides of *T. koningii* appear to be more plump, shorter and subtended at a wider angle than those of *Verticillium*, further more the verticillate arrangement of the phialides of *Trichoderma* is not as strict or as regular as in most species of *Verticillium*, because in the former genus the phialides do not normally arise at the same level. Within *Trichoderma*, the *Pachybasium*-types can be distinguished from *Trichoderma sensu stricto* in having white conidia produced from doliform phialides that are held in botryose clusters, and often with a sterile or terminally fertile extension (Samuels, 1996).

Phialospores

The overall range of variation in phialospore dimensions in *Trichoderma* may not be great; however related species can often be differentiated by slight but consistent differences in size. Bisby (1939) asserted that this character has little or no diagnostic value at the specific level. While Rifai (1969) came out with an explanation that a correlation with other morphological characters would make the phialospore reliable. He observed most of the isolates of *T. viride* had subglobose phialospores, but other isolates of this species produced constantly obovoid or ellipsoidal phialospores. Phialospore shape varies from globose, to ellipsoidal, obovoid, or short-cylindrical, with the basal end more or less tapering and truncate. The phialospore surface appeared smooth in most species in light microscope observations, although many species with apparently smooth-appearing phialospore are delicately ornamented when examined by SEM. Phialospore can also be variously roughened or verrucose in the *T. viride* aggregate, and conidia can have wing-like or bullate projections of the outer wall in two species–*T. saturnisporum* and *T. ghanense*.

Oudemans and Koning (1902) used the size of the phialospore accumulation at the tip of the phialide, the conidial head, as taxonomic evidence for separating species of *Trichoderma*. Doi *et al.* (1987) proposed the new sect. *Saturnisporum* for two species, *T. ghanense* Y. Doi, Y. Abe and Sugiy. and *T. saturnisporum* Hammill, based on the conspicuously warted, sometimes alate, phialospore.

Chlamydospores

An important aspect of sporulation almost completely disregarded in the last 50 years is the ability of *Trichoderma* to produce chlamydospores. Although chlamydospores were routinely mentioned in taxonomic papers, little had been reported on the formation and its importance (Domsch *et al.*, 1980 and Rifai, 1969). Chlamydospores are common in many species, although they tend to be uniformly globose or ellipsoidal, terminal and intercalary, smooth-walled, colourless, yellowish or greenish, and 6-15 µm diameters in most species. Some of the research had demonstrated the formation of chlamydospores by *T. hamatum*, *T. harzianum* and *T. viride* in both liquid and solid fermentation media, in sterile soil and soil extracts, and in natural plant debris and amended natural soil (Lewis and Papavizas, 1984). While these chlamydospores were characteristic of *Trichoderma* and *Hypocrea*, their form is not diagnostic of species but their presence may be (Samuels *et al.*, 1990; Bissett 1991b).

Pigments

Diffusible pigments also can be characteristic, although the colour of such pigments does not vary a great deal in *Trichoderma* (Rifai, 1965; Samuels, 1994). Strains referable to section Longibrachiatum typically have conspicuous bright greenish yellow pigments, at least when first isolated. Dull yellowish pigments were common in many species, but not very distinctive. Some species were best characterized by complete lack of pigment in reverse, whereas reddish pigments occur in reverse in a few isolates. Conidial pigments also were characteristic, varying from colourless (white in mass), to various green shades, or less often grey or brown. In some species mature conidia appear dark green in the microscopic mount, in others only pale (Rifai, 1965).

Growth

The growth rates in culture can be useful to distinguish otherwise similar species. Rifai (1969) established the growth rate of colonies of *T. hamatum* and *T. polysporum* aggregates is generally slightly slower than the other species aggregates. Differences in the linear expansion of colonies and in temperature optima can be found between isolates of different species (Rifai, 1964; Rifai and Webster, 1996).

Odour

Indistinct mouldy or musty odours are commonly produced by different strains of *Trichoderma*. Characteristic aromatic odours resembling coconut are produced commonly by strains of *T. viride*, and sometimes also by *T. atroviride*. Thus the taxonomic scheme most widely used is based on the revision of Rifai (1969), who had done a commendable work in classifying *Trichoderma* based on various morphological characters by adopting the "species aggregate" concept. He denoted the species of *Trichoderma* as species aggregates because they are made up of more then one genetic entity that cannot be distinguished with the presently used morphological characters. He recognized nine species aggregates which consist of three classical species namely *T. hamatum* (Bon.) Bain., *T. koningii* Oud., and *T. viride* Pers. Ex S.F.Gray; another one for which a new combination was proposed, *T. polysporum* (Link ex Pers.) Rifai; and five new species *viz.*, *T. longibrachiatum* Rifai, *T. piluliferum* Webster and Rifai, *T. aureoviride* Rifai, *T. harzianum* Rifai and *T. pseudokoningii* Rifai. Similarly Bissett

(1984, 1991a, 1991b, 1991c, 1992) like Rifai, adopted a morphological approach to taxonomy of *Trichoderma,* where his work perceived a noncontinuous morphological characteristics of species. However the process of classification and re-classification is still under evolution with contribution of more workers and discovery of new species and anamorph. This can be inferred by work of Samuels *et al.* (1999) who reported a new species *T. asperellum* sp. nov. which was distinguished from *T. viride* by finer conidial ornamentation, slightly ovoidal conidia, a faster growth rate, mostly paired branches, ampulliform phialides and the consistent presence of chlamydospores.

Molecular Characterization

The initial efforts for macromolecular characterization of *Trichoderma* strains and species were undertaken not by professional taxonomists, but by frustrated users of the existing taxonomy. Morphology alone in *Trichoderma* had not led to a satisfactory taxonomy or, at any rate, to a taxonomy that had been useful to many users. Characters derived from nucleic acids and enzymes were attractive because, with cladistic analysis, they seemed to offer the possibility of greater objectivity than do traditionally observed and analyzed data. Macromolecular analysis started with the use of biochemical tools such as enzyme profiling and polymorphism, antibodies production, antibiotic production, secondary metabolites etc. This later evolved into the techniques which has the ability to look at nucleic acid directly. In recent days these molecular characters have shown greater promise in defining fungal species and providing diagnostic characters, since molecular characters have advantage over biochemical characters as they do not change in response to environmental conditions and they are independent of any morphological state of the culture.

Also macromolecular analysis based on enzymes and nucleic acid have shown that certain aggregate of species are phylogenetically based (*e.g.* Okuda *et al.*, 1982; Leuchtmann *et al.*, 1996) at the same time confirming genetic diversity of the individual aggregates (Zamir and Chet, 1985; Satsz *et al.*, 1989; Meyer 1991; Meyer *et al.*, 1992; Fujiwara and Okuda, 1994; Muthumeenakshi *et al.*, 1994; Samuels *et al.*, 1994; Zimand *et al.*, 1994; Leuchtmann *et al.*, 1996).

Stasz *et al.* (1989) evaluated five aggregates species of *Trichoderma* using enzyme polymorphisms. While they concluded that morphological species are not characterized by specific alleles at single loci, or specific pattern of alleles at multiple loci, they demonstrated what they called the 'core groups' of morphological species. Despite individual strains that gave widely divergent allozymes patterns, the core group coincided well with Rifai's aggregate species. These results indicate that, in part, there is a genetic basis for the morphological aggregate species.

Latha *et al.* (2002) reported their work from BARC, where they emphasized the need for re-classification of the cultures available at MTCC and ITCC in India based on molecular methods. Samuels *et al.* (2002) identified the significant distinguishing phenotypic and genotypic character differences of various *Trichoderma* species associated with green mold epidemic of commercially grown *Agaricus bisporus*.

Thus an integrated approach involving as many characters as possible based on morphology and molecular characters would provide a better insight to the taxonomic status of *Trichoderma* spp.

Random Amplified Polymorphic DNA (RAPD)

RAPD [random amplified polymorphic DNA] procedure was used for the identification of *Trichoderma* strains. The arbitrary 10 oligonucleotide primers were used to distinguish the strains of *Trichoderma.* Of the strains identified as *T. harzianum,* 10 exhibited similar patterns. Among the strains of *T. viride,* only 3 pairs yielded similar patterns. None of the strains identified as *T. hamatum* showed any similarities. They found the isolate T-39, which is used commercially as a biological control agent against *Botrytis cinerea,* was distinguishable by this procedure (Zimand *et al.,* 1994). Fujimori and Okuda (1994) successfully used RAPD to identify duplicate *Trichoderma* strains in microbial screening.

Schlick *et al.* (1994) analyzed different patent strains and gamma-ray induced mutants of *T. harzianum* by DNA fingerprinting and PCR fingerprinting (RAPD). They were able to obtain different and discriminate fingerprint patterns for all strains and mutants investigated. Irradiation of fungi led to mutations which resulted in new fingerprint patterns. Consequently irradiation-induced mutants can be clearly distinguished from the original wild-type isolates by genomic fingerprinting which is of importance for the patent protection of fungal strains. Kuhls *et al.* (1995) used RAPD's of *T. reseei* to determine that *T. todica* (ATCC 36936), an unpublished strain patented for the production of antiviral antibiotics, is actually *T. parceramosum,* a member of Sect. *Longibrachiatum* Bissett.

Eleven strains of *T. viride,* 2 strains of the putative teleomorph *Hypocrea rufa* and 9 of several other *Trichoderma* spp. were characterized by random polymorphic DNA amplification (RAPD) fingerprinting and screened for their ability to antagonize growth of European strains of the chestnut blight causing fungus *Cryphonectria parasitica,* using a dual-culture assay. They found the best strains were in the species *T. harzianum, T. parceramosum,* a distinguishable subgroup of *T. viride* and a not named *Trichoderma* spp. Their successful application of these strains against chestnut blight in vivo was demonstrated (Arisan-Atac 1995).

Muthumeenakshi and Mills (1995) worked on DNA based techniques, including RFLP and RAPD analyses, to identify isolates of *Mycogone perniciosa, T. harzianum and Verticillium fungicola* var. *fungicola* isolates from *A.bisporus* or mushroom compost. Aggressive isolates of *T. harzianum* could be distinguished from morphologically similar non-aggressive isolates. A high level of genetic similarity was apparent within the aggressive isolate group which may indicate that outbreaks within the UK have originated from a single source. *M.perniciosa* could be separated from *M.rosae,* and analysis of *V.fungicola* isolates revealed distinct genetic pools that appear to be geographically isolated.

Turoczi *et al.* (1996) assessed the genetic differences in *T. hamatum, T. harzianum* and *T. viride* by RAPD. The RAPD patterns of *T. viride* strains were highly variable, isolates of *T. harzianum* were more uniform and *T. hamatum* showed remarkable intraspecific divergence.

Randomly amplified polymorphic DNA (RAPD) analysis and the PCR assay were used in combination with dilution plating on a semiselective medium to detect and enumerate propagules of *T. hamatum* 382, a biocontrol agent utilized in compost-amended mixes. Distinct and reproducible fingerprints were obtained upon amplification of purified genomic DNA of *T. hamatum* 382 with the random primers OPE-16, OPH-19, and OPH-20. Three amplified DNA fragments of 0.35 (OPE-16(0.35)), 0.6 (OPH.19 (0.6)), and 0.65 (OPH-20(0.65)) kb were diagnostic for *T. hamatum* 382, clearly distinguishing it from 53 isolates of four other *Trichoderma* spp. tested. Some isolates of *T. hamatum* shared these low-molecular-weight fragments with *T. hamatum* 382. However, RAPD analysis of isolates of *T. hamatum* with all three random primers used in consecutive PCR tests distinguished *T. hamatum* 382 from other isolates of *T. hamatum*. These three RAPD amplicons were cloned and sequenced, and pairs of oligonucleotide primers for each cloned fragment were designed. Use of the primers in the PCR assay resulted in the amplification of DNA fragments of the same size as the cloned RAPD fragments from genomic DNA of *T. hamatum* 382. A combination of dilution plating on a semiselective medium for *Trichoderma* spp. and PCR, with the RAPD primers OPH-19, OPE-16, and OPH-20 or the three sequence characterized primers, was used successfully to verify the presence of *T. hamatum* 382 propagules in nine different soil, compost, and potting mix samples. All 23 *Trichoderma* isolates recovered on semiselective medium from commercial potting mixes fortified with *T. hamatum* 382 were identified as *T. hamatum* 382, whereas 274 *Trichoderma* isolates recovered from the other nine samples were negative in the PCR assay. Thus, this highly specific combination of techniques allowed detection and enumeration of propagules of *T. hamatum* 382 in fortified compost-amended potting mixes. Sequence-characterized amplified region markers also facilitated the development of a very simple procedure to amplify DNA of *T. hamatum* 382 directly from fortified compost-amended potting mixes. (Abbasi, 1999).

Chen *et al.* (1999) used randomly amplified polymorphic DNA (RAPD)-PCR to estimate genetic variation among isolates of *Trichoderma* associated with green mold on the cultivated mushroom *Agaricas bisporus*. Of 83 isolates examined, 66 were sampled during the recent green mold epidemic, while the remaining 17 isolates were collected just prior to the epidemic and date back to the 1950s. *T. harzianum* biotype 4 was identified by RAPD analysis as the cause of almost 90 per cent of the epidemic-related episodes of green mold occurring in the major commercial mushroom-growing region in North America. They also found Biotype 4 was more closely allied to *T. harzianum* biotype 2, the predominant pathogenic genotype in Europe, than to the less pathogenic biotype 1 and *T. atroviride* (formerly *T. harzianum* biotype 3). No variation in the RAPD patterns was observed among the isolates within biotype 2 or 4, suggesting that the two pathogenic biotypes were populations containing single clones. Considerable genetic variation, however, was noted among isolates of biotype 1 and *T. atroviride* from Europe. Biotype 4 was not represented by the preepidemic isolates of *Trichoderma* as determined by RAPD markers and PCR amplification of an arbitrary DNA sequence unique to the genomes of biotypes 2 and 4. Goes *et al.* (2002) reported the genetic variability of *Trichoderma* isolates using the RAPD technique and their antagonistic potential against *R. solani*, as well as the relationship between their antagonistic capacities.

A duplex-PCR bioassay to detect a *T. virens* biocontrol isolate in non-sterile soil was done by development of sequence-characterized amplified region (SCAR) marker specific for the *T. virens* biocontrol isolate GV4. The marker was developed from a RAPD-PCR amplification product unique to isolate GV4. When used as a hybridisation probe in Southern blot analysis, it hybridised to the DNA of the species *T. virens* alone and not to that of other *Trichoderma* species or closely related genera *Gliocladium* and *Verticillium*. Primers designed from the sequence of the RAPD marker produced a diagnostic amplification product of 346 bp for GV4 alone, distinguishing it from all other test isolates. The duplex soil PCR assay detected GV4 down to a concentration of 10 spores g^{-1} soil in non-sterile agricultural field soil proving to be the best and most sensitive diagnostic tool. This study was the first to report the use of a duplex-PCR diagnostic bioassay for a species within the *Hypocrea/Trichoderma* genus (Dodd *et al.*, 2004).

Thus it is very clear from the above references that RAPD technique can be employed for reviewing the classification of the genus, *Trichoderma*; to differentiate the various species in this genus; to distinguish variability among the biocontrol agents; to identify the most potential biocontrol strain and development of diagnostic kit for such superior strains.

Inter Simple Sequence Repeats

Inter-simple sequence repeats (ISSRs) have been used as another method to characterize genetic variation within fungi. This technique was first demonstrated as a technique for measuring genetic diversity in plants and animals by Zietkiewicz *et al.* (1994) and then shown by Hantula *et al.* (1996) to be able to generate DNA markers in a variety of fungi as well. This technique using primers containing Microsatellite sequences and degenerate anchors at the 5′end, is highly reproducible, allows detection of interspecific and intraspecific DNA polymorphisms, and is applicable to a diverse range of fungal species (Hanulta *et al.*, 1996). His work also supported the hypothesis that there would be a greater likelihood of finding polymorphisms with ISSRs than with most other techniques, including RAPDs, because the evolutionary rate within ISSRs is considerably higher than in most other types of DNA (Charlesworth *et al.*, 1994). However, this technique has not been adopted in variability study of *Trichoderma* spp. Therefore the significant report using ISSR in variability study of other fungi are discussed below.

Vandenkoornhuyse *et al.* (2001) evaluated the genetic diversity of spores of two indigenous species of *Glomus* isolated from three soils of a long-term field experiment amended by different quantities of sewage sludge. Three populations of spores of *G. claroideum* (W2537) and three populations of spores of *Glomus* DAOM 225952 (W2538) were analyzed using a microsatellite primer and aliquots of genomic DNA were obtained from single spores (Inter Simple Sequence Repeat (ISSR) fingerprints).

39 polymorphic bands were found for *G. claroideum*, and 43 in *Glomus* DAOM 225952 and found the intraspecific diversity was high, ranging from 22 to 33 different electrophoretic types for *G. claroideum*, and 15-27 for *Glomus* DAOM 225952 depending on the population.

A study was aimed at comparing eleven biocontrol agents, nine isolates of *P. oligandrum* with two isolates of *P. paroecandrum* and *P. periplocum*, in tomato root colonization, production of molecules involved in plant resistance induction and plant growth enhancement. This was done through the testing of biochemical, morphological and molecular (ISSR) markers. Based on the distinctive features shown by the isolates of *P. oligandrum*, the molecular analysis revealed that Inter Simple Sequence Repeat (ISSR)-PCR fingerprinting showed great differences between *P. oligandrum* isolates and *P. periplocum* and *P. paroecandrum* (Floch *et al.*, 2003).

Rodrigues *et al.* (2004) established the phenotypic and genotypic variability of 18 endophytic *Guignardia* strains from different host plants (Anacardiaceae: *Anacardium giganteum, Myracrodruon urundeuva, Spondias mombin*; Apocynaceae: *Aspidosperma polyneuron*; Ericaceae: *Rhododendron* spp.; Fabaceae: *Bowdichia nitida*; Leguminosae: *Cassia occidentalis*; Rutaceae: *Citrus aurantium*) growing in different sites in Brazil was assessed by means of morphometric measurements and inter-single-sequence-repeat-anchored polymerase chain reaction (ISSR-PCR) amplifications of the DNA. Morphology of conidia and ascospores and growth rates of the Brazilian isolates corresponded well with those of *G.mangiferae*. Multiple correspondence analyses (MCA) of the ISSR-PCR data yielded three groups of strains, which did, however, not correspond either to the host or to the geographic origin. The same individual tree hosted genotypically different strains indicating multiple infections.

Little was known regarding the population biology or origin of the pathogen *Thielaviopsis basicola*, a soil-borne fungal pathogen affecting many important agricultural crops. Geldenhuis *et al.* (2004) studied 14 primer pairs using ISSR-PCR, seven of which resulted in the amplification of single polymorphic fragments in *T. basicola*. These primers would enable further studies on this economically important pathogen, and would result in an enhanced understanding of its population structure in different parts of the world.

Internal Transcribed Spacer (ITS) Region

The ITS region is perhaps the most widely sequenced DNA region in fungi. It has typically been most useful for molecular systematic at species level, and even within species (*e.g.*, to identify geographic races). Given the variability within *Trichoderma*, it is difficult to use the names assigned by Rifai (1969) or Bisset (1991) for biological control agents, since these names involve species concepts that are too broad for practical situations. In a practical biological control situation differentiation is required to define populations within a single species name. A particular strain of *T. harzianum* may be a good or bad biocontrol agent depending on the intended target and the function required. It will therefore be necessary to select the most efficient strain for each individual pathosystem. Grondona *et al.* (1997) suggested the utility of

ITS region to distinguish the functional groups within *T. harzianum* and to correlate with its biological activity so as to provide characteristics that could be used to select isolates as biocontrol agents.

Characterization of 16 biocontrol strains, previously identified as "*Trichoderma harzianum*" Rifai and one biocontrol strain recognized as *T. viride*, was carried out using several molecular techniques. A certain degree of polymorphism was detected in hybridizations using a probe of mitochondrial DNA. Sequencing of internal transcribed spacers 1 and 2 (ITS1 and ITS2) revealed three different ITS lengths and four different sequence types. Phylogenetic analysis based on ITS1 sequences, including type strains of different species, clustered the 17 biocontrol strains into four groups: *T. harzianum-T. inhamatum* complex, *T. longibrachiatum*, *T. asperellum*, and *T. atroviride-T. koningii* complex. ITS2 sequences were also useful for locating the biocontrol strains in *T. atroviride* within the complex *T. atroviride-T. koningii*. This work was done by Hermosa *et al.* (2000) to explore the synergistic effects expressed by different genotypes of *Trichoderma* strains identified as common biological control agents since it is very important to for their practical use in agriculture. Ospina–Giraldo (1998) compared the nucleotide sequences in regions of rDNA of the four biotypes of *T. harzianum* associated with mushroom production and with other species of *Trichoderma*.

Kubicek *et al.* (2000) reported the biodiversity of *Trichoderma* by collecting samples from central Russian, the urals, Siberia and Himalayan Mountains and revealed the presence of several isolates with ITS sequences identical to those of the ex type strain or established sequence variants of species of section *Trichoderma*. Dodd *et al.* (2000) reported a study on the examination of *Trichoderma* by the use of phylogenetic analysis derived from ribosomal DNA sequence data. *Trichoderma* strains that had warted conidia are traditionally identified as *T. viride*, the type species of *Trichoderma*. However, two morphologically distinct types of conidial warts (I and II) was found. Because each type corresponds to a unique mitochondrial DNA pattern, it was questioned whether *T. viride* comprises more than one species. Combined molecular data (sequences of the internal transcribed spacer 1 [ITS-1] and ITS-2 regions and of part of the 28S rRNA gene along with results of restriction fragment length polymorphism analysis of the endochitinase gene and PCR fingerprinting), morphology, physiology, and colony characteristics distinguish type I and type II as different species. Type I corresponds to "true" *T. viride*, the anamorph of *Hypocrea rufa*. Type II represents a new species, *T. asperellum*, which is, in terms of molecular characteristics, close to the neotype of *T. hamatum* (Lieckfeldt *et al.*, 1999). Similarly, Lieckfeldt and Seifert (2000) made a pioneering study where they reported the evaluation of the use of ITS sequences in the taxonomy of the Hypocreales.

Kulling *et al.* (2002) established a species phylogeny of the gene *Trichoderma*, based on the sequence analysis of multiple independent loci (ITS-1, ITS-2, 28s-rDNA) and the small mitochondrial rDNA submit and fragments from two single-copi gene loci, *viz.*, translation elongation factor 1 and endochitinase 42.

Lee and Hseu (2002) presented the adoption of multiple approaches including UP-PCR and the analysis of ITS and 5.8s sequence analysis, and DNA association for the confirmation and identification of certain characters that aid in the clarification of species within *Trichoderma* sect. *Pachybasium*.

The Teleomorphs

Teleomorphs of *Trichoderma* belong to the genus Hypocrea of ascomycete. These are characterized by the formation of fleshy, stromata in shades of light or dark brown, yellow or orange. Typically the stroma is discoidal to pulvinate and limited in extent but stromata of some species are effused, sometimes covering extensive areas. Stromata of some species (Podostroma) are clavate or turbinate. Perithecia are completely immersed. Ascospores are bicellular but disarticulate at the septum early in development into 16 part-ascospores so that the ascus appears to contain 16 ascospores. Ascospores are hyaline or green and typically spinulose. More than 200 species of Hypocrea have been described but only few have been grown in pure culture and fewer have been redescribed in modern terms.

In 1902, the specific link between *T. viride* and *Hypocrea rufa* was an accepted fact (Smith, 1902), thus establishing the generic link between *Hypocrea* and *Trichoderma.* Berkeley, as early as 1860, suspected a link between *T. viride* and some unnamed ascomycete when he enigmatically noted that the *Trichoderma* species was 'probably not autonomous'. Tulasne (1860) proved that *T. viride* and the ascomycete *Hypocrea rufa* are expressions of one life cycle. The brothers Tulasne (Tulasne and Tulasne, 1865) illustrated *T. viride* was remarkably accurate in the representation of the phialides and their disposition on the conidiophore.

In 1964 the anamorph of true *H. gelatinosa* was recognized to be a *Trichoderma* with slimy conidia (Webster, 1964) similar to *T. virens*. Ten years later Webster and Rifai published a series of papers detailing connections between *Hypocrea* species and their *Trichoderma* anamorphs (Webster, 1964; Rifai and Webster, 1966; Webster and Rifai, 1968). In addition to *H. gelatinosa*, this series established the current holomorphic concept of the common species *H. pulvinata* and *H. rufa*, and culminated in the publication of Rifai's revision of *Trichoderma* in 1969.

Hypocrea perithecia do not form in cultures of *Trichoderma* strains isolated from natural substrates, however, *Trichoderma* anamorphs occasionally develop in cultures derived from *Hypocrea* ascospores. One can question whether *Trichoderma* strains encountered in the absence of a teleomorph are parts of *Hypocrea* life cycles. Some of the aggregate species described by Rifai (1969) are *Hypocrea* in the sense that they are based on ascospore isolations from *Hypocrea* species.

The genus *Trichoderma* is defined to include anamorphs of *Hypocrea,* previously placed in *Gliocladium* and *Verticillium* having elongate phialides and irregularly branched conidiophores. Bissett made a proposal for sectional classification for *Trichoderma* recognizing the following 5 sections: section *Trichoderma,* section *Longibrachiatum,* section *Saturnisporum,* section *Pachybasium* stat. nov., and section *Hypocreanum* sect. nov. *Trichoderma lactea* sp. nov. was described, typifying section *Hypocreanum.* These sections were classified based on certain phenotypic characters like conidiophores, phialides etc (Bissett, 1991).

However, given the difficulty of understanding what constitutes a species of *Trichoderma*, it is still unclear whether *Trichoderma* strains encountered in nature are actually *Hypocrea* anamorphs. Kuhls *et al.* (1996) reported *T. reseei*, while apparently incapable of sexual reproduction and culturally distinct from *H. jecorina*, behaves exactly like any strain of *H. jecorina*.

Samuels *et al.* (1994) examined the morphological, cultural and isoenzyme characters to determine variation in the *Trichoderma* anamorph of *H. schweinitzii*, a presumably unequivocal species of *Trichoderma*, and to assess whether the anamorph of *H. schweinitzii* can be assigned to *T. longibrachiatum*, *T. reesei* or *T. pseudokoningii*. He predicted that *H. schweinitzii* comprises at least 3 distinct and taxonomically separable holomorph taxa that coincide with the geographical origin of the collections. The name *H.schweinitzii* can be applied to Northern Hemisphere and *H. jecorina* to tropical American collections; while taxonomically distinct, no name was given to a Chinese collection. More over none of the named *Trichoderma* species coincided with any of the *Hypocrea* species studied. It is concluded that *Trichoderma* species in general can be narrowly defined but that morphology alone might not suffice to allow the identification of species. The synonymy of *T. reesei* under *T. longibrachiatum* is not supported by morphological and cultural observations and by isoenzyme data and is therefore not accepted.

Kuhls *et al.* (1996) studied the relationship of the important cellulase producing asexual fungus *T. reesei* to its putative teleomorphic (sexual) ancestor *Hypocrea jecorina* and other species of the *Trichoderma* sect. *Longibrachiatum* by PCR-fingerprinting and sequence analyses of the nuclear ribosomal DNA region containing the internal transcribed spacers (ITS-1 and ITS-2) and the 5.8S rRNA gene. The differences in the corresponding ITS sequences allowed a grouping of anamorphic (asexual) species of *Trichoderma* sect. *Longibrachiatum* into *Trichoderma longibrachiatum*, *Trichoderma pseudokoningii*, and *Trichoderma reesei*. The sexual species *Hypocrea schweinitzii* and *H. jecorina* were also clearly separated from each other. *H. jecorina* and *T. reesei* exhibited identical sequences, suggesting close relatedness or even species identity. Intraspecific and interspecific variation in the PCR-fingerprinting patterns supported the differentiation of species based on ITS sequences, the grouping of the strains, and the assignment of these strains to individual species. The variations between *T. reesei* and *H. jecorina* were at the same order of magnitude as found between all strains of *H. jecorina*, but much lower than the observed interspecific variations. Identical ITS sequences and the high similarity of PCR-fingerprinting patterns indicate a very close relationship between *T. reesei* and *H. jecorina*, whereas differences of the ITS sequences and the PCR-fingerprinting patterns show a clear phylogenetic distance between *T. reesei*/*H. jecorina* and *T. longibrachiatum*. *T. reesei* is considered to be an asexual, clonal line derived from a population of the tropical ascomycete *H. jecorina*.

Turner *et al.* (1997) reported a comprehensive analysis of the phenotypic variation in RAPD patterns and geographic bias of anamorphs and teleomorphs belonging to *Trichoderma* sect. *Longibrachiatum*. They used (M13, [GTG]5 and [GACA]4) and random (V5) primers to analyse 145 isolates characteristic of *Trichoderma* sect. *Longibrachiatum* and *Hypocrea* species with anamorphs referable to that section, and to identify strains with named species of sect. *Longibrachiatum*. Ex-type strains of *T. longibrachiatum*,

T. pseudokoningii, T. citrinoviride, T. parceramosum and *T reesi* were also analysed, and the similarity coefficient of RAPD characters were used as criteria for the alignment of strains to individual species. The ex-type strains of *T. saturnisporum* and *T. ghanense* exhibited the same interspecific similarity index as the other ex-type strains of *Trichoderma* sect. *Longibrachiatum,* clustered well within it and were therefore included into this section. Of the isolates, 103 were identified as members of *Trichoderma* sect. *Longibrachiatum.* Most of the anamorphic isolates were aligned with *T. citrinoviride* or *T. longibrachiatum,* and only a few strains of *T. saturnisporum* and *T. parceramosum* occurred. No naturally occurring anamorphs with homology to the ex-type strains of *T. reesei* and *T. pseudokoningii* occurred. The geographic ranges of *T. longibrachiatum* and *T. citrinoviride* overlapped, however *T. longibrachiatum* occurred in Africa and India but not in Southeast Asia, and the reverse was true of *T. citrinoviride.* The African strains of *T. longibrachiatum* were equally distant from the ex-type strains of *T. longibrachiatum, T. parceramosum* and *T. saturnisporum.* The type strain of *T. ghanense* was identified as belonging to this group. *T. pseudokoningii,* the anamorph of *Hypocrea schweinitzii,* occurred exclusively in eastern Australia and New Zealand. Attempts to identify potential teleomorphs of the other anamorphs revealed another subgroup of *H. schweinitzii,* occurring in temperate climates in Europe and the USA, as homologous to *T. citrinoviride.* No *Hypocrea* strains were identified as teleomorphs of *T. longibrachiatum, T. parceramosum* or *T. saturnisporum.*

Chaverri *et al.* (2001) reported the species *Hypocrea virens* sp. nov. as the teleomorph of *T. virens* based on the sequence analysis of the ITS regions. Two years later Chaverri *et al.* (2003) reported the descriptions of the different species of *Trichoderma* based on the elongation of conidiophore precisely through the addition of temperature/growth information and the *Hypocrea* species that have such anamorphs are described or redescribed. And also presented the phylogenetic relationships through partial sequences of the protein coding RNA Polymerize II submit (RPB-2) and translation elongation factor (EF-1α).

Dodd *et al.* (2003) examined and compared cultural, morphological, and sequence data of cultures derived from the specimens with cultures of *T. atroviride* isolated directly from nature. Sequences from the ITS regions of the rDNA complex and partial sequence of the translation-elongation factor (EF-1α) gene helped to confirm the link between *T. atroviride* and its *Hypocrea* teleomorphs.

Potential for Biological C

Trichoderma spp. is fungi that are present in nearly all soils and other diverse habitats. In soil, they frequently are the most prevalent culturable fungi. They are favored by the presence of high levels of plant roots, which they colonize readily. Some strains are highly rhizosphere competent, *i.e.,* able to colonize and grow on roots as they develop. The most strongly rhizosphere competent strains can be added to soil or seeds by any method. Once they come into contact with roots, they colonize the root surface or cortex, depending on the strain. Thus, if added as a seed treatment, the best strains will colonize root surfaces even when roots a meter or more below the soil surface and they can persist at useful numbers up to 18 months after application. However, most strains lack this ability.

In addition to colonizing roots, *Trichoderma* spp. attack, parasitize and otherwise gain nutrition from other fungi. Since *Trichoderma* spp. grow and proliferate best when there are abundant healthy roots, they have evolved numerous mechanisms for both attack of other fungi and for enhancing plant and root growth. Several new general methods for both biocontrol and for causing enhancement of plant growth have recently been demonstrated and it is now clear that there must be hundreds of separate genes and gene products involved in these processes. A list of mechanisms follows.

- ☆ Mycoparasitism
- ☆ Antibiosis
- ☆ Competition for nutrients or space
- ☆ Tolerance to stress through enhanced root and plant development
- ☆ Solubilization and sequestration of inorganic nutrients
- ☆ Induced resistance

The soil is a complex system where processes have direct influence on crop nutrition and plant health. Unfortunately, most of the agricultural soil management practices, compact them producing poor oxygenation, low beneficial microorganism populations and metabolic disorders in plants. Those factors induce abnormal plant development and predisposition to the attack of soil borne plant pathogens as: *Sclerotium cepivorum, Sclerotium rolfsi, Sclerotinia sclerotiorum, Rhizoctonia solani, Rosellinia* spp. But also, some air borne pathogens as Colletotrichum sp., are more aggressive under those conditions. In Costa Rica some practical trails done by farmers, have shown that the addition of *Trichoderma* spp. to organic fertilizers like vermicompost, Bocashi, and other composts, has a positive effect on the structure and microbial diversity, that improves nutrient movement in plant and pathogen suppression. Vegetative material of leather fern (*Rumohra adiantiformis*) infected by *Rosellinia* spp., planted with bocashi amended with. *T. asperellum (T. harzianum)*, showed more tip root growth and new fern rhizomes. Also the plants produced higher number of fiddleheads and high quality fronds. Non-treated material showed a total loss. In some leaf vegetables as lettuce the use of vermicompost followed by *Trichoderma viride* application, inhibited *S. sclerotiorum* development on the crop, in this study sclerotia were infected and plant yields were increased. Compost previously inoculated with *Trichoderma*, inhibited 100 per cent the Botrytis and *R. solani* infection, in nurseries, and plant nutrition was improved. The observed results are attributed to the antagonistic effect of *Trichoderma* on soil and airborne pathogens, production of growth regulators, solubilization of some microelements and better mineral absorption with the development of more roots and elimination of diseased roots. As conclusion, the organic substrates inoculated with *Trichoderma* spp., besides the organic matter source, they improve the structural characteristics of soil, promote the biochemical processes, increase the level of growth regulator substances and inhibit plant diseases.

During 1970 and its early period involved a much of work on biocontrol which involved the indirect enhancement of indigenous *Trichoderma* by manipulating the environment. Yet the precise role of these indigenous populations in the biocontrol of diseases was only a matter of speculation. Indigenous *Trichoderma* has long been known to partially or completely fill the biological vacuum created by harsh soil treatments or stresses on the soil micro biota, but the implication of this phenomenon in biological control is still in question.

Bliss (1951) attributed the control of *Armillaria mellea* in citrus following carbon disulfide fumigation to the action of indigenous *Trichoderma* that rapidly colonized and reproduced in the fumigated soil, but he did not provide direct evidence that the colonizer controlled the pathogen. Ohr *et al.* (1973) provided the most convincing direct evidence of the involvement of indigenous *Trichoderma* in the biocontrol of *Armillaria mellea* in methyl bromide fumigated soil. *Trichoderma* was more resistant to methyl bromide than was *A. mellea*, and *A. mellea* weakened by methyl bromide produced less of the antibiotic that otherwise protected it against the antagonist in nonfumigated soil.

Additional conjectural evidence of the role of indigenous populations of antagonists in biocontrol included the observation of adding sulfur to soil to maintain the pH below 3.9 that controlled root rot and heart rot of pineapple in Australia. This control was attributed to a decrease in zoosporangium formation of the causal agent (*Phytophthora*) and to an increase of the acidophilic native *T. viride* (Cook and Baker, 1983). Durrell (1968), using phase contrast and electron microscopy, produced interesting photographs showing haustoria and hyphae of *T. viride* within hyphae of Phycomycetes followed by digestion of their contents. Studies on parasitism of *Sclerotium rolfsii* and *Rhizoctonia solani* by *T. hamatum* and *T. harzianum* corroborated Durrell's findings and demonstrated additional phenomena at the molecular level *viz.*, extracellular fibrillar material deposited between the interacting hyphae, the accumulation of parasite organelles in the parasitizing cells, the production of a sheath matrix that encapsules the penetrating hyphae (Elad *et al.*, 1983).

Trichoderma harzianum, in combination with other *Trichoderma* species or chemical adjuvants, had been used in control of several diseases. Some of these include: *Rhizoctonia* damping-off in radish (Lifshitz *et al.*, 1985), corn and soybean (Kommedahl *et al.*, 1981), greymould on tomato (Migheli *et al.*, 1994), grapes and strawberry (Elad *et al.*, 1995), *Colletotrichum* storage rot of apple, cucumber fruit rot caused by *Rhizoctonia solani* J. G. Kuhn (Lewis and Papavizas, 1980), takeall disease in wheat (Ghisalberti and Sivasithamparam, 1991), and a wilt-complex, predominantly caused by *Sclerotium rolfsii* Sacc., *Rhizoctonia solani* and *Fusarium oxysporum* Schldl. in lentil and chickpea (Mukhopadhyay, 1995). Howell and Stipanovic (1991) divided stains of *T. virens* into two groups on the basis of antifungal antibiotic production. One group was designated as 'Q' strains that produced gliotoxin which was effective against *Rhizoctonia solani* but inactive against *Pythium ultimum* Trow. The other groups was designated as 'P' strains, produced gliovirin and were strongly active against *P. ultimum* but were inactive against *R. solani*.

Chitinolytic enzymes from *T. harzianum* and *T. virens* are thought to be responsible for degradation of fungus cell walls and, thereby, effective in biological control of the fungal pathogen *Botrytis cinerea* Pers.: Fr. (Cruz *et al.*, 1993; Di Pietro *et al.*, 1993; Lorito *et al.*, 1993, 1994). Inbar *et al.* (1996) investigated the hyphal interactions between the mycoparasite *Trichoderma harzianum* (BAFC Cult. No. 72) and the soil borne plant pathogenic fungus *Sclerotinia sclerotiorum* in dual culture and in sterilized soil using light and scanning electron microscopy. In dual culture, *T. harzianum* hyphae grew towards and coiled around the *S.sclerotiorum* hyphae. Dense coils of hyphae of *T. harzianum* and partial degradation of the *Sclerotinia* cell wall were observed in later stages of the parasitism. In sterile soil, conidia of *T. harzianum* germinated and the developing mycelium made contact with that of *S. sclerotiorum*, forming short branches and aspersorium-like bodies which aided in holding and penetrating the host cell wall. They also developed an in vitro system to test the ability of *T. harzianum* to control *Sclerotinia* wilt in cucumber and lettuce.

Hashem (2004) worked on two hundred isolates representing 31 fungal species (20 genera) that was recovered from soybean roots. Samples were collected from 12 localities at 3 different growth stages of the crop. The most dominant species were *Aspergillus flavus, Fusarium oxysporum, Fusarium solani (Nectria haematococca), Macrophomina phaseolina and Rhizoctonia solani*. Pathogenicity tests have proved the ability of *Macrophomina phaseolina* and *Rhizoctonia solani* to infect soybean roots and produce the symptoms of damping-off and root-rot diseases. The efficacy of three antagonists (*Trichoderma harzianum, Epicoccum nigrum and Paecilomyces lilacinus*) as well as two organic compounds (Strom and F-760) was evaluated as to their control of pathogenic fungi. The biocontrol fungi significantly suppressed *Macrophomina phaseolina* and *Rhizoctonia solani invitro* and *invivo*. *Epicoccum nigrum* and *Paecilomyces lilacinus* suppressed the growth of the pathogens by producing an inhibition zone, whereas *Trichoderma harzianum* suppressed them by overgrowing. Strom and F-760 showed lower reduction effect of diseases in comparison with the antagonists.

Biocontrol by *Trichoderma* results from different mechanisms acting synergistically to achieve disease control. Those involve the competition for nutrients and living space with plant pathogenic organisms (Hjeljord and Tronsmo, 1998) the direct attack and destruction of the pathogens like antagonism and mycoparasitism (Cortes *et al.*, 1998; Kullnig *et al.*, 2000) and promotion of plant beneficial processes such as enhancement of plant growth and induction of systemic and localized resistance (Harman *et al.*, 2004).

Indian Research on *Trichoderma* (30 years)

Table 17.1: Number of Publications on *Trichoderma* for Past Thirty Years

Area	No. of Publication	
	World	India
Biocontrol potential	841	237
Molecular Biology and ISR	399	42
Formulation	141	56
Shelf Life	23	12
Delivery and Evaluation	1351	569
Species Registered	11	2
Product Available	257	23

Table 17.2: Research Areas of Important Bioagents

Area of work	*T. harz*	*T. viride*	*T. virens*	*P.fluo*	*Total*
Genetic Improvement	2	1	0	2	5
Cost-effective mass Production	15	7	4	10	36
Formulation Development	8	7	2	3	20
Formulation evaluation	80	46	10	80	216
Enhancing shelf life	5	4	2	5	16
Application delivery and equipment	10	4	1	5	20

Table 17.3: Requirement of Biocontrol Agents in India

Bioagent	*Country Need*	*Value Rs. in Crores*	*Prod. Capacity*	*Qty. Prod. (2004)*	*Sold (2004)*
Trichoderma (T)	22038	260.78	1850(8.4)	382(20.6)	326(85)
Pseudomonas (T)	11421	228.42	705(6.2)	205(29.0)	170(83)
Trichogramna (L)	176305	652.33	832(0.5)	404(48.6)	317(78)
Crysoperla (Lakhs)	459420	9.42	26(<0.1)	9(34.6)	8(89)
Cryptolemus	106	1.06	25(24.0)	16() 64.0	11(89)
(Lakhs)	2551	714.28	17(), 0.7	7() 41.2	5(71)
HaNPV (Kiloliters)	1971	542.03	6(0.3)	10(16.6)	1(100)
SINPV (Kiloliters)	341	–	–	–	–
C1GV (Kiloliters)	1341	0.0188	436(32.5)	29(6.7)	15(51)
Verticilium (T)	6117	0.098	755(12.3)	214(28.3)	195(91)
P. tilacinus (T)	525	0.01	39575.2	31(7.8)	21(67)
M. Anisoptiac (T)	3671	48.36	2(0.02)	0.58(29)	0.39(67)
Gonlozus (Lakhs)	9671	–	–	–	–
Nomuracu (T)	9082	–	–	–	–
AaNPV (KL)	505	–	–	–	–
PxGV (KL)	18	–	–	–	–

Source: Lecture by Dr. R. J. Rabindra, Director PDBC, Bangalore in Brainstorming on Integrated Pest Management and Biopesticides (26 April, 2006).

Successful Application of *Trichoderma* Species in Economically Important Agriculture Crops in India

Cereals

Rice, Wheat, Sorghum, Bajara, Maize, Pearl millet, Barley.

Pulses

Chickpea, Pigeonpea, Mothbean, Mungbean, Urdbean, Cowpea, Horsegram, Guar, Lentil, Finger millets, Soybean, Linseed.

Oilseed Crops

Sesame, Mustard, Sunflower, Groundnut, Safflower, Coconut

Fruits and Vegetables

Papaya, Guava, Citrus, Apple, Banana, Mulberry, Muskmelon, Tomato, Potato, Pea, Wingedbean, Chilli, Okra, Ginger, Onion, Cauliflower, Cabbage, French bean, Cucurbits, Elephant Foot Yam, Brinjal, Betelvine, Watermelon, Patchouli.

Ornamentals

Gerbera, Carnation, Gladiolus, Jasmine, Chrysanthemum, Balsam, Rose.

Spices

Black Pepper, Cumin, Coriander, Cardamom, Fenugreek.

Others

Cotton, Sugarcane, Sugar beet, Tea, Coffee, Tobacco, Hevea, Cassava, Tamarinds, Jojoba, Jute, Tropical Pines, Dalbergia, Neem, Acasia, Flax, Cymbopogons, Teak, Eucalyptus, Cocoa, Aubergine, Chairpine, *Prosopis juliflora.*

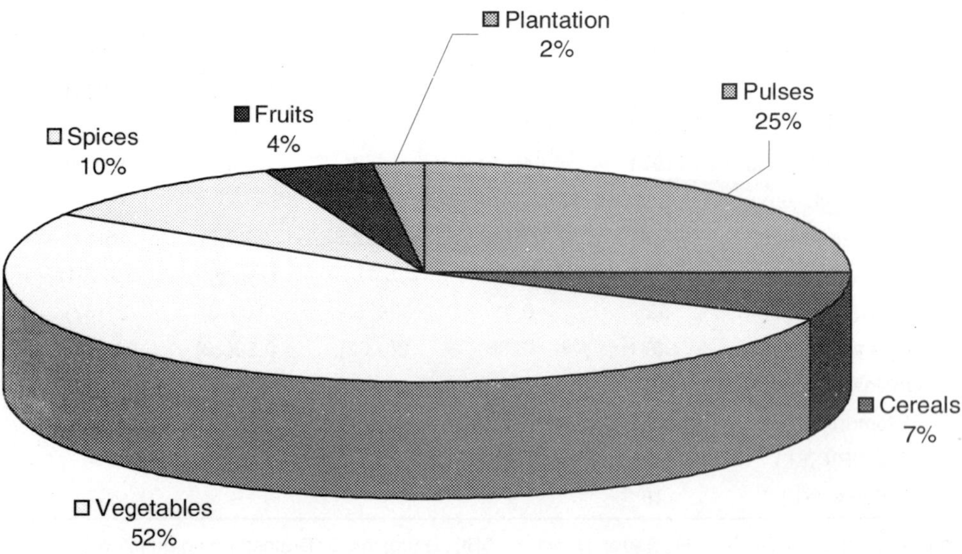

Figure 17.1: Use of Bioagents in Different Crops (Based on Indian publications)

Integration of Biocontrol with Other Control Measures

Inspite of the recent upsurge of interest in integrated pest management, examples of exploitation of biocontrol agents with other strategies for disease control are very few. Integration of biological with chemical controls has high potential for success by replacing some of the chemical treatments, the biological agent reduces both environmental pollution and the danger of the pathogen developing fungicide resistance. The regular applications of the biocontrol agent, which can establish itself in the infection court, thus, providing localization, persistence and plant growth unattainable by chemicals itself/or alone. Therefore, and due to its broad spectrum effect through a variety of mechanisms areas in which biological agent are superior to chemical components, further research should be directed toward exploiting such niches.

Chilli anthracnose caused by *Colletotrichum capsici* (Syd.) Butler and Bisby appears in Delhi region soon after heavy rainfall in first and second week of August persisting almost for 100-120 days in the field and continue in transit and storage. To break the life cycle of the pathogen at early stage of the disease development, studies on the effect of different formulations of *Trichoderma harzianum* Rifai isolates, Neemarin and neem oil were conducted on two chilli cultivars *i.e.* Pusa Sadabahar and Navjyoti. *T. harzianum* (toxin) suppressed the symptom expression, conidial germination and mycelial growth of *C. capsici* upto 100 per cent. Application of three different formulations of *T. harzianum i.e.* partially purified toxin, sorghum, and talc base (dry powder) applied in the field from the beginning *i.e.* seed, seedling followed by foliar spray in combination with Neemarin (neem product) and neem oil reduced the disease intensity upto 60.0 and 56.5 per cent in Pusa Sadabahar and Navjyoti, respectively was noticed. Similarly fruit rot was also reduced upto 63.7 and 46.8 per cent in Pusa Sadabahar and Navjyoti, respectively. The preventive spray of *T. harzianum* in combination with Neemarin and neem oil should be given early in the morning at the initial stage of the disease development (first week of August) to wash off the due deposit when the temperature ranges from 28-30^{0}C and RH 100 per cent (favorable for disease development) and should continue at 10 days interval till harvest. One pre harvest spray of *T. harzianum* two days before the harvesting should be followed to reduce the post-harvest and storage losses caused by *C. capsici*.(Sharma *et al.*, 2005)

Sclerotium rolfsii Sacc. (teleomorph *Athelia rolfsii* (Curzi) Tu and Kimbrough) is a devastating soil-borne plant pathogenic fungus with a wide host range. The disease caused by *S. rolfsii* is a severe problem of almost all crops in India and it is controlled particularly by the use of fungicides. The fungus does not produce asexual spores, over winters in soil and on plant debris as sclerotia. The purpose of this study was to characterize the field isolates of *S. rolfsii* collected from different hosts and locations and also to observe the effect of *Trichoderma harzianum* and *T. viride* isolates available in the laboratory. Cultural, morphological, mycelial compatibility, sclerotium formation and pathogenicity studies were made in ten field isolates of *S.rolfsii*. Mycelial compatibility was the major trait to differentiate the isolates, which were paired in culture in all possible combination. On the basis of intermingling and interaction zones that developed in the pairing, the isolates were assigned to mycelial compatibility groups (MCG's). Within a compatibility group, all isolates grew together when paired and the hyphae intermingled with little or no cell death. Isolates within a MCG also varied in mycelial and sclerotium morphology. These MCGs also exhibited differential response against twelve biologically effective isolates of *Trichoderma harzianum* (5) and *T. viride* (7) in the form of mycelial growth and sclerotia formation, clearly highlighting the variability in biological activity of *Trichoderma* spp. against *S. rolfsii* isolates. This research is a part of CGP NATP 399, author has worked on different strains of *Trichoderma* spp. (2002-2005), which also signifies the importance of variability in *Trichoderma*.

The economics of using biopesticides is similar to using any other pesticides in field. It has been observed that the prices of biological formulation are lesser as compared to common fungicides; therefore using biopesticides does not increase the expenditure. The main problem is the availability of pure products, which creates a confusion in the minds of farmers. Multiplication of biological agents, is possible at field level under scientific supervision possible through FYM and agriculture substrates.

Nursery management has the major role in vegetable production as diseases of seed and seedlings during germination either before or after emergence causes heavy losses. The most common means to check the disease in nurseries is by using fungicides but frequent and indiscriminate use of them often leads to atmospheric pollution and development of resistance in pathogens. Therefore, due to its economic importance a suitable, ecofriendly and economic means is needed to curb these problems in disease management. In this context, biological control is coming up as an alternative strategy for disease management, which is also eco-conscious and eco-friendly (Sharma, 2001and 2002a). Improved isolate compatible with recommended doses of fungicides *viz.*, thiram, captaf, iprodion, and mancozeb showed an increased overall growth/performance of treated tomato, cabbage, cauliflower, chilli and brinjal plants in terms of plant vigor, root colonization and disease control, indicates better performance of *Trichoderma* under solarized soils than non solarized soil. Though this isolate was also effective in non-solarized soils using the dose 2g/kg seed treatment + seedling dip +foliar spray. Rhizosphere observations based on persistence and proliferation of *Trichoderma* isolates in terms of c.f.u./g soil sample were also made. Studies to test the different nursery IPM module and cropping system for finalize suitable IPM package (nursery modules using varied four components *viz.*, solarization, seed treatment, nursery net and foliar spray of *Trichoderma* and/or iprodione) showed noticeable effective performance when the *Trichoderma* isolate used after solarization, in regular application or as in first two-crop-year under crop rotation system. The best nursery IPM module: solarization of nursery, seed treatment with captaf and *Trichoderma* (1:1), nursery covering with net and foliar spray of *Trichoderma* proved to be the best package. This can be utilized for Integrated Pest Management modules, where pesticides are used with biological methods and therefore development of pesticide tolerant/compatible fungal bioagents required.Application of *T. harzianum* in the regular cropping or in the first two year-crop show good results and persist in the rhizosphere, is able to colonize the roots and proliferate 4-5 times on existing and newly formed plant roots and rootlets well after application.

The improved isolates, pesticide tolerant in nature, *Trichoderma harzianum* Th-3 and Th-10 were compatible with recommended doses of thiram, captaf, iprodione, carbendazim and mancozeb showed an increased plant growth of treated tomato, cauliflower, chilli and brinjal plants and consistently reduced the disease caused by various pathogens. They performed better in plant growth performance under solarization as compared to non-solarized soil however; these isolates also work very well in non-solarized soil. The isolates of *T. harzianum* Th-3 and Th-10 compared for rhizosphere colonization/persistence in soil and effect on plant growth of Th-3 was found better over Th-10 in terms of population persistence, plant growth, and disease control (Sharma, 2002c and Sharma and Sain, 2003, Sharma, 2008). In continuation of this earlier study the further experiments were carried out to understand relationship between root colonization/population persistence and plant growth and to develop suitable IPM package for nursery with the use of best isolate. There are few reports available in the literature indicating marked increase in growth and greater root volumes due to use of biocontrol agent either as seed treatment or soil application. Demonstrated increased growth of several crop plants in the presence of biological control agents. These responses may be caused by a direct effect on the plant (as biofertilizer) or by control of some undiagnosed plant pathogen(s). Detailed investigation is necessary to work out the mechanism of growth promotion and to further exploit it as additional advantage of biocontrol method.

The *T. harzianum* isolates stimulate plant growth even in the apparent absence of the pathogens because it interact directly with the plant by producing plant growth promoting active metabolites without interacting with pathogens. (Windham *et al.*, 1986). Such effects of *Trichoderma harzianum* on plant growth, under solarization and under other experiments were also reported. (Sharma, 2002c and Sharma and Sain, 2003). The performance of *T. harzianum* Th-3 for rhizosphere colonization/persistence in soil and effect on plant growth of solonaceous and cruciferous crops under different cropping system was evaluated to understand the host specificity. Thus, the Th-3 isolate proved to be the best isolate to increase plant growth, disease control and persistence of the spores on the roots/rootlets of tomato crops under crop rotation but are statistically at par. However, the performance of this isolate under mono cropping and crop rotation did not show significant difference in plant growth but it differed significantly in the persistence of its population.

The time of treatment is important for control of disease and plant growth. The studies conducted for evaluation of treatment interval in crop-year did not show significant correlation between treatment interval in cropping year and plant growth. However, population persistence was found increased with regular treatment and up to some extent in alternate and first two-season crop treatment. This shows the spores remain viable in the soil and colonize the roots of crop of further coming season. Thus, the lying of viable spores in the soil/field does not help much in plant growth as compared to the direct use of bioagents on seed, seedlings in the form of seed coating or seedling dip. Which indicates that without the host root colonization *Trichoderma harzianum* could not be able to produce the active metabolites, which are behaving as growth promoting agent.

The different integrated pest management modules including various components *viz.*, seed treatment with Th-3 and captaf (1:1), foliar spray with Th-3 and/or iprodione and other cultural practices like nursery solarization and nursery covering with net proved to be the best IPM package for nursery. This shows that the best control, however, is obtained in an integrated programme using *Trichoderma* spp. with reducing doses of fungicide. Seed treatment and seedling dip with *Trichoderma harzianum* isolate proved to be the best-option along with *T. harzianum*

Th-3 and captaf, where the studies showed captaf to be the safest fungicide for *Trichoderma*. The chemical protectant can provide the short-term protection of the seed or seedling itself. A rhizosphere-competent and pesticide tolerant/resistant bioprotectent, on the other hand can colonize the entire root system and provide degree of season-long protection unattainable through acceptable levels of only chemical treatment.

Application of *Trichoderma* in Field

Like chemical pesticides *Trichoderma* formulations have been applied through seed, soil and foliage seed coating with *T. viride* and *T. harzianum* has emerged as a feasible way of delivering the antagonist for the management of plant diseases. A large number of seed, seedling, root, stem, foliar and panicle diseases have been suppressed by the use of seed treatment with antagonists mainly *T. viride* and *T. harzianum*. (Mukhopadhyay, 1995; Sharma *et al.*, 2003; 2008). For commercial purpose dry powder of *Trichoderma* are used at the rate of 3 to 10g powder per kilogram seed based on seed size and formulation. Treatment of seeds controls seed borne diseases, bulbs tuber etc. with antagonists prevents from seed-borne diseases.

Trichoderma viride and *Trichoderma harzianum* is a bio-fungicide for control of soil borne diseases like root rot, foot rot, rhizome rot, collar rot, stem rot, damping off, wilt, blight, blasts, leaf spot etc. The product developed is active on fungal genera *viz.*, *Fusarium, Sclerotium, Pythium, Rhizoctonia, Macrophomina, Phytophthora, Botrytis, Veticillium* and many other higher and lower fungi active against all soil borne pathogens.

Today, in the era of ecofriendly technology, microbes are the best alternatives for augmenting rhizosphere for sustaining soil fertility and productivity. A project entitled "On Farm Demonstration and Commercial Cultivation of *Trichoderma* as Bio-Pesticide and Growth Promoter" funded by DST is ongoing at Biocontrol Lab, Plant Pathology, which will reflect it's importance in farmers field in different soils.

References

Abbasi, P.A., Miller, S.A., Meulia, T., Hoitink, A.J. and Kim, J.M. (1999). Precise detection and tracing of *Trichoderma hamatum* 382 in compost-amended potting mixes by using molecular markers. *Appl. Environ. Microbiol.*, 65 (12): 5421-5426.

Abbott, E.V. (1926). Taxonomic studies on soil fungi. *Iowa st. Coll. J. Sci.* 1: 15-36.

Arisan-Atac, I., Heidenreich, E. and Kubicek, C.P. (1995). Randomly amplified polymorphic DNA fingerprinting identifies subgroups of *Trichoderma viride* and other *Trichoderma* spp. capable of chestnut blight biocontrol. *FEMS-Microbiol. Lett.*, 126: 249–256.

Arora, S., Sharma, Pratibha. and Madhuban Gopal (2004) Compatibility of fungicide, iprodione with *Trichoderma harzianum*. *Ann. Pl. Protec. Sci.*12: 222-223.

Berkeley, M.J. (1860). *Outlines of British Fungology*. Lovell Reeve, London. 442pp + plates.

Bisby, G.R. (1939). *Trichoderma viride* Pers. ex Fries, and notes on *Hypocrea*. *Trans. Brit. Mycol. Soc.* 23: 149-168.

Bissett, J. (1984). A revision of the genus *Trichoderma*. I. Section *Longibrachiatum* sect. nov. *Can. J. Bot*. 62: 924–931.

Bissett, J. (1991a). A revision of the genus *Trichoderma*. II. Infrageneric classification. *Can. J. Bot*. 69: 2357-2372.

Bissett, J. (1991b). A revision of the genus *Trichoderma*. III. Sect. *Pachybasium*. *Can. J. Bot*. 69: 2373-2417.

Bissett, J. (1991c). A revision of the genus *Trichoderma*. IV. Additional notes on section *Longibrachiatum*. *Can. J. Bot*. 69: 2418-2420.

Bissett, J. (1992). *Trichoderma atroviride*. *Can. J. Bot*. 70: 639-641.

Bliss, D E. (1951). The destruction of *Armillaria mellea* in citrus soils. *Phytopathology*, 41: 665-83.

Charlesworth, B., Sniegowski, P. and Stephan, W. (1994). The evolutionary dynamics of repetitive DNA in eukaryotes. *Nature*, 371: 215-220.

Chaverri, P., Castlebury, L.A., Overton, B.E. and Samuels, G.J. (2003). Species of *Hypocrea* and *Trichoderma* with green conidia, and conidiophore elongations and related species in sect. *Pachybasium*. *Mycologia*, 95: 1100–1140.

Chaverri, P., Samuels, G.f. and Stewart, E.L. (2001). *Hypocrea virens* sp. nov., the teleomorph of *Trichoderma virens*. *Mycologia*, 93: 1113-1124.

Chen, X., Romaine, C.P., Ospina-Giraldo, M.D. and Royse, D.J. (1999). A PCR-based test for the identification of *Trichoderma harzianum* biotypes 2 and 4 inciting the world-wide green mold epidemic in cultivated *Agaricus bisporus*. *Appl. Microbiol. Biotechnol*. in press.

Cook, R. J. and Baker, K. F. (1983). The nature and practice of biological control of plant pathogens. *St. Paul, Minn: Am. Phytopathol. Soc*. 539 pp.

Cortes, C., Gutierrez, A., Olmedo, V., Inbar, J., Chet, I. and Herrera-Estrella, A. (1998). The expression of genes involved in parasitism by *Trichoderma harzianum* is triggered by a diffusible factor. *Mol. Gen. Genet*. 260: 218-225.

Cruz, J de la., Hidalgo-Gallego, A., Lora, J. M., Benitez, T., Pintor-Toro, J. and Llobed, A. (1993). Isolation and characterization of three chitinases from *Trichoderma harzianum*. *Europ. J. of Bioch*. 206: 859-867.

Danielson, R. M. and Davey, C. B. (1973a). Nonnutritional factors affecting the growth of *Trichoderma* in culture. *Soil Biol. Biochem*. 5: 495-504.

Danielson, RM. and Davey, C.B. (1973b). The abundance of *Trichoderma* propagules and distribution of species in forest soils. *Soil Biol. Biochem*. 5: 485–494.

Di Pietro, A., Lorito, M., Hayes, C. K., Broadway, R. M. and Harman, G. E. (1993). Endochitinase from *Gliocladium virens*: isolation, characterization, and synergistic antifungal activity in combination with gliotoxin. *Phytopathology*, 83: 308-313.

Dodd, S.L., Crowhurst, R.N., Rodrigo, A.C., Samuels, G.J., Hill, R.A. and Stewart, A. (2000). Examination of *Trichoderma* phylogenies derived from ribosomal DNA sequence data. *Mycol. Res*. 104: 23-34.

Dodd, S.L., Lieckfeldt, E., Samuels, G.J. (2003). *Hypocrea atroviridis* sp. nov., the teleomorph of *Trichoderma atroviride*. *Mycologia*. 95 (1): 27-40.

Dodd, S-L., Hill, R.A. and Stewart, A. (2004). A duplex-PCR bioassay to detect a *Trichoderma virens* biocontrol isolate in non-sterile soil. *Soil Biol. Biochem*. 36 (12): 1955-1965.

Doi, A.Y., Abe, Y. and Sugiyama, J. (1987). *Trichoderma* sect. *Saturnisporum*, sect. nov. and *Trichoderma ghanense*, sp. nov. *Bull. Nat. Sci. Museum Ser* B (Botany). 13: 1-9.

Domsch, K.H., Gams, W. and Anderson, T.H. (1980). Compendium of soil fungi. Vol. 1. New York. Academic Press, 859pp.

Durrell, L. W. (1968). Hyphal invasion by *Trichoderma viride*. *Mycopath. Mycol. Appl.* 35: 138-44.

Elad, Y., Chet, I. and Henis. Y. (1983). Parasitism of *Trichoderma* spp. on *Rhizoctonia solani* and *Sclerotium rolfsii*–scanning electron microscopy and fluorescence microscopy. *Phytopathology*. 73: 85–88.

Elad, Y., O'Neill, T., Cohen, A. and Schtlenberg, O. (1995). Factors influencing control of gray mold by means of Trichodex (*Trchoderma harzianum* T39) under field conditions. Fifth International *Trichoderma* and *Gliocladium* Workshop, Beltsville, April 1995, abstract.

Floch, G. le., Rey, P., Bellanger, E., Benhamou, N., Guillou, A. and Tirilly, Y. (2003). Selection of *Pythium oligandrum* isolates for biocontrol in tomato soil less culture: morphological, biochemical and molecular characterization. Colloque-international-tomate-sous-abri,-protection-integree-agriculture-biologique,-Avignon,-France,-17-18-et-19-septembre-2003: 80-84.

Fritze, H. and Baath, E. (1993). Microfungal species composition and fungal biomass in a coniferous forest soil polluted by alkaline deposition. *Microbial Eco*. 25: 83-92.

Gawande, S. and Sharma, Pratibha. (2003). Changes in host enzyme activity due to induction of resistance against downy mildew in cauliflower. *Annals. Agril. Res.* 24(: 2)316-325.

Geldenhuis, M.M., Roux, J., Wingfield, M.J. and Wingfield, B.D. (2004). Development of polymorphic markers for the root pathogen *Thielaviopsis basicola* using ISSR-PCR. *Mol. Eco. Notes*. 4 (4): 547-550.

Ghisalberti, E. L. and Sivasithamparam, K. (1991). The role of secondary metabolites produced by *Trichoderma* species in biological control (abstract). *Petria*. 1: 130-131.

Gilman, J.C. (1957). A manual of soil fungi. 2nd edition. Ames, Iowa.

Goes, L.B., Costa, ABL da., Freire, LL de C., Oliveira, NT de., de C Freire LL. and Oliveira, N.T. (2002). Randomly amplified polymorphic DNA of *Trichoderma* isolates and antagonism against *Rhizoctonia solani*. *Brazil. Arch. of Biol. and Tech.*, 45: 151-160.

Grondona, I., Hermosa, M.R., Tejada, M., Gomis, M.D., Mateos, P.F., Bridge, P., Monte, E. and Garcia-Acha, J.M. (1997). Physiological and biochemical characterization of *Trichoderma harzianum*, a biological control agent against soilborne fungal plant pathogens. *Appl. Environ. Microbiol.* 63: 3189–3198.

Gutter, Y. (1957). Effect of light on sporulation of *Trichoderma viride* Pers. Ex Fr. *Bull. Res. Coun. Israel.* 5D: 273-286.

Hantula, J., Dusabenyagasani, M. and Hamelin, R. C. (1996). Random amplified microsatellites (RAMS)–A novel method for characterizing genetic variation within fungi. *Eur. J. For. Pathol.* 26: 159-166.

Harman, G.E., Howell, C.R., Viterbo, A., Chet, I. and Lorito, M. (2004). *Trichoderma* spp. opportunistic avirulent plant symbionts. *Nature Microbiol. Rev.* 2: 43-56.

Harz, C.O. (1871). Einige neue hyphomyceten Berlin's und Wien's nebst Beitragen zur Systematik derselben. *Bull. Soc. Imp. Moscow.* 44: 88-147.

Hashem, ME. (2004). Biological control of two phytopathogenic fungal species isolated from the rhizoplane of soybean (Glycine max). Assiut Univ. (Egypt). Botany Dept. Czech Mycology (Czech Republic), 56 (3-4): 223-238.

Hermosa, M.R., Grondona, I., Iturriaga, E.A., Diaz-Minguez, J.M., Castro, C., Monte, E. and Garcia-Acha, J.M. (2000). Molecular characterization and identification of biocontrol isolates of *Trichoderma* spp. *Appl. Environ. Microbiol.* 66: 1980-1998.

Hjeljord, L. and Tronsmo, A. (1998). *Trichoderma* and *Gliocladium* in Biological Control: An Overview. *In*: Harman, G.E., Kubicek, C.P. (Eds.), *Trichoderma* and *Gliocladium*. Vol. 2. Enzymes, Biological Control and Commercial Applications. Taylor and Francis Ltd., London, 131-151pp.

Howell, C. R. and Stipanovic, R. D. (1991). Antibiotic production by *Gliocladium virens* and its relation to biocontrol of seedling diseases. *Phytopathology.* 81: 1152 (abstract).

Inbar, J., Menendez, A. and Chet, I. (1996). Hyphal interaction between *Trichoderma harzianum* and *Sclerotinia sclerotiorum* and its role in biological control. *Soil. Boil. Biochem.*, 28 (6): 757-763.

Kommedahl, T., Windels, C. E., Sarbini G., and Wiley, H. B. (1981). Variability in performance of biological and fungicidal seed treatments in corn, peas, and soybeans. *Protec. Ecol.* 3: 55-61.

Kubicek, C.P., Mach, R.L., Peterbauer, C.K. and Lorito,M. (2000). *Trichoderma*. From genes to biocontrol [plant diseases]. *J. of Plant Pathology.* 83: 11-23.

Kuhls, K., Lieckfeldt, E. and Borner, T. (1995). *Microbiol. Res.* 150: 363-371.

Kuhls, K., Lieckfeldt, E., Samuels, G. J., Kovacs, W., Meyer, W., Petrini, O., Gams, W.,.

Bo¨rner, T. and Kubicek, C.P. (1996). Molecular evidence that the asexual industrial fungus *Trichoderma reesei* is a clonal derivate of the ascomycete *Hypocrea jecorina*. *Proc. Natl. Acad. Sci.* USA, 93: 7755–7760.

Kullnig, C., Szakacs, G. and Kubicek, C.P. (2000). Molecular identification of *Trichoderma* species from Russia, Siberia and the Himalaya. *Mycol. Res.* 104: 1117–1125.

Kullnig-Gradinger, C.M., Szakacs, G. and Kubicek, C.P. (2002). Phylogeny and evolution of the genus *Trichoderma*: a multigene approach. *Mycol. Res.* 106: 757–767.

Latha, J., Verma, A. and Mukherjee, P.K. (2002). PCR-Fingerprinting of some *Trichoderma* isolates from two Indian type culture collection–a need for re-identification of these economically important fungi. *Curr. Sci.* 83 (4): 372-374.

Lee, C. and Hseu, T. (2002). Genetic relatedness of *Trichoderma* sect. *Pachybasium* species based on molecular approaches. *Can. J. Microbiol.* 48: 831-840.

Lejeune, R., Nielsen, J. and Baron, G.V. (1995). Morphology of *Trichoderma reesei* QM 9414 in submerged cultures. *Biotechnol. Bioeng.* 47 (5): 609-615.

Leuchtmann, A., Petrini, O. and Samuels, G. J. (1996). Isozyme subgroups in *Trichoderma* Section *Longibrachiatum*. *Mycologia*. 88: 384-394.

Lewis, J. A. and Papavizas, G. C. (1980). Integrated control of *Rhizoctonia* fruit rot of cucumber, *Phytopathology*. 70: 85-89.

Lieckfeldt, E. and Seifert, K.A. (2000). An evaluation of the use of ITS sequences in the taxonomy of the *Hypocreales*. *Stud. in Mycol.* 45: 35-44.

Lieckfeldt, E., Samuels, G.J., Helgard, H.I. and Petrini, O. (1999). A morphological and molecular perspective of *Trichoderma viride*: is it one or two species? *Appl. Environ. Microbiol.* 65: 2418–2428.

Lifshitz, R., Lifshitz, S. and Baker, R. (1985). Decrease in incidence of *Rhizoctonia* preemergence damping-off by the use of integrated and chemical controls. *Plant Disease*. 69: 4341-4344.

Lorito, M., Harman, G. E., Hayes, C. K., Broadway, R. M., Tronsmo, A., Woo, S. L. and Di Pietro, A. (1993). Chitinolytic enzymes produced by *Trichoderma harzianum*: antifungal activity of purified endochitinase and chitobiosidase. *Phytopathology*. 84: 398-405.

Lorito, M., Hayes, C. K., Di Pietro, A., Woo, S. L. and Harman, G. E. (1994). Purification, characterization, and synergistic of a glucan 1,3-β-glucosidase and an N-acetyl-β-glucosaminidase from *Trichoderma harzianum*. *Phytopathology*, 84: 398-405.

Meyer, R. J. (1991). Mitochondrial DNAs and plasmids as taxonomic characteristics in *Trichoderma viride*. *Appl. Environ. Microbiol.* 57: 2269–2276.

Migheli, Q., Herrera-Estrella, A., Avataneo, M. and Gullino, M. L. (1994). Fate of transformed *Trichoderma harzianum* in the phylloplane of tomato plants. *Mol. Ecol.* 3: 153-159.

Moubasher, A.H. (1963). Effect of duramycin on some isolates of *Trichoderma viride*. *Nature*, London. 200: 492.

Mukhopadhyay, A. N. (1995). Biological seed treatment with *Gliocladium* and *Trichoderma* for control of chickpea and lentil wilt complex. Fifth International *Gliocladium* and *Trichoderma* workshop, Beltsville, MD, April 1995, abstract.

Muthumeenakshi, S., and P. R. Mills. (1995). Detection and differentiation of fungal pathogens of *Agaricus bisporus*. *Mushroom Sci.* 14: 603–610.

Muthumeenakshi, S., Mills, P.R., Brown, A.E. and Seaby, D.A. (1994). Intraspecific molecular variation among *Trichoderma harzianum* isolates colonising mushroom compost in the British Isles. *Microbiology*. 140: 769–777.

Niethammer, A. (1937). Die mikroskopischen Boden-Pilze. The Hague.

Ohr, H. D., Munnecke, D. E. and Bricker, J. L. (1973). The interaction of *Armillaria mellea* and *Trichoderma* spp. As modified by methyl bromide. *Phytopathology*. 63: 965-73.

Okuda, T., Fujiwara, A. and Fujiwara, M. (1982). Correlation between species of *Trichoderma* and production patterns of isonitrite antibiotics. *Agricul. and Biol. Chem.* 46: 1811-1822.

Ospina-Giraldo, M. D., Royse, D.J., Thon, M.R., Chen, X. and Romaine, C.P. (1998). Phylogenetic relationships of *Trichoderma harzianum* causing mushroom green mold in Europe and North America to other species of *Trichoderma* from world-wide sources. *Mycologia*. 90: 76–81.

Oudemans, C. A. J. A. and Koning, C. J. (1902). Prodrome d'une flore mycologique obtenue par la culture sur gelatine prepare de terre humeuse du Spanderswoud, pres de Bassum. *Arch. Neerl. Sci.* 11 (7): 266-298.

Persoon, C.H. (1794). Disposito Methodica Fungorum in Classes, Ordines, Familias et Genera. *In*: Römer, J.J. (Ed.), Neues Magazin für Botanik. Ziegler und Söhne, Zürich, 63-128.

Pratibha Sharma and S.K. Sain. (2005). Use of biotic agents and abiotic compound against damping of cauliflower caused by *Pythium aphanidermatum. Indian Phytopahology*. Vol 58(4): 395-401.

Qais, K. Zewain., P. Bhadur and Pratibha, Sharma. (2004) Effect of fungicides and neem extract on mycelial growth and myceliogenic germination of *Sclerotinia sclerotiotum*. *Indian J. Phytopahtology*. 57(1) 101-103.

Rifai, M. A. (1969). A revision of the genus *Trichoderma*. *Mycol. Papers*. 116: 1–56.

Rifai, M.A. (1964). A reinvestigation of the taxonomy of the genus *Trichoderma* Pers. Unpublished M.Sc. Thesis, University of Sheffield.

Rifai, M.A. and Webster, J. (1965). An undescribed British species of *Thuemonella*. *Trans. Brit. Mycol. Soc.* 48: 409-413.

Rodrigues, K-F., Sieber, T.N., Grunig, C.R. and Holdenrieder, O. (2004). Characterization of *Guignardia mangiferae* isolated from tropical plants based on morphology, ISSR-PCR amplifications and ITS1-5.8S-ITS2 sequences. *Brazil. Mycol. Res.*, 108: 45-52.

Sain, S.K., Gour H.N. and Pratibha Sharma (2007). Evaluation of botanicals and PGPRs against *Xanthomonas campestris* pv. *campestris*, an incitant of black rot of cauliflower, *Journal of Ecofriendly Agriculture*. 2 (2): 178-182.

Sain, S.K., H.N. Gour and P. Sharma. (2005). Biocontrol of black rot of cauliflower by plant-growth promoting rhizobacteria. *J.Mycol.Pl.Pathol* Vol. 35 (1): 99-102.

Samuels, G. J., Lieckfeldt, E. and Nirenberg, H.I. (1999). Description of *Trichoderma asperellum* sp. nov. and comparison to *Trichoderma viride*. *Sydowia*. 51: 71–88.

Samuels, G. J., Petrini, O. and Manguin, S. (1994). Morphological and macromolecular characterization of *Hypocrea schweinitzii* and its *Trichoderma* anamorph. *Mycologia*. 86: 421-435.

Samuels, G.J. (1996). *Trichoderma*: A review of biology and systematics of the genus. *Mycol. Res*. 100: 923–935.

Samuels, G.J., Dodd, S.L., Gams, W., Castlebury, L.A. and Petrini, O. (2002). *Trichoderma* species associated with the green mold epidemic of commercially grown *Agaricus bisporus*. *Mycologia*. 94: 146-170.

Samuels, G.J., Doi, Y. and Rogerson, C.T. (1990). Hypocreales. *Memoirs of the New York Botanical Garden*. 59: 6-108.

Schlick, A., Kuhls, K., Meyer, W., Lieckfeldt, E., Borner, T. and Messner, K. (1994). Fingerprinting reveals gamma-ray induced mutations in fungal DNA: implications for identification of patent strains of *Trichoderma harzianum*. *Curr. Genet*. 26: 74-78.

Sharma, Pratibha (2008). Effect of cropping system on rhizospheric competence of *T. harzianum* and growth performance of important vegetable crops. *Journal of Ecofriendly Agriculture*. 3 (2): 185-189.

Sharma, P. (2007). Mass production techniques of *Trichoderma, Kalisena* and *Pseudomonas fluorescence*, in Manual on Biomanagement strategy of Plant Pathogens 13-17 Dec. 2007, pp.17-20.

Sharma, Pratibha. and Nidhi, Sharma. (2007). Hypovirulence as tool of biological control. *Journal of Ecofriendly Agriculture*.2 (1): 46-50.

Sharma, H.K., D. Prasad, Pankaj. and Pratibha, Sharma. (2006) Bio-management of *Meliodogyne incognita* infesting Okra *Ann.pl.Prot.Sci*.14 (1) 191-193.

Sharma, Pratibha, S.R. Sharma, S.K.Sain, and A. Dhandpani. (2006). Integrated management of major diseases of cauliflower 9 Brassica oleracea var.botrytis subvar cauliflora) Indian Journal of Agricultural Sciences 76(12): 726-731.

Sharma Pratibha, Q Zewain, P. Bahadur and S. K. Sain. (2005). Effect of soil solarization on sclerotial viability of Sclerotinia sclertiorum (Lib.) de Bary of cauliflower (*Brassica oleracea* L. var. *botrytis* subvar, *cauliflower* D. C.) *Indian Journal of Agricultural Sciences* 75: 90-94.

Sharma Pratibha, L.N.Kadu and S.K. Sain. (2005). Biological management of dieback and fruit rot of chilli caused by *Colletotrichum capsici* (Syd.) Butler and Bisby. *Indian journal of plant protection*. 33(2): 226-230.

Sharma, Pratibha. (2003). Use of biocontrol agents in sustainable vegetable production, In Proceedings of International Seminar on Downsizing Technology for Rural Development at Bhubaneswer, *(Eds)* S.Khuntia, A.Parida and Vibuti N Misra. Vol.I pp 193-202.

Sharma Pratibha and P. Dureja. (2004). Evaluation of *T. harzianum* and *T. viride* isolates at BCA Pathogen Crop Interface. *J. Mycol. Pl. Pathol.* Vol. 34 (1) 47-55.

Sharma Pratibha, G. Kulshrestha, M. Gopal, and L.N. Kadu. (2004). Integrated Management of Chilli Die Back and Anthracnose in Delhi Region. *Indian Phytopathology*.57 (4) 427-434.

Sharma, P., S.R. Sharma and M. Sindhu. (2004). A detached leaf technique for evaluation of resistance in cabbage and cauliflower against three major pathogens. *Indian Phytopathology*. Vol.57 (3). 315-318.

Sharma, Pratibha. and S.K. Sain. (2004). Induction of systemic resistance in tomato and cauliflower by *Trichoderma* species against stalk rot pathogen. (*Sclerotinia sclerotiorum*). *J.Biocontrol* Vol 18(1): 21-28.

Sharma, Pratibha, (2004). Development of bioformulation from improved strains of *Trichoderma* for vegetable crops. In Proceeding of 26th Annual Conference and National Symposium of ISMPP on Advance in Fungal Diversity and Host-Pathogen Interaction at Goa University, Goa, October7-9, 2004, pp: 163-172.

Sharma, Pratibha. (2004). Integrated use of abiotic and biotic elicitors against downy mildew suppression and enzymatic activity in cauliflower in Microbial Diversity: Opportunities and Challenges pp–279-286 (Ed.) S.P.Gautam *et al.*, Shree Publishers and Distributors, New Delhi-110 002.

Sharma, Pratibha. (2004). Effect of fungicides on fungal antagonists of Sclerotinia rot of cauliflower and selection of pesticide resistant biotypes. in Eds. P. Dureja, D. B. Saxena, J. Kumar, Madhuban Gopal, S. B. Singh and R. J. Tanwar Society of Pesticide Science, IARI, India pp186-191.

Sharma, Pratibha; A. Kaur and Sain., S.K. (2004) Variability in *Sclerotium rolfsii* and differential response against *Trichoderma harzianum* and *Trichoderma viride.* In Proceeding of 26th Annual Conference and National Symposium of ISMPP on Advance in Fungal Diversity and Host-Pathogen Interaction at Goa University, Goa, October7-9, 2004, pp: 173-182.

Sharma, Pratibha and S.K. Sain (2003) Evaluation of commercial and laboratory formulations of bioagents and plant nutrients against wilt of tomato and damping off of cauliflower. *Indian J.Pl.Pathol*. Vol. 21: 105-109.

Sharma, Pratibha. and S.K.Sain. (2003). Development of suitable techniques for evaluating virulence and biocontrol activity of *Trichoderma* isolates. *Indian J.Pl.Pathol*. Vol. 21: 16-21.

Sharma, Pratibha, Sain, S.K. and James S. (2003). Compatibility Study of *Trichoderma* isolates With Fungicides against Damping-off of Cauliflower and Tomato caused by *Pythium aphanidermatum*. *Pesticide Research Journal*. 15(2): 133-138.

Sharma, Pratibha. and A.P, Singh, (2002). Multiple diseases resistance in roses against foliar and flower pathogens. *Indian Phytopathology* 55 (2): 169-172.

Sharma, Pratibha. (2002). Use of bioagents with pesticides in plant disease management. *New Agriculturist.* 13 (1): 54-60.

Sharma, Pratibha. (2002). Disease management strategies of green house crops in Crop Pest and Disease Management Challenges for the Millennium Eds.pp. 211-220 (D. Prasad and S. N. Puri Jyoti Publishers, New Delhi.

Sharma, Pratibha. (2002). Induction of systemic resistance in cauliflower to downy mildew by exogenous application of plant activator.*Annals of Plant Protection.* 10: 199-203.

Sharma, Pratibha. (2002). Production of biocontrol products at rural women site. In proceedings on the National Workshop on Role of biotechnology in Women Uplftment in the New Millenium at Agra (U.P.) 24th and 25th January, 2002, pp 41-49.

Sharma, Pratibha, L. Singh and D. Adlakha (2001). Antagonistic potential of *Trichoderma* and *Aspergillus* species on *Sclerotinia sclerotiorum* (Lib.) de Bary causing rots in cabbage and cauliflower. *Pesticides Information* 2: 41.

Singh, S.N. (1998). Effect of temperature and relative humidity on spore germination of different isolates of *Myrothecium roridum* Tode. infecting mungbean. *Ann. Pl. Protec. Sci.* 6: 19-21.

Smith, A.L. (1902). The fungi of germinating farm seeds. *Trans. Brit. Mycol. Soc.*, 1: 182-186.

Tulasne, L.R. (1860). De quelques Sphe´ries fungicoles, a' propos d'un me´moire de M. Antoine de Bary sur les *Nyctalis* (1). *Ann Sci Nat Bot, Se´r.* 13: 5–19.

Tulasne, L.R. and Tulasne, C. (1865). Selecta fungorum carpologia. Tomus Tertius. *Nectriae.–Phacidiei.–Pezizei.* Paris. xvi 1 221 p 1 Tabs. I–XVI.

Turner, D., Kovacs, W., Kuhls, K., Lieckfeldt, E., Peter, B., Arisan-Atac, I., Strauss, J., Samuels, G.J., Borner, T. and Kubicek, C.P. (1997). Biogeography and phenotypic variation in *Trichoderma* sect. *Longibrachiatum* and associated *Hypocrea* species. *Austria Mycological Research.* 101 (4): 449-459.

Turoczi, G., Fekete, C., Kerenyi, Z., Nagy, R., Pomazi, A. and Hornok, L. (1996). Biological and molecular characterization of potential biocontrol strains of *Trichoderma. J. of Basic Microbiol.*, 36 (1): 63-72.

Vandenkoornhuyse, P., Leyval, C. and Bonnin, I. (2001). High genetic diversity in arbuscular mycorrhizal fungi: evidence for recombination events. Heredity. Oxford: Blackwell Science Ltd. 87: 243-253.

Waksman, S.A. (1916). Soil fungi and their activity. *Soil Sci.* 2: 103-155.

Webster, J. (1964). Culture studies on *Hypocrea* and *Trichoderma.* I. Comparison of perfect and imperfect states of *H.gelatinosa, H.rufa,* and *Hypocrea* sp. 1. *Trans. Brit. Mycol. Soc.* 47: 75–96.

Webster, J. (1966). Culture studies on *Hypocrea* and *Trichoderma* III. *H.lactea* (5 *H.citrina*) and *H.pulvinata*. *Trans. Brit. Mycol. Soc.* 49: 297–310.

Webster, J. and Rifai, M.A. (1968). Culture studies on *Hypocrea* and *Trichoderma*. *Trans. Brit. Mycol. Soc.* 51: 511–514.

Weindling, R. and Emerson, O. H. (1936). The isolation of a toxic from the culture filtrates of *Trichoderma*. *Phytopathology*. 26: 1068-70.

Widden, P. and Abitbol, J.J. (1980). Seasonality of *Trichoderma* species in a spruce-forest soil. *Mycologia*. 72: 775-784.

Widden, P. and Scattolin, V. (1988). Competitive interactions and ecological strategies of *Trichoderma* species colonizing spruce litter. *Mycologia*. 80: 795–803.

Windham, M.T.,Elad. and Y, Baker, R.(1986). A mehanism for increased plant growth by *Trichoderma* species. *Phytopathology*. 76: 518-521.

Zamir, D. and Chet, I. (1985). Application of enzyme electrophoresis for the identification of isolates in *Trichoderma harzianum*. *Can. J. Microbiol*. 31: 578-580.

Zietkiewicz, E., Rafalski, A., and Labuda, D. (1994). Genome fingerprinting by simple sequence repeat (SSR)-anchored polymerase chain reaction amplification. *Genome*. 20: 176-183.

Zimand, G., Valinsky, L., Elad, Y., Chet, I. and Manulis, S. (1994). Use of the RAPD procedure for the identification of *Trichoderma* strains. *Mycol. Res*. 98 (5): 531-534.

Chapter 18

Vegetative Compatibility Groups (VCGs) in Fungi

Pratibha Sharma*, Susanta Banik, Mousa Najafinia and Monica Henry

Division of Plant Pathology, IARI, Pusa Campus, New Delhi – 110 012, India

ABSTRACT

Vegetative compatibility is the ability of two strains of a fungus to fuse and form a stable heterokaryon. Vegetative compatibility Groups (VCGs) study has been undertaken with respect to many filamentous fungi especially with a view to study variability within a fungal population. A great deal of work has been done in *Fusarium* spp. where naming of VCGs has been standardized. Studies with *Neurospora crassa* and *Podospora anserina* revealed many molecular and biochemical aspects of vegetative compatibility. Vegetative incompatibility, which is the inability to form a stable heterokaryon, functions as natural barrier to genetic exchange between fungal individuals. Transfer of hypovirulence factor in case of *Cryphonectria parasitica* is based on VCGs. An attempt has been made in this chapter with a view to understand the concept, importance and some recent trends in the studies of VCGs.

Keywords: *VCG, Aspergillus, Cercospora, Colletotrichum, Fusarium, Rhizoctonia, Sclerotium, Phytophthora.*

Introduction

All living organisms in nature have some systems to identify self and non-self. The obvious function of this self/non-self recognition mechanism (compatibility/

* E-mail: psharma032003@yahoo.co.in

incompatibility) is to avoid inbreeding. Compatibility systems of bacteria, fungi and plants have been studied in detail. Among animals invertebrates have a recognition system, which is involved in immunological recognition and in fertilization.

The best known compatibility systems are the pheromone system of protozoa (Vallesi *et al.*, 1995), (in) compatibility systems of the fungi (Metzenberg 1990; Begueret *et al.*, 1994; Hiscock *et al.*, 1996; Wendland *et al.*, 1995; Kothe 1996), self-incompatibility system of flowering plants (Kao and McCubbin 1996) and allorecognition systems of the invertebrate (Magor *et al.*, 1992). All of these systems are primarily involved in avoiding the harmful effects of inbreeding.

The fungal incompatibility system regulates both sexual reproduction and somatic compatibility (Vegetative compatibility). The co-existence of two different nuclei (heterokaryon or dikaryon) in the same cell is regulated by the somatic/vegetative/heterokaryon compatibility system. The first major review on this subject was presented by Raper in 1966. The somatic compatibility system which regulates self/non-self recognition during vegetative growth in filamentous fungi has been extensively reviewed (Begueret *et al.*, 1994; Glass and Kuldau 1992; Wu *et al.*, 1998).

Basidiomycetes have more complex mating type systems and (filamentous) Ascomycetes have a somatic (heterokaryon) incompatibility system in addition to the sexual mating types.

The function of the mating types is obviously avoidance of inbreeding (not allowing homozygosity). This helps to increase genetic diversity and survival chances of the species. However, in vegetative compatibility system of fungi it is only genetically similar individuals who can undergo hyphal anastomosis. Why they have to be of the same genetic type is not known clearly. One of the hypotheses is that horizontal transfer of harmful cytoplasmic genetic elements (stable RNA, mitochondria and plasmids) is reduced between incompatible (genetically dissimilar) strains and thus protecting the strains. However, the efficiency of prevention of horizontal gene transfer is dependent on the *het* (*het*erokaryon incompatibility) loci (Coenen *et al.*,1994; Debets *et al.*, 1994; Caten 1972) and *het* loci are not equally efficient in all cases.

Classification of fungal strains on the basis of pathogenicity generally cannot classify those strains, which are non-pathogenic but constitute majority of fungal microflora in the agro-ecosystem. Thus, diversity among the fungal species can be effectively brought out by VCG. This approach provides a means of characterizing subspecific groups based on the genetics of the fungus rather than on the host–pathogen interaction. In addition, vegetative compatibility allows for the characterization of the non-pathogenic portion of the population (Correll, 1991).

Vegetative Compatibility

Vegetative compatibility (also known as somatic compatibility, heterokaryon compatibility, self/non-self recognition) refers to the ability of two strains of a fungus to fuse and form a stable heterokaryon. This process necessarily involves fusion of hyphae of two strains of the fungus that is also called as anastomosis. The heterokaryon formed as a result of this process is genetically different from the parent components and may acquire new abilities such as more virulence, more survival rate in adverse situation or act as source of variation in the population of the fungus in nature. The co-existence of two different nuclei (heterokaryon or dikaryon) in the same cell is regulated by the somatic/vegetative/heterokaryon compatibility system. On the other hand, inability of two strains of a fungus to fuse and form a stable heterokaryon, which is called as vegetative incompatibility, also signifies that the natural population of a fungus is genetically diverse. More the incompatibility among the strains of a fungus more is the extent of genetic variation of the fungus. This phenomenon helps in identifying intraspecific variation present in a fungus. Moreover, study of intraspecific variation like virulence has been possible with grouping of isolates of fungi based on vegetative compatibility. Thus from the point of view of plant pathologist it is of great concern.

In nature heterokaryon formation takes place also by means of fusion of two gametes of sexually compatible strains of a fungus. The resultant dikaryon containing two genetically different nuclei may differ drastically from the haploid parents, as we know in case of Basidiomycetes like rust pathogen *Puccinia graminis tritici*. However, to what extent, vegetatively developed heterokaryon differs from parent mycelium is not known in many of the fungi wherever this phenomenon is found.

A fungal individual can be viewed as a fluid, dynamic system that is characterized by hyphal tip growth, branching and hyphal fusion (anastomosis). Hyphal anastomosis is especially important in such nonlinear systems for the purposes of communication and homeostasis. Filamentous fungi can also undergo hyphal fusion with different individuals to form heterokaryons. However, the viability of such heterokaryons is dependent upon genetic constitution at heterokaryon incompatibility (*het*) loci. If hyphal fusion occurs between strains that differ in allelic specificity at *het* loci, vegetative incompatibility, which is characterized by hyphal compartmentation and cell lysis, is induced (Glass *et al.*, 2000).

Hyphal Anastomosis

Vegetative hyphal fusion is common in filamentous fungi. Buller (1933) was the first to demonstrate the process of anastomosis. Hyphal anastomosis takes place in three distinct physiological states: pre-contact, post-contact and post-fusion. Pre-contact initiation process is under the control of diffusible chemical signal indicating that there is chemotropic movement of fungal hyphae. Physical contact is followed by growth arrest and hyphal fusion. During the process of hyphal fusion, hydrolytic enzymes take part and break down the cell wall thus forming a bridge between the two hyphae. Plasma membrane fusion and cytoplasmic mixing are the events of post-contact stage (Figure 18.1).

Vegetative Incompatibility

The terminologies vegetative compatibility and vegetative incompatibility are like two sides of the same coin. While vegetative compatibility facilitates easy exchange of genetic material through hyphal anastomosis, vegetative incompatibility is a natural genetic mechanism that limits nuclear exchange. Vegetative incompatibility, the prevention of somatic fusion between fungal individuals, occurs frequently in ascomycetes.

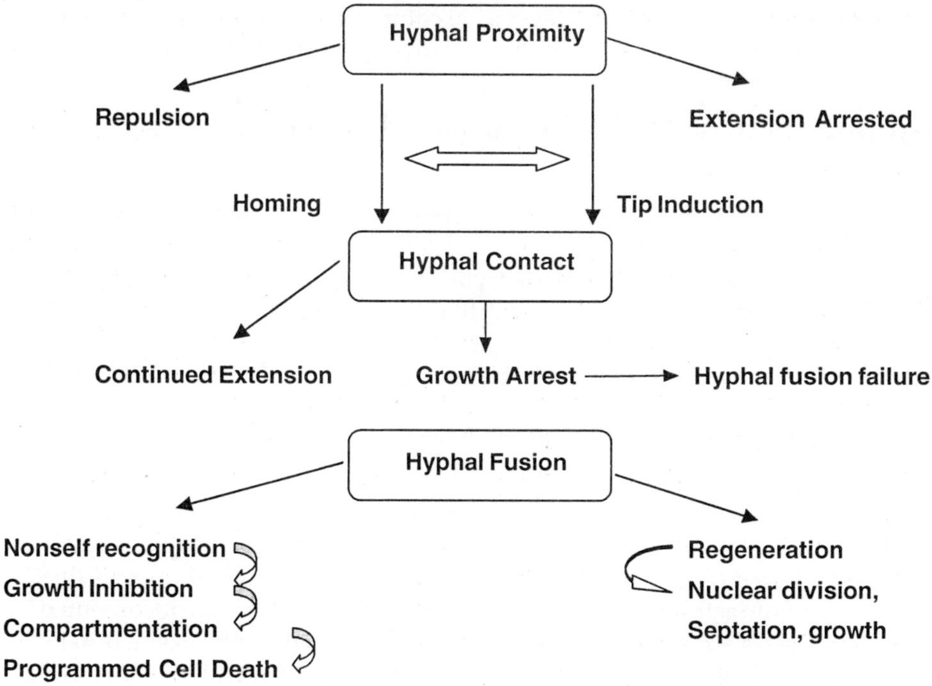

Figure 18.1: Steps in Vegetative Hyphal Fusion (Adapted from Ainsworth and Rayner 1989)

If the number of VCGs is large, selection does not lead to a significant increase in the number of VCGs. Sexual reproduction, however, may enlarge the number of VCGs, due to recombination of incompatibility genes. Strong selective pressure does give an extra increase in VCG numbers (Nauta and Hoekstra 1996).

Research workers generally take up this kind of study within their own geographical reach, say their home state or country and give nomenclature of the groupings on the basis of their study. Naturally those nomenclatures can't be globally accepted. If we think globally and work out the variability of a fungus affecting one or more crops then a clear picture may emerge and nomenclature of the groupings can be internationally standardized as considerable work has been done in grouping *Fusarium oxysporum* isolates all over the world.

Vegetative and Mycelial Compatibility

The concept of vegetative compatibility differs from mycelial compatibility in the fact that the later denotes only the ability of two strains of a fungus to fuse and form one colony while the former denotes the ability of the two strains to form a stable heterokaryon through hyphal anastomosis. Two strains of a fungus, which show mycelial incompatibility, cannot form heterokaryon due to lack of physical proximity that means they will also be vegetative incompatible. On the other hand, mycelial compatibility alone cannot assure vegetative compatibility. There may not be any hyphal fusion or anastomosis even though the mycelia of two strains may remain in contact with each other.

Morphological and Biochemical Aspects of Vegetative Incompatibility

In nature, hyphae of one fungus may come in contact with other hyphae of genetically same or different fungal individual of the same species. This may result in hyphal fusion and formation of heterokaryon. The heterokaryon thus formed may divide repeatedly and form heterokaryotic mycelium as happens in *Neurospora crassa* and *Podospora anserina*. Or else, heterokaryons are limited only to actual fusion cells and nuclei do not migrate between cells. Heterokaryons are continually reformed by repeated fusion events within the mycelium as happens in *Verticillium dahliae* (Puhalla and Mayfield 1974) and *Gibberella fujikuroi* (Puhalla and Speith 1985).

Het (heterokaryon incompatibility) genes are responsible for determining vegetative compatibility/incompatibility reaction. Hyphal fusion between strains that differ in *het* loci results in incompatibility reaction. Incompatibility reaction is manifested in terms of growth inhibition, hyphal compartmentation and death. Microscopic and ultrastructural features of incompatibility reactions documented in a few fungal species are septal plugging, vacuolization of the cytoplasm, organelle degradation, shrinkage of the plasma membrane from the cell wall (Jacobson *et al.*, 1998) and DNA fragmentation (Marek *et al.*, 1998). Some other biochemical features associated with incompatibility are decrease in RNA production, synthesis of new protein, increase in phenoloxidases, malate/NADH dehydrogenase, proteases and amino acid oxidase (Begueret and Bernet 1973; Boucherie and Bernet 1978).

Vegetative or somatic incompatibility in ascomycetes prevents the fusion of genetically different mycelia (heterogenetic incompatibility) and is usually under multigenic control by a series of bi- or multiallelic genes at vegetative incompatibility (*het* or *vic*) loci. When two mycelia in different vegetative compatibility (VC) groups meet in a substrate, they may interact with varying degrees of antagonism, as would be expected of a phenotype under multigenic control. In some cases barrage reactions may be recognized by a clear zone between the two mycelial fronts due to the lysis of the interacting cells. Sometimes incomplete interactions occur at the margins and unstable dikaryons may exist for a short time. In some cases pigments or enhanced conidia production indicates the meeting of the two mycelia.

How do heterothallic ascomycetes in different VC groups undergo sexual fusions? In the species that have been studied, separate individuals are maintained until the time of sexual reproduction when morphologically and physiologically differentiated sex organs, the ascogonia and antheridia, develop and apparently are not usually under the control of the vegetative incompatibility system.

Molecular Basis of Vegetative Incompatibility

Vegetative cell fusion (anastomosis) occurs within and also between individuals leading to cytoplasmic mixing and formation of vegetative heterokaryon, the viability of which is genetically controlled by specific loci termed *het* (for heterokaryon incompatibility) or *vic* (for *v*egetative *inc*ompatibility) loci. A *het* locus can be defined as a locus at which heteroallelism cannot be tolerated in a heterokaryon. When two fungal individuals that differ genetically at a *het* locus fuse, the resulting heterokaryotic cells are rapidly destroyed or severely inhibited in their growth as described in the above mentioned paragraph, depending on the *het* locus that is involved.

Hyphal anastomosis takes place only when a fungal hypha identifies another hypha as self. This self/non-self recognition is due to *het* locus (Tables 18.1 and 18.2). A single allelic or non-allelic difference in *het* locus is key to determine incompatibility reaction. Protein products of *het* locus participate in recognition reaction. Structural difference between the HET proteins mediates nonself recognition and trigger growth inhibition, hyphal compartmentation and cell death. Thus, structural incompatibility in the protein products of *het* locus determines the incompatibility of the participant fungal hyphae.

Table 18.1: Number of *het* loci in some Fungi

Fungus	*No. of het loci*
Aspergillus nidulans	8
Cryphonectria parasitica	7
Podospora anserina	9
Neurospora crassa	11

Table 18.2: Some *het* Genes and their Characters

Species	*Gene*	*Class of gene*	*Nature of Protein*
Neurospora crassa	*matA-1*	Allelic *het* gene	Transcriptional regulator
	mat a-1	Allelic *het* gene	Transcriptional regulator
	tol	Mediator	Leucine-rich repeat
	uet-c	Allelic *het* gene	Signal peptide
	un-24	Nonallelic *het* gene	Ribonucleotide reductase
	het-6	Nonallelic *het* gene	Region with similarity to TOL and HET-E
Podospora anserina	*het-s*	Allelic *het* gene	Prion analog
	het-e	Nonallelic *het* gene	GTP-binding domain
	het-c	Nonallelic *het* gene	Glycolipid transfer protein
	mod-D	Modifier, *het-C/E* inc	Alpha-subunit of G protein
	mod-E	Modifier, *het-R/V* inc.	Heat-shock protein
	idi-2	Induced by *het-R/V* inc	Signal peptide
	mod-A	Modifier, nonallelic inc	SH3-binding domain
	idi-1, idi-3	Induced by nonallelic inc	Signal peptide
	PspA	Induced by nonallelic inc.	Serine protease

VCG Testing in Fungi

Most fungi can utilize nitrate as a nitrogen source by reducing it to ammonium (Figure 18.2) via nitrate reductase and nitrite reductase (Garraway and Evans 1984). There are different classes of nit mutants, these classes presumably reflect mutations at a nitrate reductase structural locus (nit 1), a nitrate-assimilation pathway-specific regulatory locus (nit 3), and loci (at least) that affect the assembly of molybdenum containing cofactor necessary of necessary nitrate reductase activity (nit M) (Correll *et al.*, 1987).

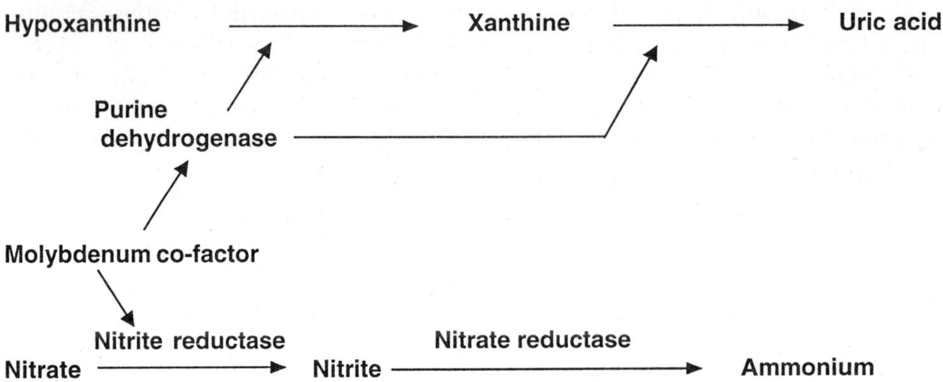

Figure 18.2: Nitrate Utilization Pathway in *Aspergillus nidulans* and *Neurospora crassa*

Vegetative compatibility of strains of fungi can be tested by pairing their nit (nitrate non-utilizing) mutants generated on potassium chlorate containing media (Correll *et al.*, 1987; Puhalla 1985). Different nit mutants can be phenotypically classified by their growth on a basal medium amended with one of several different nitrogen sources (Figure 18.3). When phyenotypically distinct nit mutants are paired on minimal containing nitrate as the sole source of nitrogen, the hyphal fusion and nutritional complementation in heterokaryotic cells will occur resulting in wild-type growth (Figure 18.4). Appearance of wild-type growth ensures the occurrence of nuclear exchange through hyphal anastomosis and thus vegetative compatibility.

However, it is not always easy to recover nit mutants from certain isolates. Moreover, weak heterokaryon formation, cross compatibility reactions and presence of vegetatively self-incompatible isolates may pose additional problems in nit mutant analysis.

VCGs in Fungi: Case Studies

Ascomycetes (Table 18.3)

Ophiostoma novo-ulmi

When the Dutch elm disease fungus *Ophiostoma novo-ulmi* invaded Europe it was a single mating type (MAT-2) and a single vegetative incompatibility (vic) type. Later the populations became diverse due to acquisition by *O. novo-ulmi* of the MAT-

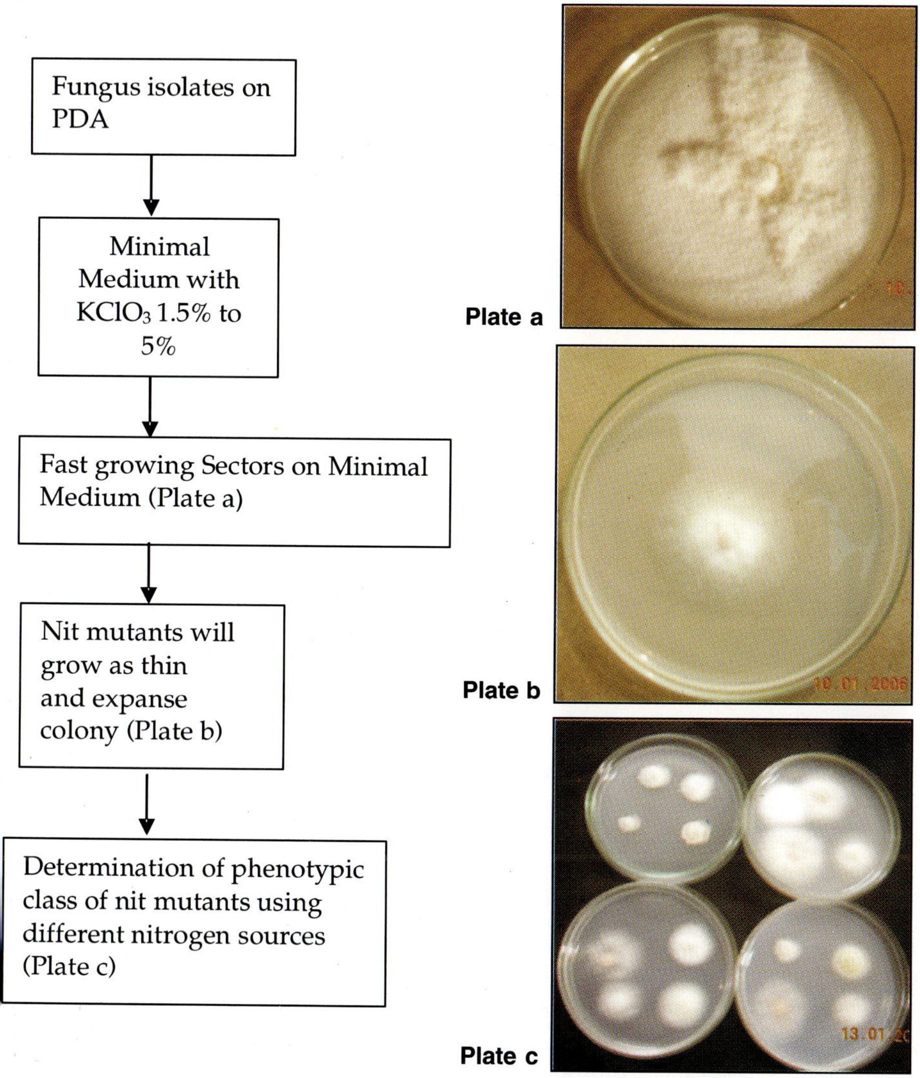

Figure 18.3: Production of Nit Mutants Using Potassium Chlorate Containing Media (Photograph: Najafinia and Sharma 2006)

1 and vic loci from another species, *O. ulmi* as evidenced from AFLP study (Paoletti *et al.*, 2006).

The diversity of vegetative compatibility (vc) types of the Dutch elm disease fungus, *Ophiostoma novo-ulmi*, in North America was recently shown to be very low compared with populations in Europe (Milgroom and Brasier 1997).

Ascochyta pisi

Vegetative compatibility of *Ascochyta pisi* was assessed using both sulfate nonutilizing and nitrate nonutilizing auxotrophic mutants. Based on the large number

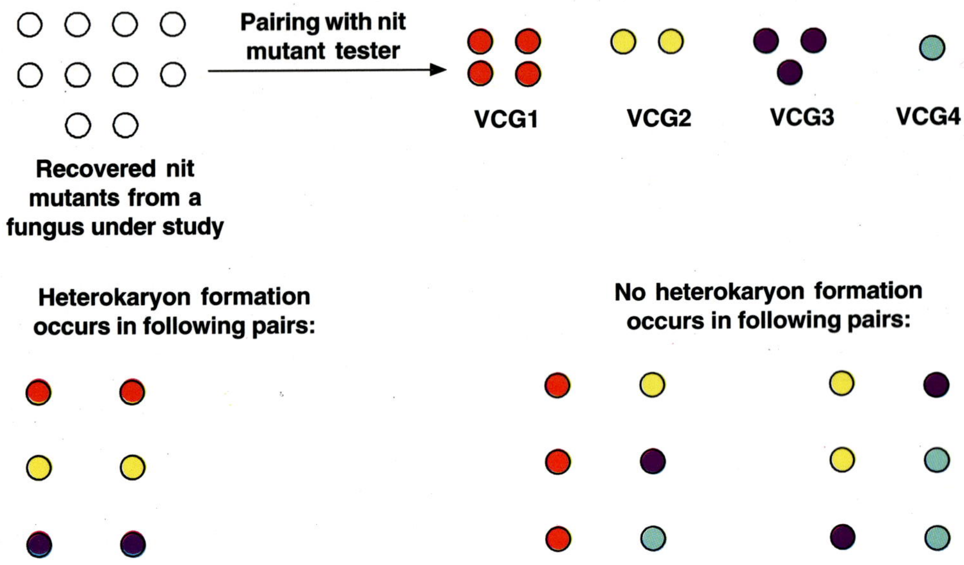

Figure 18.4: Complementation Test with Nit Mutants to Group Isolates into VCGs

of noncomplementary interactions, it was concluded that vegetative incompatibility is widespread in *A. pisi* (Furga 1997).

Table 18.3: Groups of Fungi where VCG Study has been Undertaken

Fungus Group	*Fungus*
Ascomycetes	*Rosellinia necatrix, Septoria nodorum, Ascochyta pisi, Botryotinia fuckeliana, Daldinia loculata, Neurospora crassa, Podospora anserina, Sclerotinia* sp., *Diaporthe ambigua, Monilinia fructicola, Ophiostoma novo-ulmi, Cryphonectria parasitica*
Basidiomycetes	*Ganoderma boninense, Glomus mosseae, Serpula lacrymans, Helicobasidium monpa, Amylostereum areolatum*
Deuteromycetes	*Rhizoctonia solani, Aspergillus* sp., *Cercospora kikuchii, Botrytis cinerea, Fusarium* spp, *Verticillium dahliae, Sclerotium rolfsii, Colletotrichum gloeosporioides*
Oomycetes	*Phytophthora infestans*

Diaporthe ambigua

Testing of strains of *Diaporthe ambigua* for vegetative compatibility resulted in barrage reactions indicating vegetative incompatibility. When dsRNA-containing strains of *D. ambigua* were paired with dsRNA containing strain from a different VCG they developed a broad, clear zone (Smit *et al.*,1997).

Cryphonectria sp.

In *Cryphonectria cubensis* which is the causal agent of a serious canker disease of *Eucalyptus* spp. in tropical and subtropical parts of the world it was found that vegetative incompatibility may not pose a strong barrier against virus transmission in South African isolates of *C. cubensis* (Heerden *et al.*, 2001). The workers could find that a synthetic RNA transcript corresponding to the full-length coding strand of the *C. parasitica* hypovirus (CHV1-EP713) could readily be transmitted via hyphal anastomosis to *C. cubensis*.

However, vegetative incompatibility was reported to be a barrier for the transfer of dsRNA from one isolate to another in *Cryphonectria parasitica* and the conversion of virulent strain to hypovirulent strain through hyphal anastomosis reduced with the increase in number of different *vic* genes among the donor and recipient isolates (Wang *et al.*, 2004).

The transmission of viruses between individuals of the chestnut blight fungus *C. parasitica* is controlled primarily by *vic* genes. 100 per cent Transmission occurred when donor and recipient isolates had the same *vic* genotypes (Cortesi *et al.*, 2001).

Genotypes of 64 vegetative compatibility (vc) types, controlled by six unlinked vegetative incompatibility (*vic*) loci, have been identified in the chestnut blight fungus, *Cryphonectria parasitica*. European populations have less vc type diversity than the US populations because of a combination of lower vic-allele diversity and limited recombination (Milgroom and Cortesi 1999).

Biocontrol using hypovirulence depends critically on the use of appropriate vegetative compatibility groups, which closely match the compatibility group(s) of the targeted pathogen (s). So, random application of vegetative compatibility groups will most often fail. This problem may be addressed (eventually) by application of hypovirulent strains of every possible vegetative compatibility group.

Podospora anserina

Expression of new genes during vegetative incompatibility was studied using *P. anserina* strains het-R het-V, het-r het-V and het-R het-V1. It was found that the mRNA content of fungal cells was qualitatively modified during the progress of the vegetative incompatibility reaction. Three genes induced during incompatibility were characterized and named idi genes and their protein products were small with signal peptide (Bourges *et al.*, 1998).

There was a strong correlation found between meiotic drive and vegetative incompatibility in eight different spore killer strains from the population of *P. anserina* (Gaag *et al.*, 2003).

In *P. anserina* two phenomena are associated with polymorphism at the het-s locus, vegetative incompatibility and ascospore abortion. Two het-s alleles, het-s and het-S occur naturally. The het-s allele acts as a meiotic drive element the het-s-encoded protein is a prion propagating as a self-perpetuating amyloid aggregate. When prion-infected [Het-s] hyphae fuse with [Het-S] hyphae, the resulting heterokaryotic cells die. (Dalstra *et al.*, 2003).

Neurospora crassa

The *Neurospora* sp. contributed a lot to fungal genetics and biology. The area of vegetative incompatibility is one of them (Perkins and Davis 2000).

Vegetative incompatibility reaction in *N. crassa* is regulated by genetic differences at 11 het loci, including mating type loci. RFLP study was used in progeny with crossovers in the het-6 region and a DNA transformation assay to identify two genes in a 25-kbp region that have vegetative incompatibility activity. The predicted product of one of the two genes, het-6OR, has sequence similarity to the predicted product of the het-e vegetative incompatibility gene in *P. anserina* and to the predicted product of tol, which mediates mating-type vegetative incompatibility in *N. crassa*. The second incompatibility gene was un-24OR encoding the large subunit of ribonucleotide reductase. A nonallelic interaction was suggested to occur between un-24 and het-6 and possibly other loci to mediate vegetative incompatibility in the het-6 region of *N. crassa* (Smith *et al.*, 2000).

Trans-species polymorphism has been found at loci associated with vegetative incompatibility in *N. crassa*. Several unlinked loci determined the vegetative compatibility group (VCG) of an individual. However, viable heterokaryon formation between individuals of the same VCG resulted in a fitness loss, presumably via transfer of infectious agents by hyphal fusion or exploitation by aggressive genotypes (Muirhead *et al.*, 2002).

The filamentous fungus *N. crassa* was used as a model to study the molecular mechanism of hyphal fusion. A deletion showing pleiotropic defects was identified that was restricted in its ability to undergo both self-hyphal fusion and fusion with a different individual to form a heterokaryon. Complementation with a single open reading frame (ORF) within the deletion region in this mutant, restored near wild-type growth rates, female fertility, aerial hyphae formation and hyphal fusion, but not vegetative incompatibility and wild-type conidiation pattern. This ORF was named ham-2 (for hyphal anastomois) that encodes a putative transmembrane protein that is highly conserved (Xiang *et al.*, 2002).

Rosellinia necatrix

Three genes encoding G protein alpha subunits (Rga1, Rga2, and Rga3) were cloned and characterized from the white root rot fungus, *R. necatrix*. Rga1 was found to be similar to Mod-D, which is associated with vegetative incompatibility in *P. anserina* suggesting that Rga1 is important in the vegetative incompatibility reaction in *R. necatrix* (Aimi *et al.*, 2001).

Cell death was observed in fused hypha resulted from hyphal anastomoses between incompatible strains and the nuclei of dead cell were intensified by staining with ethidium bromide. In contrast, the nuclei in a living cell were intensified by staining with acridine orange. A strain was found which did not form a barrier reaction, but which could be shown to undergo cell death and therefore showed a positive vegetative incompatibility reaction (Aimi *et al.*, 2002b).

Daldinia loculata

The local population structure of *D. loculata*, a xylariaceous species that frequently produces conidia and sexual stromata on fire-damaged deciduous host trees was determined by using both vegetative incompatibility tests and restriction enzyme analysis of PCR amplified nuclear gene fragments. Only one allele per locus was found in the mycelium isolated from the wood, suggesting that *D. loculata* grows vegetatively as haploid mycelia (Johannesson *et al.*, 2001).

Botryotinia fuckeliana

Based on the nutritional complementation on nitrate minimal medium of both the mutants it was concluded that a vegetative incompatibility system operates in *B. fuckeliana* resulting in multiple vegetative compatibility groups. Mycelial interaction zones which were formed between all parental strains when paired on NaCl-amended medium, indicated congruence between mycelial incompatibility and vegetative incompatibility (Beever and Parkes 2003).

Septoria nodorum

Widespread vegetative incompatibility in *Septoria nodorum* (*Leptosphaeria nodorum*) was reported by Newton *et al.* (1998).

Sclerotinia sp.

Isolates of *S. homoeocarpa*, the causal agent of dollar spot of turfgrass were assessed for vegetative compatibility in culture, and four types of reactions with respect to transmission of hypovirulence-associated double-stranded RNA (dsRNA) were observed. The fully incompatible reaction strongly restricted transmission of hypovirulence-associated dsRNA, whereas the partially incompatible reactions allowed limited spread between vegetative compatibility groups (VCGs), suggesting that vegetative incompatibility is not an absolute barrier to the transmission (Deng *et al.*, 2002).

Basidiomycetes

Amylostereum areolatum

Amylostereum areolatum, a fungal symbiont with woodwasp (*Sirex noctilio*), cause extensive damage to pine plantations in the Southern Hemisphere. The population diversity of the fungal isolates from South Africa, South America, Australasia and Europe was determined by vegetative incompatibility testing. All 108 South African and 26 South American isolates belonged to the same vegetative compatibility group (VCG) indicating that the South African and South American populations of *A. areolatum* share a common origin (Slippers *et al.*, 2001). Thus, VCG study also throws some light on the evolutionary aspect of a fungal individual.

Glomus mosseae

Vegetative compatibility tests performed on isolate of *Glomus mosseae* showed that six geographically different isolates were capable of self-anastomosing. Microscopic examinations revealed the manifestations of incompatible reactions like protoplasm retraction from the tips and septum formation in the approaching hyphae. The result of vegetative compatibility tests on *G. mossease* was confirmed by total protein profiles and internal transcribed spacer-restriction fragment length polymorphism profiles. Since arbuscular mycorrhizal fungi lack a tractable genetic system, vegetative compatibility tests may represent an easy assay for the detection of genetically different mycelia and an additional powerful tool for investigating the population structure and genetics of these obligate symbionts (Giovannetti *et al.*, 2003).

Helicobasidium monpa

The cytological characteristics of the vegetative incompatibility reaction in a filamentous basidiomycete, *Helicobasidium monpa* were elucidated by analysing the fluorescence emitted by ethidium bromide and acridine orange stained nuclei. Cell death was observed in fused hyphae resulted from hyphal anastomoses between strains belonging to different mycelium compatibility groups (MCG) and the nuclei of the dead cell were intensified by ethidium bromide. In contrast, the nuclei in a living cell were intensified by staining with acridine orange indicating the usefulness of ethidium bromide staining for detecting dead cells in case of *H. monpa* (Aimi *et al.*, 2002a).

Serpula lacrymans

The occurrence of geographically widespread vegetative compatibility groups (VCG) in *S. lacrymans* in Europe was demonstrated by Kauserud (2004). Among 22 heterokaryotic isolates of *S. lacrymans*, five VCG were found. The most widespread VCG included isolates from Belgium, south and central Norway, separated by more than 1500 km. No other genetic variation, measured as DNA sequence variation or ISSR polymorphisms, was detected between the investigated *S. lacrymans* isolates, whereas a considerable level of genetic variation was found among five European isolates of the sister taxon, *S. himantioides*. It was hypothesized that isolates of *S. lacrymans* have lost their ability to recognize self from nonself due to sharing of similar VC alleles, caused by a recent genetic bottleneck during the establishment in northern Europe.

Deuteromycetes

***Fusarium* spp.**

F. oxysporum is an important vascular wilt pathogen of many agriculture crops (Nelson 1981). Strains of *F. oxysporum* are often highly host specific. This specificity led Snyder and Hansen (1940), to subdivide the species into formae speciales based on a strains ability to cause disease on a particular host or group of hosts. Many formae speciales can be further subdivided into race on the basis of their virulence on differential host cultivars (Armstrong and Armstrong 1981). Although virulence has been an extremely useful characteristic for differentiating isolate of *F. oxysporum*, there are some inherent problems associated with characterizing strains based solely on pathogencity. Groupings based on host pathogen interaction (*i.e.* virulence) are dictated by the genetic makeup of the host or simply the differential cultivars one has available to distinguish strains. Moreover, virulence has been shown to be influenced by a number of variables, including temperature (Pound and Fowler 1982) host age (Hart and Endo 1981) and method of inoculation (Kraft and Haglund 1978). Furthermore, classification of strains based solely on pathogenicity precludes the characterization of non–pathogenic strains of *F. oxysporum*. Nonpathogenic strains represent a major component of the fungal microflora of agricultural soils, and an understanding of diversity in this portion of the population must be developed, if we are to fully understand diversity among the virulent strains of this species.

It is a fungus where genetic diversity has been categorized extensively (Table 18.4) by vegetative compatibility grouping in laboratories around the world. VCGs emerged as good predictors of genetic similarity, clonal lineage or both (Kistler *et al.*, 1998). Kistler *et al.* (1988) proposed a standard systemic numbering of VCGs in *Fusarium oxysporum*. In this system vegetative compatibility isolates are given a VCG code. The code number of a given VCG is composed of two parts: the first three digits correspond to host specialization, or formae speciales, and the last digit corresponds to individual VCG within the forma specialis. As a rule, non-pathogenic isolates of *F. oxysporum* should not be given four digit VCG code.

In *Fusarium oxysporum* f. sp. *vasinfectum* four phenotypic classes of nit mutants were recognized (nitA, nitB, nitC and nitD) on the basis of ability to utilize various N sources. On the basis of pairing complementary nit mutants on a minimal medium agar, 107 strains, were assigned to 4 vegetative compatibility groups (Wang *et al.*, 1996).

High degree of vegetative incompatibility (through nit mutant study) was reported among 20 isolates of *Fusarium avenaceum* [*Gibberella avenacea*], mostly from crops of white lupin or wheat. However, no relationships could be established among pathogenicity, RFLP group, RAPD group and vegetative compatibility group of the isolates of *F. avenaceum* studied (Satyaprasad *et al.*, 2000).

On a comparative study between RAPD technique and VCG analysis to find out the intraspecific genetic variability in *F. oxysporum* f. sp. *pisi* and *F. poae*, higher genetic diversity was shown on the basis of VCG analysis in both the species than the RAPD techniques. The reason for the differences in structure of populations could be the different scale of selection pressure and migration rate (Pomazi 1999).

Genetic diversity study on *Fusarium circinatum* (teleomorph=*Gibberella circinata*) through vegetative compatibility groups resulted in only eight VCGs indicating a low level of diversity. This low level of diversity indicated the prevalence of an asexually propagating population of *F. circinatum*. However, sexual reproduction is not ruled out in the Californian population of *F. circinatum* because of vegetative incompatibility between the progeny and their parents (Wikler *et al.*, 2000).

In Iran, isolates of *Fusarium graminearum* [*Gibberella zeae*] were tested for vegetative compatibility using nit mutants. Altogether 24 VCGs were reported. No correlation was found between relative virulence of the isolates and their VCGs (Naseri *et al.*, 2000).

Isolates of *Fusarium oxysporum* f. sp. *gladioli* of different geographic origin were characterized in terms of pathogenicity, vegetative compatibility and restriction fragment length polymorphisms (RFLPs). Three main RFLP groups were established which correlated with vegetative compatibility groups, but not with races. This finding establishes the reliability of VCG study as an altenative to molecular tools to find out genetic variability in fungal species (Mes *et al.*, 1994).

Among Indian isolates of *F. oxysporum* f. sp. *cucumerinum* five VCGs (arbitrarily named VCG1-A, 1-B, 1-C, 1-D and 1-E) for pathogenic isolates and three VCGs for non-pathogenic isolates of *F. oxysporum* colonizing cucumber root has been reported (Najafinia and Sharma, 2006).

Table 18.4: Some Vegetative Compatibility Groups (VCGs) of *Fusarium oxysporum*

Forma Specialis	*Host*	*Forma Specialis Code*	*Assigned Numbers for VCG*	*Number of VCG*	*Reference*
Asparagi	*Asparagus*	100-	1001 to 1008	8	Elmer and Stephens 1989
Ciceris	*Cicer*	...	...	1	Nogales *et al.*, 1993
Conglutinans	*Brassica*	010-	0101	1	Kistler *et al.*, 1991
Cubense	*Musa*	012-	0120 to 0126, 0128 to 01220	21	Ploetz 1994
Cucumerinum	*Cucumis sativus*	018-	0180	1	Katan 1996
Lycopersici	*Lycopersicon*	003-	0030 to 0033	4	Marlatt *et al.*, 1996
Melonis	*Cucumis melo*	013-	0130 to 0138	8	Jacobson and Gordon 1990; Katan *et al.*, 1994
Niveum	*Citrullus*	008-	0080 to 0082	3	Larkin *et al.*, 1990
Phaseoli	*Phaseolus*	016-	0161 to 0165	5	Woo *et al.*, 1996
Pisi	*Pisum*	007-	0070	6	Boder *et al.*, 1993; Whitehead *et al.*, 1992
Radicis-lycopersici	*Lycopersicon*	000-	0090 to 0098	8	Katan *et al.*, 1991; Rosewich *et al.*, 1997
Vasinfectum	*Gossypium*	011-	0111 to 01110	10	Fernandez *et al.*, 1994

Several patterns of VCG diversity have been identified in *Fusarium oxysporum*.

Model I, II, III and IV (Table 18.5) represents how changes in virulence could result in various patters of race-VCG diversity.

Botrytis cinerea

Botrytis cinerea (grey mould) is a major fungal pathogen of horticultural crops such as grapes, kiwifruit, berry-fruit and ornamentals in temperate regions of the world. Like most fungi, *B. cinerea* has the ability to distinguish 'self' from 'non-self'. Experiments with fusions between isolates and monosporous strains detected several types of vegetative compatibility and incompatibility reactions (Chikin and Lichachev 1997).

The *B. cinerea* homologue (Bc-hch) of Nc-het-c and Pa-hch (vegetative incompatibility loci of *Neurospora crassa* and *Podospora anserina*, respectively) was cloned and sequenced. The gene structure of Bc-hch is very close to those of Nc-het-c and Pa-hch (Fournier *et al.*, 2003).

A total of 69 VCGs have been identified in a sample of 85 isolates of *B. cinerea*; of these, 60 VCGs are represented by only a single isolate and the remaining 9 contain between 2 and 5 isolates each. These results provide a new perspective on the population structure of this important fungus. VCG analysis of progeny from a sexual cross-confirmed the hypothesis that sexual recombination generates new VCGs in addition to the parental ones.

Table 18.5: Model of Race-VCG Diversity in *Fusarium oxysporum* (Correll, 1991)

Models of VCG diversity					
I	VCG 1	→	Race A		
	VCG 2	→	Race B		
	VCG 3	→	Race C		
II	VCG 1	→	Race A		
	VCG 2	→	Race A		
	VCG 3	→	Race A		
III		↗	Race A		
	VCG1	→	Race B		
		↘	Race C		
VCG 2 →	Race A	→	Race B	→	Race C
IV				↗	VCG 1
	VCG 1	→	Race A	→	VCG 2
				↘	VCG 3
	Direction of selection pressure → Mutation				

Cercospora kikuchii

Vegetative compatibility of 58 isolates of *Cercospora kikuchii* was assessed. Two isolates were vegetatively self-incompatible. Of 56 self-compatible isolates, 16 were assigned to six multimember vegetative compatibility groups (VCGs). The other 40 isolates each belonged to a distinct VCG (Cai and Schneider 2005).

Aspergillus sp.

Pairings of complementary mutants showed that somatic compatibility between different strains was very rare in natural populations of the asexual black *Aspergillus* spp. (*A. carbonarius*, *A. japonicus*, *A. niger* and *A. tubingensis*) therefore blocking chances of horizontal transfer of mycoviruses present in natural population (Diepeningen *et al.*, 1997).

Eighty-nine isolates of *Aspergillus nige*, infecting *Welwitschia mirabilis* was studied for assessing genetic variation through RAPD marker. Although all the isolates belonged to the same vegetative compatibility group, 84 per cent had unique genotypes (Pekarek 2006).

Rhizoctonia solani

Vegetative compatibility was used as a marker to study the nature of heterokaryosis in two strains of *R. solani* AG 4. Each isolate could be placed into 1 of 2 VCGs, members of which were compatible within the group but incompatible with those of the other group and the parent strain (McCabe *et al.*, 1997).

Anastomoses between hyphae of *R. solani*, leading to both successful cell fusions and death of fused cells (vegetative incompatibility) were observed by video microscopy. Anastomosis was seen only to occur between tip cells from side branches, never main runner hyphae. Perfect hyphal fusion was only observed following pre-contact tropism between the hyphae involved (McCabe *et al.*, 1999).

Colletotrichum gloeosporioides

It was postulated that natural genetic mechanisms like vegetative incompatibility may limit nuclear exchanges in *Colletotrichum gloeosporioides* after studying nuclear condition of hyphal cells and hyphal tips of 9 *Colletotrichum gloeosporioides* [*Glomerella cingulata*] strains from olive, citrus and cherimoya (Agosteo *et al.*, 1998).

Sclerotium rolfsii

Mycelial incompatibility among 26 isolates of *S. rolfsii* [*Corticium rolfsii*] was studied. Based on mycelial compatibility, 13 vegetative incompatibility groups were identified among the isolates (Sharma *et al.*, 2002).

The sexual behaviour, clamp connection, vegetative incompatibility group and segregation of vegetative incompatibility of aquatic pathogenic strains of *S. rolfsii* [*Corticium rolfsii*], isolated from *Nelumbo nucifera*, *Trapa natans* var. *spinosa* and *Lemna minor*, have been well documented by Chand *et al.* (2003).

Oomycetes

Phytophthora infestans

Phytophthora infestans is a notorious pathogen. The self-fertility of the isolates (SFI) of *P. infestans* is relative. They convert to A1 or A2 self-sterile strains depending on the vegetative compatibility group (VCG) of SFI (SFI with VCG1 properties convert to A1 and those with VCG2 to A2). On pairing study, at the boundary of colonies of vegetatively incompatible isolates, SFI produce more oospores than at the boundary of compatible isolates. Synthesis of SFI after inoculation of solid or liquid nutrient media with a mixture of A1 and A2 isolates was not successful, possibly due to their vegetative incompatibility (Anikina *et al.*, 1997).

Vegetative incompatibility has been ascribed to be one of the reasons of limited horizontal and vertical gene transfer in tomato and potato isolates of *P. infestans*. Pairing of potato T0 with tomato T1 gave the widest barrage zone, a low number of oospores and a high proportion of abortive oospores indicating vegetative incompatibility among the isolates (Bagirova *et al.*, 1998).

Biological Importance of Vegetative Incompatibility

Heterokaryon incompatibility is a widespread phenomenon among filamentous fungi, but its biological significance is still questionable. There are two opposing views in this respect. First, it has been proposed that heterokaryon incompatibility genes exist to prevent heterokaryon formation between genetically different individuals. According to this hypothesis *het* genes determine the self/nonself recognition and preserve genetic individuality and prevent the horizontal transfer of infectious cytoplasmic elements such as senescence plasmids, mycoviruses, transposons, and debilitated organelles (Caten 1972). *het* genotype differences have been shown to limit to some extent the transfer of such cytoplasmic replicons (Anagnostakis 1982; Caten 1972; Kinsey 1990). Control of heterokaryosis may also prevent different forms of nuclear parasitism, for example, exploitation of an individual by unadapted nuclei that possess a fitness advantage in the heterokaryon (Hartl *et al.*, 1975) or resource plundering by conidia landing on an established maternal culture of a distinct individual (Debets and Griffiths 1998).

The alternative view proposes that this phenomenon is characteristic of filamentous fungi simply because they are practically the only organisms capable of natural heterokaryosis. This is also known as accident hypothesis and according to this; incompatibility system has no function in nature. The accident hypothesis postulates that *het* genes are not meant for preventing heterokaryosis and should therefore have other cellular functions in fungal biology. This hypothesis says that function of *het* gene is not solely self/non-self recognition.

Conclusion

Vegetative compatibility/incompatibility system has been studied not only in filamentous ascomycetes but in many other fungi belonging to Basidiomycetes, Deuteromycetes and Oomycetes. Where the purpose of vegetative incompatibility is to act as barrier to natural transfer of genetic or cytoplasmic materials, the role of the vic/het loci to the cellular functions is also not ruled out. Molecular and biochemical basis of self/non-self recognition are being worked out but it will need a lot of effort to bring out the complete picture. In future the VCGs can be used as taxononomic group just like race, forma specialis as VCG study can reveal a good degree of genetic variability in a natural population of a fungus. For that matter all the fungi where VCG study is in progress need to be studied thoroughly at least to the level of Fusarium spp. Moreover, a simpler and surer method to study heterokaryon formation to be identified as nit mutant study has some problems. VCG study has thrown more light on the efficacy of hypovirulent isolates of *Cryphonectria parasitica* for biological control of chestnut blight pathogen. We need to take this natural mechanism operating in many fungi seriously as it may help in determining variability of a plant pathogen, tracing its evolutionary path, geographical origin, virulence factors etc. There is also a need to know whether accident hypothesis hold true in fungi.

References

Agosteo, G.E., Pennisi, A.M. and Destri,G. (1998). Observations on the nuclear condition of *Colletotrichum gloeosporioides*. *Petria*. 8(3): 169-174.

Aimi, T., Kano, S., Wang, Q.A. and Morinaga, T. (2001). Molecular cloning of three genes encoding G protein alpha subunits in the white root rot fungus, *Rosellinia necatrix*. *Bioscience Biotechnology and Biochemistry*. 65(3): 678-682.

Aimi, T., Yotsutani, Y. and Morinaga, T. (2002a). Cytological analysis of anastomoses and vegetative incompatibility reactions in *Helicobasidium monpa*. *Current Microbiology*. 44(2): 148-152.

Aimi, T., Yotsutani, Y. and Morinaga, T. (2002b). Vegetative incompatibility in the ascomycete *Rosellinia necatrix* studied by fluorescence microscopy. *Journal of Basic Microbiology*. 42(3): 147-155.

Anagnostakis,S.L. (1982). Biological control of chestnut blight. *Science*. 215: 466-471.

Anikina, M,L., Savenkova, L.V. and D'-yakov, Y.T. (1997). Self-fertile isolates of *Phytophthora infestans*. *Biology Bulletin of the Russian Academy of Sciences*. 24(4): 334-338.

Anwar, M.M., Croft, J.H. and Dales, R.B. (1993). Analysis of heterokaryon incompatibility between heterokaryon-compatibility (h-c) groups R and GL provides evidence that at least eight *het* loci control somatic incompatibility in *Aspergillus nidulans*. *J. Gen. Microbiol*. 139: 1599-1603.

Armstrong, G.M. and Armstrong, J.K. (1978). Formae speciales and races of *Fusarium oxysporum* causing wilt of Cucurbitaceae. *Phytopathology*. 68: 19-28.

Bagirova, S.F., An, T.L. and Yu, D.T. (1998). Mechanisms of genetic isolation of *Phytophthora infestans* (Mont.) D by. specific pathogenic forms in sexual and asexual populations. *Mikologiya i Fitopatologiya*. 32(4): 47-50.

Beever, R.E. and Parkes, S.L. (2003). Use of nitrate non-utilising (Nit) mutants to determine vegetative compatibility in *Botryotinia fuckeliana* (*Botrytis cinerea*). *European Journal of Plant Pathology*. 109(6): 607-613.

Begueret, J., Turcq, B., and Clave, C. (1994). Vegetative incompatibility in filamentous fungi: het genes begin to talk. *Trends in Genetics* 10: 441.

Bodkar, L., Lewis,B.G. and Coddington,A. (1993). The occurrence of a new genetic variant of *Fusarium oxysporum* fsp. *pisi*. *Plant Pathology*. 42: 833-838.

Bourges, N., Groppi,A., Barreau,C., Clave,C. and Begueret,J. (1998). Regulation of gene expression during the vegetative incompatibility reaction in *Podospora anserina*: characterization of three induced genes. *Genetics*. 150(2): 633-641.

Cai, G. and Schneider,R.W. (2005). Vegetative compatibility groups in *Cercospora kikuchii*, the causal agent of Cercospora leaf blight and purple seed stain in soybean. *Phytopathology* 95: 257-261.

Caten, C.E. (1972). Vegetative incompatibility and cytoplasmic infection of fungi. *J. Gen. Microbiol*. 72: 221-229.

Chand, R., Verma, R., Singh,S.K., Chaurasia,S. and Lal, M. (2003). Biology of the aquatic isolates of *Sclerotium rolfsii*. *Indian Phytopathology*. 56(3): 293-294.

Coenen, A., Debets, F., and Hoekstra, R. (1994). Additive action of partial heterokaryon incompatibility (partial-het) genes in *Aspergillus nidulans*. *Current Genetics* 26: 233.

Correll, J.C. (1991). The relationship between formae speciales, races, and vegetative compatibility groups in *Fusarium oxysporum*. *Phytopathology*. 81: 1061-1064.

Correll, J.C., Klittich, C.J.R. and Leslie, J.F. (1987). Nitrate nonutilizing mutants of Fusarium oxysporum and their use in vegetative compatibility test. *Phytopathology*. 77: 1640-1646.

Cortesi,P., McCulloch,C.E., Song, H.Y., Lin,H.Q. and Milgroom,M.G. (2001). Genetic control of horizontal virus transmission in the chestnut blight fungus, *Cryphonectria parasitica*. *Genetics*. 159(1): 107-118.

Dalstra, H.J.P., Swart, K., Debets,A.J.M., Saupe,S.J. and Hoekstra, R.F. (2003). Sexual transmission of the [Het-s] prion leads to meiotic drive in *Podospora anserina*. Proceedings of the National Academy of Sciences of the United States of America. 100(11): 6616-6621.

Debets,A.J.M. and Griffiths, A.J.F. (1998). Polymorphism of *het*-genes prevents resource plundering in *Neurospora crassa*. *Mycol. Res.* 102: 1343-1349.

Debets, F.X., Yang. and Griffiths,A.J.F. (1994). Vegetative incompatibility in Neurospora: its effect on horizontal transfer of mitochondrial plasmids and senescence in natural populations. *Curr. Genet.* 26: 113-119.

Deng,F., Melzer, M.S. and Boland, G.J. (2002). Vegetative compatibility and transmission of hypovirulence-associated dsRNA in *Sclerotinia homoeocarpa*. *Canadian Journal of Plant Pathology*. 24(4): 481-488.

Diepeningen,A.D., Debets,A.J.M. and Hoekstra,R.F. (1997). Heterokaryon incompatibility blocks virus transfer among natural isolates of black *Aspergillus*. *Current Genetics*. 32(3): 209-217.

Elmer, W.H. and Stephens,C.T. (1989). Classification of *Fusarium oxysporum* f. sp. *asparagi* into vegetatively compatible groups. *Phytopathology*. 79: 88-93.

Fernandez, D., Assigbetse,K., Dubois, M.P. and Geiger,J.P. (1994). Molecular characterization of races and vegetative compatibility groups in *Fusarium oxysporum* f.sp. *vasinfectum*. *Applied Environmental Microbiology*. 60: 4039-4046.

Fournier,E., Levis,C., Fortini,D., Leroux,P., Giraud,T. and Brygoo, Y. (2003). Characterization of Bc-hch, the *Botrytis cinerea* homolog of the *Neurospora crassa* het-c vegetative incompatibility locus, and its use as a population marker. *Mycologia*. 95(2): 251-261.

Furga, W.H. (1997). Vegetative incompatibility in *Ascochyta pisi* Lib. *Acta Microbiologica Polonica*. 46(1): 75-82.

Gaag, M., Debets, A.J.M. and Hoekstra, R.F. (2003). Spore killing in the fungus *Podospora anserina*: a connection between meiotic drive and vegetative incompatibility? *Genetica*. 117(1): 59-65.

Garraway,M.O. and Evans,R.C. (1984). Fungal nutrition and physiology. *John Wiley and Sons, New York*. 401pp.

Giovannetti,M., Sbrana, C., Strani, P., Agnolucci, M., Rinaudo, V. and Avio, L. (2003). Genetic diversity of isolates of *Glomus mosseae* from different geographic areas detected by vegetative compatibility testing and biochemical and molecular analysis. *Applied and Environmental Microbiology*. 69(1): 616-624.

Glass, N.L., Jacobson, D.J., and Shiu, P.K.T. (2000). The genetics of hyphal fusion and vegetative incompatibility in filamentous ascomycete fungi. *Annual Review of Genetics*. 34: 165-186.

Glass, N.L., Kuldau, G.A. (1992). Mating type and vegetative incompatibility in filamentous ascomycetes. *Annual Review of Phytopathology* 30: 201.

Hartl, D.L., Dempster,E.R. and Brown, S.R. (1975). Adaptive significance of vegetative incompatibility in *Neurospora crassa*. *Genetics* 81: 553-569.

Hart, L.P. and Endo, R.M. (1981). The effect of time of exposure to inoculum, plant age, root development and root wounding on Fusarium yellows of celery. *Phytopathology*. 71: 77-79.

Heerden, S.W., Geletka, L.M., Preisig, O., Nuss,D.L., Wingfield,B.D. and Wingfield, M.J. (2001). Characterization of South African *Cryphonectria cubensis* isolates infected with a *C. parasitica* hypovirus. *Phytopathology*. 91(7): 628-632.

Hiscock, S.J., Kues, U. and Dickinson, H.G. (1996). Molecular mechanisms of self-incompatibility in flowering plants and fungi–different means to the same end. *Trends in Cell Biology* 6: 421.

Jacobson,D.J. and Gordon,T.R. (1990). Further investigation of vegetative compatibility within *Fusarium oxysporum* f.sp. *melonis*. *Canadian Journal of Botany*. 68: 1245-1248.

Johannesson, H., Gustafsson,M. and Stenlid, J. (2001). Local population structure of the wood decay ascomycete *Daldinia loculata*. *Mycologia*. 93(3): 440-446.

Katan, T. (1996). Vegetative compatibility groups in populations of *Fusarium oxysporum* in Israel. (Abstr.) *Phytoparasitica*. 24: 139.

Katan, T., Katan, J., Gordon, T.R. and Pozniak,D. (1994). Physiologic races and vegetative compatibility groups of *Fusarium oxysporum* f.sp. *melonis* in Israel. *Phytopathology*. 84: 153-157.

Katan, T., Zamir, D., Sarfatti, M. and Katan, J. (1991). Vegetative compatibility groups and subgroups in *Fusarium oxysporum* f.sp. *radicis-lycopersici*. *Phytopathology*. 81: 255-262.

Kao,T.H., McCubbin,A.G. (1996). How flowering plants discriminate between self and non-self pollen to prevent inbreeding. Proceedings of the National Academy of Sciences USA 93: 12059.

Kauserud, H. (2004). Widespread vegetative compatibility groups in the dry-rot fungus *Serpula lacrymans*. *Mycologia*, 96(2): 232–239.

Kistler, H.C., Momol, E.A. and Benny, U. (1991). Repetitive genomic sequences for determining relatedness among strains of *Fusarium oxysporum*. *Phytopathology*. 81: 331-336.

Kothe, E. (1996). Tetrapolar fungal mating types: sexes by the thousands. *FEMS Microbiological Reviews* 18: 65.

Kraft, J.M. and Hagland, W.A. (1978). A repraisal of the race classification of Fusarium oxysporum f. sp. pisi. *Phytopathology* 68: 273-275.

Larkin, R.P., Hopkins,D.L. and Martin,F.N. (1990). Vegetative compatibility within *Fusarium oxysporum* f.sp. *niveum* and its relationship to virulence, aggressiveness, and race. *Canadian Journal of Microbiology*. 36: 352-358.

Magor,B.G., De Tomaso. A., Rinkevich,B., Weissman,I.L. (1992). Allorecognition in colonial tunicates: protection against predatory cell lineages? *Immunological Reviews* 167: 69.

Marlatt, M.L., Correll, J.C., Kaufmann, P. and Cooper,P.E. (1996). Two genetically distinct populations of *Fusarium oxysporum* f.sp. *lycopersici* race 3 in the United States. *Plant Dis*. 80: 1336-1342.

McCabe,P.M., Gallagher,M.P. and Deacon,J.W. (1999). Evidence for segregation of somatic incompatibility during hyphal tip subculture of *Rhizoctonia solani* AG 4. *Mycological Research*. 103(10): 1323-1331.

McCabe,P.M., Gallagher,M.P. and Deacon,J.W. (1999). Microscopic observation of perfect hyphal fusion in *Rhizoctonia solani*. *Mycological Research*. 103(4): 487-490.

Mes,J.J., Van,D.J., Roebroeck,E.J.A., Egmond,E.V., Aartrijk,J.V., Boonekamp,P.M. (1994). Restriction fragment length polymorphisms, races and vegetative compatibility groups within a worldwide collection of *Fusarium oxysporum* f.sp. *gladioli*. *Plant Pathology*. 43 (2): 362–370.

Metzenberg, R.L. (1990). The role of similarity and difference in fungal mating. *Genetics* 125: 457.

Milgroom,M.G. and Brasier,C.M. (1997). Potential diversity in vegetative compatibility types of *Ophiostoma novo-ulmi* in North America. *Mycologia*. 89(5): 722-726.

Milgroom,M.G. and Cortesi,P. (1999). Analysis of population structure of the chestnut blight fungus based on vegetative incompatibility genotypes. Proceedings of the National Academy of Sciences of the United States of America. 96(18): 10518-10523.

Muirhead,C.A., Glass, N.L. and Slatkin, M. (2002). Multilocus self-recognition systems in fungi as a cause of trans-species polymorphism. *Genetics*. 161(2): 633-641.

Mylyk, O.M. (1975). Heterokaryon incompatibility genes in *Neurospora crassa* detected using duplication-producing chromosome rearrangements. *Genetics* 80: 107-124.

Najafinia, M. and Sharma, P. (2006). Vegetative compatibility groups and pathogenic variability among isolates of *Fusarium oxysporum* f.sp. *cucumerinum causing* wilt in India. International meeting on Biotic and Abiotic Stress Responses in Plants, 11-13 December, ICGEB, New Delhi.

Naseri, B., Alizadeh, A., Saidi, A. and Safaee, N. (2000). Population diversity *of Fusarium graminearum* based on vegetative compatibility groups (VCGs) and its relationship to virulence of isolates. *Iranian Journal of Plant Pathology*. 36(3/4): 261-280.

Nauta, M.J. and Hoekstra, R.F. (1996). Vegetative Incompatibility in Ascomycetes: Highly Polymorphic but Selectively Neutral? *Journal of Theoretical Biology*, 183(1): 67-76.

Nelson, P.E. (1981). Life cycle and epidemiology of *Fusarium oxysporum*. In: ME Mace, AA Bell and CH Beckman, eds. Fungal Wilt Disease of Plants. *Academic Press. New York,* 51-80pp.

Newton, A.C., Osbourn, A.E. and Caten, C.E. (1998). Heterokaryosis and vegetative incompatibility in *Stagonospora nodorum*. *Mycologia*. 90(2): 215-225.

Nogales-Moncada, A.M., Perez-Artes,E. and Jimenez-Diaz, R.M. (1993). VCG diversity within *Fusarium oxysporum* f. sp. *ciceris* (Abstr.) Page 169 in: Abstr. Int. Congr. Plant Pathol, 6th. National Research Council Canada, Ottawa.

Paoletti, M., Buck, K.W. and Brasier,C.M. (2006). Selective acquisition of novel mating type and vegetative incompatibility genes via interspecies gene transfer in the globally invading eukaryote *Ophiostoma novo-ulmi*. *Molecular Ecology*. 15(1): 249-262.

Pekarek, E., Jacobson,K. and Donovan,A. (2006). High Levels of Genetic Variation Exist in *Aspergillus niger* Populations Infecting *Welwitschia mirabilis*. *Hook Journal of Heredity*. 97(3): 270-278.

Perkins, D.D. and Davis,R.H. (2000). *Neurospora* at the millennium. *Fungal Genetics and Biology*. 31(3): 153-167.

Ploetz, R.C. (1994). Panama disease: Return of the first banana menace. *Int. J. Pest Manage.* 40: 326-336.

Pomazi, A. (1999). Population genetics analysis in fungi. Vegetative incompatibility and clonal multiplication in the genus *Fusarium*. *Mikologiai Kozlemenyek*. 38(1/3): 69-78.

Pound, G.S. and Fowler,D.L. (1953). Fusarium wilt of radish in Wisconsin. *Phytopathology*. 43: 277-280.

Puhall, J.E. (1985). Classification of strains of *Fusarium oxysporum* on the basis of vegetative compatibility. *Can. J. Bot.* 63: 179-183.

Puhalla, J.E. and Mayfield,J.E. 1974. The mechanism of heterokaryotic growth in *Verticillium dahliae*. *Genetics*. 76: 411-422.

Puhalla, J.E. and Spieth, P.T. (1985). A comparison of heterokaryosis and vegetative incompatibility among variety of *Gibberella fujikuroi* (*Fusarium moniliforme*). *Exp. Mycol.* 9: 39-47.

Raper, J.R. (1966). Genetics of Sexuality in Higher Fungi. New York, Ronald Press.

Rosewich, U.L., Pettway, R.E., Kistler,H.C. and Katan,T. (1997). Morphological and molecular characterization of isolates of *Fusarium oxysporum* f. sp. *radicis-lycopersici* from Florida (Abstr.) Page 83 in: abstr. Fungal Genet. Conf., 19th Fungal Genetics Stock Center, Kansas City, KS.

Sharma, B.K., Singh,U.P. and Singh,K.P. (2002). Variability in Indian isolates of *Sclerotium rolfsii*. *Mycologia*. 94(6): 1051-1058.

Satyaprasad, K., Bateman, G.L. and Ward, E. (2000). Comparisons of isolates of *Fusarium avenaceum* from white lupin and other crops by pathogenicity tests, DNA analyses and vegetative compatibility tests. *Journal of Phytopathology*. 148(4): 211-219.

Slippers, B., Wingfield,M.J., Coutinho,T.A. and Wingfield,B.D. (2001). Population structure and possible origin of *Amylostereum areolatum* in South Africa. *Plant Pathology*. 50(2): 206-210.

Smit, W.A., Wingfield, B.D. and Wingfield, M.J. (1997). Vegetative incompatibility in *Diaporthe ambigua*. *Plant Pathology*. 46(3): 366-372.

Smith, M.L., Micali, O.C., Hubbard,S.P., Mir-Rashed, N., Jacobson, D.J. and Glass, N.L. (2000). Vegetative incompatibility in the het-6 region of *Neurospora crassa* is mediated by two linked genes. *Genetics*. 155(3): 1095-1104.

Snyder, W.C. and Hansen, H.N. (1940). The species concept in *Fusarium*. *Am. J. Bot.* 27: 64-67.

Vallesi, A., Giuli,G., Bradshaw,R.A., and Luporini,P. (1995). Autocrine mitogenic activity of pheromones produced by the protozoan ciliate Euplotes raikovi. *Nature* 376: 522.

Wang, K.R., Meng, A.Z. and Bu, Z.G. (1996). Vegetative compatibility among isolates of *Fusarium oxysporum* f. sp. *vasinfectum* from diseased tissues of cotton plants. *Journal of Nanjing Agricultural University*. 19(1): 30-33.

Wang, K.R., Zhou, E.X., Jiang, A.P. and Cheng, G.Y. (2004). Transmission of dsRNA in *Cryphonectria parasitica* and it's affecting factors. *Scientia Silvae Sinicae*. 40(1): 185-188.

Wendland, J., Vaillancourt, L.J., Hegner, J., Lengeler,K.B., Laddison, K.J., Specht, C.A., Raper, C.A., and Kothe, E. (1995). The mating-type locus B alpha 1 of *Schizophyllum commune* contains a pheromone receptor gene and putative pheromone genes. *EMBO Journal* 14: 5271.

Whitehead, D.S., Coddington, A. and Lewis, B.G. (1992). Classification of races by DNA polymorphism and vegetative compatibility grouping in *Fusarium oxysporum* f. sp. *pisi*. *Physiol. Mol. Plant Pathol*. 41: 295-305.

Wikler, K., Gordon, T.R., Clark, S.L., Wingfield, M.J. and Britz, H. (2000). Potential for outcrossing in an apparently asexual population of *Fusarium circinatum*, the causal agent of pitch canker disease. *Mycologia*. 92(6): 1085-1090.

Woo, S.L., Zoina, A., Del, sorbo. G., Lorito, M., Nanni, B., Scala, F. and Noviello,C. (1996). Characterization of *Fusarium oxysporum* f. sp. *phaseoli* by pathogenic races, VCGs, RFLPs and RAPD. *Phytopathology*. 86: 966-973.

Wu,J., Saupe, S.J., and Glass, N.L. (1998). Evidence for balancing selection operating at the het-c heterokaryon incompatibility locus in a group of filamentous fungi. *Proceedings of the National Academy of Sciences USA* 95: 12398.

Xiang, Q.J., Rasmussen, C. and Glass, N.L. (2002). The ham-2 locus, encoding a putative transmembrane protein, is required for hyphal fusion in *Neurospora crassa*. *Genetics*. 160(1): 169-180.

Chapter 19

Molecular Methods for Detection of Plant Pathogen with Special Reference to DNA Based Methods

*Kahekashan Tanveer, Sairose S. Lalani and M.M.V. Baig**

Department of Botany and Biotechnology,
Yeshwant Mahavidyalaya, Nanded – 431 602, India

ABSTRACT

The fundamental aspect of studying any diseases and its management lies in the detection, identification and differentiation of microbial phytopathogens–viruses, viroids, phytoplasmas, bacteria and fungi. The conventional methods of isolation, microbiological morphological characterization are time consuming and difficult may derive indecisive results. The advent of modern molecular techniques. On the other hand can provide accurate consistent and reproducible result rapidly. Because of the distinct advantage over conventional method, biochemical, immunoassay, nucleic acid based methods are preferred. The advantage of these molecular methods lies in the identification of obligate parasites and fungi, which are otherwise difficult to cultivate, or takes long time. Some of the nucleic acid methods are described in the paragraphs to follow.

Keywords: *Immunoassays, ELISA, RFLP, T-RFLP, Autoradiography.*

* E-mail: mmvbaig@gmail.com

Introduction

The correct and exact identification of causal organism of plant diseases can lead to successful disease control. Symptom based diagnostics of diseases and their causative agents by symptoms are not so reliable. This can be attributed to the similarity of in many diseases and physiological disorders caused by abiotic factors. Beside plant pathogens can cause asymptomatic disease or disease with indistinct symptoms as non reliability of diagnosis using indicator plants or differential cultivars (Moricca *et al.*, 1998). In pesticidal disease control methods, the treatments with pesticide are less efficient when applied at the stage of symptom expression.

Identification of bacterial and fungal pathogens using isolation in pure culture and the testing thereafter by microbiological methods are cumbersome and time consuming. The diagnostic methods of plant viruses by means of electron microscopy are time-consuming as well as laborious. These techniques have low productivity and are difficult to be made automated. Apart from these problems, the obligate parasites are difficult or almost impossible to culture *in vitro e.g.* Smuts rusts, viroids, viruses, phytoplasmas etc. (Ward *et al.*, 2004). Thus the need to overcome all these problems of conventional methods of diagnostics in plant pathology has led to the development of newer methods of plant pathogen detection and identification. However, these new methods basically differ from the conventional techniques but can supplement with conventional methods or in some cases can replace conventional techniques.

The modern technologies, especially, the molecular based methods used in plant pathology for diagnostics of plant pathogens are extremely wide. Some of them are improved or modified conventional plant pathological methods including certain cultural, histo-chemical or biochemical techniques, and electron microscopy while some of these techniques have been adapted from other sciences such as molecular biology, biochemistry, or immunology. The development of novel methods for detection and identification of plant pathogens is a continuous active process and most of these techniques have inherent advantages and disadvantages (Schaad and Frederick, 2002; Rementeria *et al.*, 2004; Moricca *et al.*, 2004).

This article focuses on diagnostic methods based on DNA based methods of molecular biology. The specific examples of application of the molecular methods for detection and identification of viruses, bacteria and fungi are limited to the most important pathogens and are mostly mentioned without references to the original papers. The objective is not to provide a comprehensive list of phytopathogens or other microorganisms that are detected and/or characterized with these techniques. The purpose is rather to focus on the principle of each method, in order to demonstrate how the new molecular approaches, whether developed in the field of plant pathology, medical science, industry, or environmental microbiology, can be successfully applied to solve practical problems in different biological systems. We thus aim at demonstrating how the new technologies can overcome problems inherent with conventional identification of plant pathogens (Gilles *et al.*, 2000).

Immunoassays

Immunodiagnostic assays have been applied for detection, differentiation and quantification of microbial pathogens. They have been found to be highly specific, sensitive, simple, rapid and cost effective and can be automated for large scale applications. Hence, the immunoassays have largely replaced conventional analytic methods which are time-consuming, cumbersome and expensive (Narayanasamy 2001, 2005). Improvements and modifications, however, have been made to suit different host-pathogen combinations. The advancements during the recent years have focused the attention of the researchers on the need for applying rapid, reliable and sensitive methods for the detection of microbial pathogens and diagnosis of diseases caused by them.

Immunoassays primarily depend on the visualization of the specific binding of antibody to the antigen, directly or indirectly. The formation of precipitate or precipitin lines indicating the binding between reactants may be seen. These tests require large volumes of the reactants and longer time to provide results which may be inconclusive. On the other hand, the assays requiring labeling of antibodies are more sensitive and rapid with possibility of automation for large scale application. The antibodies may be labeled with enzymes such as alkaline phosphatase or fluorescent dyes.

Among the techniques using labeled antibodies, Enzyme-Linked Immunosorbent Assay (ELISA) has been most widely applied for studying various aspects of plant pathogen interactions, in addition to detection and characterization of microbial pathogens. Three formats of ELISA *viz.*, the Double Antibody Sandwich (DAS)-ELISA, the Triple Antibody Sandwich (TAS)-ELISA and the Plate Trapped Antibody (PTA)-ELISA are frequently employed for detection and disease diagnosis. Other immunoassays such as Direct Tissue Blot Immunoassays (DTBIA) and Immunosorbent Electron Microscopy (ISEM) for *in situ* detection of viruses have also been employed. The development of monoclonal antibody (MAB) technology has remarkably enhanced the specificity of immunoassays, resulting in the differentiation of strains, biotypes or races of microbial pathogens more precisely than by using polyclonal antibodies (PABs).

Enzyme-Linked Immunosorbent Assay (ELISA)

ELISA has been frequently employed using PABs specific against viral nucleoprotein, due to the ease in the purification step, in preference to other methods that rely on detection of complete virus particle with envelop which require a different enveloped particle preparation (Hsu *et al.*, 2000; Takeuchi *et al.*, 2001; Tanina *et al.*, 2001). *Calla lilies* (*Zantedeschia* spp.) were infected by a new tospovirus. By using ELISA, this new virus was found to be distantly related to *Watermelon mottle silver virus* (WSMoV) (Chen *et al.*, 2005). By using PABs and MABs specific to *Calla lily chlorotic spot virus* (CCSV) and WSMoV in indirect ELISA and immuno-blotting assays, CCSV was shown to be a distinct member in the genus Tospovirus. It was suggested that CCSV is a new species belonging to WSMoV serogroup (Lin *et al.*, 2005). The presence of *Eriwinia chrysanthemi* in potato stems and tubers was detected by using a specific MAB in the triple antibody ELISA test (Singh *et al.*, 2000). Melanins derived from 1,8-dihydroxy naphthalene (DHN) have an important role in the pathogenicity and survival of fungal pathogens like *Alternaria altetnata*. Competitive inhibition-ELISA format revealed that the phage-display antibody (scFV) M1 bound specifically to 1,8 DHN that was located in the septa and outer (primary) walls of wild type *A. alternate* conidia. It is possible to detect melanized fungal pathogens in different plant tissues by using M1 antibody (Carzaniga *et al.*, 2002).

Diagnostics Based on Nucleic Acid Analysis

The progress of molecular biology in the 1980s–1990s allowed the development of specific, sensitive, and rapid methods for plant disease diagnostics. The most important advantage of the molecular methods in comparison with immunochemical techniques for detection and identification of plant pathogens is that they can be used for analysis of genomic information *i.e.* the most conservative properties of a pathogen. The application of nucleic acid based detection method allows researchers to overcome a number of disadvantages encountered in immunological based techniques. The insufficient specificity expressed by cross-reactivity leading to false positive results and insufficient sensitivity in cases where the target pathogen is present in too little quantity. The important aspect associated with these molecular methods is very little amount of samples (microlitres) can be analyzed by molecular methods with high accuracy. An important advantage of some nucleic acid methods is the combination of fast detection and identification of plant pathogens.

There is a range of molecular methods adopted in various areas of plant pathology, *e.g.* for development of genetic characteristic of species, *form species*, stains, races, and isolates, studying the population structure of pathogenic species, identification of pathogenicity genes, investigation of plant resistance to diseases, toxinogenesis, early stages of host–parasite interactions, etc. (Narayansamy 2001, 2005). In the recent decade, the number of phytopathology studies with these techniques has been growing.

There are many methods that are most important for plant pathogen detection and identification and are currently used for this purpose. They may all be divided into two groups: the probing methods based on nucleic acid hybridization with DNA/RNA probe sequences and those based on polymerase chain reaction (PCR) amplification of nucleic acid sequences. The techniques of the first group were developed before the PCR methods and were alternatives to PCR for pathogen identification before discovery of PCR. Nowadays the probing methods are used in conjunction with PCR. In some cases, PCR combined with probe enhances the diagnostic potential of PCR. Depending on whether the target nucleotide sequence is known or partially or fully unknown, a particular version of a method within each group may be employed. Plant pathogen diagnostics involving restriction fragment length polymorphism (RFLP) will also be briefly discussed. The PCR, nucleic acid hybridization, and RFLP are the subjects covered in huge and rapidly growing number of publications, including molecular biology textbooks, explaining the theoretical aspects of these techniques.

DNA-Based Methods

Nucleic acid-based (primarily DNA-based) diagnostic tools have been widely adopted for applications in research labs, regulatory labs, and diagnostic clinics.

DNA-based diagnostic have increased the accuracy of pathogen identifications and decreased the time for most diagnoses. These technologies are likely to be further adopted in the near future. A long list of DNA-based technologies appropriate for invasive species monitoring have been reviewed and include: Heteroduplex polymorphism assay (HPA) (Gil-Lamaignere *et al.*, 2003), Single-Strand Conformational Polymorphism and Denaturing Gradient Gel Electrophoresis (SSCP and DGGE) (Masoud *et al.*, 2004; Noll and collins, 1987; Plachu *et al.*, 2005), Polymerase Chain Reaction/Restriction Fragment Length Polymorphism (PCR/RFLP), Species-Specific PRC (SSP), DNA barcoding (COX fingerprinting), quantitative PCR (qPCR), Shotgun barcoding, Terminal-Restriction Fragment Length Polymorphism (T-RFLP), and Phylogenetic Oligonucleotide microArray (POA) (Darling and Blum, 2007).

Southern Blotting

Southern blotting was named after Edward M. Southern who developed this procedure at Edinburgh University in the 1970s. In this, DNA molecules are transferred from an agarose gel onto a membrane. It allows to determine the molecular weight of a restriction fragment, to measure relative amounts in different samples and to locate a particular sequence of DNA within a complex mixture in plant pathogen identification. Nucleic acid-based techniques involving hybridization with and amplification of nucleic acid sequences have been shown to be useful to study various aspects of microbial pathogens and their interaction with host plants, in addition to detection differentiation and quantification of pathogens and their strains/races/pathotypes: (*i*) production of disease-free seeds and propagative materials; (*ii*) prevention of introduction of new pathogens by domestic and international quarantines; (*iii*) field surveys to determine incidence and distribution in a geographical location/country; (*iv*) detection and resolution of disease complexes; (*v*) identification of additional hosts to determine the manner of pathogen perptuation; (*vi*) studying the nature of pathogen-vector relationships particularly in the case of viruses; (*vii*) screening to assess the levels of disease resistance in genotypes, cultivars and breeding lines; (*viii*) studying of the mechanisms of disease resistance of plants following incorporation of genes or induced by inducers of resistance and (*ix*) functions of trangenes of pathogen, plant and animal origins.

For example, Southern Blotting could be used to locate a particular gene within an entire genome. A sample of DNA (genomic or other source) containing fragments of different sizes is digested with a restriction enzyme or mechanical shearing and separated by gel electrophoresis and transferred from an agarose gel onto a membrane which is then incubated with a probe which is single-stranded DNA segment of few to several nucleotides while polyacrylamide is preferred for smaller fragments. Very large DNA fragments of upto 1000-2000 kb are seperated in agarose gel with pulsed electrical field or field inversion. This probe will form base pairs with its complementary DNA sequence and bind to form a double-stranded DNA molecule. This is called as hybridization reaction. The probe is labeled before hybridization either radioactively or enzymatically (*e.g.* alkaline phosphatase or horseradish peroxidase). Genomic DNA is digested with one or more restriction enzymes, and the resulting fragments are separated according to size by electrophoresis through an agarose gel. The DNA is then denatured *in situ* and transferred from the gel to a solid support (usually nitrocellulose filter or nylon membrane). The relative positions of the DNA fragments are preserved during their transfer to the filter. The DNA attached to the filter is hybridized to radiolabeled DNA or RNA, and autoradiography is used to locate the positions of bands complementary to the probe.

The tissue–print hybridization, involves the transfer of viral nucleic acid from infected plant tissues directly on to nitrocellulose or nylon membrane followed by hybridization of the printed membrane with radioactive or non-radioactive chemiluminescent DIG-labelled probes. This technique is useful for studying the localization pattern of the virus in specific host plant tissues. The DIG-labelled probes could be employed for the detection and differentiation of viruses.

The use of digoxigenin (DIG)-labelled cDNA probes has substantially enhanced the sensitivity of detection of viriods, phytopathogens, viral (Navaez *et al.*, 2000), bacterial and fungal pathogens. DIG-labelled cDNA probes were useful for the detection of *potato spindle tuber viriod* (PSTVd) (Welnicki and Hiruki 1992), *sweet potato witches broom phythoplasma* (Ko and Lin 1994), *Cucumber mosaic virus* (CMV) (Kiranmai *et al.*, 1998), *Pseudomonas campestris pv. phaseolicola* (Schaad *et al.*, 1989) and *Pythium spp.* (Klassen *et al.*, 1996).

Autoradiography

Autoradiography is the direct exposure of film by β-particles or γ-rays. It is a procedure for localizing and recording a radiolabeled compound within a solid sample, which involves the production of an image in a photographic emulsion. Nucleic acid-based techniques are either based on the specificity by which nucleic acids (DNA or RNA) hybridize to form double-stranded molecules or detection of similarities between nucleic acids by restriction enzymes to cleave DNA into fragments at or near a defined recognition sequence. Specific sequences can be identified, labeled and employed as a probe for hybridization with nucleic acid from target organism. Different hybridization methods such as colony and dot-blot hybridization have been performed for detection and qualitative differentiation to distinguish groups of plant pathogens, when a specific probe is available. Nucleic acid probes may be labeled with radioactive (P^{32} or S^{35}) or non-radioactive (biotin, digoxigenin, fluorescein, enzyme, steroid antigens) marker. The hazards associated with handling radioisotopes, despite the greater sensitivity of probes labeled with radioactive markers, have resulted in wider application of non-radioactive markers. Among the non-radioactive labels, biotin and digoxigenin labeled deoxyuridine triphosphate (Dig-dUTP) are incorporated into the DNA more frequently. Biotin can be readily cross-linked to deoxyribonucleotides and DNA with biotin is able to form stable complexs with avidin. Enzymes such as peroxidase and alkaline phosphatase can be attached to avidin. The avidin enzyme conjugate is then used to detect the presence of biotin colorimertrically. The digoxenin labeled hybrids can be detected by ELISA, by using an anti-digoxigenin antibody-ALP conjugate.

RNA is the nucleic acid of most plant viruses. Diagnostic protocols for these viruses often require a preliminary reverse transcription step to generate a DNA sequence with which to conduct the array of amplification-based and/or restriction enzyme-based methods. The challenge of high cost per diagnosis associated with many nucleic acid based technologies is a significant issue for many plant diagnostic clinics processing routine plant samples. The high cost of the instrumentation involved in these DNA based methods had hindered application of these technologies for field diagnostics (*e.g.*, thermocyclers, DNA sequencers MicroArray platform and readers etc.) (Healy *et al.*, 2004; Wise *et al.*, 2007). However, more emphasis on development of methods make it feasible, user friendly and workable at field level.

As compared to human and related pathogens relatively very few plant pathogen species and about 50 fungal genome (http:/fungal.genome.duke.edu) sequence are deposited in public repositories. This has restricted the development of species specific, pathovar specific or strain specific probes to support the diagnosis of some of devastating pathogens besides hindering the development of forensic protocols to assist law enforcement agencies.

Restriction Fragment Length Polymorphism (RFLP)

The RFLP analysis proceeds with isolation of DNA from pathogens/plant tissues and is "cut" by specific bacterial enzymes, restriction endonucleases that recognize characteristic nucleotide sequences. The restriction fragments are separated by electrophoresis in agarose gel, transferred to nitrocellulose membrane with the Southern blotting, then are hybridized with probes which are usually DNA fragments of the size 500–2000 bases, then analyzed according to the band position in the gel or with the autoradiography. Random products of restriction of genomic DNA can serve as probes. However, more often specially selected clones of unique and rarely repeating DNA sequences, used for drawing up molecular maps of the genome, are used as probes. RFLP technique to identify different strains in cereals (Alizadeh *et al.*, 1997), in rice plant tissues (Ura *et al.*, 1998), papaya (Gibb *et al.*, 1996) and rye (Nicholson *et al.*, 1994). The major advantages of RFLP-analysis are its universality and good reproducibility.

The RFLP-analysis requires considerable amounts of DNA. Depending on the genome dimension, electrophoresis is carried out with 2–15 μg of the mixture of restriction fragments is loaded in each well in agarose gel and hence cannot be used when quantity of material is very less. Beside, this method requires extra pure DNA free from protein, RNA, carbohydrates, and other substances that can interfere with restriction or have an adverse effect on electrophoretic separation of the restriction fragments. Maintenance of a large RFLP-clone library and procurement of reagents and materials is time and resource consuming but safety measures for handling of radioactive materials are also expensive. Owing to all these reasons RFLP-analysis is used in diagnostic laboratories less often than PCR, regardless of all its indisputable advantages.

PCR Methods

The development of Polymerase chain reaction (PCR) in 1983 by Kerry Mullis has revolutionized most of the DNA based techniques. "Science" regarded PCR as one of the most remarkable discoveries of that decade. Practical applications of the method were first reported in 1985. PCR spread worldwide at a lightning speed. The number of publications on the studies where PCR is used as a method of scientific and applied research is recorded high. However, the primary application of the method was in microbiological diagnostics and became a preferred tool for detection and identification of plant pathogens (Dietgen *et al.*, 2002).

The importance of PCR for diagnostic task follows from the fact that it enables repeated finding and amplification *in vitro* of specific DNA sequences (or cDNA transcripts) present in trace amount among a huge number of other polynucleotides. This provides extremely sensitive detection and identification of a target pathogenic microorganism in complex biological or environmental matrix. PCR allows detection of about 10 bacterial cells in a sample. PCR is based on the process of natural DNA replication which includes untwining of the DNA double helix, separation ("unzipping") of the DNA strands, and complementary assembling of both strands by the cell enzyme DNA-dependent DNA polymerase. The DNA replication cannot begin at any point but only in fixed starting blocks, short double-stranded areas. Marking the DNA site specific only for the given species (but not for other species) with such blocks allows repeated reproduction (amplification) of this site.

PCR process includes the mixture of two oligonucleotide probes–primers forward and reverse are added to the DNA sample melted by temperature in a very small tube. The primers are short oligonucleotides of approximately 16–35 bases in length which serves as precursors for synthesis of the chosen DNA site. They are complementary to the sequences on the left and right ends of the fragment and are oriented so as to constrain the completion of the new DNA strand. The primers added to the sample being analyzed binds to the DNA sequence to which they are complementary. After binding (annealing) of two primers, the second strand of the specific DNA fragment is generated, extending from the primers, with the help of a thermostable DNA polymerase, Taq polymerase (isolated from the thermophilic bacterium *Thermus aquaticus* and having the activity optimum at 70–72°C). The newly synthesized DNA fragments serve as matrix for synthesis of the new strands in the next amplification cycle, *i.e.* chain reaction takes place.

The number of copies grows in logarithmic succession and after 25 amplification cycles (from 30 seconds to a few minutes each) about 100 fragment copies are produced. In an automated thermal cycler, 2–3 hours are sufficient for 30–40 cycles, and they generate more than a million copies of DNA or complementry DNA (cDNA) fragments, from the length as short as 50 bp to over 10,000 bp. This results in production of enough DNA for visual registration of the reaction results after electrophoresis.

Classical PCR-analysis comprises of three basic stages: sample preparation the isolation of DNA or RNA, PCR, and detection of the product–amplified DNA fragment. The reaction mixture consists of the following four main components: (i) DNA sample of the segment to be amplified; (ii) a pair of primers; (iii) Taq DNA polymerase; and (iv) four types of deoxyribonucleotide triphosphate (dNTP). Apart from these component, Mg2+ ions must be present in reaction buffer.

The synthesized product of polymerase reaction (amplicon) can be detected by various ways. One of the simplest, efficient, and common method is electrophoretic separation of the amplified DNA with subsequent detection of the fragments with intercalating fluorescent probes, such as ethidium bromide. In addition to ethidium bromide, PCR products can be detected in gel by staining with silver. The difference of several nucleotides between the amplified fragments is found with electrophoresis in agarose gel.

Prior to electrophoresis, the amplified DNA may be "cut" with a restriction endonuclease to study RFLP of the amplified product. Detection of the amplified DNA by hybridization with different labelled probes is also used (Cocolin *et al.*, 2006). Detection methods using microtitre plate format where the reaction products are revealed by ELISA (PCR–ELISA) have become common. Combination of PCR and ELISA has been used for analysis of plant pathogens. PCR-ELISA was found to be useful for the detection of *Wheat streak mosaic virus* (Jones *et al.*, 2005), *Plum pox virus* (PPV), *Cherry leaf roll virus* (CLRV), *Citrus tristeza virus* (CTV), *Prunus necrotic ringspot virus* (PNRSV) and *Tomato ring spot virus* (ToRSV) (Olmos *et al.*, 1997). There are also other methods to analyze the amplification reaction product. They include high performance liquid chromatography (HPLC), capillary electrophoresis, and mass spectroscopy.

In modern plant pathology, PCR is applied not only for plant pathogen diagnostics but also in diversity, taxonomy and phylogeny of plant pathogens including viroids, viruses, phytoplasmas, bacteria, fungi, oomycetes, and nematodes. This technique is widely used for monitoring plant diseases, and also for detection of their causal organism in vegetative part of plants and seeds and fruits in storage. Subsequent PCR reaction is required to obtain the best yeilds, the best specificity and the highest sensivity (Yu and Pauls, 1993; Zinger *et al.*, 2007). The PCR protocols have been elaborated for many important plant pathogens.

Modifications of PCR Method

Reverse Transcription-Polymerase Chain Reaction (RT-PCR)

The ordinary polymerase reaction cannot facilitates the identification of RNA and hence not suitable for diagnostics of viroids and RNA-containing viruses. The limitation of *Taq* polymerase which cannot catalyze DNA synthesis on the RNA template is evident here. In practice, the task of detection and identification of these pathogens is accomplished by an additional enzyme, RNA-dependent DNA polymerase (reverse transcriptase). The reaction catalyzed by this enzyme leads to formation of single-stranded cDNA fragments that are further amplified by Taq polymerase. RT-PCR is successfully used for detection of viral and viroid pathogens. In the nested PCR technique, two PCRs are performed. In the first reaction, the amount of template for the second reaction is increased, making the detection effective particularly for the virus occuring in low titres or when inhibitors of DNA polymerase are present in the host tissue extract. In the technique, a combination of degenerate deoxyinosine (dI)-substitued primers amplified part of the RNA-RNA polymerase (RdRP) domain, followed by a semi-nested PR amplification that increased the sensitivity of virus detection is employed. If viral genomic dsRNA or viral RNA is not available RNA extracts from infected plants can be used for testing. *Mycospharella graminicola* (Ray *et al.*, 2003), *Alternaria brassicicola* (Schenk *et al.*, 2003), *Potato black ringspot virus* (PBRSV) and *Cherry leaf roll virus* (CLRV) were detected efficiently (Maliogka *et al.*, 2006). It is well known that detection of viroids present in infected plants in minute amounts is difficult. Generally, these causal organisms are diagnosed by hybridization analysis or return gel electrophoresis, where these two methods are 100 and 1000 times less sensitive respectively, as compared to RT-PCR. In RT-PCR isolation of 1–100 pg of total nucleic acid from the plant tissue is sufficient. The important advantage of RT-PCR for diagnosis of bacteria, fungi, or oomycetes is its ability not only to detect pathogens but also to confirm their viability. The reverse transcription can be also accomplished by DNA polymerase of the other thermophile, *Thermus thermophilus.* This enzyme (Tth polymerase) catalyzes synthesis of a single-strand complementary DNA on RNA in the presence of manganese ions. Interestingly, after addition of magnesium ions and manganese ion chelators in the reaction medium, the enzyme is able to catalyze an ordinary DNA-dependent DNA polymerase reaction. This enzyme can be used in diagnostics of viroids and RNA-containing viruses to synthesize simultaneously a single-stranded DNA, complementary to RNA of the target pathogen, and amplify the DNA fragments.

RAPD-PCR

The modern plant pathology possesses a variety of PCR-based diagnostic tools hence the researcher must clearly envisage the sequences that could serve as target for DNA assay for a plant pathogen in question. This necessitates the availability of information of genome sequence of plant pathogen. The correct choice and design of the primers is fundamental requirement for success of any PCR analysis. The primer choice is the first stage of PCR analysis based diagnosis. Owing to small genomes of viruses and viroids complete genome sequence is available in databases and appropriate primers can be easily designed. On the other hand a little information is available regarding the genomes of bacteria and fungi. But the volume of data is increasing day by day. The techniques based on screening of random regions of DNA have been developed for these pathogens and general approaches for selection of particular known DNA target fragments are available. Generally DNA encoding ribosomal RNA (rDNA) is used as a target sequence for bacteria and fungi. There are several facts that make rDNA suitable for diagnostic purpose. Many copies of rDNA are present in each cell which ensures the sensitivity of detection. The genes are present in all organisms and contain highly conserved 5.8 S regions that give rDNA universal applicability whereas internal transcribed spacer (ITS) regions are highly variable regions. The conserved regions can be used to design universal primers for the group detection of microorganisms within a taxon, while the presence of variable regions allows finding distinctions between races, strain, and isolates (Jepersen *et al.*, 2000; Kasuga and Mitchelson, 2000; Cadez *et al.*, 2002). Species specific ribosomal gene amplification can be achieved by careful defination of selective primers, providing a sensitive assay for the detection of fungi in mixed cellular environments. For example, Feau and colleagues (2005) created specific assays for the popular pathogenic fungi *Septoria musia*, *S. populicola* and *S. populi* based on variants of the internal transcribed spacer (ITS), without detectable amplification of products from 12 other *Septoria species* or other fungal species collected form poplar trees, thus permitting selective detection of these fungal speices on poplars. Other target sequences that are used for detection of fungi are beta-tubulin genes which are connected with resistance to fungicides. DNAs contained in bacterial plasmids and pathogenicity-associated genes usually serve as the sources of the target fragments.

One of the techniques used if the target nucleotide sequence is unknown is random amplified polymorphic DNA PCR (RAPD-PCR), or arbitrary primed polymerase chain reaction (AP-PCR). The RAPD-PCR is usually applied, alone or together with RFLP, in studying DNA polymorphism, in gene mapping, and in population and evolutionary biology. The RAPD-PCR is important for plant pathogen diagnostics as it enables screening the sequences specific for closely related species, strains, races, and isolates, and differentiate them (Chen *et al.*, 2007).

Apart from the described PCR techniques, where two primers encompassing the amplified sequence are used, RAPD-PCR involves annealing of single primer. The primer binds to the random complementary sequences of the genomic DNA and after amplification, the RAPD-PCR product generated are of arbitrary length that is partially or completely homologous to the arbitrarily primed sequence at both ends. The DNA polymorphism due to insertions, deletions, and base substitutions affects the generation of the RAPD-PCR product that is evident by presence or absence of bands in gel after RAPD-PCR. This method provides the possibility of amplifying gene products from many organisms analyzed where the banding pattern is specific for a particular organism. Many different primers have to be tested to identify a band that is specific for a target. Specific bands can be used for synthesis of highly specific primers (Milczarski *et al.*, 2007; Grimmer *et al.*, 2007).

However one limitation of the technique is that all organisms, from bacteria to mammals, are able to generate RAPD fingerprints, thus pure fungal cultures or highly purified DNA sources are required for such analysis, as contamination of the target genomic DNA may invalidate the RAPD fingerprinting (Lockhart *et al.*, 2005). This poses a strong limitation to plant disease diagnosis, where identification of the target pathogens from infected tissues or other environmental samples is needed.

When RAPD fingerprinting is used with pure fungal isolates the high information content may be used for detailed and precise identification of genotypes, sufficient for large scale medical or epidemiological classification of fungal species important to human health such as *Candida sp., Aspergillus fumigatus* and other anthropophilic dermatophytes (Lockhart *et al.*, 2005; Song *et al.*, 2006). RAPD fingerprinting has been used widely for the identification of strains of fungi important to quality in food production (Cocolin *et al.*, 2006; Chen *et al.*, 2007; Walczak *et al.*, 2007) and for fungi important in food spoilage (Lopandic *et al.*, 2006).

Lastly it should be noted that PCR is not the only amplification diagnostic technique. Some plant pathogens have been detected with ligase chain reaction (LCR) which is based on the ability of DNA-dependent DNA-ligase to ligate a DNA strand in the presence of adenosine triphosphate (ATP) and Mg^{++} ions, at break of the phosphodiesteric bond (Wu and Wallace, 1989). A characteristic feature of DNA ligase work is high specific activity in ligation of single-stranded ruptures at the template which constitutes the second complementary strand, and low specific activity in simultaneous ligation of two ruptures in both strands or rupture in single-stranded DNA. Implementation of LCR requires finding two pairs of primers complementary to each other and to the initially chosen fragment of the matrix (for instance, DNA of some causative agent), as "head to tail" arrangement in direction from 5? to 3? end. As early as after the second LCR cycle, the reaction mix accumulates the product which is a ligated double-stranded DNA fragment, structurally identical to the four primers used. It is characteristic that even a one-nucleotide error in the place of annealing leads to a negative result. Therefore, LCR is promising for enhanced detection of plant pathogens and revealing the point mutations in the wild types of causative agents.

Advanced and Future Technologies

The development of methods based on the molecular characterization of the plant pathogen by the amplification of specific sequences of nucleic acids has advanced through the years owing to the availability of on-line databases and of entire sequenced genomes of different microorganisms. Inspite of these development of molecular methodologies, they are not yet applied extensively for routine diagnosis of plant pathogens because of technical and economic limitations such as the risks related to sample contamination, the interpretation of results and the reliability of the extraction procedures in terms of quantity and purity (Mumford *et al.*, 2006). The advancement in this direction could be PCR amplification based methods for detection of many pathogens at a time. (Cacciola and Faedda 2007). RT PCR is still expensive method and is confined to few specialized laboratory. Some technologies are still in a development stage: Li *et al.* (2005) studied a bar code system based on DNA nanostructures called dendrimers marked with fluorescent dyes that will permit the unambiguous identification of a pathogen. The very large number of host plant species and the even larger number of pathogens that cause diseases in those plant species makes it impossible for any one diagnostician to be an expert in the diagnosis of all diseases in all hosts. One of the benefits of a plant diagnostic network is providing the ability to access specific expertise when a diagnostician encounters plant diseases or pathogens with which they have no experience. Most local plant diagnostic clinics lack the capacity to process a lot of samples quickly. This leads to a host of problems including a decline in sample quality and the unnecessary quarantine of uninfected plants.

Conclusions

The proper identification of diseases and the causal agents leads to appropriate control and eradication measures of plant diseases which are either introduced accidentally or deliberately. Therefore, diagnosis is one of the most important aspects for a plant pathologist. For this purpose proper infrastructures required for diagnosis should developed and special attention should be devoted in the future to upgrade those serving poor farmers in developing countries. The consequences of the climate change on the disease development could be even more devastating in the absence of adequate facilities and trained people.

References

Alizadeh, A., Arlat, M., Sarrafi, A., Boucher, C. A. and Barrault, G. (1997). Sensitive detection of pea seedborne mosaic potyvirus by dot blot and tissue print hybridization assays. *Austrial Agriculture Res.* 49: 191-197.

Cacciola, S.O., Faedda, R. (2007). Nanotecnologie e diagnostica fitopatologica. Informatore Fitopatologico–*La Difesa delle Piante* 57 (11): 12–20.

Cadez, N., Raspor, P., de Cock A. W., Boekhout T. and Smith, M. T. (2002). Molecular identification and genetic diversity within species of the genera Hansenia spora and Kloeckera. *FEMS Yeast Research.* 1: 279-289.

Carzaniga, R., Fiocca D., Bowyer, P. and Oconnell, R. J. (2002). Localization of melanin in conidia of Alternaria alternate using phage antibodies. *Molecular Plant Microbe Interaction.* 15: 216-224.

Chen, C. C., Chen, T. C., Lin, Y. H., Yeh,S. D. and Hsu, H. T. (2005). A chlorotic spot disease on *calla lilies* (Zantedeschia spp.) is caused by a *tospovirus* serologically but distantly related to *Watermelon silver mosaic virus. Plant Disease.*89: 440-445.

Chen, Q. H., Wag, Y. C., Li, A. N., Zhang, Z. G. and Zheng, X. B. (2007). Molecular mapping of two cultivar-specific avirulence genes in the rice blast fungus *Magnaporthe grisea. Molecular Genetics and Genomics.* 277: 139-148.

Coclin, L., Urso, R., Rantsiou, K., Cantoni, C. and Comi, G. (2006). Dynamics and characterization of yeast during natural fermentation of Italian sausages. *FEMS Yeast research.* 6: 692-701.

Darling, J.A. and Blumm, M.J. (2007). DNA-based methods for monitoring invasive species: a review and prospectus. *Biological Invasions.*9: 751–765.

Dietgen, R. G. (2002). Application of PCR in Plant Virology In: Khan J A., Dijkstra J (ads) Plant viruses as molecular pathogens. The Haworth Press, Inc., New York, USA, pp 471-500.

Feau, N., Weiland, J. E., Stanosz, G. R. and Bernier, L. (2005). Specific and sensitive PCR-based detection of *Septoria musiva, S. populicola and S. populi,* the causes of leaf spot and stem canker on poplars. *Mycological Research.* 109: 1015-1028.

Gibb, K. S., Persley, D. M., Schneider, B. and Thomas, J. E. (1996). Phytoplasmas associated with papaya disases in Austrilia. *Plant Dis.* 80: 174-178.

Gil-Lamaignere, C., Roilides, E., Hacker, J. and Muller, F. M. (2003). Molecular typing for fungi-a critical review of the possibilities and limitations of currently and future methods. *Clinical Microbiology and infections.* 9: 172-185.

Gilles, T., Evans, N., Fitt, B. D. L. and Jeger, M. J. (2000). Epidemiology in relation to forecasting light leaf spot (*Pyrenopeziza brassicae*) severity on winter oilseed rape (*Brassica napus*) in the UK. *European Journal of plant Pathology.* 106: 593-605.

Grimmer, M. K., Trybush, S., Hanley, S., Francis, S. A., Karp, A. and Asher, M. J. (2007). An anchored linkage map for sugar beet based on AFLP, SNP and RAPD markers and QTL mapping of a new source of resistance to *Beet necrotic yello vein virus. Theoritical and applied genetics.* 114: 1151-1160.

Healy, M., Reece, K., Walton, D., Huong, J., Shah, K. and Kontoyiannis, D. P. (2004). Identification to the species level and differentiation between strains of *Aspergillus* clinical isolates by automated repetitive-sequence-based PCR. *Journal of Clinical Microbiology*. 44: 3299-3305.

Hsu, H. T., Ueng, P. P.,Chu,F. H., Ye, Z. and Yeh, S. D. (2000). Serological and molecular characterization of a high temperature recovered virus belonging to *tospovirus* serogroup IV. *Journal of Gen Plant Pathology*. 66: 167-175.

Jepersen, L., Van der Kuhle, A. and Peterson, K. M. (2000). Phenotypic and genetic diversity of *Saccharomyces* contaminants isolated from lager breweries and their phylogenetic relationship with brewing yeast. *International Journal of Food Microbiology*. 60: 43-53.

Jones, R. A. C., Cults, B. A., Mackie, A. E. and Dwyer, G. I. (2005). Seed transmission of *wheat streak mosaic virus* shown unequivally in wheat. *Plant Disease*. 89: 1048-1050.

Kasuga, T., and Mitchelson, K. R. (2000). Interesterility group differentiation in *Heterobasidion annosum* using ribosomalIGS1 region polymorphism. *European Journal of Forest Pathology*, 30: 329-344.

Kiranmai, G., Satyanarayana, T. and Sreenivasulu, P. (1998). Molecular cloning and detection of *cucumber mosaic cucumovirus* causing infectious chlorosis disease of banana using DNA probes. *Current Science* 74: 356-359.

Klassen, G. R., Balcerzak, M. and de Cock, A.W. (1996). 5S ribosomal for eight species of *Pythium*. *Phythopathology*. 86: 581-587.

Li, Y. Hong Cu, Y.T. and Luo, D. (2005). Multiplexed detection of pathogen DNA with DNA-based fluorescence nanobarcodes. *Nature Biotechnology*. 23: 885–889.

Lin, Y. H., Chen, T. C., Hsu, H. T., Liu, F. L., Chu, F. H., Chen, C. C., Lin, Y. Z. and Yeh, S. D. (2005). Serological comparision and molecular characterization for verification of *Calla lily chlorotic spot virus* as a new *tospovirus* species belonging to *Watermelon silver mottle virus* serogroup. *Phytopathology*.95: 1482-1488.

Lockhart, S. R., Pujol, C., Dogson, A. R. and Soll, D. R. (2005). Deoxyribonucleotide acid fingerprinting methods for *Candida* species. *Methods in molecular Medicine*. 118: 15-25.

Lopandic, K., Zelger, S., Banszky, L.K., Eliskases-Lechner, F. and Prillinger, H. (2006). Identification of yeasts associated with milk products using traditional and molecular techniques. *Food Microbiology*. 23: 341-350.

Maliogka, V., Dovas, C. I., Leseman, D. E., Winter, S. Katis, N. I. (2006). Molecular identification, reverse transcription-PCR, detection host reaction and specific cytopathology of *Arilichoke yellow ringspot virus* infecting onion crops. *Phytopathology*. 96: 622-629.

Masoud, W., Cesar, L. B., Jesperson, L. and Jakobsen, M. (2004). Yeast involved in fermentation of *Coffee arabica* in East Africa determined by genotyping and by direct denaturating gradient gel electrophoresis. *Yeast*. 21: 549-556.

Milczarski, P., Banek-Tabor, A., Lebiecka, K., Stojalowski, S., Myskow, B. and Masojc, P. (2007). New genetic map of rye composed of PCR-based molecular markers and its alignment with reference map of the DS2 x RXL10 intercross. *Journal of Applied Genetics*.48: 11-24.

Moricca, S., Hantula, J. and Muller, M. M. (2004). Genomic and genetic perspectives on fungal endophytes. In: Ragazzi A., Moricca S. and Dellavalle I. (Eds.), Endophytism in forest trees (pp. 51-71). Firenze: Accademia Italiana di Scianze Forestali.

Moricca, S., Ragazzi, A., Kasuga, T. and Mitchelson, K. R. (1998). Detection of *Fusarium oxysporum f.sp. vasinfectum* in cotton tissue by polymerase chain reaction. *Plant pathology*. 47: 486-494.

Mumford, R., Boonham, N., Tomlinson, J., Barker, I. (2006). Advances in molecular phytodiagnostics–NISC (National Invasive Species Council) (2001) Meeting the invasive species challenge. National Invasive Species Council, Washington DC, 24 May 2002, www.invasivespecies.gov. Accessed on 10 December 2007.

Narayanasamy, P. (2001). Plant pathogen detection an disease diagnosis, 2nd edition Marcel Dekker, Inc., New York.

Narayanasamy, P. (2005). Immunology in plant health and its impact on food safety. The Haworth Press, Inc., New York.

Narvaez, G., Slimane, S. B., Ayllon, M. A., Rubio, L., Guerri, J. and Moreno, P. (2000). A new procedure to differentiate *Citrus tristeza virus* isolate by hybridization with digoxigenin labelled cDNA probes. *Journal of Virology Meth*. 85: 83-92.

Nicholson, P., Rezanoor, H. N., Hollins, T. N. (1994). The identification of a pathotype specific DNA probe for the R type of *Pseudocercosporella herpotrichoides*. *Plant Pathology* 43: 694-700.

Noll, W. W. and Collins, M. (1987). Detection of human DNA polymorphisms with a simplified denaturing gradient gel electrophoresis technique. Proceedings of the National Academy of the Science, USA. 84: 3339-3343.

Olmos, A., Cambra, M., Dasi, M A., Candresse, T., Esteban, O., Gorris, M. T., Asensio, M. (1997). Simultaneous detection and typing of *plum pox virus* (PPV)isolates by hemmi-nested-PCR and PCR-ELISA. *Journal of Virology Method*.68: 127-137.

Plachy, R., Hamal, P. and Raclavsky, V. (2005). Mc RAPd as a new approacg to rapid and accurate identification of pathogenic yeasts. *Journal of Microbiological Methods*. 60: 107-113.

Ray, S., Anderson, J. M., Urmeev, F. I. and Goodwin, S. B. (2003). Rapid induction of a protein disulfide isomerase and defense-related genes in wheat in response to hemibiotrophic pathogen *Mycosphaerella graminicola*. *Plant Molecular Biology*. 53: 601-714.

Rementeria, A., Vivanco, A. B., Cadaval, A., Ruesga, M. T., Breana, S., Pomton, J., Quindos, G. and Garaizar, J. (2004). Typing fungal isolates: molecular methods and computerized analysis. *Methods in Molecular Biology*, 268: 117-125.

Schaad, N. W., Azad, H., Peet, R C. and Panopolus, N. J. (1989). Identification of *Pseudomonas syringae pv phaseolicola* by a DNA hybridization probe. *Phythopathology* 79: 903.

Schaad, N. W. and Frederick, R. D. (2002). Real-time PCR and its application for rapid plant disease diagnostics. *Canadian J. Plant Pathol.* 24: 250-258.

Schenk, P. M., Kazan, K. and Manneres, J. M. (2003). Systemic gene expression in Arabdiosis during an incompatible interaction with *Alternaria brassicola. Plant physiology*.132: 999-1010.

Singh, U., Trevors, C. M., de Boer, S. H. and Janse, J. D. (2000). Fimbrial-specific monoclonal antibody based ELISA for European potato strains of *Erwinia chrysanthemi* and comparison to PCR. *Plant Diseases*.84: 443-448.

Song, X., Sun, J., Store, G., Eribe, E. R., Hansen, B. F. and Olsen, I. (2006). Genotypic relatedness of yeast in thrush and denture stomatitis. *Oral Microbiol Immunology.* 21: 301-308.

Takeuchi, S., Okuda, M., Hanada, K., Kawada, Y. and Kameya-Iwaki, M. (2001). Spotted wilt disease of cucumber (*Cucumis sativus*) caused by *Melon Yellow spot virus. Japan J. Phythopath.* 67: 46-51.

Tanina, K., Inoue K., Date, H., Okuda, M., Hanada, K., Nasu, H. and Kasuyama, S. (2001). Necrotic spot disease of cineraria caused by Impatiens *necrotic spot virus. Japan J. Phythopath.* 67: 42-45.

Ura, H., Mutsumoto, M., Iiyama, K., Furuya, N. and Matsuyama, N. (1998). PCR-RFLP analysis for distinction of *Burkholderia* species, causal agent of various rice diseases. *Proc Assoc Plant Protec Kyushu*. 44: 5-8.

Walczak, E., Czaplinska, A., Barszczewski. W., Wilgosz, M., Wojtatowicz, M. and Robak, M. (2007). RAPD with microsatellite as a tool for differentiation of *Candida* genus yeast isolated in brewing. *Food Microbiology*. 24: 305-312.

Ward, E., Foster, S. J., Fraaije, B. A. and McCrrtney, H. A. (2004). Plant pathogen diagonostics: Immunological and nucleic acid approaches. *Annals of applied Biology*. 145: 1-16.

Welnicki, M. and Hiruki, C. (1992). Highly sensitive digoxigenin-labelled DNA probe for detection of potato spindle tuber viriod. *J. Virology Meth.* 39: 91-99.

Wise, M. G., Healy M., Reece, K., Smith, R., Walton, D., Dutch, W. (2007). Species identification and strain differentiation of clinical *Candida* isolates using the Diversilab system of automated repetitive sequence-based PCR. *J.Medical Microbiol.* 56: 778-787.

Wu, D.Y. and Wallace, R, B. (1989). The ligation amplification reaction (LAR)—amplification of specific DNA sequences using sequential rounds of template-dependent ligation. *Genomics.* 4 (4): 560-569.

Yu, K. and Pauls, K. P. (1993). Optimization of DNA extraction PCR procedures for random amplified polymorphic DNA (RAPD) analysis in plants. In: H. G.

Griiffen and A M. Griffin (Eds.), PCR technology current innovations (pp. 193-200). Boca Raton, FL: CRC Press.

Zinger, L., Gury, J., Giraud, F., Krivobok, S., Gielly, L., Taberlet, P. and Geremia, R. A. (2007). Improvements of polymerase chain reaction and capillary electrophoresis single-strand conformation polymorphism methods in microbial ecology: toward a high-throughput method for microbial diversity studies in soil. *Microbial ecolog.* 54: 203-216.

Index

AFLP 58, 160
Aspergillus 336, 337
Autoradiography 360, 365

Biological control 40, 302
Bipolaris sorokiniana 18

Cassava 1
Cercospora 336
Characterization 302
Chitinase 40, 44
Cloning 204
Coconut 131
Colletotrichum 336, 352

DAC-ELISA 146
DAS-ELISA 146
Detection techniques 81
DIBA 131, 146
DNA hybridization 204
DNA support media 123

ELISA 84, 131, 146, 360

F. oxysporum 159, 160
F. oxysporum f .sp. *gladioli* 105
F. solani 105
Frogskin disease 1
Fusarium 336, 348

Ganoderma lucidum 131
Gene bank 5
Genetic diversity 74
Genome sequencing 58
Ginger 236
Gladiolus hybridus 105
Glucanase 40, 45

HMA 204

I-CZE 146
Immunoassays 360

ISEM 81, 146
Isozyme 58, 236

Macrophomina phaseolina 73
Manihot esculenta 1
Molecular 236, 302
Molecular techniques 249
Molecular variability 19, 105, 191
Morphological 302
Morphological variability 186
mtDNA 58

NAH 87, 94, 204

P. infestans 58
PAC-ELISA 146
Pathogen 81
PCR 2, 19, 81, 88, 146, 204, 366
Phytophthora 336, 352
Phytoplasma 1, 204, 249
Plant growth promotion 302
Plasmopora halstedii 178
Pogostimon patchouli 105
Population dynamics 58
Protease 45
Pyricularia grisea 191
Pythium aphanidermatum 236

RAPD 19, 74, 105, 134, 160, 178, 191, 236
rDNA 204
Re-melt 123
Resistance 191
RFLP 4, 8, 11, 58, 160, 204, 366
Rhizoctonia 336, 352
Rhizome rot 236
RT-PCR 1, 90, 146, 368

Sclerotium 336, 347
Serology 81, 204
SNP 58
Spot blotch 19
SSR 58

Transgenic 40
T-RFLP 360
Trichoderma 40, 302
Tris-acetate 123
Tris-borate 123

URP 19

Variability 236
VCG 336, 337
Virus 145, 249

Wheat 18, 19